Water Quality Engineering and Wastewater Treatment

Water Quality Engineering and Wastewater Treatment

Editors

Yung-Tse Hung
Hamidi Abdul Aziz
Issam A. Al-Khatib
Rehab O. Abdel Rahman
Mario GR Cora-Hernandez

MDPI • Basel • Beijing • Wuhan • Barcelona • Belgrade • Manchester • Tokyo • Cluj • Tianjin

Editors
Yung-Tse Hung
Cleveland State University
USA

Hamidi Abdul Aziz
Universiti Sains Malaysia
Malaysia

Issam A. Al-Khatib
Graduate Studies Birzeit University
Palestine

Rehab O. Abdel Rahman
Atomic Energy Authority of Egypt
Egypt

Mario GR Cora-Hernandez
University of Maryland
University College
USA

Editorial Office
MDPI
St. Alban-Anlage 66
4052 Basel, Switzerland

This is a reprint of articles from the Special Issue published online in the open access journal *Water* (ISSN 2073-4441) (available at: https://www.mdpi.com/journal/water/special_issues/Water_Quality_Wastewater).

For citation purposes, cite each article independently as indicated on the article page online and as indicated below:

LastName, A.A.; LastName, B.B.; LastName, C.C. Article Title. *Journal Name* **Year**, *Volume Number*, Page Range.

ISBN 978-3-0365-1022-4 (Hbk)
ISBN 978-3-0365-1023-1 (PDF)

Contents

About the Editors

Yung-Tse Hung, Ph.D., P.E., DEE, F-ASCE, has been a Professor of Civil Engineering at Cleveland State University, Cleveland, Ohio, USA, since 1981. He is a Fellow of the American Society of Civil Engineers. He has taught at 16 universities in 8 countries. His primary research interests and publications have been involved with biological wastewater treatment, industrial water pollution control and industrial waste treatment, and municipal wastewater treatment. Prof. Hung has over 400 publications and presentations on water and wastewater treatment. He received his BSCE and MSCE degrees from Cheng-Kung University, Taiwan, and his PhD from the University of Texas at Austin. He is Editor-in-Chief of *International Journal of Environment and Waste Management*, Editor-in-Chief of *International Journal of Environmental Engineering*, and Editor-in-Chief of *International Journal of Environmental Pollution Control and Management*.

Hamidi Abdul Aziz is a Professor at the School of Civil Engineering, Engineering Campus, Universiti Sains Malaysia. His research interests are in water supply and water treatment, municipal wastewater treatment, industrial waste treatment, biological waste treatment, water and wastewater treatment plant design, water pollution control, water quality engineering.

Issam A. Al-Khatib is a Professor at the Institute of Environmental and Water Studies Faculty of Graduate Studies, Birzeit University, West Bank, Palestine. His research interests are in water recourses management and quality, environmental assessment, wastewater management, advocacy, coordination and networking.

Rehab O. Abdel Rahman is a Professor at the Hot Lab. & Waste Manag. Center, Atomic Energy Authority of Egypt, Cairo, Egypt. Her research interests are in radioactive waste management, wastewater treatment, natural and synthetic materials assessments, pollution control.

Mario GR Cora-Hernandez is Adjunct Associate Professor, School of Undergraduate Studies, Sciences Program, University of Maryland University College, Maryland, USA. His research interests are in industrial and municipal wastewater treatment, waste treatment, industrial hygiene, environmental management and sustainability, and air pollution control.

Editorial

Water Quality Engineering and Wastewater Treatment

Yung-Tse Hung [1], Hamidi Abdul Aziz [2,3,*], Issam A. Al-Khatib [4], Rehab O. Abdel Rahman [5] and Mario GR Cora-Hernandez [6]

1 Department of Civil and Environmental Engineering, Cleveland State University, Cleveland, OH 44115, USA; y.hung@csuohio.edu
2 School of Civil Engineering, Engineering Campus, Universiti Sains Malaysia, Pulau Pinang 14300, Malaysia
3 Solid Waste Management Cluster, Engineering Campus, Universiti Sains Malaysia, Pulau Pinang 14300, Malaysia
4 Institute of Environmental and Water Studies, Faculty of Graduate Studies, Birzeit University, Birzeit P.O. Box 14, West Bank, Palestine; ikhatib@birzeit.edu
5 Hot Lab. & Waste Manag. Center, Atomic Energy Authority of Egypt, Inshas, Cairo P.O. No. 13759, Egypt; alaarehab@yahoo.com
6 School of Undergraduate Studies, Sciences Program, University of Maryland University College, 3501 University Blvd East, Adelphi, MD 20783, USA; mariocora2000@yahoo.com
* Correspondence: cehamidi@usm.my

Citation: Hung, Y.-T.; Abdul Aziz, H.; Al-Khatib, I.A.; Abdel Rahman, R.O.; Cora-Hernandez, M.G. Water Quality Engineering and Wastewater Treatment. *Water* **2021**, *13*, 330. https://doi.org/10.3390/w13030330

Received: 13 January 2021
Accepted: 27 January 2021
Published: 29 January 2021

Wastewater treatment is crucial to prevent environmental pollution. Wastewater sources include domestic households, municipal communities, or industrial activities. Wastewater that is discharged to the environment must be treated to prevent pollution to the environment. However, wastewater remains one of the major pollutants of our inland waterways. To satisfy the tighter regulatory requirements, the implementation of more advanced design in wastewater treatment technologies is required. Treatment of wastewater usually includes physical, chemical, and biological processes. Today's wastewater treatments are much more technologically advanced than they were in previous years. Many centralized mechanized treatments are run via a computer system, and they are run more efficiently. In this Special Issue, we attempted to discuss and address the state-of-the-art of wastewater quality, treatment, and its management.

This special issue is composed of 18 innovative papers and reviews that address water quality engineering and wastewater treatment. The study areas in which these topics are developed include wastewater treatment from acid mine drainage, municipal wastewater, landfill leachate, groundwater, greywater, industrial wastewater, and urban wastewater. The issue also covers the degradation mechanism of one of the nonsteroidal anti-inflammatory medications most widely used, diclofenac (DIC), by an MnO_2 catalyst. In addition, the issue also present papers on eutrophication, wastewater de-nitrification, micropollutants treatment, nanoparticles application in wastewater treatment, and a paper each for a new ecotoxicity measuring tool by using optical camera and inactivation and loss of solar irradiation infectivity of Enterovirus 70. The three review papers include the use of natural polymers' modification in wastewater treatment of toxicant dye compounds, metallic iron for environmental remediation using metallic iron (Fe^0) as a reactive agent, and the utilization of ionizing radiation in wastewater purification.

The papers in the Special Issue are summarized as follows:

Biological sulfate reduction (BSR) has been recognized as a favorable option for the purification of acid mine drainage. The influence of temperature, pH, and hydraulic retention time (HRT) on BSR has been examined in downflow mode packed bed reactors [1]. They concluded that HRT and temperature had a powerful interaction; however, the effect of pH was negligible. In addition, due to higher flow rates, a decrease in HRT had a positive effect on the rate of sulfate reduction. On the other hand, it had a deleterious effect on the performance of sulfate remediation, most probably caused by substrates being washed out.

During the wastewater treatment process, Lange et al. [2] examined the behavior of nanoparticles $nTiO_2$ and $nCeO_2$ in wastewater treatment modes (at lab-scale) to model the mass flow of anthropogenic nanoparticles (NP) in the wastewater treatment process. The findings indicate that $nTiO_2$ and $nCeO_2$ are adsorbed to at least 90% of the sludge. In addition, the findings show that there are steps that entail a shift in the shape of the NP in the effluent during the passage of the treatment method, as the NP in the effluent was found to be partly lesser than in the added solution. This statement was denoted, especially for $nCeO_2$, and may be caused by dissolution action or higher particles sedimentation during the transit of the treatment procedure.

One of the water species used to assess ecotoxicity is *Lemna minor* (Lesser duckweed-Angiospermae, Lemnaceae). Haffner et al. [3] tested a cheaper process for a *Lemna minor* bioassay assessment by computer and machine vision. The aim was to use a digital camera and a framework software instrument to design a methodology for image analysis. In this paper, instead of counting individual leaves, they suggested using computer vision to calculate and compare the area covered by the leaves and compared the modern procedure with the ordinary one. They concluded that the toxic effect was more meaningful when examining the leaf area instead of the number of leaves. In addition, errors resulting from a human element are removed using the computer vision-based approach during leaf counting.

The degradation of Diclofenac (DIC) by tunnel-structured MnO_2 based on solution pH with excellent oxidative and catalytic capacities was investigated by Hu et al. [4]. DIC can be effectively oxidized in an acidic medium by γ-MnO_2, and at alkaline and neutral conditions, the removal rate decreased significantly. The developed model could successfully match the kinetics of DIC degradation and demonstrated the regulation of electron transfer at acidic environments and the complex precursor structure control mechanism within neutral to alkaline environments, by which the pH level corresponds precisely to the percentage of distribution of ionized DIC species for two mechanisms. 5-iminoquinone DIC, hydroxyl-DIC, and 2,6-dichloro-N-o-tolylbenzenamine were the major oxidation products with a strong dynamic hydroxylation route in the tunnel-structured Mn-oxide.

The ability of *Moringa oleifera* (MO) seeds as a coagulant to remove turbidity, biochemical oxygen demand (BOD), and chemical oxygen demand (COD) from urban wastewater was investigated by Adelodun et al. [5]. Their outcomes indicated that employing a MO dose of 150 mg/L, the maximal reduction in turbidity, BOD, and COD was 94%, 69%, and 58%, respectively. They recommended this low-cost and natural MO coagulant be used for the safe removal of turbidity, BOD, and COD from urban wastewater.

Remediation of metal contamination from groundwater and greywater supplies in Riyadh, Saudi Arabia, was investigated by Alomar et al. [6]. Ultrasonic power before adsorption was also explored to measure the distribution of renewable carbon from mixed waste sources (RC-MWS) as an adsorbent and to improve the water treatment system. From the actual water samples under study, the renewable carbon adsorbent exhibited a greater adsorption potential of Pb(II), Zn(II), Cu(II), and Fe(II). The increased adsorption method demonstrated the greatest efficiency at a pH of 6, room temperature, and 60 min contact time.

Thaher et al. [7] examined the local population's views of On-site GreyWater Treatment Plants (GWTPs) for rural community wastewater management in Palestine. They found that the reasons for adopting GWTPs included the use of GWTPs in irrigation, the reduction in the frequency and financial effects of cesspit discharges, water scarcity, the reduction in possible risks of groundwater contamination, the decrease in the water bill and improved hygiene, and funds availability for their use. The reuse of treated greywater in irrigation has also been approved by the Islamic religion. In GWTP management, women play a major role. The majority (70%) of GWTP recipients were pleased. Operation and maintenance took little effort, with just a mean of 0.4 h of working per week. Odor pollution, insect infestation, implementing agency limitations in doing follow-up and monitoring, system failures triggered by insufficient beneficiary expertise in service and maintenance and deficiency of system awareness, and health issues and doubts concerning the quality of crops irrigated by treated greywater were among the obstacles to the implementation of the GWTPs. In rural areas, house on-site greywater management systems were acceptable; therefore, an adequate system is needed to deal with wastewater and replace cesspits and their hazardous environmental, groundwater, and public health implications.

Alkhudhiri et al. [8] used air gap membrane distillation (AGMD) to remove heavy metals from synthetic industrial wastewater specimens comprising mercury (Hg), arsenic (As), and lead (Pb). The results demonstrated that TF200 and TF450 showed excellent reductions at a wide range of concentrations, which exceeded 96% for heavy metal ions. Furthermore, the pH value did not have a major impact on the efficiency of metal reduction. Energy usage was controlled at diverse pore sizes of the membrane, and it was identified to be almost unrelated to the pore size of the membrane and class of metal.

In an urban river in North China, Bai et al. [9] tested various sequentially constructed wetlands for contaminated water. From April to October 2016, the monitoring results showed that chemical oxygen (COD), ammonia nitrogen (NH_4-N), total nitrogen (TN), total phosphate (TP), and suspended solids (SS) could be efficiently removed by multiple wetland ecosystems at average elimination rates of 75%, 81%, 71%, 78%, and 92%, respectively. Of all methods, the floating-bed wetlands exhibited the highest elimination rate of SS (80%), which could effectively stop the blockage of sub-surface flow wetlands. The NH_4-N, TN, and TP were efficiently eliminated by the sub-surface flow wetland, and the contribution rates were 79%, 65%, and 82%. The surface wetland flow will further clean the TN, and the reduction in the TN will approach 23%. The general expense of this investment in environmental engineering was $12,000. Building and operation costs were $120 and $0.02 per tonne of wastewater.

The influence of inundation and eutrophication on the production of the *Wedelia Trilobata* (WT) trait over its congener native *Wedelia Chinensis* was investigated by Azeem et al. [10] to understand the terrestrial plant reaction when affected by a riparian zone. It was found that both species survive and grow well under submergence and eutrophication, but high submergence and eutrophication provide superior conditions for WT to flourish. Environmental modeling suggested that artificial disruption and climate change, which was approximately 1/3 to 1/5 and 1/6 to 1/3 of traditional sewage treatment, will ensure submergence and eutrophication, respectively.

Hamid et al. [11] investigated the pattern of a unique zeolite supplemented by the electrocoagulation method (ZAEP) by an aluminum electrode to treat high-strength ammonia concentration (3471 mg/L) from saline (15.36 ppt) landfill leachate in nature. It was observed that up to 71% of ammonia was removed, with the ideal working treatment operation as follows: 105 g/L zeolite dosage, the current density of 600 A/m^2, 60 min electrolysis duration, and pH 8.20. The results indicate that ZAEP treatment is a sustainable solution for concentration landfill leachate without the use of auxiliary salinity.

A laboratory-scale aerobic-methane oxidation bioreactor (MOB)-anoxic device was developed by Le et al. [12], which combined MOB with the aerobic-anoxic de-nitrification method and assessed its possibility to eliminate nitrogen in wastewater treatment plants (WWTPs). Based on three months of continuous running employing real wastewater, the total nitrogen elimination was 76%, close to the performance of a tertiary-advanced WWTP, and the total phosphorus reduction reached 84%.

Mojiri et al. [13] based their research on the treatment of pharmaceuticals micropollutant. The goal was to eliminate diclofenac (DCF), ibuprofen (IBP), and naproxen (NPX) from the water by cross-linked magnetic chitosan/activated biochar (CMCAB). Two point four one milligram per liter (96%) of DCF, 2.47 mg/L (99%) of IBP, and 2.38 mg/L (95%) of NPX were eliminated at a pH of 6.0 and an initial micropollutant (MP) concentration of 2.5 mg/L. Eventually, desorption experiments demonstrated that cross-linked magnetic chitosan/activated biochar could be used for a minimum of eight adsorption–desorption series.

Solar Irradiation Inactivation and Loss of Enterovirus 70 Infectivity Enterovirus 70 (EV70) is a novel viral pathogen that could be present in the final effluent. For this, solar irradiation as a low-cost natural disinfection technique was studied by Jumat and Hong [14] to alleviate possible concerns. EV70 reduced its infectivity in the presence of sunlight in PBS, effluent, and chlorinated effluent, respectively, by 1.7 log, 1.0 log, and 1.3 log. The decrease in EV70 infectivity was consistent with a decrease in the viral binding capacity of Vero cells. Furthermore, a genome sequencing analysis exhibited five non-synonymous nucleotide displacements in irradiated viruses following 10 days of infection in Vero cells, leading to amino acid substitutions.

Aziz et al. [15] used a mixture of polyaluminium chloride (PACl) as a coagulant and *Dimocarpus longan* seed powder (LSP) as a coagulant to help treat landfill leachate by introducing a coagulation–flocculation process. The highest reductions in COD, SS, and color were 69%, 100%, and 99%, respectively, when LSP was employed as a coagulant aid with PACl. The PACl dose was reduced from 5 to 2.75 g/L when LSP was applied as a coagulant aid. The cost estimate for using PACl as a sole coagulant and using LSP as a flocculant indicated a decrease of about 40% in the cost of using just PACl.

Ishak et al. [16] reviewed the use of diverse forms of natural and modified polymers to remove wastewater toxicant dyes discharged by the dye industry. Even though modified polymers are favored for dye treatment because of their biodegradability and non-toxic nature, high amounts of polymers are needed, which is expensive. To treat dyes from wastewater, surface-modified polymers are more effective. A study of 80 recently published research work showed that modified polymers have an excellent capacity to eliminate dye and, therefore, have high applicability in the treatment of industrial wastewater.

Hu et al. [17] performed a significant review of the abundant literature on the environmental remediation and water purification metallic iron (Fe^0) as a reactive agent. The requirement to characterize Fe^0 materials in relation to intrinsic reactivity and performance is important for the design and running of Fe^0/H_2O systems. A deeper knowledge of long-term corrosion of related Fe^0 materials at site-specific environments is envisaged to assist in the final design of low-cost, appropriate, and effective Fe^0/H_2O remediation techniques. A wide range of household and small-size water treatment plants are currently being used for Fe^0-based systems, which includes rainwater harvesting methods for the supply of drinking water, decentralized domestic wastewater treatment, urban stormwater treatment, agricultural and industrial wastewater purification, and as philter media for constructed wetlands.

Abdel Rahman and Hung [18] reviewed ionizing radiation technology in the decomposition of bio-refractory organic pollutants and the disinfection of various wastewater effluents. Compared to other disinfection technologies that are not influenced by periodic changes in the effluent constituents, ionizing radiation technology provides cheap, effective, and safer operations and reduces the production of secondary toxic intermediates. Owing to increasingly strict regulatory standards and the upgrading of operating procedures, the operating safety of industrial irradiators has been increased, leading to a reduction in the likelihood of incidents from 10^{-2} to 10^{-4} a^{-1}.

Author Contributions: Conceptualization, Y.-T.H. and I.A.A.-K.; writing—original draft preparation, H.A.A. and R.O.A.R.; writing—review and editing, M.G.C.-H. and Y.-T.H. All authors have read and agreed to the published version of the manuscript.

Funding: This research received no external funding.

Acknowledgments: The first draft of this paper was prepared by Hamidi Abdul Aziz, who is the corresponding author of this paper. Thanks to all guest editors of this special issue. Thanks to the journal's editors, as well as to the authors who contributed to the Special Issue with their articles. Finally, a special thank goes to the anonymous reviewers, who have made a significant contribution to improving the quality of the articles.

Conflicts of Interest: The authors declare no conflict of interest.

References

1. Mukwevho, M.J.; Maharajh, D.; Chirwa, E.M.N. Evaluating the effect of ph, temperature, and hydraulic retention time on biological sulphate reduction using response surface methodology. *Water* **2020**, *12*, 2662. [CrossRef]
2. Lange, T.; Schneider, P.; Schymura, S.; Franke, K. The fate of anthropogenic nanoparticles, $nTiO_2$ and $nCeO_2$, in wastewater treatment. *Water* **2020**, *12*, 2509. [CrossRef]
3. Haffner, O.; Kučera, E.; Drahoš, P.; Cigánek, J.; Kozáková, A.; Urminská, B. *Lemna minor* bioassay evaluation using computer image analysis. *Water* **2020**, *12*, 2207. [CrossRef]
4. Hu, C.Y.; Liu, Y.J.; Kuan, W.H. pH-dependent degradation of diclofenac by a tunnel-structured manganese oxide. *Water* **2020**, *12*, 2203. [CrossRef]
5. Adelodun, B.; Ogunshina, M.S.; Ajibade, F.O.; Abdulkadir, T.S.; Bakare, H.O.; Choi, K.S. Kinetic and prediction modeling studies of organic pollutants removal from municipalwastewater using moringa oleifera biomass as a coagulant. *Water* **2020**, *12*, 2052. [CrossRef]
6. Alomar, T.S.; Habila, M.A.; Alothman, Z.A.; AlMasoud, N.; Alqahtany, S.S. Evaluation of groundwater and greywater contamination with heavy metals and their adsorptive remediation using renewable carbon from a mixed-waste source. *Water* **2020**, *12*, 1802. [CrossRef]
7. Thaher, R.A.; Mahmoud, N.; Al-Khatib, I.A.; Hung, Y.T. Reasons of acceptance and barriers of house on-site greywater treatment and reuse in Palestinian rural areas. *Water* **2020**, *12*, 1679. [CrossRef]
8. Alkhudhiri, A.; Hakami, M.; Zacharof, M.P.; Homod, H.A.; Alsadun, A. Mercury, arsenic and lead removal by air gap membrane distillation: Experimental study. *Water* **2020**, *12*, 1574. [CrossRef]
9. Bai, X.; Zhu, X.; Jiang, H.; Wang, Z.; He, C.; Sheng, L.; Zhuang, J. Purification effect of sequential constructedwetland for the polluted water in urban river. *Water* **2020**, *12*, 1054. [CrossRef]
10. Azeem, A.; Sun, J.; Javed, Q.; Jabran, K.; Du, D. The effect of submergence and eutrophication on the trait's performance *Ofwedelia Trilobata* over its congener native *Wedelia Chinensis*. *Water* **2020**, *12*, 934. [CrossRef]
11. Hamid, M.A.A.; Aziz, H.A.; Yusoff, M.S.; Rezan, S.A. Optimization and analysis of zeolite augmented electrocoagulation process in the reduction of high-strength ammonia in saline landfill leachate. *Water* **2020**, *12*, 247. [CrossRef]
12. Kim, I.T.; Lee, Y.E.; Yoo, Y.S.; Jeong, W.; Yoon, W.H.; Shin, D.C.; Jeong, Y. Development of a combined aerobic–anoxic and methane oxidation bioreactor system using mixed methanotrophs and biogas for wastewater de-nitrification. *Water* **2019**, *11*, 1377. [CrossRef]
13. Mojiri, A.; Kazeroon, R.A.; Gholami, A. Cross-linked magnetic chitosan/activated biochar for removal of emerging micropollutants fromwater: Optimization by the artificial neural network. *Water* **2019**, *11*, 551. [CrossRef]
14. Jumat, M.R.; Hong, P.Y. Inactivation and loss of infectivity of Enterovirus 70 by solar irradiation. *Water* **2019**, *11*, 64. [CrossRef]
15. Aziz, H.A.; Rahim, N.A.; Ramli, S.F.; Alazaiza, M.Y.D.; Omar, F.M.; Hung, Y.T. Potential use of *dimocarpus longan* seeds as a flocculant in landfill leachate treatment. *Water* **2018**, *10*, 1672. [CrossRef]
16. Ishak, S.A.; Murshed, M.F.; Akil, H.M.; Ismail, N.; Rasib, S.Z.M.; Al-Gheethi, A.A.S. The application of modified natural polymers in toxicant dye compounds wastewater: A review. *Water* **2020**, *12*, 2032. [CrossRef]
17. Hu, R.; Yang, H.; Tao, R.; Cui, X.; Xiao, M.; Amoah, B.K.; Cao, V.; Lufingo, M.; Soppa-Sangue, N.P.; Ndé-Tchoupé, A.I.; et al. Metallic iron for environmental remediation: Starting an overdue progress in knowledge. *Water* **2020**, *12*, 641. [CrossRef]
18. Abdel Rahman, R.O.; Hung, Y.T. Application of ionizing radiation in wastewater treatment: An overview. *Water* **2020**, *12*, 19. [CrossRef]

Article

Evaluating the Effect of pH, Temperature, and Hydraulic Retention Time on Biological Sulphate Reduction Using Response Surface Methodology

Mukhethwa Judy Mukwevho [1,2,*], Dheepak Maharajh [3] and Evans M. Nkhalambayausi Chirwa [1]

1 Department of Chemical Engineering, Water Utilisation and Environmental Engineering Division, University of Pretoria, Private Bag X20, Hatfield, Pretoria 0002, South Africa; evans.chirwa@up.ac.za

2 Mineral Processing Division, Mintek, Private Bag X3015, Randburg 2125, South Africa

3 YRD2 Solutions, 1093 Raslouw Glen Estate, Centurion 0157, South Africa; dheepakmaharajh@gmail.com

* Correspondence: mukhethwam@mintek.co.za; Tel.: +27-11-709-4666

Received: 27 June 2020; Accepted: 21 September 2020; Published: 23 September 2020

Abstract: Biological sulphate reduction (BSR) has been identified as a promising alternative for treating acid mine drainage. In this study, the effect of pH, temperature, and hydraulic retention time (HRT) on BSR was investigated. The Box–Behnken design was used to matrix independent variables, namely pH (4–6), temperature (10–30 °C), and HRT (2–7 days) with the sulphate reduction efficiency and sulphate reduction rate as response variables. Experiments were conducted in packed bed reactors operating in a downflow mode. Response surface methodology was used to statistically analyse the data and to develop statistical models that can be used to fully understand the individual effects and the interactions between the independent variables. The analysis of variance results showed that the data fitted the quadratic models well as confirmed by a non-significant lack of fit. The temperature and HRT effect were significant ($p < 0.0001$), and these two variables had a strong interaction. However, the influence of pH was insignificant ($p > 0.05$).

Keywords: acid mine drainage; sulphate reduction; sulphate reducing bacteria; response surface methodology

---

1. Introduction

Acid mine drainage (AMD) is a widespread problem that is considered the most important pollution problem caused by mining industries worldwide. AMD is formed when a sulphide-bearing mineral comes into contact with oxygen and water during or after the closure of mining operations. This oxidation process leads to the formation of sulphuric acid, which further reacts with the sulphide mineral and other exposed minerals and leaches out toxic heavy metals such as lead, cadmium, and arsenic [1–3]. Pyrite is the most common pathway for AMD formation, and its oxidation is shown in Equation (1) [4].

$$4FeS_2 + 15O_2 + 14H_2O \rightarrow 4Fe(OH)_3 + 8SO_4^{2-} + 16H^+ \quad (1)$$

AMD is characterised by low pH, a high concentration of sulphate, and high concentrations of heavy metals such as iron, manganese, arsenic, zinc, copper, and aluminum. In South Africa, sulphate is considered one of the major contributors to water quality issues for mining operations as it is typically above 2000 mg/L in AMD. As a result, the maximum sulphate discharge levels should be less than 600 mg/L [5].

Conventionally, AMD is treated by neutralisation using lime or calcium carbonate, which precipitate metals and increase the pH but do not effectively reduce sulphate concentration in the mining effluent to levels suitable for discharge. This method is costly and produces large sludge volumes that are difficult to dispose of. Due to this, more research on AMD treatment has been done over the years, and biological sulphate reduction (BSR) has been identified as a promising alternative treatment for AMD. BSR is a process where sulphate is metabolically converted to sulphide by sulphate-reducing bacteria (SRB), and it simultaneously increases pH and precipitate metals under anaerobic conditions [6,7]. This process requires an electron donor and uses sulphate as a terminal electron acceptor. Electron donors, also known as substrates, that have been used include simple organic compounds such as ethanol, methanol, and butyrate [8–10], and complex organic compounds such as manure, food waste, woodchips, sludge, and hay [11–15]. Simple organic compounds are preferred as they are readily available; however, they are expensive [16]. Most studies have been leaning towards using complex compounds for BSR as they are considered to be cost-effective. SRB oxidises the organic matter, denoted by CH_2O in Equation (2), to produce alkalinity and hydrogen sulphide, which binds to the metals and precipitates them as metal sulphides, as shown in Equation (3).

$$2CH_2O + SO_4^{2-} \overset{SRB}{\rightarrow} 2HCO_3^- + H_2S \tag{2}$$

$$H_2S + M^{2+} \rightarrow MS + 2H^+ \tag{3}$$

where M represents metals.

The performance of BSR is highly dependent on the availability of organic substrate, pH, temperature, and hydraulic retention time (HRT). Most known SRBs are mesophilic, and they perform optimally at neutral pH. Low pH and high pH suppress and inhibit SRBs, respectively [17–19], whereas low temperatures slow down the metabolic activity of SRBs [20]. Due to the sensitivity of SRB to temperature and pH, most research was done at neutral pH and temperatures greater than 20 °C [17,21–23]. HRT affects the rate at which sulphate is removed from AMD. Short retention times are known to washout biomass, whereas long retention times may lead to the depletion of organic matter if complex organic compounds are used [10,11].

Response surface methodology (RSM) is a collection of mathematical and statistical techniques based on the fit of a polynomial equation to the experimental data with the objective of statistically predicting and understanding the system's behaviour [24]. RSM was developed for the simplification of multivariable experimental design enabling the reduction of the number of experiments that are required to identify ideal variables for a process. An advantage of using RSM includes less time required for experimentation due to reduced experimental runs and therefore a cost reduction of materials and reagents [24,25].

In the literature, studies have been focused on increasing the pH of mining influent while simultaneously precipitating dissolved heavy metals [15,26–30]. However, the main focus of this study was primarily on the reduction of sulphate in AMD. Although previous studies have used response surface methodology to investigate how different factors affect biological sulphate reduction [25,31], as far as the authors are aware, there is no published work on how pH, HRT, and temperature affects biological sulphate reduction using response surface methodology. This study forms part of the ongoing BSR work at Mintek, which has a pilot plant running at a coal mine in Mpumalanga, South Africa. This study aimed to investigate the effects of operational factors—namely pH, temperature and HRT—on sulphate reduction efficiency and sulphate reduction rate using response surface methodology. Additionally, the purpose of this study was to develop mathematical models that can be used to predict how the pilot plant would behave when factors are changed within the investigated range. This will help operate the pilot plant such that the sulphate discharge standard is met. The lab-scale reactors were set up to mimic the pilot plant.

2. Materials and Methods

2.1. Reactor Set-Up

In the current study, three laboratory-scale water-jacketed reactors were operated in parallel in a downflow mode. The schematic diagram of the lab-scale reactors is shown in Figure 1. The lab-scale packed bed reactors contained the following components: feed and effluent buckets, peristaltic pumps (Watson-Marlow Fluid Technology Group, Johannesburg, South Africa), water-jacketed columns, and a PolyScience Whispercool® (PolyScience, Niles, IL, USA) heater/chiller for temperature control. Each reactor consisted of a base section that functioned as a stand and housed a conical section with an outlet at the bottom of the cone. Above the conical section was a perforated plate with 1 cm holes. The reactors were 1 m in height and 0.15 m in internal diameter, and they had a total working volume of 8 L. A piece of silicone tubing, with an outer and inner diameter of 1 cm and 0.7 cm respectively, was connected at the bottom of the cone. The tube was routed up the column to near the top edge. A T-piece was fitted at the top of the tube to assist in fluid level adjustment. The water jacket of the reactor was connected in closed circuit with a heater/chiller unit with a built-in pump that recirculated water through the water jacket of the column. The feed water for each reactor was stored in 10 L or 25 L plastic buckets, where it was pumped to the top of the column using a variable flowrate Watson Marlow 120 series peristaltic pump. The reactor overflow was collected in a 15 L bucket for each column.

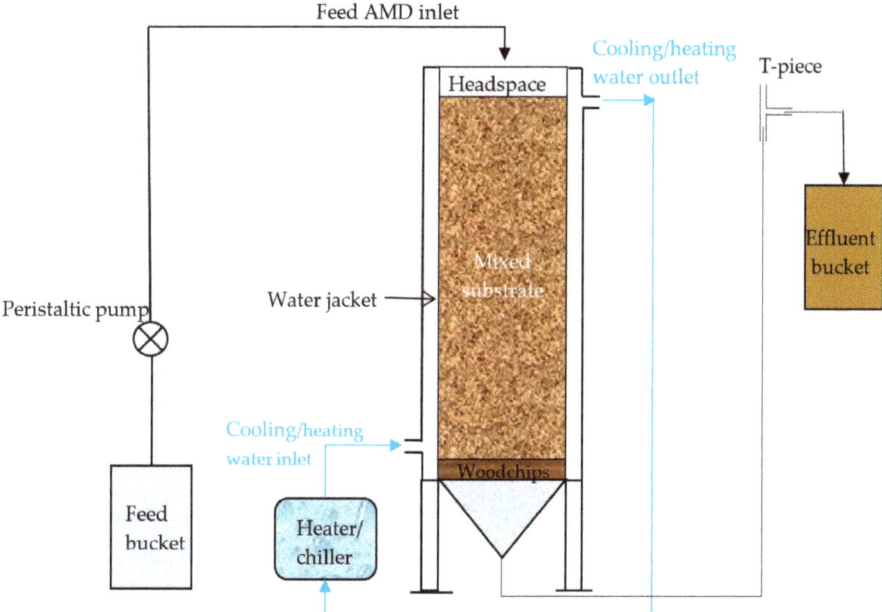

Figure 1. Schematic diagram of the reactors.

2.2. Substrates

Initially, the three lab-scale reactors were packed with 30% woodchips, 30% wood shavings, 20% hay, 10% lucerne straw, and 10% cow manure measured by volume. A mixture of the above-mentioned substrates was blended and loaded into the reactors. Woodchips, wood shavings, hay, and lucerne straw are cellulosic compounds that can contribute as substrates, although their contribution is small [32,33]; hence, they were used as a support for the biofilm. Above the perforated plate, a 2–3 cm layer of woodchips was evenly spread to prevent the holes on the perforated plate from blocking when the substrates migrated downwards. Cow manure purchased from Lifestyle, Johannesburg, South Africa and lucerne pellets purchased from Milmac Feeds, Fourways, South Africa were used as the main substrates. Then, 186 mL (128.07 g) of cow manure and 186 mL (63.69 g) of lucerne pellets were added on top of the reactor packing and replenished once every week.

2.3. *AMD and Inoculum*

The AMD used for all experiments was collected from a coal mine in eMalahleni, Mpumalanga province, South Africa. The raw AMD was characterised by pH less than 3 and a sulphate concentration ranging between 2500 mg/L and 5200 mg/L. The anaerobic mixed culture used was collected from one of the reactors that were operating at the Mintek's pilot plant at the coal mine. The pilot plant had been operating for 10 months at HRT varying between 5 days and 7 days, influent pH > 5 with sulphate reduction efficiency above 90%, sulphide concentration varying between 200 mg/L and 700 mg/L, and it was packed with the same mixture as that used in lab-scale reactors. The lab-scale reactors were inoculated with a mixture of mine water adjusted using hydrated lime to pH approximately 6.5 (70% *v*/*v*) and the inoculum (30% *v*/*v*). For the duration of the study, the flow rate varied between 1.14 L/day and 4 L/day with a sulphate loading rate between 0.36 g/L/day and 2.6 g/L/day. Hydrated lime was used for all pH adjustments.

2.4. *Sampling and Analysis*

The effluent pH was measured immediately after the samples were taken. Samples were collected using a beaker; then, the pH meter was immersed into the sample, and the reading was recorded once stabilised. A Metrohm pH sensor (Metrohm, Herisau, Switzerland) was used for pH measurements, and it was calibrated for pH 4 and pH 7 buffer solutions before analysis.

For sulphate analysis, a turbidimetric method was used to measure the influent and effluent sulphate concentration. This was achieved by using a Merck Spectroquant® Prove 300 (Merck, Darmstadt, Germany). All the samples were filtered using 0.22 μm membrane syringe filters before analysis to prevent interferences from suspended solids. Samples were analyzed immediately after collection.

The potentiometric determination of hydrogen sulphide using 0.1 M $AgNO_3$ was used to determine the total sulphide concentration in the effluent. A Metrohm Titrando (Metrohm, Herisau, Switzerland) was used for sulphide titrations using $AgNO_3$.

2.5. *Experimental Design*

Design-Expert® (version 11.1.2.0, Stat Ease Inc., Minneapolis, MN, USA), a statistical tool that helps with the design of experiments (DoE), was used to design the experiments. The Box–Behnken model with 3 centre points was used for the design. A total of 16 experiments were conducted. The effect of three factors—namely, pH ranging from 4 to 6 in 1 unit steps, temperature ranging from 10 °C to 30 °C in steps of 10 °C, and HRT ranging from 2 days to 7 days in steps of 2.5 days—was studied. The pH range was selected after it was found in the literature that most SRB perform better near neutral pH, and that at pH less than 4, SRB are suppressed and therefore affecting their performance [17,34]. The temperature range was selected after considering the average temperatures at eMalahleni throughout the year, and HRT was selected based on previous studies done at Mintek. The levels of the chosen variables in the design of experiments are shown in Table 1. The sulphate reduction efficiency and sulphate reduction rate were the corresponding response variables, as shown in Table 2.

Table 1. Box–Behnken design for three factors in experimental design.

Code	Factors	Factor Range and Levels (Coded)		
		−1	0	1
A	pH	4	5	6
B	Temperature (°C)	10	20	30
C	HRT (days)	2	4.5	7

Table 2. Experimental runs and obtained responses.

Run	Independent Variables			Response Variables	
	A: pH	B: Temperature (°C)	C: HRT (days)	Sulphate Reduction Efficiency (%)	Sulphate Reduction Rate (mol/m³/day)
1	5	20	4.5	76.94	4.57
2	5	30	7	98.73	7.66
3	4	30	4.5	90.34	5.70
4	5	30	2	65.61	9.75
5	6	20	2	50.28	8.97
6	5	10	7	41.74	1.66
7	5	20	4.5	84.46	4.99
8	4	10	4.5	22.76	1.67
9	6	30	4.5	96.87	6.17
10	6	20	7	97.56	5.63
11	4	20	2	55.17	9.89
12	4	20	7	93.43	5.35
13	5	20	4.5	84.38	4.98
14	6	10	4.5	30.38	1.88
15	6	30	7	98.40	7.54
16	5	30	4.5	96.57	6.46

2.6. Statistical Analysis

Response surface methodology [24] was used to understand the interactions between the independent variables. This was achieved by fitting the experimental data into a polynomial quadratic equation to obtain regression coefficients, as shown in Equation (4).

$$Y = b_0 + \sum b_i x_i + \sum b_{ii} {x_i}^2 + \sum b_{ij} x_i x_j \quad (4)$$

where Y is the response variable, b_0 is the constant term, b_i is the linear coefficient, b_{ii} is the quadratic coefficient, b_{ij} is the interaction coefficient, and x_i and x_j are the values of the coded variables. In this study, the sulphate reduction efficiency (%) and sulphate reduction rate (mol/m^3/day) were chosen as response variables and therefore were fitted into Equation (4). Analysis of variance (ANOVA) was used to evaluate the validity and significance of the fitted model. The coefficient of determination R^2, adjusted R^2, predicted R^2, lack of fit, adequate precision, F-value, and *p*-value were used to further evaluate the quality and accuracy of the model. In the present study, the significance level was set at 0.05.

3. Results and Discussion

3.1. Statistical Analysis

The data obtained from the 16 experiments that were conducted were fitted into polynomial quadratic equations as shown in Equations (5) and (6) in terms of coded factors.

$$\begin{aligned}\text{Sulphate reductionefficiency} &= +\,82.86 + 2.65 \times A + 33.79 \times B + 21.31 \times C - 0.7714 \times AB \\ &-0.6874 \times AC - 5.67 \times BC + 1.08 \times A^2 - 23.64 \times B^2 + 10.51 \times C^2\end{aligned} \quad (5)$$

$$\begin{aligned} \text{Sulphate reductionrate} \\ &= +4.92 + 0.2257 \times A + 2.11 \times B\, 1.97 \times C + 0.05 \times AB \\ &-0.1838 \times AC + 0.8878 \times BC - 0.1347 \times A^2 - 0.8936 \times B^2 \\ &+ 2.63 \times C^2 \end{aligned} \quad (6)$$

The reliability, quality, and accuracy of the fitted quadratic models were evaluated using analysis of variance (ANOVA), as shown in Table 3. The significance of the models is confirmed by high F-values and low p-values [35]. The models were significant as confirmed by low probability values of less than 0.0001 and high F-values of 101.70 for sulphate reduction efficiency and 221.37 for sulphate reduction rate. The reported F-values imply that there is only a 0.01% chance that the differences could be due to noise. For this study, the lack of fit for both models was insignificant, which shows that the data fitted the models well.

Table 3. ANOVA results for the fitted quadratic model.

Response	Source of Variation	Sum of Squares	df	Mean Square	F-Value	p-Value
Sulphate reduction efficiency (%)	Model	10,155.47	9	1128.39	101.70	<0.0001 [1]
	A-pH	59.11	1	59.11	5.33	0.0604 [2]
	B-Temperature	6519.62	1	6519.62	587.61	<0.0001 [1]
	C-HRT	2442.99	1	2442.99	220.19	<0.0001 [1]
	AB	2.65	1	2.65	0.2389	0.6424 [2]
	AC	2.11	1	2.11	0.1897	0.6784 [2]
	BC	72.74	1	72.74	6.56	0.0429 [1]
	A^2	3.68	1	3.68	0.3316	0.5856 [2]
	B^2	1808.36	1	1808.36	162.99	<0.0001 [1]
	C^2	370.19	1	370.19	33.37	0.0012 [1]
	Residual	66.57	6	11.10		
	Lack of Fit	29.32	4	7.33	0.3937	0.8060 [2]
	Pure Error	37.25	2	18.62		
	Correction Total	10,222.04	15			
Sulphate reduction rate (mol/m^3/day)	Model	101.69	9	11.30	221.37	<0.0001 [1]
	A-pH	0.5037	1	0.5037	1.49	0.2617 [2]
	B-Temperature	25.50	1	25.50	499.63	<0.0001 [1]
	C-HRT	20.96	1	20.96	410.67	<0.0001 [1]
	AB	0.0093	1	0.0093	0.1822	0.6844 [2]
	AC	0.1504	1	0.1504	2.95	0.1369 [2]
	BC	1.78	1	1.78	34.89	0.0010 [1]
	A^2	0.0577	1	0.0577	1.13	0.3287 [2]
	B^2	2.58	1	2.58	50.60	0.0004 [1]
	C^2	23.09	1	23.09	452.42	<0.0001 [1]
	Residual	0.3062	6	0.0510		
	Lack of Fit	0.1926	4	0.0481	0.8472	0.6045 [2]
	Pure Error	0.1137	2	0.0568		
	Correction Total	102.00	15			

df—degree of freedom; [1] Significant; [2] Not significant.

Fit statistics are shown in Table 4. The coefficient of determination R^2 is a statistical parameter that measures how well the data fits the line. Adjusted R^2 is a version of R^2 that is always smaller than R^2, and predicted R^2 measures the predictive accuracy of the model [25]. A model is considered well fitted when the R^2 value is greater than 0.8 [25]. R^2, adjusted R^2, and predicted R^2 were found to be 0.9935, 0.9837, and 0.9716 for sulphate reduction efficiency and 0.9970, 0.9925, and 0.9853 for sulphate reduction rate. The difference between the predicted and adjusted R^2 should be less than 0.2 for the model to be considered well fitted and able to make satisfactory predictions. For this study, predicted and adjusted R^2 were in agreement with this. Adequate precision measures the signal-to-noise ratio, and a ratio greater than 4 is desirable. The values of adequate precision were 29.36 and 46.21 for sulphate reduction efficiency and sulphate reduction rate, respectively, which indicates an adequate signal.

Table 4. Fit statistics.

	R^2	Adjusted R^2	Predicted R^2	Adequate Precision
Sulphate reduction efficiency (%)	0.9935	0.9837	0.9716	29.36
Sulphate reduction rate (mol/m^3/day)	0.9970	0.9925	0.9853	46.21

The diagnostic section provides plots that can be used to further validate the accuracy of the model. The normal probability plots illustrated in Figure 2 show that the residuals are normally distributed as the points are closer to the line. Residuals vs. predicted, as shown in Figure 3, proved the models' quality by having random scatters that are evenly distributed above and below the horizontal axis [25]. The correlation between predicted and actual values is shown in Figure 4. The clustering of the values along the diagonal line confirms that the model is accurate and robust [25,36].

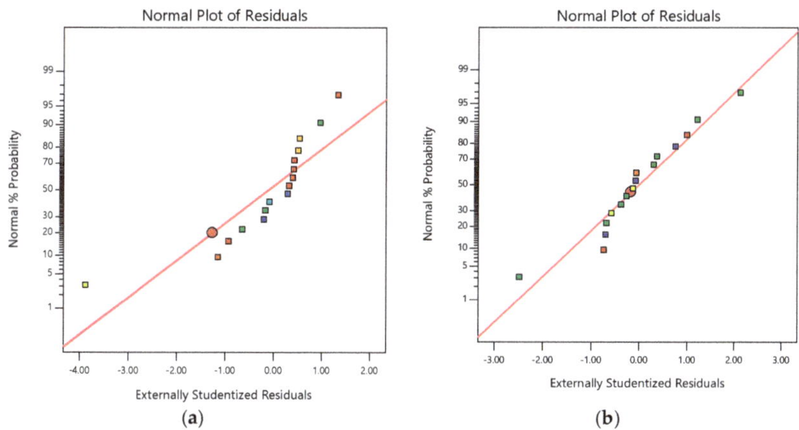

Figure 2. Normal plot of residuals for (**a**) sulphate reduction efficiency and (**b**) sulphate reduction rate.

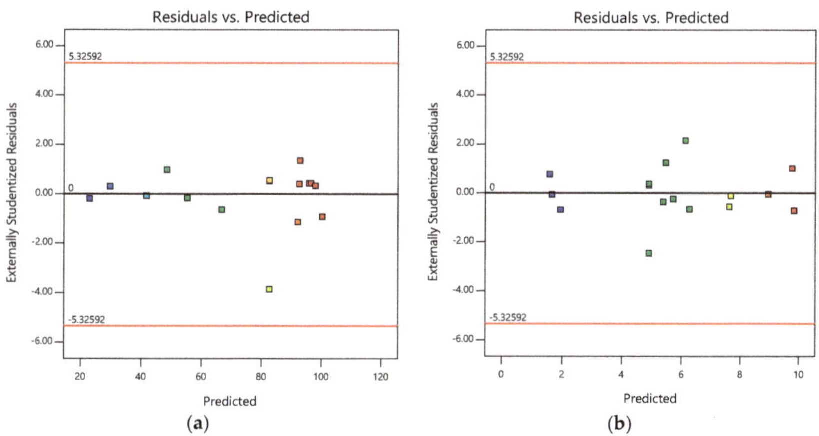

Figure 3. Plot of residuals vs. predicted for (**a**) sulphate reduction efficiency and (**b**) sulphate reduction rate.

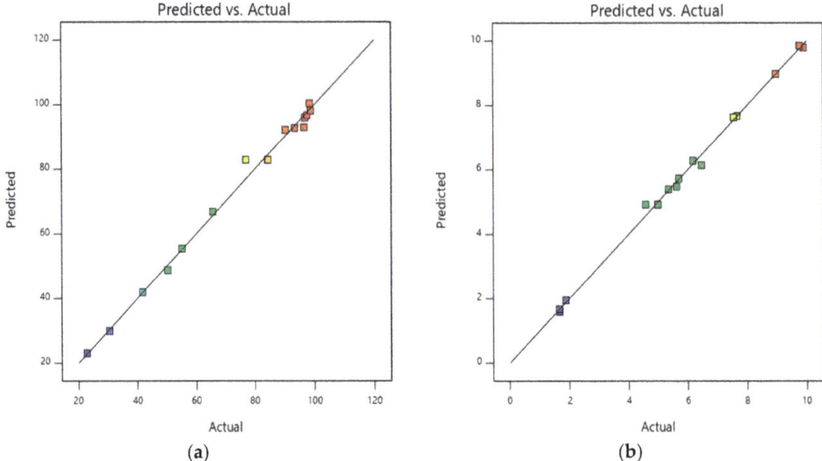

Figure 4. Correlation between predicted and actual values for (**a**) sulphate reduction efficiency and (**b**) sulphate reduction rate.

3.2. *Effect of Individual Factors*

The effect of individual factors on the responses is shown in this section. One factor is changed at a time while keeping other factors at the centre point. The steeper the slope, the more sensitive a response is to the factor.

3.2.1. pH

Figure 5a,b shows the effect of pH on sulphate reduction efficiency and sulphate reduction rate, respectively. From the graphs, both responses slightly increase with an increase in pH from pH 4 to pH 6; however, the increase is not significant. This is evident in Table 3, where pH was insignificant for both responses. This shows that SRBs were not suppressed at an initial pH of 4, which is considered too low for SRB to grow [37]. This could imply that lower costs need to be expended for pH adjustment, and this could have a positive impact on the process's operating expenses. Sulphate reduction at pH approximately 4 was highly impacted in some studies [38–40], which may be because the reactor pH was controlled [39]. However, Jong and Parry [34] found a sulphate reduction efficiency of above 80% when the reactor's influent pH was 4.07. For this study, only the influent pH was controlled, and the average effluent pH, for experiments that had an influent pH of 4, was above 6, as shown in Figure 6. Jong and Parry [34] also observed an effluent pH of above 6. As a result, there are uncertainties about the pH at which sulphate reduction occurs if only the influent pH is controlled, as the effluent pH is always higher [41].

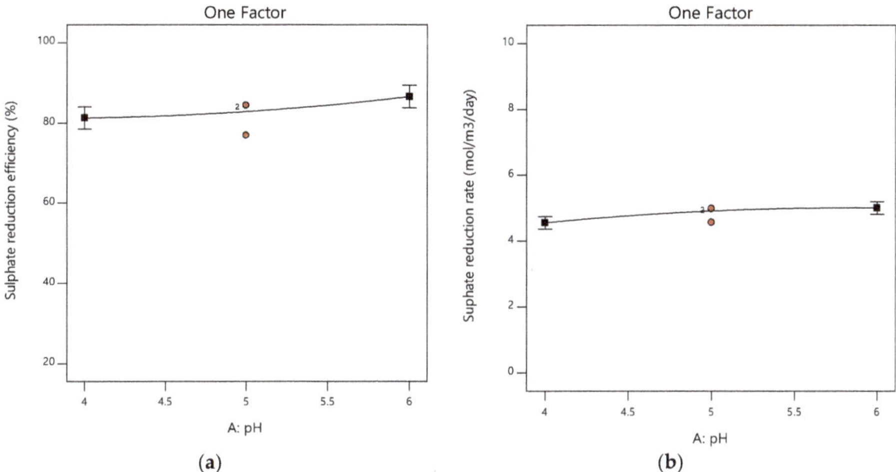

Figure 5. pH effect on (**a**) sulphate reduction efficiency and (**b**) sulphate reduction rate at 20 °C and hydraulic retention time (HRT) of 4.5 days.

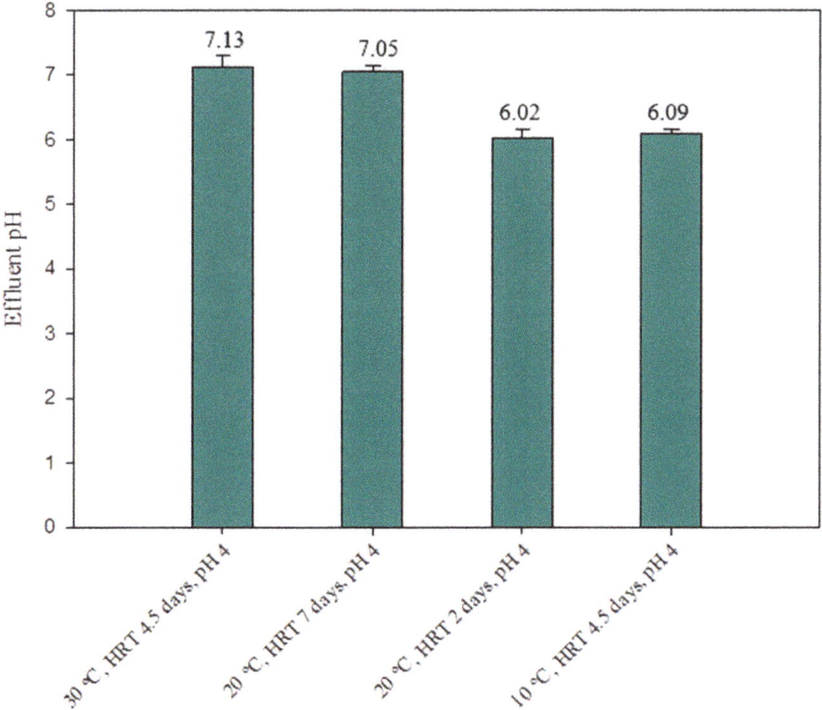

Figure 6. Average effluent pH at different conditions with an influent pH of 4.

3.2.2. Temperature

The effect of temperature is depicted in Figure 7a,b. Decreasing temperature from 30 to 20 °C with HRT and pH at the centre point had a minimal impact on SRB activity, which was expected, as 20 °C is in a range that supports the growth and activity of SRB, as also observed in other studies [42–44]. A further decrease in temperature from 20 to 10 °C slowed down the metabolic activity significantly. Sheoran et al. [10] suggested that sulphate reduction is likely to decrease by 50% at temperatures lower than 10 °C compared to sulphate reduction at 20 °C. Similarly, the same effect can be observed from the graphs. The sulphate reduction efficiency and sulphate reduction rate decrease by more than 50% with a decrease in temperature from 20 to 10 °C. According to the graphs, the sulphate reduction efficiency decreases from approximately 80% to about 25%, and the sulphate reduction rate decreases from approximately 5 to 2 mol/m^3/day when the temperature is decreased from 20 to 10 °C with HRT and pH kept constant at the centre point.

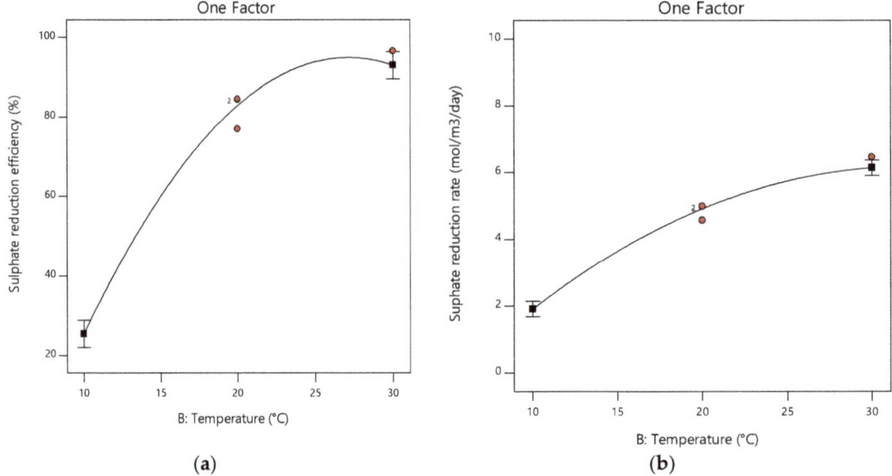

Figure 7. Temperature effect on (**a**) sulphate reduction efficiency and (**b**) sulphate reduction rate at pH 5 and HRT of 4.5 days.

3.2.3. HRT

Figure 8a shows that the sulphate reduction efficiency response increases as the HRT increases from 2 to 7 days. Conversely, the sulphate reduction rate decreases with increasing HRT from 2 to approximately 5 days, followed by a slight increase with a further increase in HRT to 7 days, as shown in Figure 8b. Although sulphate reduction efficiency was observed to decrease with decreasing HRT, the sulphate reduction rate increased with a decrease in HRT due to higher feed rates. A longer HRT resulted in higher sulphate reduction efficiency but lower sulphate reduction rates. Similar observations were made in earlier studies [45–48].

Figure 9a,b shows the effect of HRT over time at 30 °C and pH 5. When HRT was decreased from 7 to 4.5 days, there was only a slight decrease in both the sulphate reduction efficiency and sulphate reduction rate. An interesting observation was made when the HRT was decreased from 4.5 days to 2 days. At a HRT of 2 days, both the sulphate reduction efficiency and sulphate reduction rate increased rapidly upon replenishing the substrates. This was followed by a decrease after a maximum was reached. Although some studies show that HRT leads to a decrease in sulphate reduction due to SRB washout [49,50], it is presumed in this study that the decrease in sulphate reduction efficiency and sulphate reduction rate was a result of substrates washout [51]. This was evident when both responses improved upon the replenishing of substrates. A study done by Poinapen et al. [52] using primary sewage sludge (PSS) as a substrate demonstrated that a decrease in HRT did not have an impact on sulphate reduction. This was because the PSS was fed into the reactor together with the synthetic AMD; therefore, a decrease in HRT implied that the PSS loading was increasing. In other words, the substrate loading was increased with a decrease in HRT, which was not the case in this study. Hence, it is presumed that the substrates were washed out quicker.

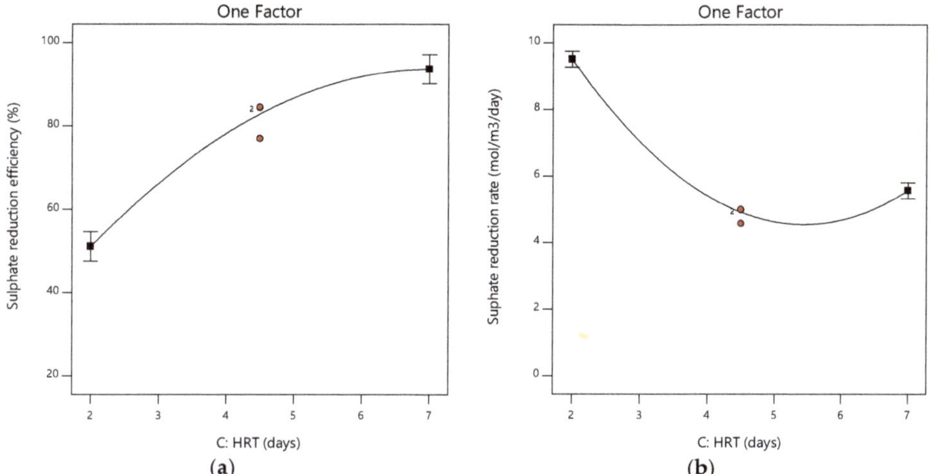

Figure 8. HRT effect on (**a**) sulphate reduction efficiency and (**b**) sulphate reduction rate at 20 °C and pH 5.

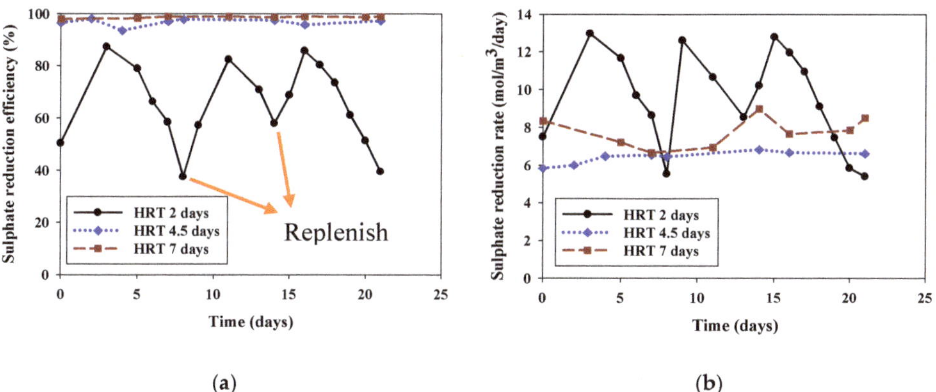

Figure 9. Effect of HRT over time at 30 °C and pH 5 for (**a**) sulphate reduction efficiency and (**b**) sulphate reduction rate.

3.3. Interactions between Factors

The interactive effects between pH, temperature, and HRT on sulphate reduction efficiency and sulphate reduction rate are shown in this section. The interaction between temperature and pH illustrated in Figure 10a,b shows that both the sulphate reduction efficiency and sulphate reduction rate decrease with a decrease in temperature. Furthermore, it is clear from the graphs that pH does not affect both responses, especially at low temperatures. At maximum temperature, the responses slightly increase with an increase in pH.

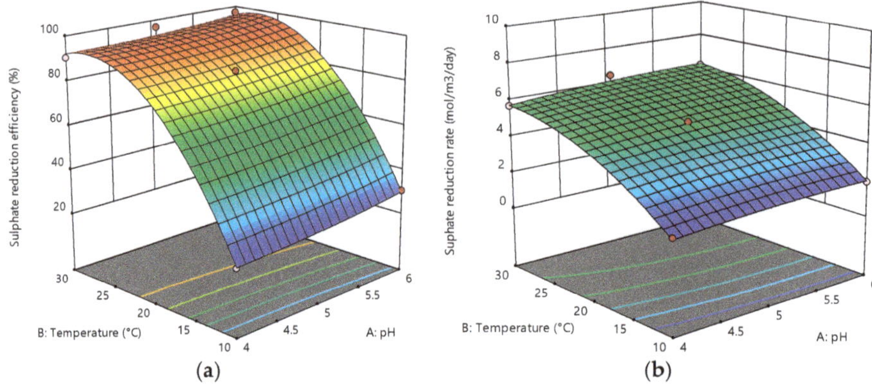

Figure 10. Interactive effects between temperature and pH on (**a**) sulphate reduction efficiency and (**b**) sulphate reduction rate.

Figure 11a,b shows the interactive effects between HRT and pH on sulphate reduction efficiency and sulphate reduction rate. From the graphs, both responses are not impacted by pH at all HRTs. However, HRT is shown to have different effects on the responses. The sulphate reduction efficiency increases with an increase in HRT, whereas the overall trend for the sulphate reduction rate is that it increases with decreasing HRT. The graphs in Figure 10a,b and Figure 11a,b show that the interactions between pH and temperature (AB) and between pH and HRT (AC) were not strong, which is proven by ANOVA analysis in Table 3.

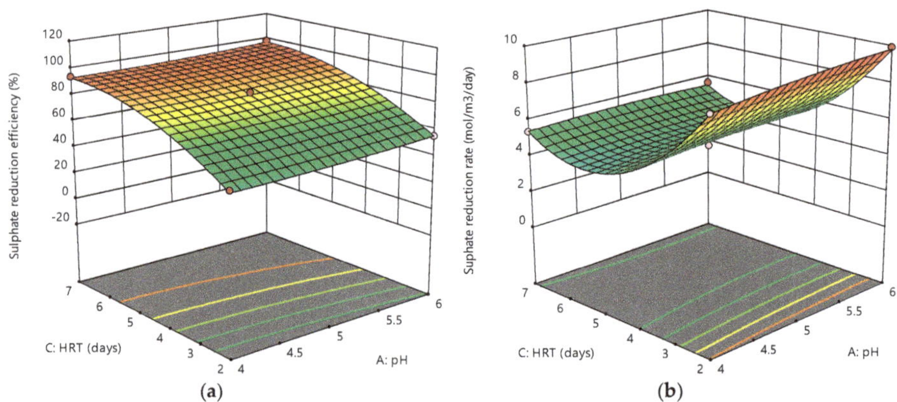

Figure 11. Interactive effects between HRT and pH on (**a**) sulphate reduction efficiency and (**b**) sulphate reduction rate.

Strong interactions were observed between HRT and temperature, as depicted in Figure 12a,b. In Figure 12a, the sulphate reduction efficiency increases with a simultaneous increase in temperature and HRT. However, temperature had more impact on the sulphate reduction efficiency compared to HRT. For example, at 30 °C, the sulphate reduction efficiency decreased from almost 100% at an HRT of 7 days to just above 60% at an HRT of 2 days. On the other hand, at an HRT of 7 days, it decreased to approximately 40% with a decrease in temperature from 30 to 10 °C. Conversely, the increase in sulphate reduction rate caused by a decrease in the HRT was greater at all temperatures than that caused by an increase in temperature at all HRTs, as shown in Figure 12b. The highest sulphate reduction rate was approximately 10 mol/m^3/day, which was observed at 30 °C and an HRT of 2 days

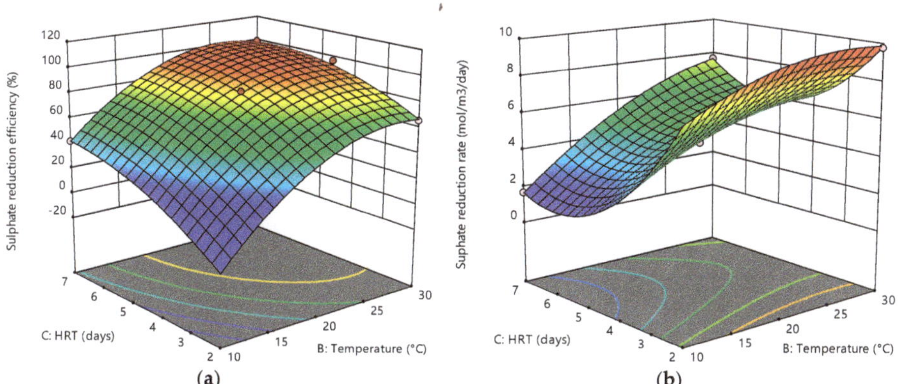

Figure 12. Interactive effects between HRT and temperature on (**a**) sulphate reduction efficiency and (**b**) sulphate reduction rate.

In this section, it was shown that pH had an insignificant interaction with both temperature and HRT. However, HRT and temperature were shown to have strong interactions. The information obtained and the mathematical models developed will be used to evaluate the performance of the pilot plant. For example, due to the difficulty in controlling temperature in an open system, the models developed will be used to determine at what HRT and pH the pilot plant should operate during winter and summer seasons to compensate for drops in sulphate reduction associated with temperature changes. Considering that pH had no effect within the range investigated, it presumed that the only parameter that will be controlled is the HRT. However, further tests will be done on the pilot plant to confirm this.

3.4. Optimisation

RSM does not only help with understanding the behaviour of systems, but it is also used as a decision-making tool by evaluating the consequences of different scenarios [53]. The numerical optimisation section in design expert allows one to maximise the desirability function. The desirability level varies from 0 to 1, with level 0 indicating that one of the responses is outside the specified limit and a level closer to 1 indicating that the corresponding factor combination is closer to optimal [53]. The optimisation process was carried out with the goal to maximise the sulphate reduction efficiency simultaneously with the sulphate reduction rate by minimising the pH and setting the temperature at 10 °C, 20 °C, and 30 °C. The HRT was set to be in range between 2 days and 7 days. At 10 °C, a 41.89% sulphate reduction efficiency and 1.67 mol/m^3/day sulphate reduction rate can be achieved at pH 5 and HRT of 7 days. However, the desirability was low at 0.068. This could mean that more retention time, beyond that which was investigated in this study, will be required to achieve a higher sulphate reduction efficiency and sulphate reduction rate. At 20 °C, a sulphate reduction efficiency of 92.56% and sulphate reduction rate of 5.06 mol/m^3/day can be achieved at pH 4 and HRT 6.7 days, and the desirability was 0.756. A sulphate reduction efficiency of 94.46% and sulphate reduction rate of 5.65 mol/m^3/day can be achieved at 30 °C, pH 4, and HRT 4.8 days, and the desirability was 0.8. This shows that at higher temperatures, the pilot plant can operate at lower HRTs.

3.5. Sulphide

Hydrogen sulphide is produced during the reduction of sulphate, as shown in Equation (2), and it is known for its toxicity. Hydrogen sulphide causes problems such as odour, corrosion, and sulphate reduction inhibition [54,55]. The sulphide concentration for all experiments in this study ranged between 114 and 798 mg/L. Due to the high sulphide concentrations observed, there are further tests that are currently being done at Mintek for a downstream process that will use sulphide oxidising bacteria to oxidise sulphide to elemental sulphur (S^0), as shown in Equation (7). The oxidation of sulphide to elemental sulphur is a result of incomplete oxidation. The complete oxidation of sulphide results in the formation of sulphate, as shown in Equation (8). Therefore, it is recommended that the ratio of sulphide to oxygen be kept at 2:1 to prevent complete oxidation to sulphate [56].

$$HS^- + 1/2O_2 \rightarrow S^0 + OH^- \quad (7)$$

$$HS^- + 2O_2 \rightarrow SO_4^{2-} + H^+ \quad (8)$$

4. Conclusions

In the current study, RSM was used to statistically analyse the data. ANOVA results showed that the sulphate reduction efficiency and sulphate reduction rate models were significant and adequate, as proven by statistical indexes including lack of fit, coefficient of variation, and adequate precision. Individually, the pH effect was insignificant for both responses, and therefore, its interaction with other independent variables was also not significant. However, there was a strong interaction between HRT and temperature. Additionally, a decrease in HRT impacted the sulphate reduction rate positively due to increased flow rates. Conversely, it had a negative impact on the sulphate reduction efficiency, which was likely due to substrates washout. This study developed mathematical models that were found to be statistically significant. These models can be used as decision-making tools by using them to predict how the process will react to different conditions within the investigated range. This will help adjust controllable factors such as pH and HRT when the temperature fluctuates.

Author Contributions: Conceptualisation, E.M.N.C. and D.M.; methodology, M.J.M., E.M.N.C. and D.M.; formal analysis, M.J.M.; investigation, M.J.M.; resources, E.M.N.C. and D.M.; data curation, M.J.M.; writing—Original draft preparation, M.J.M.; writing—Review and editing, E.M.N.C. and D.M.; supervision, E.M.N.C. and D.M.; project administration, M.J.M. All authors have read and agreed to the published version of the manuscript.

Funding: This research received no external funding.

Acknowledgments: The authors gratefully acknowledge Mintek for providing facilities to conduct the research.

Conflicts of Interest: The authors declare no conflict of interest.

References

1. Equeenuddin, S.M.; Tripathy, S.; Sahoo, P.K.; Panigrahi, M.K. Hydrogeochemical characteristics of acid mine drainage and water pollution at Makum Coalfield, India. *J. Geochem. Explor.* **2010**, *105*, 75–82. [CrossRef]
2. Mccarthy, T.S. The impact of acid mine drainage in South Africa. *S. Afr. J. Sci.* **2011**, *107*, 1–7. [CrossRef]
3. Akcil, A.; Koldas, S. Acid Mine Drainage (AMD): Causes, treatment and case studies. *J. Clean. Prod.* **2006**, *14*, 1139–1145. [CrossRef]
4. Kaksonen, A.H.; Puhakka, J.A. Sulfate Reduction Based Bioprocesses for the Treatment of Acid Mine Drainage and the Recovery of Metals. *Eng. Life Sci.* **2007**, *7*, 541–564. [CrossRef]
5. Arnold, M.; Mariekie, G.; Ritva, M. Technologies for sulphate removal with valorisation options. In Proceedings of the IMWA 2016 Annual Conference: Mining Meets Water—Conflicts and Solutions, Leipzig, Germany, 11–15 July 2016; pp. 1343–1345.
6. Arnold, D.E. Diversion wells—A low-cost approach to treatment of acid mine drainage. In Proceedings of the Twelfth West Virginia Surface Mine Drainage Task Force Symposium, Morgantown, WV, USA, 3–4 April 1991.
7. Johnson, D.B.; Hallberg, K.B. Acid mine drainage remediation options: A review. *Sci. Total Environ.* **2005**, *338*, 3–14. [CrossRef]

8. Liamleam, W.; Annachhatre, A.P. Electron donors for biological sulfate reduction. *Biotechnol. Adv.* **2007**, *25*, 452–463. [CrossRef]
9. Luptakova, A.; Macingova, E. Alternative substrates of bacterial sulphate reduction suitable for the biological-chemical treatment of acid mine drainage. *Acta Montan. Slovaca* **2012**, *17*, 74.
10. Sheoran, A.S.; Sheoran, V.; Choudhary, R.P. Bioremediation of acid-rock drainage by sulphate-reducing prokaryotes: A review. *Miner. Eng.* **2010**, *23*, 1073–1100. [CrossRef]
11. Jiménez-Rodríguez, A.M.; Durán-Barrantes, M.D.L.M.; Borja, R.; Sánchez, E.; Colmenarejo, M.; Raposo, F. Biological sulphate removal in acid mine drainage using anaerobic fixed bed reactors with cheese whey as a carbon source. *Lat. Am. Appl. Res.* **2010**, *40*, 329–335.
12. Logan, M.V.; Reardon, K.F.; Figueroa, L.A.; Mclain, J.E.; Ahmann, D.M. Microbial community activities during establishment, performance, and decline of bench-scale passive treatment systems for mine drainage. *Water Res.* **2005**, *39*, 4537–4551. [CrossRef]
13. Mannucci, A.; Munz, G.; Mori, G.; Lubello, C. Factors affecting biological sulphate reduction in tannery wastewater treatment. *Environ. Eng. Manag. J.* **2014**, *13*, 1005–1012. [CrossRef]
14. Neale, J.W.; Muller, H.H.; Gericke, M.; Muhlbauer, R. Low-Cost Biological Treatment of Metal- and Sulphate-Contaminated Mine Waters. *IMWA* **2017**, *1*, 453–460.
15. Seyler, J.; Figueroa, L.; Ahmann, D.; Wildeman, T.; Robustelli, M. Effect of solid phase organic substrate characteristics on sulfate reducer activity and metal removal in passive mine drainage treatment systems. In Proceedings of the National Meeting of the American Society of Mining and Reclamation and the 9th Billings Land Reclamation Symposium, Billings, MT, USA, 3–6 June 2003.
16. Zagury, G.; Neculita, C.; Bussiere, B. Passive treatment of acid mine drainage in bioreactors: Short review, applications, and research needs. In Proceedings of the 60th Canadian Geotechnical Conference & 8th Joint CGS/IAH-CNC Groundwater Conference, Ottawa, ON, Canada, 21–24 October 2007.
17. Bijmans, M.; Buisman, C.; Meulepas, R.; Lens, P. Sulfate reduction for inorganic waste and process water treatment. *Compr. Biotechnol.* **2011**, *6*, 435–446. [CrossRef]
18. Chaiprapat, S.; Preechalertmit, P.; Boonsawang, P.; Karnchanawong, S. Sulfidogenesis in Pretreatment of High-Sulfate Acidic Wastewater Using Anaerobic Sequencing Batch Reactor and Upflow Anaerobic Sludge Blanket Reactor. *Environ. Eng. Sci.* **2011**, *28*, 597–604. [CrossRef]
19. Moon, C.; Singh, R.; Veeravalli, S.; Shanmugam, S.; Chaganti, S.; Lalman, J.; Heath, D. Effect of COD:SO_4^{2-} Ratio, HRT and Linoleic Acid Concentration on Mesophilic Sulfate Reduction: Reactor Performance and Microbial Population Dynamics. *Water.* **2015**, *7*, 2275–2292. [CrossRef]
20. Doshi, S.M. *Bioremediation of Acid Mine Drainage Using Sulfate-Reducing Bacteria*; US Environmental Protection Agency, Office of Solid Waste and Emergency Response and Office of Superfund Remediation and Technology Innovation: Washington, DC, USA, 2006; p. 65.
21. Lettinga, G.; Pol, L.H.; Koster, I.; Wiegant, W.; De Zeeuw, W.; Rinzema, A.; Grin, P.; Roersma, R.; Hobma, S. High-rate anaerobic waste-water treatment using the UASB reactor under a wide range of temperature conditions. *Biotechnol. Genet. Eng. Rev.* **1984**, *2*, 253–284. [CrossRef]
22. Salo, M.; Bomberg, M.; Grewar, T.; Seepei, L.; Mariekie, G.; Arnold, M. Compositions of the Microbial Consortia Present in Biological Sulphate Reduction Processes During Mine Effluent Treatment. *IMWA* **2017**, *1*, 14–21.
23. Visser, A.; Gao, Y.; Lettinga, G. Effects of short-term temperature increases on the mesophilic anaerobic breakdown of sulfate containing synthetic wastewater. *Water Res.* **1993**, *27*, 541–550. [CrossRef]
24. Bezerra, M.A.; Santelli, R.E.; Oliveira, E.P.; Villar, L.S.; Escaleira, L.A. Response surface methodology (RSM) as a tool for optimization in analytical chemistry. *Talanta* **2008**, *76*, 965–977. [CrossRef]
25. Najib, T.; Solgi, M.; Farazmand, A.; Heydarian, S.M.; Nasernejad, B. Optimization of sulfate removal by sulfate reducing bacteria using response surface methodology and heavy metal removal in a sulfidogenic UASB reactor. *J. Envion. Chem. Eng.* **2017**, *5*, 3256–3265. [CrossRef]
26. Cabrera, G.; Pérez, R.; Gomez, J.; Abalos, A.; Cantero, D. Toxic effects of dissolved heavy metals on Desulfovibrio vulgaris and Desulfovibrio sp. strains. *J. Hazard. Mater.* **2006**, *135*, 40–46. [CrossRef] [PubMed]
27. Dvorak, D.H.; Hedin, R.S.; Edenborn, H.M.; Mcintire, P.E. Treatment of metal-contaminated water using bacterial sulfate reduction: Results from pilot-scale reactors. *Biotechnol. Bioeng.* **1992**, *40*, 609–616. [CrossRef] [PubMed]

28. Macingova, E.; Luptakova, A. Recovery of metals from acid mine drainage. *Chem. Eng.* **2012**, *28*, 109–114. [CrossRef]
29. Ňancucheo, I.; Johnson, D.B. Selective removal of transition metals from acidic mine waters by novel consortia of acidophilic sulfidogenic bacteria. *Microb. Biotechnol.* **2012**, *5*, 34–44. [CrossRef]
30. Utgikar, V.P.; Harmon, S.M.; Chaudhary, N.; Tabak, H.H.; Govind, R.; Haines, J.R. Inhibition of sulfate-reducing bacteria by metal sulfide formation in bioremediation of acid mine drainage. *Environ. Toxicol.* **2002**, *17*, 40–48. [CrossRef]
31. Dev, S.; Roy, S.; Das, D.; Bhattacharya, J. Improvement of biological sulfate reduction by supplementation of nitrogen rich extract prepared from organic marine wastes. *Int. Biodeterior. Biodegrad.* **2015**, *104*, 264–273. [CrossRef]
32. Choudhary, R.P.; Sheoran, A.S. Comparative study of cellulose waste versus organic waste as substrate in a sulfate reducing bioreactor. *Bioresour. Technol.* **2011**, *102*, 4319–4324. [CrossRef]
33. Kuyucak, N.; St-Germain, P. In situ treatment of acid mine drainage by sulfate reducing bacteria in open pits: Scale-up experiences. In Proceedings of the International Land Reclamations and Mine Drainage Conference and the Third International Conference on the Abatement of Acidic Drainage, Pittsburgh, PA, USA, 24–29 April 1994.
34. Jong, T.; Parry, D.L. Microbial sulfate reduction under sequentially acidic conditions in an upflow anaerobic packed bed bioreactor. *Water Res.* **2006**, *40*, 2561–2571. [CrossRef]
35. Sun, Y.; Wei, J.; Zhang, J.P.; Yang, G. Optimization using response surface methodology and kinetic study of Fischer–Tropsch synthesis using SiO_2 supported bimetallic Co–Ni catalyst. *J. Nat. Gas Sci. Eng.* **2016**, *28*, 173–183. [CrossRef]
36. Talib, N.A.A.; Salam, F.; Yusof, N.A.; Ahmad, S.A.A.; Sulaiman, Y. Optimization of peak current of poly (3,4-ethylenedioxythiophene)/multi-walled carbon nanotube using response surface methodology/central composite design. *RSC Adv.* **2017**, *7*, 11101–11110. [CrossRef]
37. Kolmert, Å.; Johnson, D.B. Remediation of acidic waste waters using immobilised acidophillic sulfate-reducing bacteria. *J. Chem. Technol. Biotechnol.* **2001**, *76*, 836–843. [CrossRef]
38. Elliott, P.; Ragusa, S.; Catcheside, D. Growth of sulfate-reducing bacteria under acidic conditions in an upflow anaerobic bioreactor as a treatment system for acid mine drainage. *Water Res.* **1998**, *32*, 3724–3730. [CrossRef]
39. Sen, A.M.; Johnson, B. Acidophilic sulphate-reducing bacteria: Candidates for bioremediation of acid mine drainage. *Process Metall.* **1999**, *9*, 709–718. [CrossRef]
40. Tsukamoto, T.; Killion, H.; Miller, G. Column experiments for microbiological treatment of acid mine drainage: Low-temperature, low-pH and matrix investigations. *Water Res.* **2004**, *38*, 1405–1418. [CrossRef]
41. Lopes, S.I.C. Sulfate Reduction at Low pH in Organic Wastewaters. Ph.D. Thesis, Wageningen University, Wageningen, The Netherlands, 2007.
42. Ferrentino, R.; Langone, M.; Andreottola, G. Temperature effects on the activity of denitrifying phosphate accumulating microorganisms and sulphate reducing bacteria in anaerobic side-stream reactor. *J. Environ. Bio. Res.* **2017**, *1*, 1.
43. Greben, H.; Bologo, H.; Maree, J. The effect of different parameters on the biological volumetric and specific sulphate removal rates. In Proceedings of the WISA Biennial Cinference 2002, Durban, South Africa, 19–23 May 2002.
44. Marais, T.; Huddy, R.; Van Hille, R.; Harrison, S. The Effect of Temperature on the Kinetics of Sulphate Reduction and Sulphide Oxidation in an Integrated Semi-Passive Bioprocess for Remediating Acid Rock Drainage. In Proceedings of the 11th ICARD|IMWA| MWD Conference—Risk to Opportunity, Pretoria, South Africa, 10–14 September 2018.
45. Gibert, O.; De Pablo, J.; Cortina, J.L.; Ayora, C. Chemical characterisation of natural organic substrates for biological mitigation of acid mine drainage. *Water Res.* **2004**, *38*, 4186–4196. [CrossRef] [PubMed]
46. Glombitza, F. Treatment of acid lignite mine flooding water by means of microbial sulfate reduction. *Waste Manag.* **2001**, *21*, 197–203. [CrossRef]
47. Greben, H.; Maree, J. The effect of reactor type and residence time on biological sulphate and sulphide removal rates. In Proceedings of the WISA Biennial Conference 2000, Sun City, South Africa, 28 May–1 June 2000.
48. Mallelwar, P.C. Effect of Linoleic Acid and Hydraulic Retention Time on Anaerobic Sulfate Reduction in High Rate Reactors. Master's Thesis, University of Windsor, Windsor, ON, Canada, 2013.

49. Isa, Z.; Grusenmeyer, S.; Verstraete, W. Sulfate reduction relative to methane production in high-rate anaerobic digestion: Microbiological aspects. *Appl. Environ. Microbiol.* **1986**, *51*, 580–587. [CrossRef]
50. Sipma, J.; Osuna, M.B.; Lettinga, G.; Stams, A.J.; Lens, P.N. Effect of hydraulic retention time on sulfate reduction in a carbon monoxide fed thermophilic gas lift reactor. *Water Res.* **2007**, *41*, 1995–2003. [CrossRef]
51. Mukwevho, M.; Chirwa, E.; Maharajh, D. Performance of Biological Sulphate Reduction at Low pH, Low Temperature and Low Hydraulic Retention Time. *Chem. Eng. Trans.* **2020**, *79*, 439–444. [CrossRef]
52. Poinapen, J.; Wentzel, M.W.; Ekama, G. Biological sulphate reduction with primary sewage sludge in an upflow anaerobic sludge bed (UASB) reactor. Part 1: Feasibility study. *Water SA* **2009**, *35*, 525–534. [CrossRef]
53. Kasim, M.S.; Harun, N.H.; Hafiz, M.S.A.; Mohamed, S.B.; Mohamad, W.N.F.W. Multi-Response Optimization of Process Parameter in Fused Deposition Modelling by Response Surface Methodology. *Int. J. Recent Technol.* **2019**, *8*, 327–338. [CrossRef]
54. Greben, H.; Maree, J.; Eloff, E.; Murray, K. Improved sulphate removal rates at increased sulphide concentration in the sulphidogenic bioreactor. *Water SA* **2005**, *31*, 351–358. [CrossRef]
55. Valdés, F.; Muñoz, E.; Chamy, R.; Ruiz, G.; Vergara, C.; Jeison, D. Effect of sulphate concentration and sulphide desorption on the combined removal of organic matter and sulphate from wastewaters using expanded granular sludge bed (EGSB) reactors. *Electron. J. Biotechnol.* **2006**, *9*, 371–378. [CrossRef]
56. Marais, T.S.; Huddy, R.; Harrison, S.T.L.; Van Hille, R. Demonstration of simultaneous biological sulphate reduction and partial sulphide oxidation in a hybrid linear flow channel reactor. *J. Water Process Eng.* **2020**, *34*, 101143. [CrossRef]

Article

The Fate of Anthropogenic Nanoparticles, $nTiO_2$ and $nCeO_2$, in Waste Water Treatment

Thomas Lange [1], Petra Schneider [2,*], Stefan Schymura [3] and Karsten Franke [3]

1 AUD Analytik- und Umweltdienstleistungs GmbH, Jagdschänkenstr. 52, D-09117 Chemnitz, Germany; thomas.lange@aud-chemnitz.de

2 Department Water, Environment, Civil Engineering and Safety, Magdeburg-Stendal University of Applied Sciences, Breitscheidstrasse. 2, D-39114 Magdeburg, Germany

3 Helmholtz-Zentrum Dresden-Rossendorf e.V., Institut für Ressourcenökologie, Forschungsstelle Leipzig, Permoserstrasse 15, D-04318 Leipzig, Germany; s.schymura@hzdr.de (S.S.); k.franke@hzdr.de (K.F.)

* Correspondence: petra.schneider@h2.de

Received: 13 July 2020; Accepted: 3 September 2020; Published: 9 September 2020

Abstract: Wastewater treatment is one of the main end-of-life scenarios, as well as a possible reentry point into the environment, for anthropogenic nanoparticles (NP). These can be released from consumer products such as sunscreen or antibacterial clothing, from health-related applications or from manufacturing processes such as the use of polishing materials ($nCeO_2$) or paints ($nTiO_2$). The use of NP has dramatically increased over recent years and initial studies have examined the possibility of toxic or environmentally hazardous effects of these particles, as well as their behavior when released. This study focuses on the fate of $nTiO_2$ and $nCeO_2$ during the wastewater treatment process using lab scale wastewater treatment systems to simulate the NP mass flow in the wastewater treatment process. The feasibility of single particle mass spectroscopy (sp-ICP-MS) was tested to determine the NP load. The results show that $nTiO_2$ and $nCeO_2$ are adsorbed to at least 90 percent of the sludge. Furthermore, the results indicate that there are processes during the passage of the treatment system that lead to a modification of the NP shape in the effluent, as NP are observed to be partially smaller in effluent than in the added solution. This observation was made particularly for $nCeO_2$ and might be due to dissolution processes or sedimentation of larger particles during the passage of the treatment system.

Keywords: synthetic nanoparticles; $nTiO_2$ and $nCeO_2$; waste water treatment; sp-ICP-MS nanoparticle tracking

1. Introduction

Recent global challenges, such as climate change adaptation and the transition to green energy, can only be overcome using innovative new technologies such as nanotechnology. At the same time, research on risks of these new materials for humans and the environment must be promoted. The advancing development of synthetic nanoparticles (NP) continuously produces new types of materials. NP are materials in which 50 percent or more of the particles have one or more dimensions between 1 nm and 100 nm, often exhibiting vastly different properties than bulk materials. These NPs can be divided into various groups according to their origin and respective properties. For example, there are carbon NPs (fullerenes, carbon nanotubes), metal NPs (Ag, Au), metal oxide NPs (TiO_2, CeO_2), and polymeric NPs [1]. Because of their special antibacterial, photocatalytic, mechanical, electronic and biological properties, NPs made of silver (nAg), titanium dioxide ($nTiO_2$), zinc oxide (nZnO) and cerium oxide ($nCeO_2$) are used as substantial components of personal care products, pharmaceuticals, paints, electronic devices, energy storage, coatings and new environmental engineering technologies [2–4]. The widespread use of such NPs leads to their inevitable arrival in domestic sewage treatment plants [5].

It is predicted that wastewater treatment (WWT) serves as an important sink for NPs released from consumer products. Therefore, wastewater treatment plants (WWTPs) are key factors controlling entry paths of NPs into the environment and the food chain. The distribution of NPs between the WWTP effluents, sludge and cleared water, determines the NP mass flow and thus controls the expected dose to the environment and humans. The need for the investigation of NP behavior and properties in sewage sludge lead to various research efforts on the topic. For example, Brar et al. [6] investigated NP behavior in different WWTP process stages and ultimately also in sewage sludge, but at the same time, found that no methodological studies were carried out to determine the NP presence and removal during various WWT processes or their presence in wastewater at all. Furthermore, DiSalvo et al. [7] addressed the knowledge gap on NP behavior in the sewer network and in wastewater treatment. Brar et al. [6] and Yamaguchi [8] gave a general overview of the origin of various NP types and their source products. More recent studies went beyond just looking at the source products and were related to the NM life cycle [5]. Some authors go further and, in addition to the manufacturing mechanism, consider the regeneration and reusability of NM in wastewater treatment [9,10] to exploit their positive effects during treatment [11,12]. However, other authors also describe cumulative and/or combination effects of NP in the wastewater-sewage sludge pathway [13]. The impairment of COD and ammonium degradation by stabilized silver NP in sewage treatment plants was investigated by Hou et al. [14] and Jarvie et al. [15], for example, with silicon NP in waste water. Kim et al. [16,17] and Gartiser et al. [18] considered $nTiO_2$. Limbach et al. [19] and Yang et al. [20] dealt with the degradation of oxidic NP in model sewage treatment plants. Oleszczuk et al. [21] focused on the effects of applied sewage sludge on different types of plants. Soils and sediments act as a NM sink in the environment [22,23], whereby an essential NM entry path consists of the agricultural use of sewage sludge [18,24–28]. Fundamental for the understanding of the interaction of NM are reactivity and bioavailability [29–34]. Various experimental approaches attempt to determine key parameters such as surface charge, degradation, geochemical milieu, interaction with natural colloids, parameters influencing hydrodynamic conditions (e.g., pore size, roughness, flow velocity), and to derive statements on mobility and bioavailability [23,35–45].

Besides the emerging risks of NPs released through WWTP effluents, the effects of NPs on the WWT itself are reason for concern. Most municipal WWTPs depend on an activated sludge process that degrades waste water components with the help of microorganisms [46]. Activated sludge contains microbial eukaryotes, including protozoa, fungi and metazoans, and various types of bacteria that are responsible for metabolic functions, e.g., the oxidation of organic compounds, the removal of nitrogenous pollutants and phosphates. Several authors refer to toxic effects of different NPs on different microorganisms which may pose potential environmental hazards and effect the effectiveness of WWT. It was shown that Ag-NPs penetrate the membrane wall of Escherichia coli and other gram-negative bacteria. In addition, growth experiments with nitrifying bacteria showed a strong inhibition caused by Ag-NPs [47–49]. NPs show toxicity to many species, including bacteria, algae, invertebrates and vertebrates [22,50–55]. While studies have shown that NPs are toxic to single species, the complexity of an activated sludge community might not respond to NPs in the same way as single species systems. Therefore, little is known about the impact of NPs on complex microbial communities that are effective in degrading waste in the activated sludge, or if microorganisms are able to remove them [56]. A sudden increase in the NP concentration in the feed water (shock load) poses a toxicity risk even for the beneficial microorganisms contained in the sludge, and it might take months to recover the performance of the treatment plant [56,57]. Furthermore, the availability of the organic substances to microorganisms can decrease due to competitive adsorption, and thus organic substances might remain untreated by the microorganisms [58,59]. This might lead to a reduced treatment efficiency of the sewage treatment plant, which consequently poses the risk of potentially pathogenic microbes remaining in the treated water.

The ecological risks that are expected with increasing NP use cannot yet be adequately assessed in many areas [60,61]. Although previous research shows that initial fears that NM are an inherent risk to humans and the environment have not been confirmed, long-term, low-dose and interaction effects have not yet been adequately investigated. The low expected environmental NP concentrations ranging from ng/L to µg/L (water) or µg/kg (soils) [26] make the investigation of the consequences of an environmental impact considerably more difficult, because of the challenge of sensitive detection. Furthermore, the complexity of the matrices involved leads to a considerable experimental effort [62], leading, for example, to a considerable uncertainty regarding NP behavior in soils, although soil is one of the main sinks for released NP [63]. A resulting trend towards studies with high concentrations and a lack of characterization can actually lead to a significant reduction in the informative value of this work up to the complete meaninglessness of the results [64]. The main entry path for the control of the entry of synthetic NM in soils is the waste water purification process, which is considered a preliminary NM sink, and at the same time represents their release into the environment via sewage sludge from sewage treatment plants [26]. While the water clarification process is the main entry path for such NM into the environment, the subsequent main exposure path for humans is the potential NP absorption introduced into the soil through their absorption in plants and thus their entry into the food chain [60]. So far, there is no country yet that has comprehensive legislation to deal with NM, particularly in wastewater treatment processes.

Here, we report on our studies concerning the fate $nTiO_2$ and $nCeO_2$, both widely used in consumer products, in WWT. In light of the studies referenced above, the main goal was to investigate the influence of the NP on the effectiveness of the WWT process, to gain valuable information on the distribution of the NP between the WWTP effluents, and to establish single-particle mass spectroscopy (ICP-MS) [65] as a measurement modality to gain a detailed insight in the fate of the NP. Only few studies exist that applied sp-ICP-MS as determination approach for NP in water or wastewater treatment samples, particularly for $nCeO_2$ [66], $nTiO_2$ [67,68], and nAg NP, that are indicators for medical NP residues in wastewater samples [67–70]. Thus, the scope of the present study was to further investigate the feasibility of the sp-ICP-MS for the determination of nCe and nTi in wastewater, and to suggest recommendations for the monitoring of municipal wastewater samples.

2. Materials and Methods

2.1. NM Measuring Technique

Conventional ICP-MS can be used to determine element concentrations with very high sensitivity. In general, samples are digested and transformed into a homogenous solution. This solution is injected into an inductively coupled plasma that ionizes the dissolved species which are then analyzed by the mass spectrometry. The uniform distribution of ions in the analyzed solution results in a constant MS signal, which is recorded at specified intervals ("dwell times") at a specified frequency. In contrast to that, in sp-ICP-MS, the solution is diluted to such a degree and the dwell time of the measurements is reduced, such that only a single particle is measured at each step. NP enter the MS individually as a cluster of ions from the plasma. This results in a time-resolved accumulation of detected ions giving rise to peaks, which provide information about the concentration (peak frequency), size (peak height) and the type of nanoparticles (mass-charge ratio). Ions which originated from non-NP-species still reach the detector when these transient signals are recorded and contribute to the background signal. The height of this element characteristic background signal has a significant influence on the detection limit for the nanoparticles. The relevant factor for detecting single particles is the speed: for sp-ICP-MS analysis, continuous data acquisition at a dwell time is substantially smaller than for conventional ICP-MC, and the fundamental instrumental requirement for NP counting and sizing. For the NP measurements in the present investigation, we used an Elan DRCII ICP-MS (PerkinElmer, Waltham, MA, USA). The assessment of the ICP-MS results was done through the Single Particle Calculation tool (SPC) that was developed by the Food Safety Research group from Wageningen, The Netherlands [71,72].

2.2. Experimental Setup

The investigation workflow is illustrated in Figure 1. Two behrotest® KLD4N/SR (behr Labor-Technik GmbH, Düsseldorf, Germany) two-stage lab-scale wastewater treatment systems (three tanks: denitrification, activation and post-treatment, see Figure 2) were operated according to OECD TG 312 [23]. Both were inoculated with activated sludge from a municipal sewage treatment plant in Chemnitz-Heinersdorf (Chemnitz, Germany) and fed with standardized, synthetic wastewater. Synthetic wastewater was generated according to OECD Test 303A. The system has a total volume of 10.5 L. The operating parameters were set to hydraulic retention time between 4 to 5 h and a sludge age of around 12 days. The oxygen content in the aeration tank was controlled in real time between 1.5 and 2.0 mg/L. The feed was pumped from the feed tank (1) into the denitrification vessel (3) by a peristaltic pump. From there, the sludge reached the activation tank (2) via an overflow, which was aerated by an air pump and a frit at the bottom of the tank (4). The activated sludge reached the clarifier via a settling pipe (5). The sludge recirculation pumped the sludge back into the denitrification vessel. Clarified water flowed into the drain tank via an overflow. Fresh sludge from the Chemnitz sewage treatment plant was used for each inoculation, and background concentrations of the relevant elements, and the media used (activated sludge, wastewater, cleared water) were determined. The mass-charge ratio *m/z* = 48 was used for titanium dioxide and *m/z* = 139 for cerium dioxide.

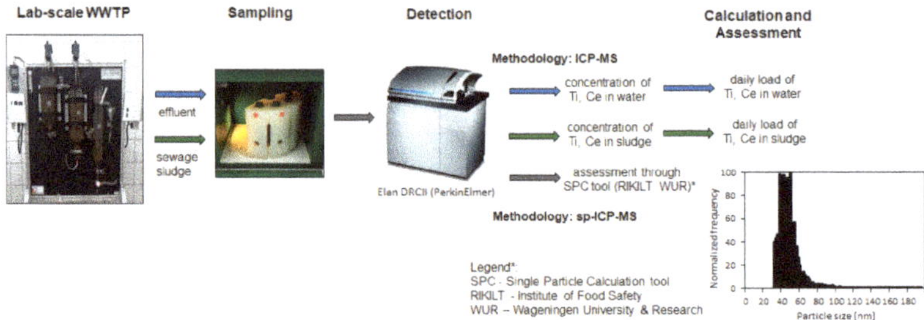

Figure 1. Investigation strategy for $nCeO_2$ and $nTiO_2$ on the pathway wastewater-sewage sludge.

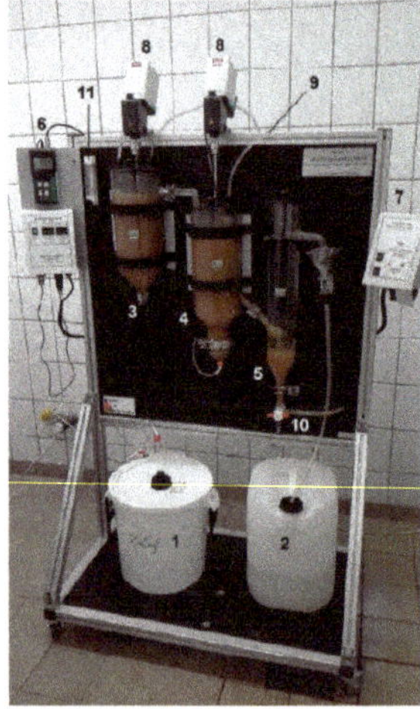

Figure 2. Lab-scale wastewater treatment systems: 1—storage tank for feed water, 2—storage tank for clarified drain water, 3—denitrification tank, 4—activation tank, 5—secondary clarification tank, 6—oxygen meter and control panel oxygen entry control, 7—control panel sludge return, 8—agitators, 9—sludge recirculation, 10—sludge recirculation, 11—volume flow measurement.

The monitoring of the performance included the daily measurement of the pH value in the aeration tank and the acid capacity in the secondary clarifier. With each change of feed, the parameters electrical conductivity, pH, carbon, nitrogen and phosphorus content in the inlet and outlet were determined. The dry matter content in the aeration and secondary clarification was monitored every working day. Two NP experiments were carried out with different concentrations using a single NP addition and one experiment using continuous particle addition, was performed. Cerium dioxide was added at 3 mg or 380 µg once and 3.95 µg/day continuously. Titanium dioxide was added at 50 mg or 63 mg once and 1 mg/day continuously. The sludge and drainage samples were taken on a weekly basis, digested using acid-assisted microwaves, and then measured using ICP-MS on the *m/z* 48 for titanium and 139 for cerium. The measurement results were converted into daily loads and summed up. In addition, selected run-off samples were analyzed without digestion using sp-ICP-MS to compare the particle size and number with the initial suspension.

2.3. Operational Setting of the Lab-Scale Wastewater Treatment System

To establish the optimal operational setting and to enhance the treatment efficiency of the lab-scale wastewater treatment system, the first test runs were performed using the wastewater of the Chemnitz-Heinersdorf municipal WWTP. Chemnitz has approx. 246,000 inhabitants. The sewage treatment plant has a total size of 400,000 population equivalents. The daily wastewater volume averages 71,400 m^3/day. The canal system has a length of around 1000 km; more than 600 km are mixed wastewater and rain water, more than 190 km are wastewater and approximately 160 km are rainwater. The connection to the sewage treatment plant is 96.6 percent. Beside residential areas, commercial centers are also connected to the treatment plant. The plant is operated as anaerobic sludge stabilization and the biological cleaning as denitrification with subsequent nitrification. The inlet to the sewage treatment plant is characterized by an average COD content of around 398 mg/L, an average BOD_5 of 196 mg/L, and a total Kjeldahl nitrogen of 38 mg/L. In the outlet, there is an average of only 21 mg/L COD. Fresh sludge from the Chemnitz sewage treatment plant was used for the inoculation of the lab-scale wastewater treatment system. On the 9th (system 1) and 3rd trial day (system 2), the performance of both recirculation pumps was increased to 100 percent, in order to improve the denitrification performance and thus raise the pH in the activation vessel. The final operating settings and sludge retention times are shown in Table 1.

Table 1. Operational parameters of the lab-scale sewage treatment systems.

Operational Parameter	System 1	System 2
Hydraulic retention time	3.0 h	2.4 h
Volume flow inlet pump	0.5 L/h	0.5 L/h
Volume flow recirculation pump	1.7 L/h	2.4 L/h
Volume flow draw-back pump	0.5 L/h	0.4 L/h

The return pumps (7) were operated in batch mode (8 min on, 32 min off). This setting was made together with the increase in performance of the pumps (previously 8 min on, 22 min off). Less sludge with a low pH should be pumped back in order to prevent a further drop in the pH in the aeration. A sludge age of 12 days was targeted for both systems. For this purpose, excess sludge was removed daily via the return pump. In both systems, an oxygen content of approx. 1.5–2 mg/L was set up in the aeration tank. The lower limit of the automatic oxygen control (6) was set to 1.3 mg/L, the upper limit to 1.4 mg/L. The diaphragm pump flow rate was set to approx. 20 L/h (11). The performance of the stirrer motors (8) was reduced to 10 percent in the denitrification and in the activation vessel. With this setting, the targeted carbon removal rate on average of minimum 90 percent for dissolved organic carbon (DOC) and total organic carbon (TOC) was achieved in both lab-scale wastewater treatment systems. The obtained optimized operational setting was used for the further operation of the lab-scale wastewater treatment system with synthetic wastewater, based on the OECD guideline 303.

2.4. Input: Synthetic Wastewater

Synthetic wastewater based on the OECD guideline 303 served as the inlet into the lab-scale systems. Moreover, 8 g peptone from casein, 5.5 g meat extract, 1.5 g urea, 1.4 g dipotassium hydrogen phosphate, 0.35 g sodium chloride, 2 g $CaCl_2 \cdot 2\ H_2O$ and 0.1 g $MgSO_4 \cdot 7\ H_2O$ were used. The components were dissolved in 1 L of ultrapure water (≥18.2 MΩ/cm, Evoqua Water Technologies). This concentrate was divided into two fillings, since 500 mL of concentrate were diluted with 25 L of deionized water in the feed tank. 10 mL of methanol were added to each replenishment. The feed tank was emptied, cleaned and refilled after two days. The dry matter was determined daily from activated sludge and excess sludge from day 1. The sample of the old feed (rest of the two-day-old feed) and the sample of the new feed were withdrawn via the feed hose. The drain sample was withdrawn using a pipette on the water surface of the secondary clarifier. Furthermore, the pH value in the aeration tank was determined every day according to DIN EN 38404-5:2009-07, and the acid capacity in the drain according to DIN 38409-7:2005-12. 50 mL of sample were used to measure the acid capacity. With each change of inlet, the parameters pH value, TOC/DOC/TNb according to DIN EN 1484:1997-08 and total phosphorus according to DIN EN ISO 6878:2004-09 were determined for the inlet and outlet. These series of measurements started with the first change of inflow on the second day.

2.5. NM Investigation Methodology

Endurance tests were then carried out with $nTiO_2$ and $nCeO_2$ from ready-to-use suspensions (US Research Nanoparticles Inc., Houston, TX, USA, each 30–50 nm). Two types of particle addition were used:

Single NP addition: To determine the blank value levels, the inoculum was sampled twice and the system (sludge and drain) 7 times in 4 days and averaged. On the fifth day after test start, 62.7 mg of titanium dioxide (anatase, 15 wt%, particle size 30 to 50 nm; US Research Nanomaterials Inc., Houston, TX, USA) and 380 µg of the cerium dioxide NP (NM-212, particle size 28.4 ± 10.4 nm, Joint Research Center, Institute for Health and Consumer Protection, Ispra, Italy) were added as a pure water suspension in the denitrification basin. The suspensions were ultrasonicated with ice cooling for 30 min and shaken vigorously before the addition in order to destroy agglomerates.

Continuous NP addition: To determine the blank value level, the inoculum was sampled twice and the system (sludge and drain) 11 times in 10 days. From day 13 after commissioning the system, approx. 1 mg/day $nTiO_2$ was added to the denitrification basin as a pure water suspension. This was achieved by adding approximately 1.9 mg of $nTiO_2$ suspension to each new feed. The suspension was ultrasonicated with ice cooling before the addition, in order to destroy agglomerates. A sample of the stock solution was taken once a week immediately after the addition to check the concentration. At the same time, two solutions for two future inlets were taken.

2.6. Sampling and Analysis

To determine the titanium load in the sludge, 10 mL of excess sludge was removed every week via return pump. The sludge was dried in a ceramic dish at 105 °C and then acidified with 5 mL of nitric acid (67 percent, suprapur; Merck KGaA, Darmstadt, Germany). This solution was transferred to a microwave vessel and mixed with 5 mL of nitric acid and 2 mL of hydrochloric acid. For sampling, 50 mL of sample were removed from the water surface of the secondary clarifier every week. From the digestion, 10 mL of sample with 3 mL of nitric acid and 2 mL of hydrogen peroxide (30 percent, e.g., Merck KGaA) were subjected to a microwave treatment. The isotope Titan 48 (47.948 amu) was examined. Rhodium (10 µg/L; 102.905 amu) was used as the internal standard. The device settings of the Elan DRC II (PerkinElmer) used can be found in Table 2.

Data evaluation: the measured values were converted into a solid concentration through the dry substance weight. The activated sludge mass was calculated using the dry substance and the total volume of the activated sludge and denitrification vessel. The secondary clarifier sludge mass was calculated using the dry matter of the excess sludge and the sludge volume of the secondary clarifier. The titanium load was calculated using those figures and the solids concentration. The accumulated masses of NP in aeration, recirculation, clarifier and sludge return resulted in the daily NP load (mg).

Table 2. Mass spectroscopy (ICP-MS) settings.

Parameter	Setting
Injection tube	2 mm, quartz
Sprayer	concentric
Sprayer chamber type	cyclone
Scan mode	Peak hopping
Dwell time	150 ms
Sweeps	20
Readings	1
Replicates	3
Detector mode	Pulse
Nebulizer Gas Flow	0.76 L/min Ar
Auxiliary Gas Flow	1.2 L/min Ar
Plasma Gas Flow	16 L/min Ar

3. Results

3.1. Operational Monitoring of the Lab-Scale Sewage Treatment Systems

The structure of the sludge changed from coarse-flaky dark brown to fine-flaky ocher in the course of the experiments. As this change was observed with every system run, it can be assumed that the sludge adapted to the changed composition of the standardized wastewater. Organic carbon degradation was always higher than 90 percent. The removal of nitrogen fluctuated in the range of 40 to 80 percent. Since the experimental set-up did not contain phosphorous removal processes, phosphorus degradation was only possible biochemically.

3.2. Discharge Loads of Titanium NP-Spike Particle Addition

The series of measurements of the singe particle addition of $nTiO_2$ began five days after the system was commissioned. The mean blank level before addition was 635.8 mg/kg in the sludge and 36.6 µg/L in the aqueous phase. The blank value of the deionized water was 0.6 µg/L. The experiment was terminated after 37 days because the titanium concentration in the excess sludge fell below the blank value level (618.0 mg/kg). In Figure 3, the titanium loads are plotted against the test days. The first vertical line marks the day of the NP addition; the second vertical line marks the day of the undershoot. Until the NP addition, the titanium load in the sludge steadily dropped to 12.9 mg, while the load in the aqueous phase remained almost constant. After the particle addition, the load in the excess sludge increased to a maximum of 71.6 mg on the 7th day of the experiment. An increase in the titanium load in the aqueous phase was also observed, but a maximum of 4.5 mg was reached on the 6th day of the experiment. In the following test days, the titanium load in the excess sludge fell continuously until day 14 and then fluctuated between 40.5 mg and 17.9 mg by the 28th test day. Until the test end, the values decreased continuously to 6.8 mg. After the maximum load in the aqueous phase, the values decreased to 0.4 mg by the 19th day of the experiment, and fluctuated between 0.2 mg and 1.3 mg by the last day of the experiment. The titanium load in the aqueous phase was almost constant. From the first to the 33rd day of the trial, 74.0 mg of titanium left the system via excess sludge and drainage. As expected, the load in the sludge and in the aqueous phase reached its maximum after the addition. The load in the sludge sank as the trial days progressed, as titanium was removed from the system by removing excess sludge. The load in the aqueous phase decreased faster when the water retention time was shorter than that of the sludge. This load consequently left the system faster. The small fluctuations in the aqueous phase are due to an almost homogeneous particle distribution in the water. The NPs are not homogeneously distributed in the sludge.

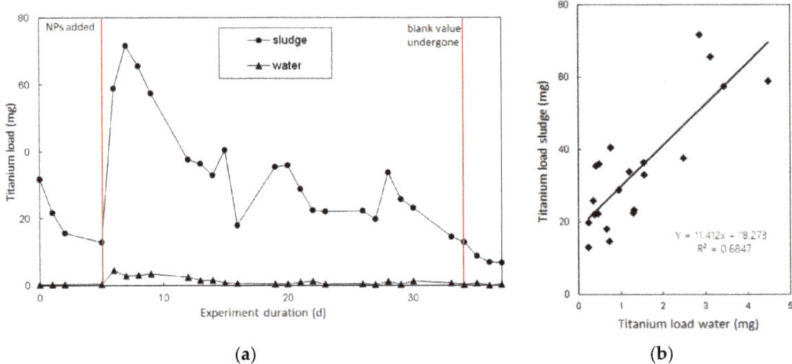

Figure 3. Discharge loads of titanium nanoparticles (NP) after single particle addition on test day 5. (**a**) Titanium load over experiment duration, (**b**) correlation between titanium load in water and sludge.

Compared to Gartiser et al. [18] with R^2 = 0.84, there was a lower correlation of R^2 = 0.68. The adsorption was determined with 90 percent in the sludge and 9 to 10 percent in the aqueous phase, compared to 95 percent in the sludge and 3 to 4 percent in the aqueous phase in Gartiser et al. [18]. The reason for the lower correlation were the different test procedures. Gartiser et al. [18] metered constant titanium dioxide into the system (similar to system 2), and the loads increased in the course of the experiment. With the single particle addition, the loads decreased steadily due to washing, which brought the sludge load closer to the water load. Calculating the adsorption ratio from the absolute mass gave a ratio of 96 percent in the sludge and 4 percent in the aqueous phase. In principle, this experiment confirms the fractionation processes observed in the literature.

3.3. Discharge Loads of Titanium NP-Continuous Particle Addition

The titanium load in the excess sludge and in the aqueous phase are shown in Figure 4, as well as the cumulative NP addition. The vertical line marks the beginning of the continuous NP addition. Before the particle addition, the sludge loads fluctuated between 8.9 mg and 24.8 mg (σ = 5.9 mg), while the loads in the aqueous phase remained almost constant. After the addition, there were minor fluctuations in the sludge, from 7.3 mg to 17.1 mg (σ = 2.4 mg). The constancy in the aqueous phase remained. Up to the last day of the test, approximately 26 mg of $nTiO_2$ was added to the system.

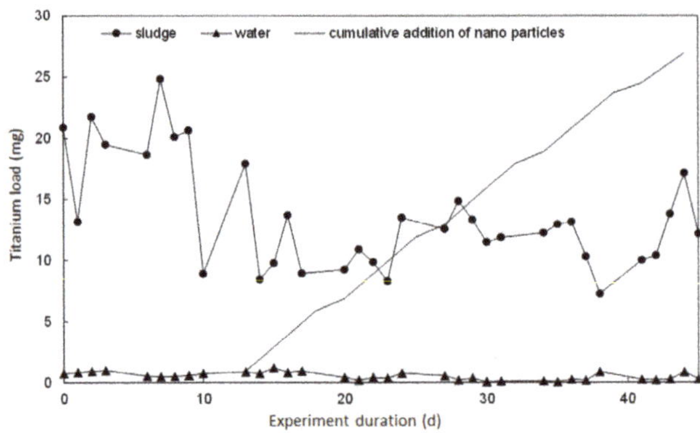

Figure 4. Discharge loads of titanium NP after continuous particle addition.

3.4. $nTiO_2$ Characteristics of Added NP Suspension and Effluent

Furthermore, we analyzed the $nTiO_2$ scan of the added NP suspension and the effluent. The height of the peaks corresponds to the size of the particles and the frequency to the quantity. Even if a direct comparison is not possible due to different dilutions, it can be concluded that the particles are both smaller and fewer in the effluent. The background level of the ionic non-NP-species value is in the range from 0 to approx. 30,000 cps. Most of the individual peaks end in a range from 100,000 cps to 280,000 cps. Two peaks exceeded this range. This observation is an indication of processes during wastewater treatment that contribute to the partial or complete dissolution of the NP.

3.5. Discharge Loads of Cerium NP-Spike Particle Addition

Figure 5 shows the cerium load in the sludge and the outlet within the test period. The cerium load of sludge increases at the 5th day of the experiment. A short-term increase in cerium load to over 300 μg was found at day 6 and 7. The loads then fell until day 12. The loads fluctuated between 90 and 220 μg between day 13 and day 27. From trial day 28, the loads dropped continuously to a value of 30 μg. The discharge loads are approximately constant between 1 and 35 μg over the test period. All loads are well below 50 μg. The discharge loads were always lower than the mud loads. The concentration of the sludge blank value is 1.79 mg/kg; the blank value of the drain was 0.15 μg/L.

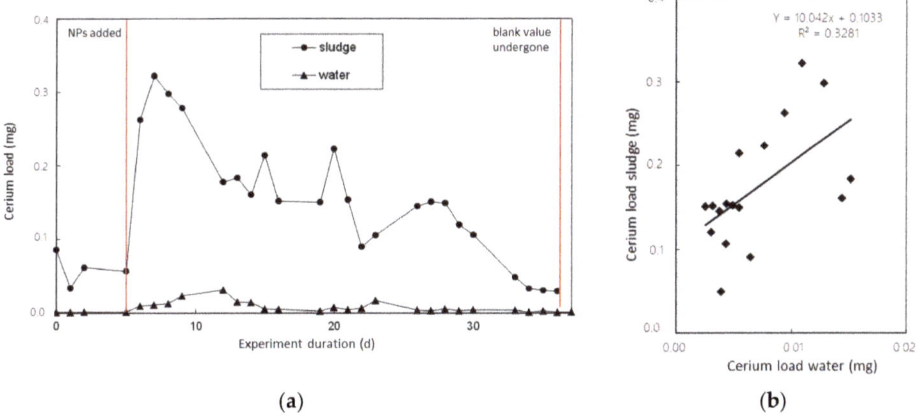

Figure 5. Discharge loads of cerium NP after single particle addition on test day 5. (**a**) Cerium load over experiment duration, (**b**) correlation between cerium load in water and sludge.

The maximum concentration was reached immediately after the NP addition, since at this point, all the particles were in the system and hardly any sludge has been removed. In the following test days, the $nCeO_2$ load in the sludge decreased more and more. This was done by taking daily samples to determine the dry matter of the excess sludge and by withdrawing excess sludge to maintain the sludge age. The significant drop in sludge loads from test day 7 to 12 could be due to the fact that excess sludge was withdrawn on test days 7, 8 and 9. This relationship was also observed from day 20 to 22. The blank level was reached both in the sludge and in the effluent at day 34. It can be concluded that the cerium dioxide NP were removed from the system by daily sampling, the removal of excess sludge and by the water running off. As a result, it was found that 5 percent of the $nCeO_2$ was in the effluent and thus 95 percent was accumulated in the sludge. When comparing the results with other studies carried out, it was found that similar results were achieved.

3.6. Discharge Loads of Cerium NP-Continuous Particle Addition

Figure 6 shows both the cerium load in the sludge and in the effluent. It is noticeable that the cerium load in sludge fluctuated between 280 and 410 µg by the seventh day of the experiment. Then, there was a drop to 125 µg. No trend can be seen between days 8 and 34; the loads fluctuated between 45 and 200 µg. An increase in cerium load can be observed after day 35 with one exception. The loads in the effluent fluctuated between 1 and 30 µg up to day 17. The discharge loads were almost constant between 0.5 and 7 µg after day 20.

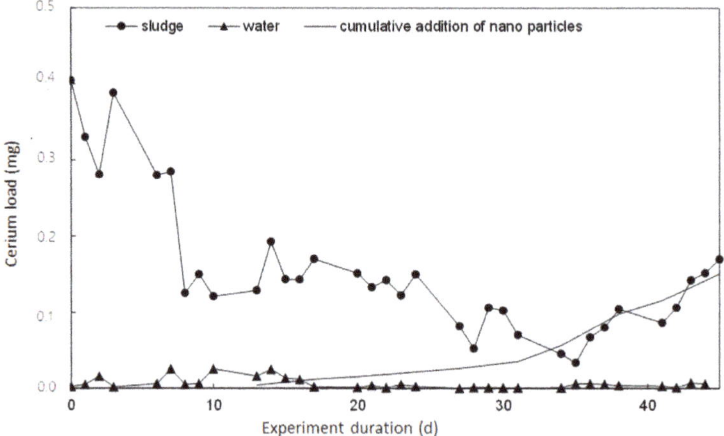

Figure 6. Discharge loads of cerium NP after continuous particle addition.

The concentration of the sludge blank value was 8.81 mg/kg, the blank value of the drain is 1.02 µg/L. The sludge loads decreased despite the addition of the nanoparticles by the 35th day of the experiment. The reason for this is that the cerium load that leaves the lab-scale wastewater treatment system via the effluent, by removing excess sludge and taking daily samples, is higher than the 3.95 µg added per day. The fluctuations between test days 0 and 13 are more pronounced than the fluctuations between test days 14 and 35. As a result, it was found that 4 percent of the $nCeO_2$ remained in the effluent and thus 96 percent in the sludge.

3.7. $nCeO_2$ Characteristics of Added NP Suspension and Effluent, Particle Size Distribution

The NP characteristic is shown in Figure 7 as the $nCeO_2$ scan of the added NP suspension and the effluent. Figure 7a shows the intensity of the stock solution of the $nCeO_2$ before they are added to the lab-scale sewage treatment systems. The time used is given in seconds and the intensity in cps. The background contributes to the intensity range 0 to 80,000 cps. The majority of the particles are detected between 100,000 and 300,000 cps. Four peaks have an intensity in the range of 350,000 to 600,000 cps. Figure 7b shows the intensity of the $nCeO_2$ in the course of experiment day 33. The time is again given in seconds and the intensity in cps. Signals in the range 0 to 2000 cps are formed by the non-NP-background. The majority of the particles are between 3000 and 10,000 cps. Six peaks are above an intensity of 20,000 cps. The $nCeO_2$ used for the test can be detected in the effluent after addition. As is visible from Figure 7, from left (stock solution) to right (effluent), the particle number concentration decreases (expressed through the decreased number of peaks), and the particle size decreases (expressed through a significantly smaller scale of intensity). This leads to the conclusion that after the addition to the system, the NPs are observed to be significantly smaller than in the stock solution (before the addition of Figure 7). The implications for dissolution processes in the course of wastewater treatment are even clearer than for $nTiO_2$.

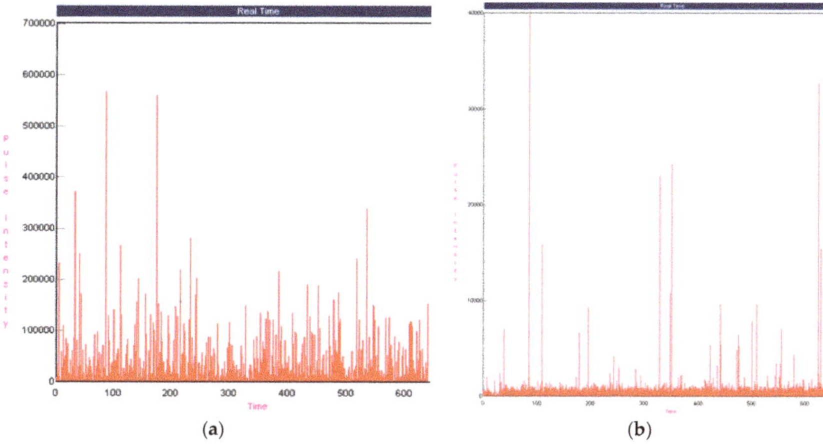

Figure 7. $nCeO_2$ scan of the added NP stock solution (a) and the effluent (b).

Since the background level of cerium is negligible in contrast to that of titanium, this comparison points to dissolution processes during the passage through the system leading to an increase in the baseline.

4. Discussion and Conclusions

Scope of the investigation was to investigate the fate of synthetic NP on the pathway wastewater-sewage sludge exemplified by $nTiO_2$ and $nCeO_2$. Furthermore, the feasibility of sp-ICP-MS as standard operation procedure for NP tracking in wastewater was investigated. Even though there is still room for further development, as was stated by Mozhayeva and Engelhardt [73], the present study confirmed the feasibility of the procedure. In this context, the investigation approach is of fundamental and application-oriented importance for research and industry, since the results provide fundamental knowledge for consulting in wastewater disposal. Particularly in the area of municipal wastewater treatment, the topic is met with great interest. This applies to both information on the operating modes of the wastewater treatment systems that contain NPs and the possibilities of verifying NPs in wastewater. In addition, the results obtained can be used to draw conclusions about expected procedural adjustments to WWTPs.

The investigation results show that both $nTiO_2$ and $nCeO_2$, are adsorbed, to a substantial extent, in the sludge. For both NM, the experiments resulted in an adsorption ratio of at least 90 percent in the sludge. The results for $nTiO_2$ show that the adsorption ratio in the sludge during a single particle addition is even higher; up to 96 percent. Thus, 4 percent of the NP remained in the aqueous phase. In case of continuous particle additions, 90 percent of the added $nTiO_2$ were adsorbed to the sludge, while 9 to 10 percent remained in the aqueous phase. The adsorption process mechanism of the different application procedures is subject to future investigations. For example, it might be assumed that the adsorption efficiency is higher during a single particle application. Thus, the available adsorption capacity might be used more comprehensively. The results also indicate that during a continuous particle addition, the maximum adsorption capacity is reached sooner. For the practical wastewater treatment, it can be concluded that a permanent $nTiO_2$ presence in the wastewater might reduce the NP adsorption capacity over time. In such a case, the remaining $nTiO_2$ could pass the wastewater treatment process. However, for $nCeO_2$ NP, a difference in the adsorption capacity during single or continuous application was not observed. This may be due to the much lower concentration used in the CeO2 experiments. Independently from the type of $nCeO_2$ addition, 4 to 5 percent of the $nCeO_2$ remained mobile in the effluent, while 96 percent were accumulated in the sludge. In case of the application of such sludge in agriculture, agglomerated metal concentrations might accumulate in crops. Investigations by the Bavarian State Office for the Environment (LfU) [74] and by the ETH Zurich [75] gave comparable results using a Hitachi S300N scanning electron microscope and X-ray diffraction and IR diffuse reflectance spectrum as NP detection methodology, respectively. Both studies can be considered as validation for the applicability of the sp-ICP-MS approach for NP detection in wastewater. Even though the scope of the respective studies was similar, the investigation approach for NP detection was different. Generally speaking, if sp-ICP-MS is further developed and comprehensively applied, it could become a standard method for NP monitoring in communal wastewater systems, particularly as the majority of monitoring labs are usually already equipped with ICP-MS technology. The further development of the sp-ICP-MS methodology might enable them to perform NP monitoring in wastewater with an acceptable affordability, once real wastewater samples can be determined in a reliable way. The lab-scale investigations with synthetic wastewater are a first step in that direction. Future investigations will have to look also on radio-labelled NPs to identify the process behavior of NPs in real wastewater. A further question in the beginning of the investigations was if $nTiO_2$ and $nCeO_2$ might have a negative impact on the microbial community and lead to disturbances of the wastewater treatment performance. Even though microbiological measurements were not performed during the lab-scale tests, it can be implicitly concluded that this is not the case, as a decrease of the degradation of organic carbon or the removal of nitrogen was not observed during the experiments. No negative effects on the microbial sludge degradation behavior due to $nTiO_2$ and $nCeO_2$ presence were observed. This confirms literature observations [64,74], which refers to the photocatalytic activity of $nTiO_2$ only to UV light [76]. They are not toxic in the opaque sludge dispersion of the lab-scale sewage treatment systems. However, due to the lower pH value, the constantly higher temperature and the synthetic wastewater, there are different conditions in the lab-scale sewage treatment systems than in the Chemnitz WWTP. The study of the Bavarian State Office for the Environment (LfU) [74] also showed that NP (nAg, nZnO, nCuO, $nTiO_2$) have no significant influence on the performance of microbiology in activated sludge.

Furthermore, the results indicate that there are processes that lead to a modification of the NP shape in the effluent, as NP are observed to be significantly smaller than in the stock solution before the addition. Two mechanisms can be assumed to play a role in the shift towards smaller particle size distributions during WWT. First, a slow dissolution of the particles, which should be more pronounced for CeO_2 than for the relatively inert TiO_2, and second, a preferential sedimentation of larger particles in the final clarifier. This has further implications for the risk assessment of consequent NP release, considering that smaller size NP (which pose the highest risks) might not be cleared from waste water to the same extent as larger particles.

Author Contributions: Conceptualization, P.S. and T.L.; methodology, P.S. and T.L.; validation, T.L., formal analysis, P.S., S.S. and K.F.; investigation, T.L.; writing—original draft preparation, P.S.; writing—review and editing, T.L., S.S. and K.F.; visualization, T.L.; supervision, K.F.; project administration, T.L.; funding acquisition, P.S. and T.L. All authors have read and agreed to the published version of the manuscript.

Funding: This research was funded by the National Ministry of Education and Research of Germany under the project "NanoSuppe-Behavior of engineered nanoparticles in the pathway wastewater-sewage sludge-plant, using the examples TiO_2, CeO_2, MWCNT and quantum dots", grant number 03X0144C.

Acknowledgments: Thanks to Marcel Neugebauer for conducting the sp-ICP-MS lab experiments.

Conflicts of Interest: The authors declare no conflict of interest.

References

1. Khan, I.; Saeed, K.; Khan, I. Nanoparticles: Properties, applications and toxicities. *Arab. J. Chem.* **2019**, *12*, 908–931. [CrossRef]
2. Mei, H.Y.; Man, C.; Bo, H.Z. Effective removal of Cu(II) ions from aqueous solution by amino-functionalized magnetic nanoparticles. *J. Hazard. Mater.* **2010**, *184*, 392–399.
3. Colvin, V.L. The potential environmental impact of engineered nanomaterials. *Nat. Biotechnol.* **2003**, *21*, 1166–1170. [CrossRef]

4. Wager, S.; Gondikas, A.; Neubauer, E.; Hofmann, T.; von der Kammer, F. Spot the Difference: Engineered and Natural Nanoparticles in the Environment—Release, Behavior, and Fate. *Angew. Chem. Int. Ed.* **2014**, *53*, 12398–12419.
5. Weber, C.H.M.; Chiche, A.; Krausch, G.; Rosenfeldt, R.; Ballauff, M.; Harnau, L.; Göttker-Schnetmann, I.; Tong, Q.; Mecking, S. Single Lamella Nanoparticles of Polyethylene. *Nano Lett.* **2007**, *7*, 2024–2029. [CrossRef]
6. Brar, S.K.; Verma, M.; Tyagi, R.D.; Surampalli, R.Y. Engineered nanoparticles in wastewater and wastewater sludge—Evidence and impacts. *Waste Manag.* **2010**, *30*, 504–520. [CrossRef]
7. DiSalvo, R.M., Jr.; McCollum, G.R. Evaluating the Impact of Nanoparticles on Wastewater Collection and Treatment Systems in Virginia. In Proceedings of the Water JAM 2008, Virginia Beach, VA, USA, 7–11 September 2008.
8. Yamaguchi, A.; Mashima, Y.; Iyoda, T. Reversible Size Control of Liquid-Metal Nanoparticles under Ultrasonication. *Angew. Chem.* **2015**, *127*, 13000–13004. [CrossRef]
9. Hendren, C.O.; Badireddy, A.R.; Casman, E.; Wiesner, M.R. Modeling nanomaterial fate in wastewater treatment: Monte Carlo simulation of silver nanoparticles (nano-Ag). *Sci. Total Environ.* **2013**, *449*, 418–425. [CrossRef]
10. Ali, I.; Naz, I.; Khan, Z.M.; Sultan, M.; Isalm, T.; Abbasi, I.A. Phytogenic magnetic nanoparticles for wastewater treatment: A review. *RSC Adv.* **2017**, *7*, 40158–40178. [CrossRef]
11. Lu, H.; Wang, J.; Stoller, M.; Wang, T.; Bao, Y.; Hao, H. An Overview of Nanomaterials for Water and Wastewater Treatment. *Adv. Mater. Sci. Eng.* **2016**, *2016*, 1687–8434. [CrossRef]
12. Samanta, H.S.; Das, R.; Bhattachajee, C. Influence of Nanoparticles for Wastewater Treatment—A Short Review. *Austin Chem. Eng.* **2016**, *3*, 1036.
13. Sheng, Z.; Van Nostrand, J.D.; Zhou, J.; Liu, Y. Contradictory effects of silver nanoparticles on activated sludge wastewater treatment. *J. Hazard. Mater.* **2018**, *341*, 448–456. [CrossRef] [PubMed]
14. Hou, L.; Li, K.; Ding, Y.; Li, Y.; Chen, J.; Wu, X.; Li, X. Removal of silver nanoparticles in simulated wastewater treatment processes and its impact on COD and NH4 reduction. *Chemosphere* **2012**, *87*, 248–252. [CrossRef] [PubMed]
15. Jarvie, H.P.; Al-Obaidi, H.; King, S.M. Fate of Silica Nanoparticle in Simulated Primary Wastewater Treastment. *Environ. Sci. Technol.* **2009**, *43*, 8622–8628. [CrossRef] [PubMed]
16. Kim, J.Y.; Choi, S.B.; Noh, J.H.; Hunyoon, S.; Lee, S.; Noh, T.H.; Frank, A.J.; Hong, K.S. Synthesis of CdSe-TiO_2 nanocomposites and their applications to TiO_2 sensitized solar cells. *Langmuir* **2009**, *25*, 5348–5351. [CrossRef]
17. Kim, B.; Murayama, M.; Colmane, B.P.; Hochella, M.F., Jr. Characterization and environmental implications of nano- and larger TiO_2 particles in sewage sludge, and soils amended with sewage sludge. *J. Environ. Monit.* **2012**, *14*, 1129–1137. [CrossRef] [PubMed]
18. Gartiser, S.; Flach, F.; Nickel, C.; Stintz, M.; Damme, S.; Schaeffer, A.; Erdinger, L.; Kuhlbusch, T.J.A. Behavior of nanoscale titanium dioxide in laboratory wastewatertreatment plants according to OECD 303 A. *Chemosphere* **2014**, *104*, 197–204. [CrossRef]
19. Limbach, L.K.; Bereiter, R.; Müller, E.; Krebs, R.; Gälli, R.; Stark, W.J. Removal of Oxide Nanoparticles in a Model Wastewater Treatment Plant: Influence of Agglomeration and Surfactants on Clearing Efficiency. *Environ. Sci. Technol.* **2008**, *42*, 5828–5833. [CrossRef]
20. Yang, Y.; Zhang, C.; Hu, Z. Impact of metallic and metal oxide nanoparticles on wastewater treatment and anaerobic digestion. *Environ. Sci. Process. Impacts* **2013**, *15*, 39–48. [CrossRef]
21. Oleszczuk, P.; Hollert, H. Comparison of sewage sludge toxicity to plants and invertebrates in three different soils. *Chemosphere* **2011**, *83*, 502–509. [CrossRef]
22. Johansen, A.; Pedersen, A.L.; Jensen, K.A. Effects of C60 fullerene nanoparticles on soil bacteria and protozoans. *Environ. Toxicol. Chem.* **2008**, *27*, 1895–1903. [CrossRef] [PubMed]
23. Nickel, C.; Gabsch, S.; Hellack, B.; Nogowski, A.; Babick, F.; Stintz, M.; Kuhlbusch, T.A.J. Mobility of coated and uncoated TiO_2 nanomaterials in soil columns–Applicability of the tests methods of OECD TG 312 and 106 for nanomaterials. *J. Environ. Manag.* **2015**, *157*, 230–237. [CrossRef] [PubMed]
24. Ganzleben, C.; Hansen, S.F. *Environmental Exposure to Nanomaterials—Data Scoping Study: Final Report*; Milieu: Brussels, Belgium, 2012.
25. Gladkova, M.M.; Terekhova, V.A. Engineered nanomaterials in soil: Sources of entry and migration pathways. *Mosc. Univ. Soil Sci. Bull.* **2013**, *68*, 129–134. [CrossRef]

26. Gottschalk, F.; Sonderer, T.; Scholz, R.W.; Nowack, B. Modeled environmental concentrations of engineered nanomaterials (TiO_2, ZnO, Ag, CNT, fullerenes) for different regions Environ. *Sci. Technol.* **2009**, *43*, 9216–9222. [CrossRef] [PubMed]
27. Johnson, A.C.; Bowes, M.J.; Crossley, A.; Jarvie, H.P.; Jurkschat, K.; Jürgens, M.D.; Lawlor, A.J.; Park, B.; Rowland, P.; Spurgeon, D.; et al. An assessment of the fate, behaviour and environmental risk associated with sunscreen TiO_2 nanoparticles in UK field scenarios. *Sci. Total Environ.* **2011**, *409*, 2503–2510. [CrossRef] [PubMed]
28. Yang, X.; Gondikas, A.P.; Marinakos, S.M.; Auffan, M.; Liu, J.; Hsu-Kim, H.; Meyer, J.N. Mechanism of Silver Nanoparticle Toxicity Is Dependent on Dissolved Silver and Surface Coating in Caenorhabditis elegans. *Environ. Sci. Technol.* **2012**, *46*, 1119–1127. [CrossRef]
29. Barton, L.E.; Therezien, M.; Auffan, M.; Bottero, J.-Y.; Wiesner, M.R. Theory and Methodology for Determining Nanoparticle Affinityfor Heteroaggregation in Environmental Matrices UsingBatch Measurements. *Environ. Eng. Sci.* **2014**, *31*, 421–427. [CrossRef]
30. Cornelis, G.; Hund-Rinke, K.; Van den Brink, N.W.; Nickel, C. Fate and Bioavailability of Engineered Nanoparticles in Soils: A Review. *Rev. Environ. Sci. Technol.* **2014**, *44*, 2720–2764. [CrossRef]
31. Dunphy Guzmán, K.A.; Taylor, M.R.; Banfield, J.F. Environmental Risks of Nanotechnology: National Nanotechnology Initiative Funding, 2000–2004. *Environ. Sci. Technol.* **2006**, *40*, 1401–1407. [CrossRef]
32. Hotze, E.M.; Phenrat, T.; Lowry, G.V. Nanoparticles aggregation: Challanges to understanding transport and reactivity in the envioenment. *J. Environ. Qual.* **2010**, *39*, 1909–1924. [CrossRef]
33. Jiang, J.; Oberdörster, G.; Biswas, P. Characterization of size, surface charge, and agglomerationstate of nanoparticle dispersions for toxicological studies. *J. Nanopart. Res.* **2008**, *11*, 77–89. [CrossRef]
34. De Santiago-Martín, A.; Constantin, B.; Guesdon, G.; Kagambega, N.; Raymond, S.; Galvez Cloutier, R. Bioavailability of engineered nanoparticles in soil systems. *J. Hazard. Toxic. Radioact. Waste* **2016**, *20*, B4015001. [CrossRef]
35. Bayat, A.E.; Junin, R.; Shamshirband, S.; Chong, W.T. Transport and retention of engineered Al2O3, TiO_2 and SiO_2 nanoparticles through various sedimentary rocks. *Sci. Rep.* **2015**, *5*, 14264. [CrossRef] [PubMed]
36. Cornelis, G.; Thomas, C.D.M.; McLaughlin, M.J.; Kirby, J.K.; Beak, D.G.; Chittleborough, D. Retention and Dissolution of Engineered Silver Nanoparticles in Natural Soils. *Soil Sci. Soc. Am. J.* **2012**, *76*, 891–902. [CrossRef]
37. Fang, J.; Shan, X.; Wen, B.; Lin, J.; Owens, G. Stability of titania nanoparticles in soil suspensions and transport in saturated homogeneous soil columns. *Environ. Pollut.* **2009**, *157*, 1101–1109. [CrossRef]
38. French, R.A.; Jacobson, A.R.; Kim, B.; Isley, S.L.; Penn, R.L.; Baveye, P.C. Influence of Ionic Strength, pH, and Cation Valence on Aggregation Kinetics of Titanium Dioxide Nanoparticles. *Environ. Sci. Technol.* **2009**, *43*, 1354–1359. [CrossRef]
39. Gogos, A.; Knauer, K.; Bucheli, T.D. Nanomaterials in plant protection and fertilization: Current state, foreseen applications, and research priorities. *J. Agric. Food Chem.* **2012**, *60*, 9781–9792. [CrossRef]
40. Al-Salim, N.; Barraclough, E.; Burgess, E. Quantum dots transport in soil, plants and insects. *Sci. Total Environ.* **2011**, *409*, 3237–3248. [CrossRef]
41. Simonin, M.; Martins, J.M.F.; Uzu, G.; Vince, E.; Richaume, A. Combined Study of Titanium Dioxide Nanoparticle Transport and Toxicity on Microbial Nitrifying Communities under Single and Repeated Exposures in Soil Columns. *Environ. Sci. Technol.* **2016**, *50*, 10693–10699. [CrossRef]
42. Sun, P.; Zhang, K.; Fang, J.; Lin, D.; Wang, M.; Han, J. Transport of TiO2 nanoparticles in soil in the presence of surfactants. *Sci. Total Environ.* **2015**, *527–528*, 420–428. [CrossRef]
43. Sun, P.; Shijirbaatar, A.; Fang, J.; Owen, G.; Lin, D.; Zhang, K. Distinguishable Transport Behavior of Zinc Oxide Nanoparticles in Silica Sand and Soil Columns. *Sci. Total Environ.* **2015**, *505*, 189–198. [CrossRef]
44. Tourinho, P.S.; Cornelis, A.; Van Gestel, M.; Lofts, S.; Svendsen, C.; Soares, A.M.V.M.; Lourero, S. Metal-based Nanoparticles in Soil: Fate, Baheviour, and Effects on Soil. *Environ. Toxicol. Chem.* **2012**, *31*, 1679–1692. [CrossRef] [PubMed]
45. Zhao, L.; Peralta-Videa, J.R.; Varela-Ramirez, A.; Castillo-Michel, H.; Li, C.; Zhang, J.; Aguilera, R.J.; Keller, A.A.; Gardea-Torresdey, J.L. Effect of surface coating and organic matter on the uptake of CeO_2 NPs by corn plants grown in soil: Insight into the uptake mechanism. *J. Hazard. Mater.* **2012**, *225–226*, 131–138. [CrossRef] [PubMed]

46. Kaegi, R.; Voegelin, A.; Sinnet, B.; Zuleeg, Z.; Hagendorfer, H.; Burkhardt, M.; Siegrist, H. Behavior of metallic silver nanoparticles in a pilot wastewater treatment plant. *Environ. Sci. Technol.* **2011**, *45*, 3902–3908. [CrossRef] [PubMed]
47. Sondi, I.; Salopek-Sondi, B. Silver nanoparticles as antimicrobial agent: A case study on E. coli as a model for Gram-negative bacteria. *J. Colloid Interface Sci.* **2004**, *275*, 177–182. [CrossRef] [PubMed]
48. Morones, J.R.; Elechiguerra, J.L.; Camacho, A.; Holt, K.; Kouri, J.B.; Tapia Ramirez, J.; Yacaman, M.J. The bactericidal effect of silvernanoparticles. *Nanotechnology* **2005**, *16*, 2346–2353. [CrossRef] [PubMed]
49. Adams, L.K.; Lyon, D.Y.; McIntosh, A.; Alvarez, P.J.J. Comparative toxicity of nano-scale TiO_2, SiO_2 and ZnO water suspensions. *Water Sci. Technol.* **2006**, *54*, 327–334. [CrossRef]
50. Huang, Y.; Liang, Y.; Rao, Y.; Zhu, D.; Cao, J.; Shen, Z.; Ho, W.; Lee, S.C. Environment-Friendly Carbon Quantum Dots/$ZnFe_2O_4$ Photocatalysts: Characterization, Biocompatibility, and Mechanisms for NO Removal. *Environ. Sci. Technol.* **2017**, *51*, 2924–2933. [CrossRef]
51. Wang, B.; Feng, W.-Y.; Wang, T.-C.; Jia, G.; Wang, M.; Shi, J.-W.; Zhang, F.; Zhaoa, Y.-L.; Chai, Z.-F. Acute toxicity of nano- and micro-scale zinc powder in healthy adult mice. *Toxicol. Lett.* **2006**, *161*, 115–123. [CrossRef]
52. Exbrayat, J.-M.; Moudilou, E.N.; Lapied, E. Harmful Effects of Nanoparticles on Animals. *J. Nanotechnol.* **2015**, *2015*, 861092. [CrossRef]
53. Heinlaan, M.; Ivask, A.; Blinova, I.; Dubourguier, H.C.; Kahru, A. Toxicity of nanosized and bulk ZnO, CuO and TiO_2 to bacteria Vibrio fischeri and crustaceans Daphnia magna and Thamnocephalus platyurus. *Chemosphere* **2008**, *71*, 1308–1316. [CrossRef]
54. Kasemets, K.; Ivask, A.; Dubourguier, H.-C.; Kahru, A. Toxicity of nanoparticles of ZnO, CuO and TiO_2 to yeast Saccharomyces cerevisiae. *Toxicol. In Vitro* **2009**, *23*, 1116–1122. [CrossRef] [PubMed]
55. Wu, F.; Harper, B.J.; Harper, S.L. Comparative dissolution, uptake, and toxicity of zinc oxide particles in individual aquatic species and mixed populations. *Environ. Toxicol. Chem.* **2019**, *38*, 591–602. [CrossRef] [PubMed]
56. Goyal, D.; Zhang, X.J.; Rooney-Varga, J.N. Impacts of single-walled carbon nanotubes on microbial community structure in activated sludge. *Lett. Appl. Microbiol.* **2010**, *51*, 428–435. [CrossRef] [PubMed]
57. Boon, N.; Top, E.M.; Verstraete, W.; Siciliano, S.D. Bioaugmentation as a tool to protect the structure and function of an activated-sludge microbial community against a 3-chloroaniline shock load. *Appl. Environ. Microbiol.* **2003**, *69*, 1511–1520. [CrossRef]
58. Henriques, I.D.S.; Love, N.G. The role of extracellular polymeric substances in the toxicity response of activated sludge bacteria to chemical toxins. *Water Res.* **2007**, *41*, 4177–4185. [CrossRef]
59. Roco, M.C. The long view of nanotechnology development: The National Nanotechnology Initiative at 10 years. *J. Nanopart. Res.* **2001**, *13*, 427–445. [CrossRef]
60. Yang, J.; Cao, W.; Rui, Y. Interactions between nanoparticles and plants: Phytotoxicity and defense mechanisms. *J. Plant Interact.* **2017**, *12*, 158–169. [CrossRef]
61. Von der Kammer, F.; Ferguson, P.L.; Holden, P.A.; Masion, A.; Rogers, K.R.; Klaine, S.J.; Koelmans, A.A.; Horne, N.; Unrine, J.M. Analysis of engineered nanomaterials in complex matrices (environment and biota): General considerations and conceptual case studies. *Environ. Toxicol. Chem.* **2012**, *31*, 32–49. [CrossRef]
62. Pan, B.; Xing, B. Applications and implications of manufactured nanoparticles in soils: A review. *Soil Sci.* **2012**, *63*, 437–456. [CrossRef]
63. Krug, H. Nanosicherheitsforschung–sind wir auf dem richtigenWeg? *Angew. Chem.* **2014**, *126*, 12502–12518. [CrossRef]
64. Durenkamp, M.; Pawlett, M.; Ritz, K.; Harris, J.A.; Neal, A.L.; McGrath, S.P. Nanoparticles within WWTP sludges have minimal impact on leachate quality and soil microbial community structure and function. *Environ. Pollut.* **2016**, *211*, 399–405. [CrossRef]
65. Thomas, R.; Stephan, C. Single-Particle ICP-MS: A Key Analytical Technique for Characterizing Nanoparticles. *Spectroscopy* **2017**, *32*, 12–25.
66. Donovan, A.R.; Adams, C.D.; Ma, Y.; Stephan, C.; Eichholz, T.; Shi, H. Detection of zinc oxide and cerium dioxide nanoparticles during drinking water treatment by rapid single particle ICP-MS methods. *Anal. Bioanal. Chem.* **2016**, *408*, 5137–5145. [CrossRef]

67. Georgantzopoulou, A.; Almeida Carvalho, P.; Vogelsang, C.; Tilahun, M.; Ndungu, K.; Booth, A.M.; Thomas, K.; Macken, A. Ecotoxicological Effects of Transformed Silver and Titanium Dioxide Nanoparticles in the Effluent from a Lab-Scale Wastewater Treatment System. *Environ. Sci. Technol.* **2018**, *52*, 9431–9441. [CrossRef]
68. Donovan, A.R.; Adams, C.D.; Ma, Y.; Stephan, C.; Eichholz, T.; Shi, H. Single particle ICP-MS characterization of titanium dioxide, silver, and gold nanoparticles during drinking water treatment. *Chemosphere* **2016**, *144*, 148–153. [CrossRef]
69. Chang, Y.J.; Shih, Y.H.; Su, C.H.; Ho, H.C. Comparison of three analytical methods to measure the size of silver nanoparticles in real environmental water and wastewater samples. *J. Hazard. Mater.* **2017**, *322*, 95–104. [CrossRef]
70. Théoret, T.; Wilkinson, K.J. Evaluation of enhanced darkfield microscopy and hyperspectral analysis to analyse the fate of silver nanoparticles in wastewaters. *Anal. Methods* **2017**, *9*, 3920–3928. [CrossRef]
71. Pace, H.E.; Rogers, N.J.; Jarolimek, C.; Coleman, V.A.; Higgins, C.P.; Ranville, J.F. Determining transport efficiency for the purpose of counting and sizing nanoparticles via single particle inductively coupled plasma mass spectrometry. *Anal. Chem.* **2011**, *83*, 9361–9369. [CrossRef]
72. Mitrano, D.M.; Lesher, K.; Bednar, A.; Monserud, J.; Higgins, C.P.; Ranville, J.F. Detecting nanoparticulate silver using single-particle inductively coupled plasma–mass spectrometry. *Environ. Toxicol. Chem.* **2012**, *31*, 115–121. [CrossRef]
73. Mozhayeva, D.; Engelhard, C. A critical review of single particle inductively coupled plasma mass spectrometry—A step towards an ideal method for nanomaterial characterization. *J. Anal. At. Spectrom.* **2020**. [CrossRef]
74. Bayerisches Landesamt für Umwelt (LfU). Umweltrelevante Eigenschaften Synthetischer Nanopartikel–Abschlussbericht, Analysis of Engineered Nanomaterials in Complex Matrices (Environment and Biota): General Considerations and Conceptual Case Studies Investigation on Likely Effects of Ag, TiO2, and ZnO Nanoparticles on Sewage Treatment. 2013. Available online: https://www.lfu.bayern.de/analytik_stoffe/nanopartikel/doc/abschlussbericht_nanoprojekt.pdf (accessed on 1 June 2020).
75. Burkhardt, M.; Zuleeg, S.; Kägi, R.; Sinnet, B.; Eugster, J.; Bollen, M.; Siegrist, H. Verhalten von nanosilber in Kläranlagen und dessen Einfluss auf die Nitrifikationsleistung in Belebtschlamm. *Umweltwiss. Schadst. Forsch.* **2010**, *22*, 529–540. [CrossRef]
76. Wang, S.; Liu, Z.; Wang, W.; Youa, H. Fate and transformation of nanoparticles (NPs) in municipal wastewater treatment systems and effects of NPs on the biological treatment of wastewater: A review. *RSC Adv.* **2017**, *7*, 37065–37075. [CrossRef]

Article

Lemna minor Bioassay Evaluation Using Computer Image Analysis

Oto Haffner [1,*], Erik Kučera [1], Peter Drahoš [1], Ján Cigánek [1], Alena Kozáková [1] and Barbora Urminská [2]

[1] Faculty of Electrical Engineering and Information Technology, Slovak University of Technology in Bratislava, 841 04 Bratislava, Slovakia; erik.kucera@stuba.sk (E.K.); peter.drahos@stuba.sk (P.D.); jan.ciganek@stuba.sk (J.C.); alena.kozakova@stuba.sk (A.K.)

[2] Faculty of Chemical and Food Technology, Slovak University of Technology in Bratislava, 812 37 Bratislava, Slovakia; b.urminska@gmail.com

* Correspondence: oto.haffner@stuba.sk

Received: 17 June 2020; Accepted: 3 August 2020; Published: 5 August 2020

Abstract: This article deals with using computer vision in the evaluation of the *Lemna minor* bioassay. According to the conventional method, the growth of *Lemna minor* mass is determined from the number of leaves grown. In this work, instead of counting individual leaves, we propose measuring the area occupied by the leaves using computer vision and compare the new approach with the conventional one. The bioassay is performed according to the ISO 20079 standard as a 168 h growth inhibition test; the aim of the experiment was to quantify the negative effects on the vegetative growth using two parameters—the number of leaves and the area occupied by the leaves. The method based on image processing was faster and also more precise since it enabled us to detect the negative effect of the tested substance on leave size, not only on their number. It can be concluded that the toxic effect has shown to be more significant when considering the leaves area rather than the number of leaves. Moreover, mistakes caused by human factor during leaves counting are eliminated using the computer vision based method.

Keywords: *Lemna minor* bioassay; visual system; computer vision; water pollution assessment; bioindicators

1. Introduction

It is necessary to develop more and more advanced wastewater treatment methods. Novel possibilities of environment monitoring are constantly advancing, which leads to the improvement of toxicant monitoring methods in threatened water. As a relevant approach for the evaluation of the impact of this water on ecosystems, ecotoxicological tests are used. Pollution in wastewater treatment plants is not always removed perfectly, it is only lowered to acceptable levels. It must be judged thoroughly whether treated water has a sufficiently low level of ecotoxicity and if there will be any effect on the organisms in it [1,2].

The *Lemna minor* (lesser duckweed—Angiospermae, Lemnaceae) is one of the water organisms used to measure ecotoxicity. It is a freshwater plant which can be found in most countries all over the world, mainly in lowlands and foothill areas in stagnant and slow-flowing waters. The plant's body consists of a relatively long root and 2 to 5 mm oval shape leaves which can float on the water surface. Mostly, the plant occurs in colonies of two to five leaves. A small single plant can itself reproduce every three days under ideal conditions in nutrient-rich waters [3]. *Lemna minor* bioassays are quite common and easy to perform. The test takes 7 days, where the plant is in the tested samples and by evaluation of the growth inhibition, the effects of substances on vegetation growth are quantified.

During the test of ecotoxicity, the biostimulation or the inhibition of plant growth in the tested samples is monitored for seven days. The objective of the test is to measure the effects of the tested substance on the growth of the *Lemna minor* plant. The basic measured indicator is the number of insoles (leaves). In addition, a second indicator must be measured—either the area overgrown with insoles, dry matter, or chlorophyll. For the legitimacy of the tests, the average increase of the number of leaves has to increase seven times in reference samples. For a standard substance, the level of EC50 (effective concentration causing a 50% inhibition) must be from 5.5 to 10 $g \cdot L^{-1}$ and the pH should not change by more than 1.5 unit [4].

The expansion of image analysis algorithms has allowed semi-automatic or fully automatic measurements. Computer systems can evaluate the changes and describe their nature, both qualitatively and quantitatively. It is important to develop a method to extract data encoded in a graphical form, usually created on the features of predefined characteristics. To acquire information from living organisms in the form of digital images, various progressive evaluation tools are used [5–7], including plant species classification using a deep convolutional neural network [8].

Computer vision includes methods for acquiring, analyzing, and understanding images. Applications of computer vision range from computers or robots that can comprehend the world around them through artificial intelligence up to industrial machine vision. Computer vision covers the core technology of automated image analysis which is used in many fields. Machine vision represents a process of combining automated image analysis with other methods and technologies to provide automated inspection or robot guidance in industrial environments and applications [9–12].

In classical bioassays using the *Lemna minor*, dry biomass of plants from the control and the tested samples is compared. However, this is an invasive approach that does not allow one to continue the experiment because the plants are dried for biomass measurements. A non-invasive approach is based on manually counting the number of leaves. The leaf size is not taken into account in this approach. A more accurate and reliable approach to defining the biomass growth rate can be comparison of the area occupied by the leaves in the experimental and control samples. This non-invasive approach allows one to continue the experiment. Computer image analysis, algorithms, and methods improve the objectivity of data collection and evaluation [13,14].

The visual evaluation of the number of leaves is done manually by laboratory workers. This process can be long and loaded with human error. There are devices with a partially automated process for *Lemna minor* experiment evaluation that are owned by big scientific institutions dealing with wastewater management. These devices are often not available for smaller institutions or scientific teams because of the high price.

In the work [15], authors deal with using computer vision in the *Lemna minor* bioassay. Calibration of their method showed a strong correlation between the frond area and the fresh biomass weight. Based on it, several experiments on the reaction of the *Lemna minor* to various detergent solutions (Brij 59 and Brij 38) were performed. Experiments have shown that non-ionic detergents changed the plant surfaces. However, in the ISO 20079:2005 standard [4], the dry biomass weight is considered, whereas in [15], the fresh biomass is used as an indicator for assay evaluation.

In this article, we focus on testing a low-cost method for a *Lemna minor* bioassay evaluation based on computer and machine vision, realizable by a research worker using reliable tools. The objective is to design a methodology for image analysis using a digital camera and a framework software tool. This method was based on the ISO 20079 standard as a 168 h growth inhibition test with the aim to quantify the effects on vegetative growth in terms of the number of leaves compared to leaves-occupied area. The number of leaves is considered as the basic indicator, however this is old and inaccurate according to current modern technological possibilities; growth reaction of the *Lemna minor* can be evaluated more accurately by the leaves-occupied area using computer vision algorithms.

2. Experiment

A test with the *Lemna minor* was performed according to the ISO 20079 standard (Water quality—Determination of the toxic effect of water constituents and waste water on duckweed (*Lemna minor*)—Duckweed growth inhibition test) as a 168 h growth inhibition test [4]. The observed endpoint was growth inhibition by the effect of the samples during a 7-day exposure (168 h). The aim of the test was to quantify the negative effects on vegetative growth in two parameters—the number of leaves and the leaves-occupied area. Then, it was possible to compare these two evaluation approaches.

The potassium dichromate $K_2Cr_2O_7$ was used as the reference substance for experiments; the concentrations used were: 5, 10, 20, 40, 80, and 160 mg·L^{-1}. Tests on reference substances serve to verify the quality and sensitivity of organisms and thus, their usability for the test. For $K_2Cr_2O_7$, the EC50 reference value is in the range of 5.5 to 10.0 mg·L^{-1}.

For the purpose of testing, the modified Steinberg medium was used. This medium provides all the necessary nutrients for the *Lemna minor* growth. The Steinberg medium composition was according to [4] and can be seen in Table 1.

Table 1. Ingredients of the modified Steinberg medium.

Macronutrients			Micronutrients		
Number	**Compound**	**c (g·L^{-1})**	**Number**	**Compound**	**c (mg·L^{-1})**
	KNO_3	17.5	**IV.**	H_3BO_3	120
I.	KH_2PO_4	4.5	**V.**	$ZnSO_4 \cdot 7\,H_2O$	180
	K_2HPO_4	0.63	**VI.**	$Na_2MoO_4 \cdot 2\,H_2O$	44
II.	$MgSO_4 \cdot 7\,H_2O$	5.0	**VII.**	$MnCl_2 \cdot 4\,H_2O$	180
III.	$Ca(NO_3)_2 \cdot 4\,H_2O$	14.75	**VIII.**	$FeCl_3 \cdot 6\,H_2O$	760
				EDTA	1500

For 1 L of medium, 20 mL of each solution of macronutrients (I–III) and 1 mL of each solution of micronutrients (IV–VIII) were pipetted. The remaining volume was distilled water. At the beginning of the test, approximately 9 beads were placed into each beaker with 100 mL of sample (3 parallel assays for each $K_2Cr_2O_7$ concentration and 6 assays with a clean modified Steinberg medium as a control) and the beakers were covered with a transparent foil and kept under constant light for 7 days. After a week, the test was evaluated and an EC50 value was determined from both the number of leaves and the area occupied by the leaves.

The growth rate was calculated as follows:

$$r = \frac{\ln x_{tn} - \ln x_{t1}}{t_n} \times 100, \quad (1)$$

Where x_{tn} is the final number of leaves (or the final area), x_{t1} is the initial number of leaves (or the initial area), t_n is the duration of the test.

The growth inhibition was calculated as follows:

$$I_r = \frac{(r_c - r_t)}{r_c} \times 100, \quad (2)$$

where r_c is the growth rate of the reference sample, r_t is growth rate of tested sample.

In order to evaluate the assay and determine the EC, the dependence of percentual growth inhibition on the sample concentration logarithm was plotted. From the regression line equation, the EC50 value was calculated.

3. Visual System for the *Lemna minor* Bioassay Evaluation

The main idea for using a visual system was the low-cost solution. Our proposed visual system consists of a camera holder, a light holder, a camera, a light, a voltage source, and a computer with image acquisition software. The whole visual system can be seen in Figure 1.

Figure 1. Visual system for the *Lemna minor* bioassay evaluation.

The holders are built using the building kit Merkur (Merkurtoys Ltd., Hradec Králové, Czechia). The lights are cheap LED spot lights (Vakesun Industrial Co., Ltd., Shenzhen, China) with constant luminosity and light diffusors made of paper. The camera used in the experiment is Logitech HD Webcam C270 (Logitech Inc., California, CA, USA), with its own image acquisition software Logitech Webcam Software (Logitech Inc., California, CA, USA). The sensing parameters of the camera were set manually to receive focused and well-exposed images (neither under- nor over-exposed), see Figure 2.

Figure 2. Well exposed (**left**) and overexposed (**right**) images.

4. Image Processing

The experiment images have to be processed after image acquisition. In this article, we compare the conventional evaluation method based on leaves counting and the new method based on the leaves-occupied area measurement. To measure the leaves-occupied area, we used the number of pixels representing the leaves. At first, we had to check whether the leaves overlap. If so, it was necessary to detach them from each other. Then, we had to select the area occupied by the leaves. This can be done automatically using sophisticated computer vision algorithms.

In our previous work [16], we proposed the system for visual evaluation of the *Lemna minor* bioassays. In this paper, we will just describe the main algorithm of the proposed system. The block diagram of the algorithm is depicted in Figure 3.

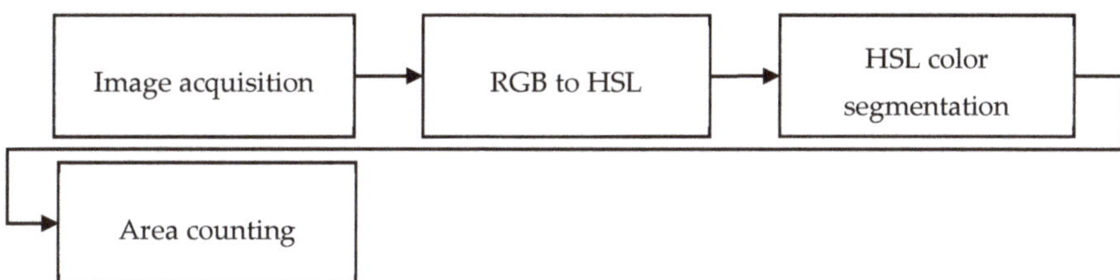

Figure 3. The flow chart of the leaves-occupied area counting algorithm.

After the image acquisition, the image was converted from the RGB (red, green, blue) image color space to the HSL (hue, saturation, value) color space. In the HSL color space, it is easy to determine which pixels of the image have a green hue (the leaves), thus the HSL color segmentation can be done. After the color segmentation, we can see the white mask in the image, which determines the area occupied by the leaves (Figure 4).

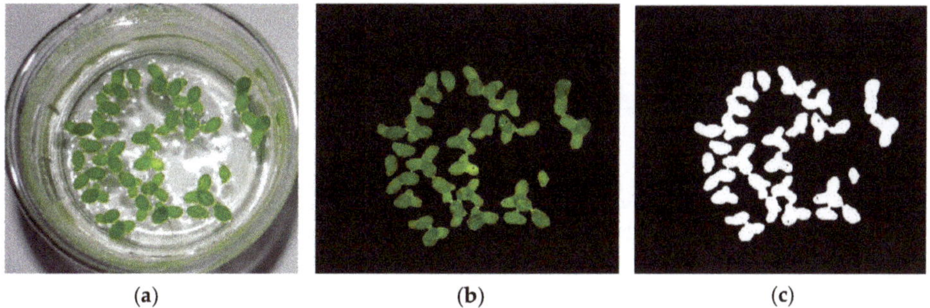

Figure 4. Original image (**a**), image of segmented leaves (**b**), image of the mask of the leaves (**c**).

The result obtained from the selected area is a new binary image where the background is black and the leaves are white (Figure 4c). It is then very easy to count the number of pixels (the white ones) from the binary image using the ready-to-use tool in the image analysis framework. The algorithm was implemented in the NI Vision Assistant software which offers various pre-prepared machine vision algorithms. We used the simple area counting using a module called Particle Analysis.

5. Results

The numerical results of the tests after 168 h exposure are shown in Tables 2 and 3. Unfortunately, we were not able to maintain the recommended temperature of 24 °C during the test, which caused lower growth rate values in the control than required by the ISO standard. However, it was still possible to compare the two methods of evaluation. The method using image processing showed significantly higher values of growth inhibition, especially at the lowest tested concentrations. A box and whisker plot representation of the result is in Figures 5–8. There are significant differences in growth rate and growth rate inhibition, especially at concentration values 5 and 10 mg·L^{-1}. This could be attributed to the fact that low concentration of the toxic substance did not yet have much impact on the number of leaves grown but rather on their size. The differences in concentrations between 20–160 mg·L^{-1} were not so significant. In both evaluation methods, higher $K_2Cr_2O_7$ concentrations caused higher growth rate inhibition (number of leaves and pixels). In case of lower toxic concentrations, using the area counting method may lead to more accurate measurements.

Table 2. Results of the test on the *Lemna minor* obtained from the counted number of leaves.

Concentration of $K_2Cr_2O_7$ c(mg·L^{-1})	Average Initial Number of Leaves x_{t1} and Stdev [1]		Average Number of Leaves at the End of the Test x_{t1} and Stdev [1]		Growth Rate r (d^{-1}) and Stdev [1]		Growth Rate Inhibition I_r (%) and Stdev [1]	
0	9.8	0.6	34.2	6.0	17.9	2.56	0.0	0.0
5	9.7	1.5	23.7	3.5	12.8	1.5	28.4	8.6
10	9.3	1.2	15.7	3.1	7.4	1.7	58.6	9.9
20	9.7	0.6	11.7	1.2	2.7	2.2	85.0	12.6
40	9.7	1.2	10.7	0.6	1.4	1.4	92.1	8.0
80	9.7	2.1	9.7	2.1	0.0	0.0	100.0	0.0
160	10.7	0.6	11.3	0.6	0.9	0.8	95.1	4.2

[1] standard deviation.

Table 3. Results of the test on the *Lemna minor* obtained from the leaves-occupied area measurement (image analysis).

Concentration of $K_2Cr_2O_7$ c (mg·L^{-1})	Average Initial Area of Leaves x_{t1} (px) and Stdev [1]		Average Area of Leaves at the End of the Test x_{t1} (px) and Stdev [1]		Growth Rate r (d^{-1}) and Stdev [1]		Growth Rate Inhibition I_r (%) and Stdev [1]	
0	60,860	5178	184,916	27,675	15.9	2.9	0.0	0.0
5	57,898	7923	102,413	8980	8.1	2.3	54.4	14.2
10	52,490	1522	76,473	5582	5.4	1.4	69.9	9.0
20	58,796	4043	76,109	1887	3.7	0.8	79.4	5.3
40	54,943	7595	66,054	7714	2.6	0.3	85.3	2.0
80	53,853	10,592	56,148	9645	0.6	0.6	96.7	4.15
160	58,422	7606	60,745	10,299	0.6	1.7	96.9	10.7

[1] standard deviation.

Figures 9 and 10 depict the evaluation of EC50 by both methods. Only the values of growth rate inhibition that are lower than 100% have to be used for the calculation, therefore the regression line consists only of these values (black points on the graph). According to the leaves counting method, the EC50 had a value of 8.50 mg·L^{-1}, which lies within the expected range. According to the image processing method, the EC50 was 3.05 mg·L^{-1}, which is slightly below the mentioned range. A lower value of EC50 means higher toxicity. It can be concluded that the toxic effect was more significant in the case of the leaves-occupied area than in the case of the number of leaves.

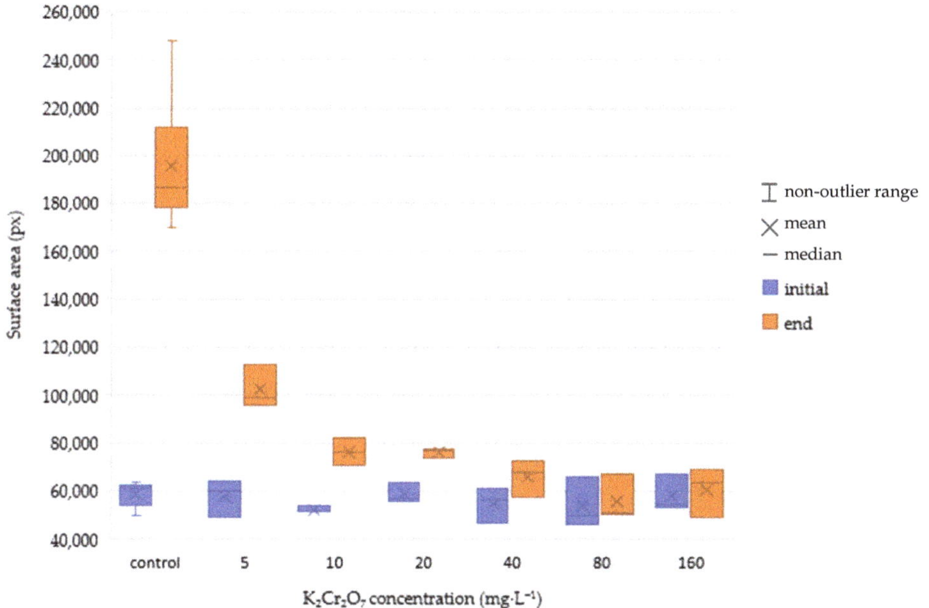

Figure 5. Box and whisker plot of the initial and end area of the *Lemna minor* leaves.

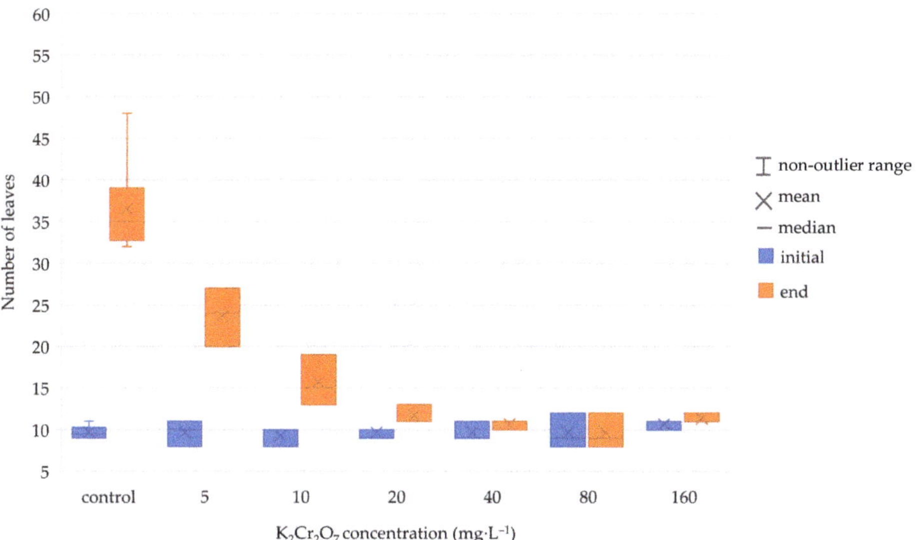

Figure 6. Box and whisker plot of the initial and end number of the *Lemna minor* leaves.

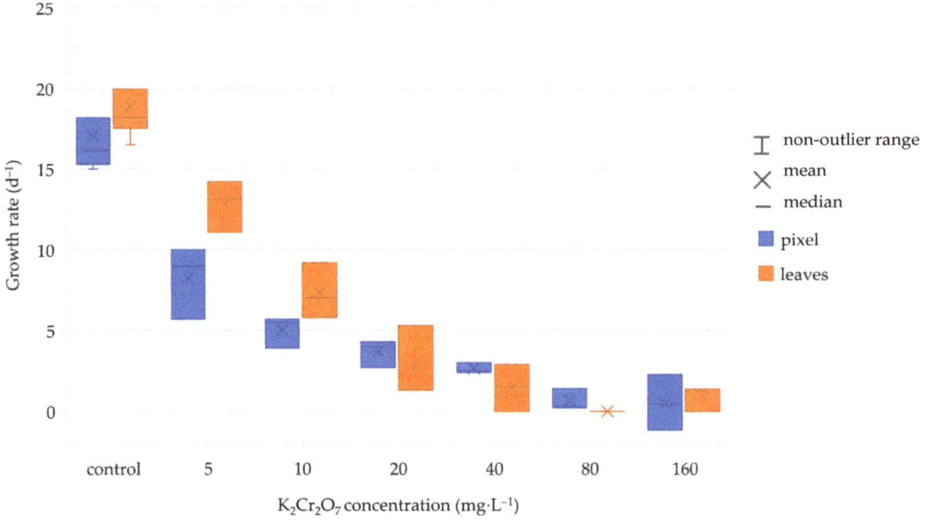

Figure 7. Box and whisker plot of the growth rate for pixels and the number of leaves.

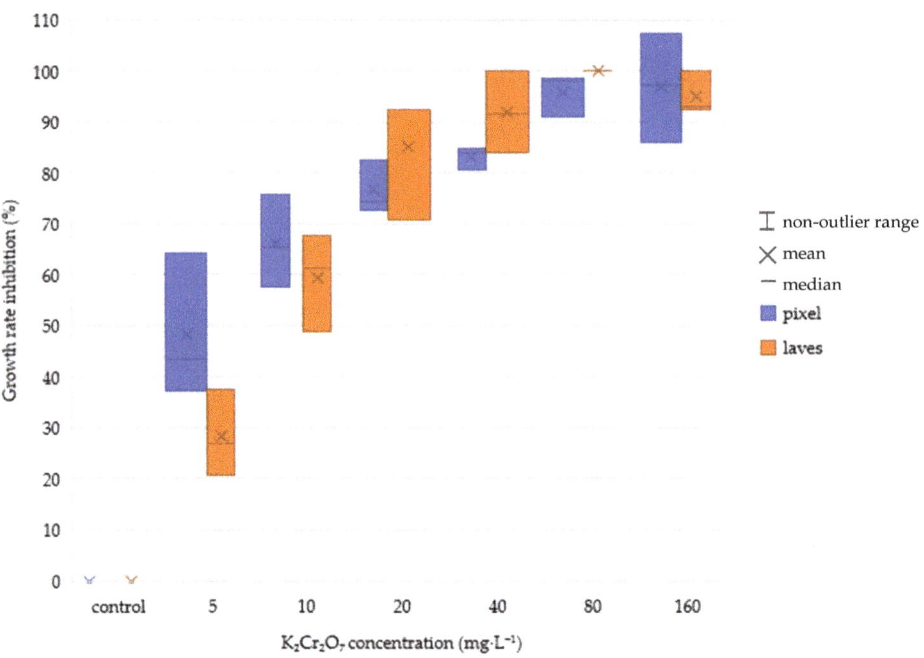

Figure 8. Box and whisker plot of the growth rate inhibition for pixels and the number of leaves.

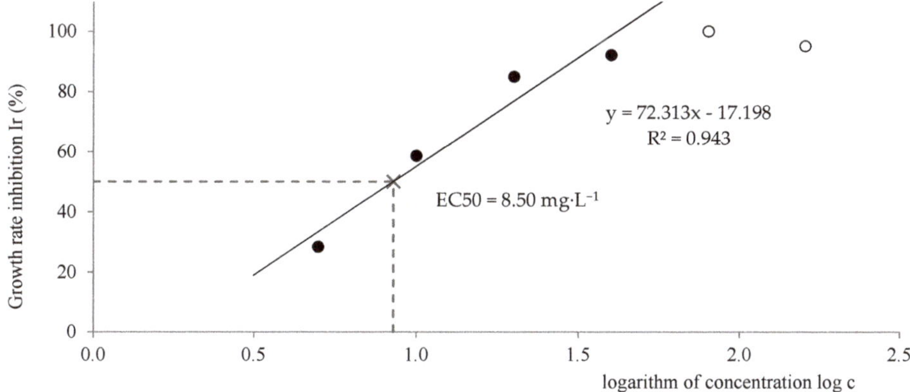

Figure 9. Graphical results of the test on the *Lemna minor* from the counted number of leaves.

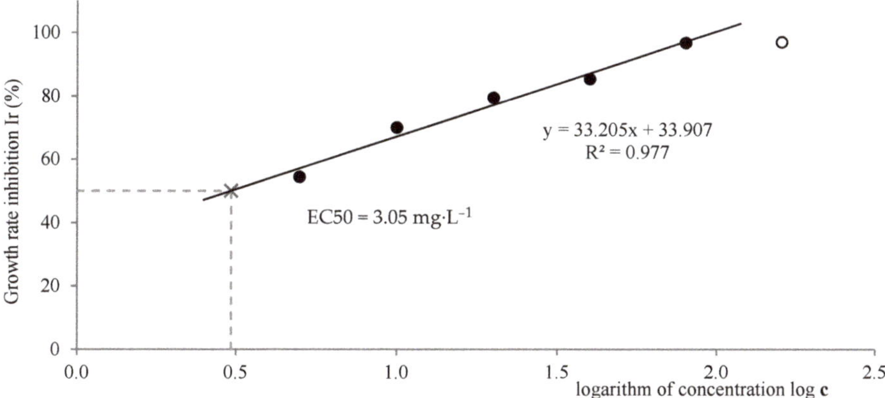

Figure 10. Graphical results of the test on the *Lemna minor* from area measurement (image analyzing).

6. Discussion

In this study, the image processing-based method was used for the *Lemna minor* bioassay evaluation. According to the ISO 20079: "Water quality—Determination of the toxic effect of water constituents and waste water on duckweed (*Lemna minor*)—Duckweed growth inhibition test" [4], the main indicator of the assay evaluation is the number of leaves. However, our study shows that the number of leaves is not as precise an indicator as the leaves-occupied area and can cause significant differences mainly for lower toxic concentrations.

The image processing approach was faster and more precise since it enabled us to detect the negative effect of the tested substance on the size of the leaves, not just on their number. Also, the errors caused by the human factor during the leaves counting were eliminated when using the image processing. On the other hand, overlapping leaves can cause a false negative result (smaller occupied area) if overlooked. Therefore, it is necessary to carefully check and detach the overlapping leaves before processing. More details of our proposed and used low-cost, reliable method and algorithm can be found in previous works [16–18].

In the work [15], authors showed a strong correlation between frond area and fresh biomass weight of the *Lemna minor*. Based on this, computer image analysis was used to measure the plant surface area. It was confirmed that the *Lemna minor* reacts to a toxic environment (pollution by detergents) by changing the surface area, thus modern techniques such as computer image analysis can be used to evaluate assays. A small disadvantage of the study [15] is that the weight of the fresh biomass is not considered as a second indicator, but the dry biomass weight as required in the ISO 20079 standard [15]. However, results of experiments confirm that computer image analysis can provide important support in the bioassays. Technical details of the proposed method are not described in depth.

Both our results and the results of [15] confirm that the application of modern technologies, such as computer image processing, and in combination with reliable low-cost hardware for bioassays evaluation, may lead to new standards for quantified and more objective evaluation of the negative influence of toxins on bioindicators. This also leads to the need to perform similar experiments and confirm the need for new, more accurate metrics. However, such experiments require the cooperation of seemingly different scientific disciplines. The synergy of such disciplines is rare, resulting in a low number of similar research works.

In the following research, our aim is to also include color evaluation (hue) in the image processing, which can further improve the sensitivity of the test since many toxic substances cause color changes of the leaves (e.g., a white color is caused by necrosis and a yellowish color by chlorosis). This effect is hard to quantify when using the standard method of counting leaves and can be easily overlooked, especially if the leaves growth has not yet been affected by the toxic substance.

7. Patents

The system for visual evaluation of the *Lemna minor* bioassays mentioned in Section 4 and in our previous work [16] is protected by the utility model number 8230 by the Industrial Property Office of the Slovak Republic [18].

Author Contributions: Conceptualization, O.H. and B.U.; methodology, O.H.; software, E.K.; validation, E.K., J.C., and B.U.; formal analysis, B.U.; resources, B.U.; data curation, J.C.; writing—original draft preparation, B.U.; writing—review and editing, O.H. and A.K.; visualization, J.C.; supervision, P.D. and A.K.; project administration, E.K.; funding acquisition, P.D. and A.K. All authors have read and agreed to the published version of the manuscript.

Funding: This work has been supported by the Cultural and Educational Grant Agency of the Ministry of Education, Science, Research and Sport of the Slovak Republic, KEGA 038STU-4/2018 and KEGA 016STU-4/2020, by the Slovak Research and Development Agency APVV-17-0190.

Conflicts of Interest: The authors declare no conflict of interest.

References

1. Mendonça, E.; Picado, A.; Paixão, S.; Silva, L.; Barbosa, M.; Cunha, M. The Role of Ecotoxicological Evaluation in Changing the Environmental Paradigm of Wastewater Treatment Management. In Proceedings of the 6th Dubrovnik Conference on Sustainable Development of Energy, Water and Environment Systems (SDEWES 2011), Dubrovnik, Croatia, 25–29 September 2011; pp. 7–11. [CrossRef]
2. Bundschuh, M. The Challenge: Chemical and ecotoxicological characterization of wastewater treatment plant effluents. *Environ. Toxicol. Chem.* **2014**, *33*, 2407. [CrossRef]
3. Kuznetsova, T.; Politaeva, N.; Smyatskaya, Y.; Ivanova, A. Lemna Minor Cultivation for Biofuel Production. *IOP Conf. Ser. Earth Environ. Sci.* **2019**, *272*, 022058. [CrossRef]
4. ISO 20079:2005. *Water Quality—Determination of the Toxic Effect of Water Constituents and Wastewater on Duckweed (Lemna Minor)—Duckweed Growth Inhibition Test*; ISO: Geneva, Switzerland, 2005.
5. Nowakowski, K.; Boniecki, P.; Tomczak, R.J.; Raba, B. Identification process of corn and barley kernel damages using neural image analysis. In Proceedings of the Third International Conference on Digital Image Processing (ICDIP 2011), Chengdu, China, 15–17 April 2011.
6. Sozzani, R.; Busch, W.; Spalding, E.P.; Benfey, P.N. Advanced imaging techniques for the study of plant growth and development. *Trends Plant Sci.* **2014**, *19*, 304–310. [CrossRef] [PubMed]
7. Nowakowski, K.; Boniecki, P.; Dach, J. The identification of mechanical damages of kernels basis on neural image analysis. In Proceedings of the 2009 International Conference on Digital Image Processing, Bangkok, Thailand, 7–9 March 2009.
8. Dyrmann, M.; Karstoft, H.; Midtiby, H.S. Plant species classification using deep convolutional neural network. *Biosyst. Eng.* **2016**, *151*, 72–80. [CrossRef]

9. Rodrigues, M.; Kormann, M.; Schuhler, C.; Tomek, P. An intelligent real time 3D vision system for robotic welding tasks. In Proceedings of the 2013 9th International Symposium on Mechatronics and its Applications (ISMA), Amman, Jordan, 9–11 April 2013; pp. 1–6.
10. Laganieère, R. Over 50 recipes to master this library of programming functions for real-time computer vision. In *OpenCV 2 Computer Vision Application Programming Cookbook;* Packt Publishing: Birmingham, UK, 2011; ISBN 9781849513241.
11. Neves, A.C.; González, I.; Leander, J.; Karoumi, R. A new approach to damage detection in bridges using machine learning. In *Lecture Notes in Civil Engineering;* Springer International Publishing: Cham, Switzerland, 2018.
12. Wierzbicki, D. Calibration Low-Cost Cameras with Wide-Angle Lenses for Measurements. *J. Autom. Mob. Robot. Intell. Syst.* **2018**, *12*, 45–51. [CrossRef]
13. Vaverková, M.D.; Elbl, J.; Koda, E.; Adamcová, D.; Bilgin, A.; Lukas, V.; Podlasek, A.; Kintl, A.; Wdowska, M.; Brtnický, M.; et al. Chemical Composition and Hazardous Effects of Leachate from the Active Municipal Solid Waste Landfill Surrounded by Farmlands. *Sustainability* **2020**, *12*, 4531. [CrossRef]
14. Rahaman, M.M.; Chen, D.; Gillani, Z.; Klukas, C.; Chen, M. Advanced phenotyping and phenotype data analysis for the study of plant growth and development. *Front. Plant Sci.* **2015**, *6*. [CrossRef] [PubMed]
15. Mazur, R.; Szoszkiewicz, K.; Lewicki, P.; Bedla, D. The use of computer image analysis in a Lemna minor L. bioassay. *Hydrobiologia* **2018**, *812*, 193–201. [CrossRef]
16. Haffner, O.; Kucera, E.; Urminska, B.; Kozak, S. Proposal of system for visual evaluation of lemna minor bioassays. In Proceedings of the 29th International Conference on Cybernetics and Informatics, K and I 2018, Lazy pod Makytou, Slovakia, 31 January–3 February 2018.
17. Haffner, O.; Kucera, E.; Kozak, S.; Urminska, B. Application of Information Technologies for Visual Evaluation of Lemna. *Int. J. Inf. Technol. Appl.* **2018**, *7*, 85–97.
18. Haffner, O.; Kučera, E.; Urminska, B. System for Visual Evaluation of Lemna Minor Growth Inhibition in Ecotoxicity Tests: Utility Model (in Slovak). Available online: https://wbr.indprop.gov.sk/WebRegistre/UzitkovyVzor/Detail/145-2017 (accessed on 16 June 2020).

Article

pH-Dependent Degradation of Diclofenac by a Tunnel-Structured Manganese Oxide

Ching-Yao Hu [1], Yu-Jung Liu [2] and Wen-Hui Kuan [3,4,*]

1 School of Public Health, Taipei Medical University, Taipei 110, Taiwan; cyhu@tmu.edu.tw
2 Graduate Institute of Environmental Engineering, National Taiwan University, Taipei 106, Taiwan; yujungliu77@gmail.com
3 Department of Safety, Health and Environmental Engineering, Ming Chi University of Technology, New Taipei 24301, Taiwan
4 Chronic Diseases and Health Promotion Research Center, Chang Gung University of Science and Technology, Chiayi 61363, Taiwan
* Correspondence: whkuan@mail.mcut.edu.tw; Tel.: +886-2-29089899 (ext. 4653); Fax: +886-29041914

Received: 25 June 2020; Accepted: 3 August 2020; Published: 5 August 2020

Abstract: The mechanism of diclofenac (DIC) degradation by tunnel-structured γ-MnO_2, with superior oxidative and catalytic abilities, was determined in terms of solution pH. High-performance liquid chromatography with mass spectroscopy (HPLC–MS) was used to identify intermediates and final products of DIC degradation. DIC can be efficiently oxidized by γ-MnO_2 in an acidic medium, and the removal rate decreased significantly under neutral and alkaline conditions. The developed model can successfully fit DIC degradation kinetics and demonstrates electron transfer control under acidic conditions and precursor complex formation control mechanism under neutral to alkaline conditions, in which the pH extent for two mechanisms exactly corresponds to the distribution percentage of ionized species of DIC. We also found surface reactive sites (S_{rxn}), a key parameter in the kinetic model for mechanism determination, to be exactly a function of solution pH and MnO_2 dosage. The main products of oxidation with a highly active hydroxylation pathway on the tunnel-structured Mn-oxide are 5-iminoquinone DIC, hydroxyl-DIC, and 2,6-dichloro-N-o-tolylbenzenamine.

Keywords: diclofenac (DIC); pH-dependent degradation mechanism; reactive site; tunnel-structured manganese oxide; γ-MnO_2

1. Introduction

Diclofenac (DIC), one of the most commonly used nonsteroidal anti-inflammatory drugs (NSAIDs) worldwide, is discharged in large amounts from wastewater treatment plants because of its high hydrophilic nature [1] and low biodegradability [2]. Thus, DIC is widely found in the aquatic environment in a range from ng/L to μg/L and is one of the most frequently detected pharmaceutical and personal care products in water [3,4]. No evidence suggests that DIC is harmful to humans; however, it might be toxic to aquatic organisms and harmful to embryos, infants, children, and adults with low immunity and being sensitive to pharmaceuticals [5–8]. Most of the evidence was focused on its adverse effects on the aquatic and terrestrial organisms, which might cause ecological damage [9–11]. Besides, the transformation products of diclofenac might be more toxic than diclofenac [12,13], which needs to be investigated further.

Manganese oxides (Mn-oxides) are effective natural oxidants for organic pollutants including phenols [14], chlorophenol [14–17], aliphatic amines [18] and anilines [16,19] in soils and sediments. In past decades, Mn-oxides have been used to remove antibacterials and compounds with phenolic and fluoroquinolonic moieties [20,21], triazine, aromatic *N*-oxides [22], tetracyclines [23], and estrogenic compounds such as the synthetic hormone 17R-ethinylestradiol [24,25] and as an alternative treatment for wastewater or groundwater containing DIC because Mn-oxides are cheap and operated-friendly [26,27]. These compounds may be endocrine disruptors (EDS) and precursors of harmful disinfection byproducts such as haloacetonitriles, haloacetamides, and nitrogenous heterocyclic [28].

In general, Mn-oxides are classified into a layer phase and tunnel structure with edge-sharing and corner-sharing octahedral MnO_6, respectively. Studies have commonly used layer-structured birnessite because of its high sorption capacity for target pollutants to remove organic pollutants [15,20,24]. Recently, tunnel-structured Mn-oxides have received considerable attention because of their excellent catalytic oxidative capacity for organic pollutant degradation; however, their mechanisms and feasibility remain unclear and must be investigated [29,30]. γ-MnO_2, a tunnel-structured Mn-oxide, normally contains a combination of pyrolusite (1 × 1 tunnel) and ramsdellite (1 × 2 tunnel) and has been confirmed to be environmentally friendly without apparent cell toxicity [31,32]. To the best of the authors' knowledge, this is the first study investigating the action of γ-MnO_2 on DIC degradation.

The ionized and acid form of a weak acid such as DIC significantly alters its adsorptive behavior between solution and solid surface, i.e., pollutants adsorption onto oxides from water and analytes separation in high-performance liquid chromatography (HPLC) column. The ionization of a weak acid such as DIC for which forms of the acid exist under different pH values [33]. The extent of adsorption is, as with anions of a weak acid, strongly dependent on pH and favored by lower pH values [34]. In HPLC if the pH of the sample solution and the eluent is not well-matched with each other and around the pKa of the organic acid deformed peaks will appear and then mislead the HPLC analysis conclusions [35,36]. Therefore, the pH could influence the mechanism of the antibiotic interaction with the manganese oxide while those issues have not been systematically addressed.

This study investigated the oxidative degradation of DIC on γ-MnO_2 suspensions by varying the key operating parameter pH, which highly influenced the surface features and redox potential of Mn-oxides [37] and the charge density of the chemical form of DIC [38]. Therefore, degradative mechanisms, intermediates, and final products were investigated in terms of pH and compared with the performances of other structures of Mn-oxides presented in the literature.

2. Materials and Methods

2.1. Materials

The sodium diclofenac (CAS 15307-86-5) with a purity of >98.5% was purchased from Sigma-A. All other chemicals used in this study were of analytical grade and purchased from Sigma A (St. Louis, MO, USA), J. T. Baker (Phillipsburg, NJ, USA) and Riedel-deHean (Seelze, Germany). Mn-oxide purchased from Tosoh was characterized as tunnel-structured (or molecular sieve) γ-MnO_2 (JCPDS 14-0644, PANalytical X'Pert Pro MRD diffractometer) with the Brunauer–Emmett–Teller (BET) specific surface area of 45.6 m^2 g^{-1} (Micrometrics ASAP, USA, 2010) and pH_{zpc} of 5.1 (Dispersion Technology, USA, DT-1200).

2.2. Batch Experiments

Experiments were conducted as a function of pH (4–9). For each batch system, various amounts of a γ-MnO_2 suspension solution were added to 15-mL glass centrifuge tubes. In the solution, 0.005 M NaH_2PO_4 and NaH_2BO_3 were added as a buffer. Various proportions of 0.1 M HCl and NaOH were used to adjust pH to the designed value within a controlled range (±0.07). After each reaction course, the solution pH was remeasured to confirm that it remained within the controlled range. For simplification, the pH value is indicated as the designed value in the following paragraphs. The initial concentration of sodium diclofenac (CAS 15307-86-5) prepared was 100 μM, which was confirmed to be completely soluble, with a water solubility of $10^{-5.1}$–$10^{-1.78}$ M [39,40]. The centrifuge tubes were covered with an aluminum foil to prevent light exposure. The suspensions were rotated at 25 °C through end-over-end rotations at 10 rpm for a specific time in kinetic trials and 24 h in thermodynamic tests. All experiments were conducted in duplicate. Moreover, controls (no MnO_2 powder) were established using a similar preparatory process to account for sorption on glass tubes and other reactions in the solution.

2.3. Sample Preparation and Analysis

After reactions, the suspensions were centrifuged (Pico 17, Thermo Scientific, Waltham, MA, USA) at 8000 rpm for 40 min, and then the supernatant was quantified using high-performance liquid chromatography (HPLC, L-7200, Hitachi, Japan) with a diode array (DAD) detector (L-7455, Hitachi) at 270 nm. Chromatographic separation was conducted using an RP-18 column (150 μm × 4.6 μm and an internal diameter of 5 μm) purchased from Mightysi with an eluent comprising 60% acetonitrile and 40% acidified water (25 mM phosphoric acid). The flow rate and injection volume were 1 mL min^{-1} and 20 μL, respectively.

2.4. Identification of Oxidation Products

Major oxidation products were identified using HPLC with mass spectrometry (HPLC–MS). The HPLC–MS system comprised an Agilent 1100 Series liquid chromatography system (LC, Agilent, Palo Alto, CA, USA) with a CTC PAL auto-sampler (CTC Analytica, Carrboro, NC, USA) separation module interfaced with an API 4000 triple quadrupole mass spectrometer (Applied Biosystems AB/MDS Sciex, Foster City, CA, USA). The LC column was a Luna Polar RP (150 mm × 2.1 mm internal diameter) column purchased from Phenomenex (Torrance, CA, USA). The flow rate and injection volume were 0.5 mL min^{-1} and 10 μL, respectively. An HPLC gradient was established by mixing two mobile phases: acetonitrile and deionized water, with 10 mM formic acid. Chromatographic separation was achieved with the following gradient: 0 to 1 min: 0% acetonitrile; 1–5 min: linear-gradient to 100% acetonitrile; 5–10 min: 100% acetonitrile; 10 to 10.1 min: 0% acetonitrile; and 10.1–15 min: 0% acetonitrile. Mass spectrometer parameters operated in a positive ion mode were as follows: curtain gas, 20 psi; ion source gas 1, 30 psi; ion source gas 2, 40 psi; source temperature, 500 °C; entrance potential, 10 V; and nebulizer current, 5 μA, and the interface heater was turned on. Positive ions in the range of 100–500 m/z were scanned at a cycle time of 1 s. The data obtained were processed with Analyst 1.4.2 software.

3. Results and Discussions

3.1. DIC Degradation Kinetics on the Tunnel-Structured Mn-Oxide

Figure 1 presents the kinetic data of DIC degradation on the tunnel-structured Mn-oxide denoted as dot symbols. Interfacial reactions between DIC and γ-MnO_2 were highly pH-dependent and initially involved a rapid removal of DIC, followed by gradual slowdown and eventual approach to a plateau. In the acidic medium (pH 4–6), the gradual slowdown period was longer than that in neutral and alkaline conditions (pH 7–9) within the tested pH range. The complicated and multistep reactions between the organic micropollutant interface and Mn-oxides result in limitations of kinetics studies; therefore, only initial reaction rates have been explored in most studies and have been generally characterized with a pseudo-first-order degradation model [27,41]. However, the pseudo-first-order kinetics may not satisfy conditions for the later stage of the interfacial reaction. In general, the interfacial reaction can be initiated by the formation of a precursor complex between the Mn(IV) of oxide surface and target organic pollutants, subsequently followed by electron transfer within the precursor complex, redox product formation (Equations (1)–(5)). Either formation of the precursor complex (Equation (1)) or electron transfer within the precursor complex (Equation (2)) is likely to be the rate-limiting step [42]. The formation of redox products, including the surface Mn(III) and Mn(II) (Equation (4)), and further oxidization or combination of organic radicals to form products (Equation (5)), was rapid because of the unstable nature of intermediates.

$$\equiv Mn^{IV} + DIC \underset{k_{-1}}{\overset{k_1}{\rightleftarrows}} \equiv Mn^{IV} \cdots DIC \tag{1}$$

$$\equiv Mn^{IV} \cdots DIC \underset{k_{-2}}{\overset{k_2}{\rightleftarrows}} \equiv Mn^{III} \cdots DIC \tag{2}$$

$$\equiv Mn^{III} \cdots DIC \underset{k_{-3}}{\overset{k_3}{\rightleftarrows}} \equiv Mn^{III} + DIC \tag{3}$$

$$\equiv Mn^{III} \overset{fast}{\rightarrow} \equiv Mn^{II} \tag{4}$$

$$DIC\cdot \overset{fast}{\rightarrow} products \tag{5}$$

An integrated kinetic model [42] was applied to examine the DIC reaction over γ-MnO_2. The kinetic equation can be expressed as follows:

$$\frac{d[\equiv Mn^{IV} \cdots DIC]}{dt} = k_1[\equiv Mn^{IV}][DIC] - k_{-1}[\equiv Mn^{IV} \cdots DIC] - k_2[\equiv Mn^{IV} \cdots DIC] \tag{6}$$

Total reactive surface sites (S) for DIC degradation can be represented as follows:

$$S = [\equiv Mn^{IV}] + [\equiv Mn^{IV} \cdots DIC] + [\equiv Mn^{III} \cdots DIC\cdot] + [\equiv Mn^{III}] + [\equiv Mn^{II}] \tag{7}$$

Both $\equiv Mn^{III}$—DIC and $\equiv Mn^{III}$ are negligible in Equation (7) because of their high instability. The $\equiv Mn^{II}$ concentration can be calculated with the concentration difference between parent DIC at initial (C_0) and specific (C) times on account of two electrons transferred from parent DIC to Mn-oxide [43]. To verify the rate-limited step and degradation mechanism, $k_1C >> k_{-1} + k_2$ and $k_{-1} + k_2 >> k_1C$, respectively, were assumed for electron transfer control and precursor complex formation control, and the analytical solution of Equation (6) for electron transfer control kinetic model is as follows:

$$C = (C_0 - S) + Se^{-k_{et}t}, \tag{8}$$

where k_{et} equals to k_2 and represents the rate constant of the electron transfer control mechanism.

The analytical solution for the precursor complex formation control model is as follows:

$$C = \frac{S - C_0}{\frac{S}{C_0}e^{k_{pf}(S-C_0)t} - 1}, \tag{9}$$

where k_{pf} equals to $\frac{k_1k_2}{k_{-1}+k_2}$ and denotes the rate constant of the precursor formation control mechanism.

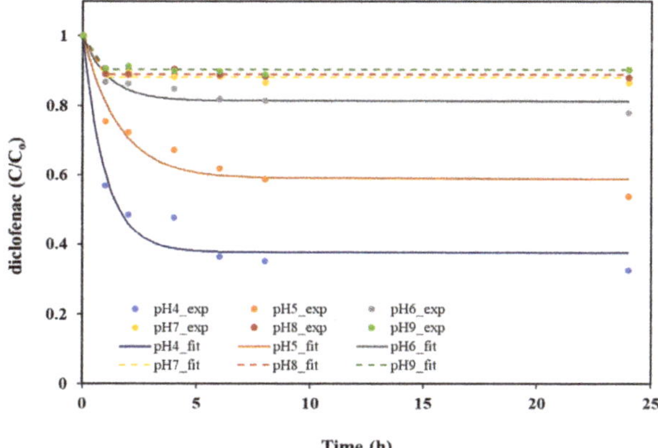

Figure 1. Diclofenac (DIC) oxidation by Mn-oxide. The initial γ-MnO_2 dosage ($[MnO_2]_0$) was (100 mg/L) (1150 μM) and the initial DIC concentration ($[DIC]_0$) was 100 μM. Dotted symbols represent the experimental data, lines indicate the fitting model, solid lines denote the electron-transfer control mechanism model, dash lines indicate the precursor complex-formation control mechanism model.

3.2. Correlation between pH and Oxidative Kinetic Constants

The pH of the solution shifted the degradation kinetics from electron-transfer control to precursor complex-formation control (Figure 1). The electron-transfer control mechanism model successfully described degradation evolution with time under acidic conditions (pH 4–6), whereas under neutral-to-alkaline conditions (pH 7–9), the precursor complex-formation control mechanism was highly fitting to the experimental data (high r-value in Table 1). As Table 1 indicates, the pHs (7–9) with precursor complex-formation control mechanism exactly correspond to the DIC existing as 100% ionized species. Figure S1 (supplementary materials) displayed the DIC species distribution versus solution pH based on the calculation of DIC's pKa 4.15 [43].

When pH was higher than 7, the ionized species account for 100% of DIC in solution. Since the formation of precursor complex of DIC with the γ-MnO_2 surface is coupled with a release of OH^- ions [34], the anionic species of DIC confront with the competition of OH^- ions for the surface sites under pH 7–9. Thus, adsorption was not favored by higher pH value (7–9) and the precursor complex formation becomes the control mechanism.

Moreover, the rate constant (k_{et} or k_{pf}) and surface reaction site (S) for DIC degradation decreased when pH increased (Table 1). The inverse relationship between k and pH for DIC could be partially attributed to a decrease in the reduction potential of MnO_2 with an increase in pH (Equation (10)).

$$MnO_2 + 4H^+ + 2e^- \leftrightarrow Mn^{2+} + 2H_2O \; E_H = E_H^0 + 0.0296\, log\frac{\{MnO_2\}\{H^+\}^4}{\{Mn^{2+}\}}. \tag{10}$$

Table 1. Fitting constants (k and S) of the kinetic model under various pH.

pH	k_{et} or k_{pf}	S (μM)	r	Kinetic Model	Ionized Species of DIC (%) †
4.0	0.998	62.3	0.96	et *	41.5
5.0	0.623	41.1	0.94	et	87.6
6.0	0.887	18.6	0.91	et	98.6
7.0	0.590	11.8	0.93	pf #	100
8.0	0.588	11.0	0.97	pf	100
9.0	0.541	9.7	0.96	pf	100

* et: electron-transfer-limited mechanism; # pf: precursor complex-formation-limited mechanism; † calculated by the DIC's pKa 4.15.

In addition to reducing the potential, solution pH alters the amounts of surface reactive sites (S, Table 1). Under acidic conditions, a large amount of S was expected because of the relatively strong affinity of anionic DIC (pKa = 4.15 [44]) species on the surface of net positively charged MnO_2. Consequently, electron transfer was limited against sufficient active reaction sites for DIC within the tested acidic pH (4–6). When pH increased, electrostatic attraction between the net negatively charged surface and anionic DIC species decreased. Furthermore, OH^- strongly competed against DIC for surface-bound Mn(IV). A lower amount of S at higher pH represented insufficient active reaction sites for DIC attachment, leading to the removal mechanism to shift to precursor complex-formation control mechanism. In addition, each component was actually derived from the initial dosage of MnO_2 (Equation (7)); therefore, surface reaction sites were presumed to be functions of initial dosage of MnO_2 and solution pH:

$$S = [MnO_2][H^+]^n. \tag{11}$$

To present the H^+ concentration as pH, the log form can be written as follows:

$$logS = \log[MnO_2] - n\,pH. \tag{12}$$

To investigate the influence of pH on S and the kinetic mechanism shift, the log S values extracted from Table 1 and the log value of original MnO_2 loading (log 1150 = 3.06) were plotted as a function of pH (Figure 2). The influencing order (n) of pH was determined as the slope of the straight line of correlation and was equal to 0.24. The correlation coefficient (r) of 0.95 corroborates the presumption that the amounts of surface reactive sites are the function of solid oxidant loadings and solution pH.

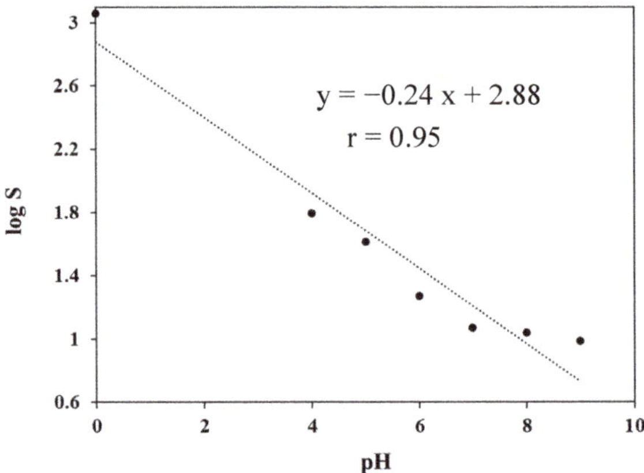

Figure 2. Surface active sites (S) relative to solution pH.

Compared with the Mn-oxide dosage yield (total mole of DIC removal per mole of MnO_2 dosage), that for DIC removal by employing γ-MnO_2 in this study was 0.07, which falls in relatively higher than 4.49×10^{-4}–0.14 reported in studies on DIC degradation using other structured Mn-oxides [27,45,46], at 24 h under similar pH conditions. Despite varying the DIC concentration and Mn-oxide dosage from μM to mM in these studies, degradation efficiencies could be compared in a unified manner when the oxide dosage yield was introduced. The remarkable differences in dosage yields indicated that the structure of Mn-oxides substantially influenced their degradative capacity toward DIC, and striking differences were observed for their sorption, oxidative, catalytic, and electrochemical properties [47–49]. The higher oxide dosage yield of γ-MnO_2 could be ascribed to the higher amounts of more flexible corner-shared MnO_6 sites dominated in Mn-oxide bulk, which may facilitate oxidation for target pollutant degradation [49–51].

3.3. Identification of Oxidation Products Using HPLC–MS

HPLC–MS was used to determine the M/Z ratio of parent DIC, oxidation intermediates (reaction time of 2 h), and products (reaction time of 24 h), and Figures 3–5, respectively, present their MS chromatograms. Figure 3a has a peak with a very pronounced tailing and this phenomenon should be due to the pH mismatch effect mentioned in a previous study [35]. This fact should not affect the identification of the oxidation intermediates and products because the pronounced tail did not appear after reaction (because the concentration of DIC decreased significantly) and most of the compounds did not appear in this region. System peaks were observed in Figure 4 (Figure 4a,b) and Figure S2. This phenomenon reflected some compounds which are strongly absorbed to the stationary phase were generated after reaction [36]. These compounds cannot be identified using this effluent procedure.

Under neutral-to-alkaline conditions, no intermediates were detected at a reaction interval of 2 h, and compared with acidic conditions, fewer oxidative products were obtained at 24 h according to MS analysis results (Figure S2). Because of the relatively low degradation of DIC under neutral and alkaline conditions, the MS analyses of intermediates and final products were mainly conducted under the acidic condition.

Because of the ionic nature of DIC, two electrospray ionization (ESI) methods, ESI+ and ESI−, were employed to study degradation products, and Table 2 presents the results. Under acidic conditions, three intermediates (I_1, I_2, and I_3) were formed after 2 h of the reaction, and four final products (F_1, F_2, F_3, and F_4) were obtained after a day of the reaction.

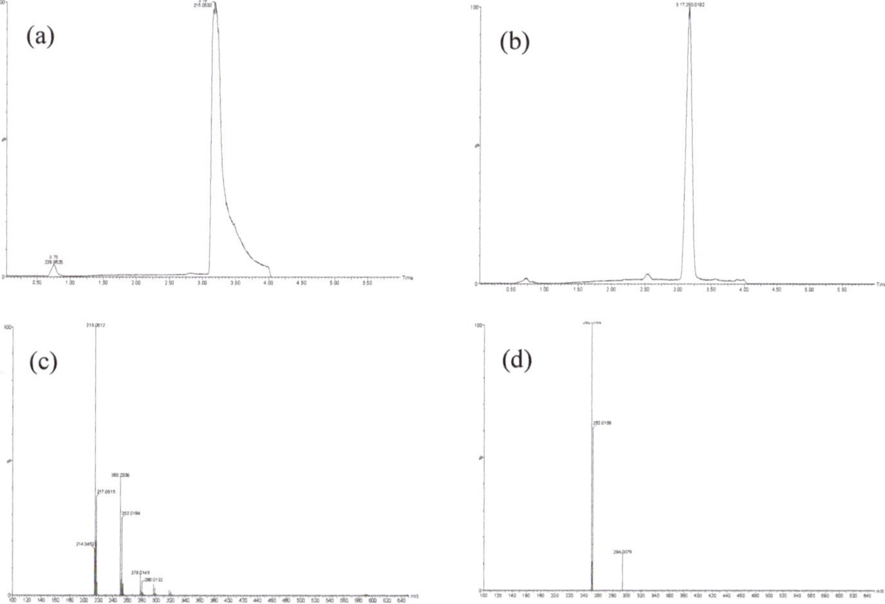

Figure 3. High-performance liquid chromatography (HPLC)– mass spectroscopy (MS) chromatographic patterns of DIC standard. (**a**) Total ion chromatogram (TIC) in the ESI+ mode, (**b**) TIC in the ESI− mode, (**c**) MS patterns in the ESI+ mode, and (**d**) MS patterns in the ESI− mode.

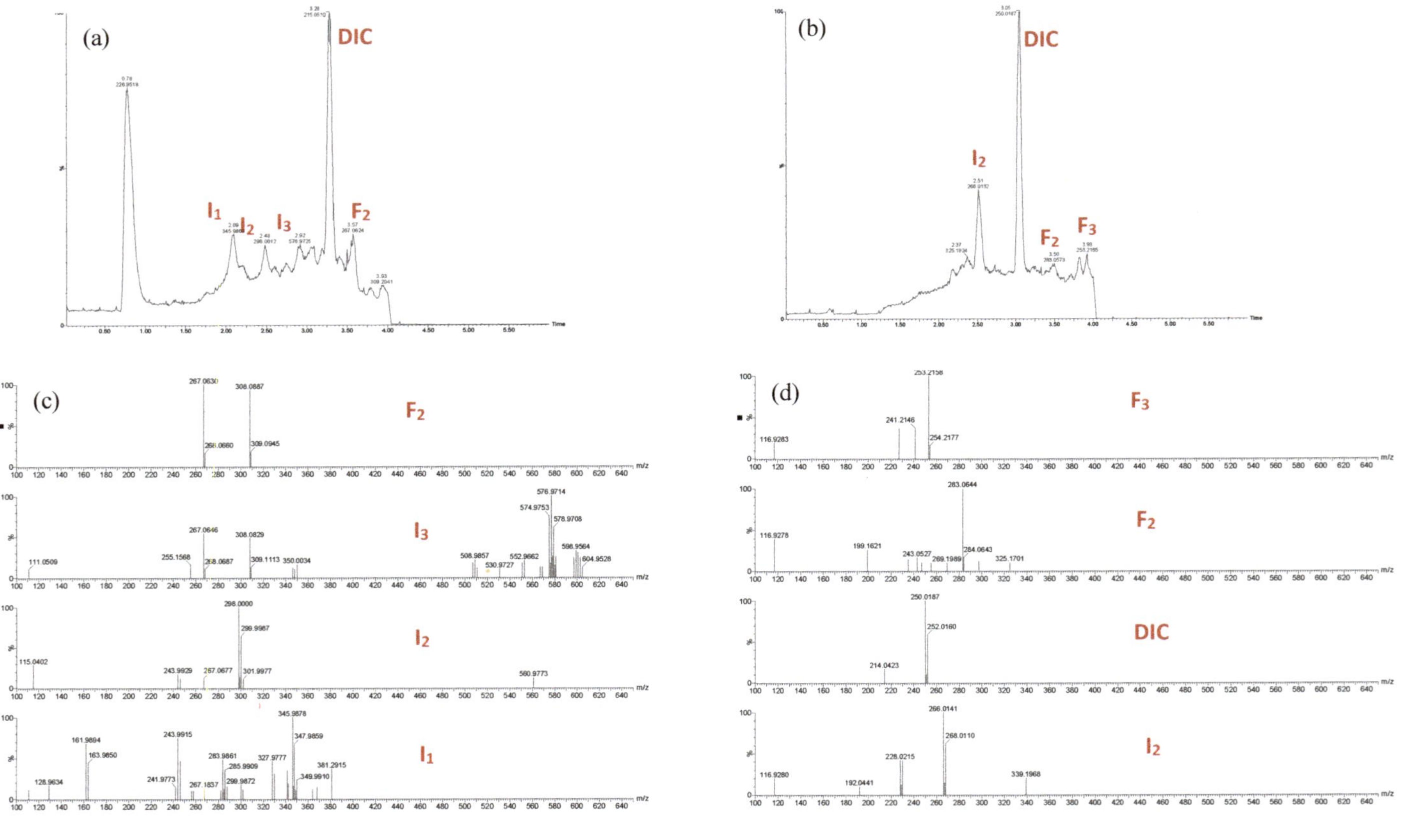

Figure 4. HPLC–MS chromatographic patterns of degradation intermediates. (**a**) TIC in the ESI+ mode, (**b**) TIC in the ESI− mode, (**c**) MS patterns in the ESI+ mode, and (**d**) MS patterns in the ESI− mode under a pH of 5 for a reaction time of 2 h with an initial MnO_2 loading of 200 mg/L.

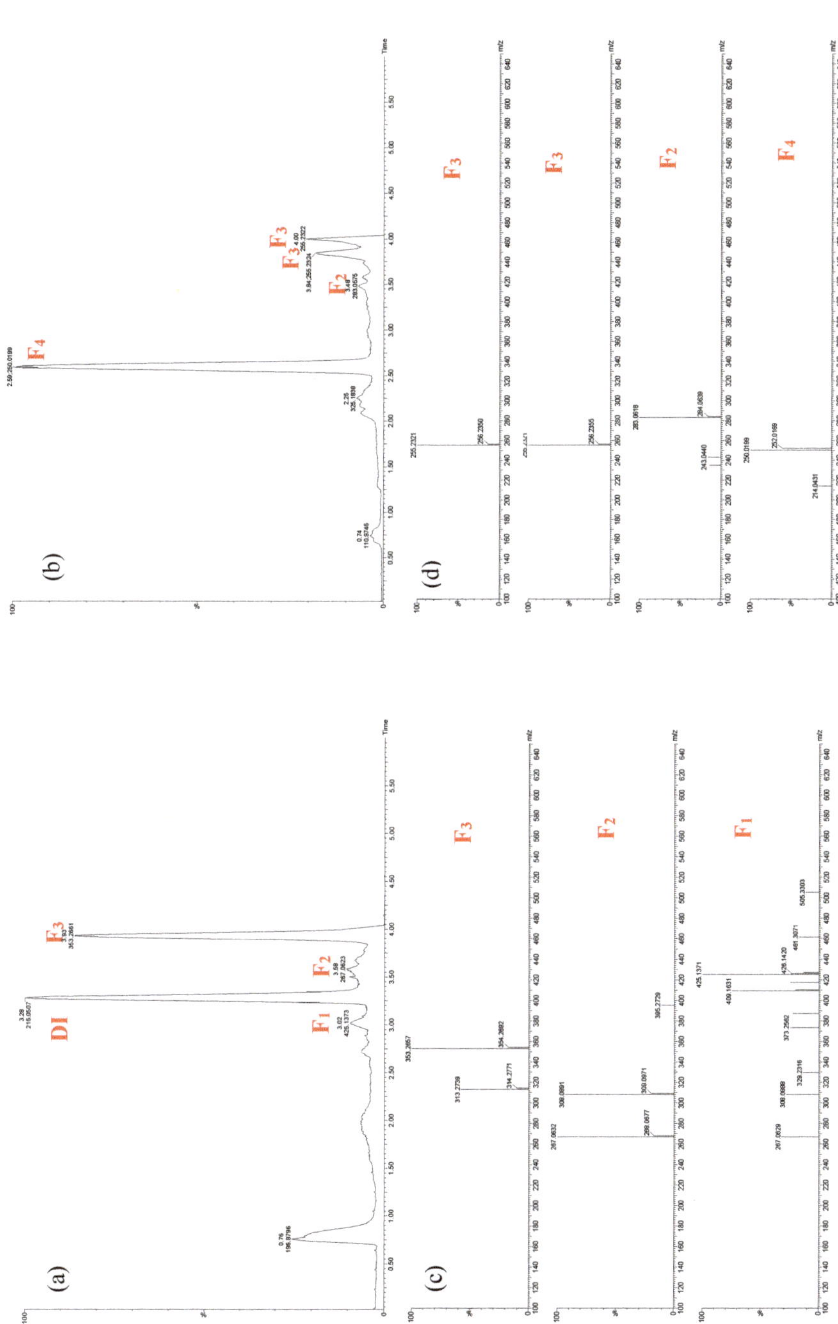

Figure 5. HPLC–MS chromatographic patterns of degradation products. (**a**) TIC in the ESI+ mode, (**b**) TIC in the ESI− mode, (**c**) MS patterns in the ESI+ mode, and (**d**) MS patterns in the ESI− mode under a pH of 5 for a reaction time of 24 h with an initial MnO_2 loading of 200 mg/L.

According to Monteagudo et al. [52,53], I_1 (RT = 2.09, m/z = 346) correspond totri-hydroxyl-DIC (m/z = 346) or di-hydroxyl-DIC (m/z = 328). I2 (RT = 2.48, m/z = 298) should be a hydrolyzed-decarboxylated DIC (296 − 14 + 16 = 298). The molecular weight of I_3 (RT = 2.92, m/z = 597) is considerably higher than that of DIC. Moreover, I_3 exhibited numerous isotopic peaks, and its intensity ratio of (M + 1)/Z to (M + 3)/Z was approximately 3:4, which revealed that these compounds contained four chlorine atoms. Thus, I_3 should be a dimmer of 5-iminoquinone DIC (m/z = 308) and another intermediate. This finding indicated that polymerization or dimerization, which was found during the reaction of other aromatic compounds with Mn-oxides, may occur during DIC degradation by γ-MnO_2. Similar results were reported by Huguet et al. [26].

F1 (RT = 3.02, m/z = 503) is a new product, and its molecular weight is substantially higher than that of DIC. Therefore, it should be a transformation product of I_3. F_2 (RT = 3.58, m/z = 308) and F_3 (RT = 3.93, m/z = 312) correspond to 5-iminoquinone DIC and hydroxyl-DIC, respectively, which have been reported in literature [26]. The peak of hydroxyl-DIC (F_3) split into two and the m/z ratio (255) in negative mode was considerably lower than the m/z ratio (312) in the positive mode. The split of the peak could be attributed to the different sites of the hydroxyl group of the compound (structure isomers) leading to different hydrophilicity, and the observed difference of m/z ratio for F_3 between in positive mode and negative mode should result from the carbon chain ($-CH_2COOH$, M = 57) broke during ionization. F_4 (RT = 2.59, m/z = 250) should be a decarboxylation product of DIC (2,6-dichloro-N-o-tolylbenzenamine), which was reported by Martínez et al. [54].

The intensity of F_3 was much higher than that of F_2, and multiple hydroxyl intermediates (I_1) were found. According to studies, decarboxylation, hydroxylation, and dimerization are the three main pathways of DIC transformation by Mn-oxides [26]. The pathways of DIC transformation by γ-MnO_2 are the same as those of birnessite or other natural manganese oxides [26], and compared with the layer-structured birnessite that is widely used in studies, hydroxylation of DIC by γ-MnO_2 was more active than that through other pathways. This phenomenon could corroborate that the large amounts of highly flexible corner-shared MnO_6 may provide abundant reactive hydroxyl groups and facilitate oxidation for target pollutant degradation [50,51]. Therefore, the dimerization products of DIC obtained through γ-MnO_2 are highly hydrophilic and can be detected without extraction. Hydroxylation intermediates were not detected after 1 day because they were oxidized to smaller or hydrophilic compounds due to further hydroxylation.

Table 2. MS data of intermediate and final products.

Compound	RT (min)	Model	m/z	Possible Structure	Ref
DIC	3.18	ESI+	**296**/298 (3/2) 278/280, 250/252, 215/217	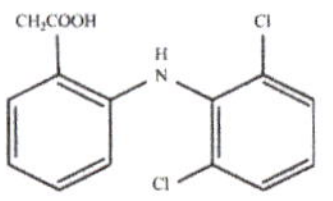	
		ESI−	**250**/252 (3/2)		
I1	2.09	ESI+	**346**/348 (3/2) 328/330, 284/286, 244/246, 162/164	Tri or Di—hydroxyl DIC	[52,53]
I2	2.48	ESI+	**298**/300 (3/2) 267, 244	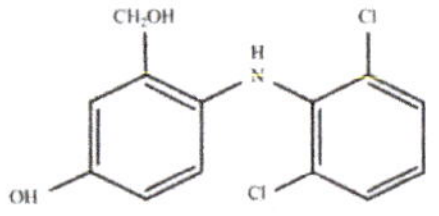	
		ESI−	**266**/264 (3/2) 228/230		
I3	2.92	ESI+	**597**/599 (3/4) 575/577 (3/4), 551/553 (3/4), 507/509 (3/4), 308, 267, 255	Dimer	

Table 2. *Cont.*

Compound	RT (min)	Model	m/z	Possible Structure	Ref
F1	3.02	ESI+	**503** 461, 425, 409, 373, 329, 308, 267	Dimer	
F2	3.58	ESI+	**308**, 267	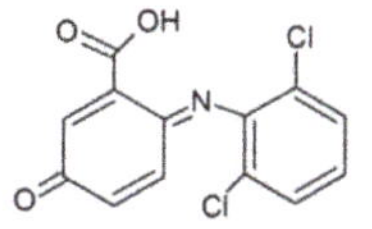	[26,55]
		ESI−	**283**		
F3	3.93	ESI+	**312**	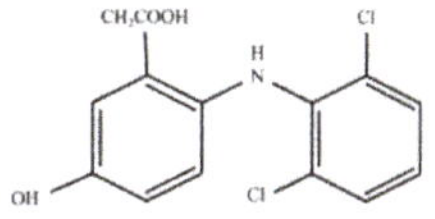	[26,55]
		ESI−	**255**		
F4	2.59	ESI−	**250/252 (3/2)** 214	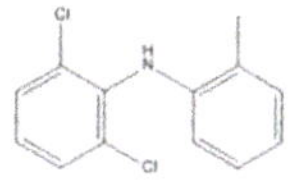	[54]

RT: retention time; bold number: parent ion.

4. Conclusions

This study demonstrated that the pH of media highly influences DIC oxidative degradation on the tunnel-structured Mn-oxide (γ-MnO_2). The reduction potential of Mn-oxide, the number of surface reactive sites (S), and electrostatic affinity between DIC and γ-MnO_2 increase with a decrease in pH value. Consequently, the electron-transfer control mechanism model successfully described degradation evolution with time under acidic conditions (pH = 4–6). While under neutral-to-alkaline conditions (pH = 7–9), the precursor complex-formation control mechanism was highly fitting to the experimental data. At pH 7–9 the anionic species account for 100% DIC in solution and hence confront with the competition of OH^- ions for the complex formation on the γ-MnO_2 surface. In contrast, the acid form of DIC with a substantial ratio under pH 4–6 is favored for the surface complex formation with less competition. The results of the analysis of oxidative intermediates and products by using HPLC–MS revealed decarboxylation, hydroxylation, and dimerization as the three main pathways of DIC transformation by γ-MnO_2. Although the oxidation products obtained by γ-MnO_2 are similar to those obtained by other Mn-oxides, hydroxylation of DIC by γ-MnO_2 is more active than other pathways because of an abundance of flexible corner-shared MnO_6 for target pollutant degradation.

Supplementary Materials: The following are available online: http://www.mdpi.com/2073-4441/12/8/2203/s1, Figure S1: DIC species distribution versus solution. HA and A^- represent the acid and ionized form of DIC, respectively. The black and gray line were calculated based on pKa = 4.15 of DIC [44]. When pH higher than 7, the ionized species (A^-) accounts for 100% of DIC in solution pH. Figure S2: LC–MS chromatographic patterns of degradation intermediates (a) TIC in ESI^+ mode (b) MS patterns in ESI^+ mode under pH 7.0 for reaction time = 24 h with initial MnO_2 loading 200 mg.

Author Contributions: C.-Y.H. conceived and designed the experiments. Y.-J.L. performed the experiments. C.-Y.H. and W.-H.K. analyzed the data and wrote the paper. All authors have read and agreed to the published version of the manuscript.

Funding: This research received the funding from Ministry of Science and Technology of the Republic of China, Contract No. NSC 98-2221-E-038 -004, MOST 105-2221-E-131-001-MY3 and MOST 108-2221-E-131-024-MY3.

Conflicts of Interest: The authors declare no conflict of interest.

References

1. Ternes, T.A.; Herrmann, N.; Bonerz, M.; Knacker, T.; Siegrist, H.; Joss, A. A rapid method to measure the solid-water distribution coefficient (K-d) for pharmaceuticals and musk fragrances in sewage sludge. *Water Res.* **2004**, *38*, 4075–4084. [CrossRef] [PubMed]
2. Joss, A.; Zabczynski, S.; Gobel, A.; Hoffmann, B.; Loffler, D.; McArdell, C.S.; Ternes, T.A.; Thomsen, A.; Siegrist, H. Biological degradation of pharmaceuticals in municipal wastewater treatment: Proposing a classification scheme. *Water Res.* **2006**, *40*, 1686–1696. [CrossRef] [PubMed]
3. Camacho-Munoz, D.; Martin, J.; Santos, J.L.; Aparicio, I.; Alonso, E. Occurrence, temporal evolution and risk assessment of pharmaceutically active compounds in Donana Park (Spain). *J. Hazard. Mater.* **2010**, *183*, 602–608. [CrossRef] [PubMed]
4. Comeau, F.; Surette, C.; Brun, G.L.; Losier, R. The occurrence of acidic drugs and caffeine in sewage effluents and receiving waters from three coastal watersheds in Atlantic Canada. *Sci. Total Environ.* **2008**, *396*, 132–146. [CrossRef] [PubMed]
5. Schwab, B.W.; Hayes, E.P.; Fiori, J.M.; Mastrocco, F.J.; Roden, N.M.; Cragin, D.; Meyerhoff, R.D.; D'Aco, V.J.; Anderson, P.D. Human pharmaceuticals in US surface waters: A human health risk assessment. *Regul. Toxicol. Pharmacol.* **2005**, *42*, 296–312. [CrossRef]
6. Sidlova, P.; Podlipna, R.; Vanek, T. Cytotoxic pharmaceuticals in the environment. *Chem. Listy* **2011**, *105*, 8–14.
7. Tauxe-Wuersch, A.; de Alencastro, L.F.; Grandjean, D.; Tarradellas, J. Occurrence of several acidic drugs in sewage treatment plants in Switzerland and risk assessment. *Water Res.* **2005**, *39*, 1761–1772. [CrossRef]
8. Weber, S.; Khan, S.; Hollender, J. Human risk assessment of organic contaminants in reclaimed wastewater used for irrigation. *Desalination* **2006**, *187*, 53–64. [CrossRef]
9. Lonappan, L.; Brar, S.K.; Das, R.K.; Verma, M.; Surampalli, R.Y. Diclofenac and its transformation products: Environmental occurrence and toxicity—A review. *Environ. Int.* **2016**, *96*, 127–138. [CrossRef]
10. Oaks, J.L.; Gilbert, M.; Virani, M.Z.; Watson, R.T.; Meteyer, C.U.; Rideout, B.A.; Shivaprasad, H.L.; Ahmed, S.; Iqbal Chaudhry, M.J.; Arshad, M.; et al. Diclofenac residues as the cause of vulture population decline in Pakistan. *Nature* **2004**, *427*, 630–633. [CrossRef]
11. Taggart, M.A.; Senacha, K.R.; Green, R.E.; Jhala, Y.V.; Raghavan, B.; Rahmani, A.R.; Cuthbert, R.; Pain, D.J.; Meharg, A.A. Diclofenac residues in carcasses of domestic ungulates available to vultures in India. *Environ. Int.* **2007**, *33*, 759–765. [CrossRef] [PubMed]
12. Schmitt-Jansen, M.; Bartels, P.; Adler, N.; Altenburger, R. Phytotoxicity assessment of diclofenac and its phototransformation products. *Anal. Bioanal. Chem.* **2007**, *387*, 1389–1396. [CrossRef] [PubMed]
13. Stülten, D.; Zühlke, S.; Lamshöft, M.; Spiteller, M. Occurrence of diclofenac and selected metabolites in sewage effluents. *Sci. Total Environ.* **2008**, *405*, 310–316. [CrossRef]
14. Stone, A.T. Reductive dissolution of manganese(III/Iv) oxides by substituted phenols. *Environ. Sci. Technol.* **1987**, *21*, 979–988. [CrossRef] [PubMed]
15. Zhang, H.C.; Huang, C.H. Oxidative transformation of triclosan and chlorophene by manganese oxides. *Environ. Sci. Technol.* **2003**, *37*, 2421–2430. [CrossRef] [PubMed]
16. Park, J.W.; Dec, J.; Kim, J.E.; Bollag, J.M. Effect of humic constituents on the transformation of chlorinated phenols and anilines in the presence of oxidoreductive enzymes or birnessite. *Environ. Sci. Technol.* **1999**, *33*, 2028–2034. [CrossRef]
17. Ulrich, H.J.; Stone, A.T. The oxidation of chlorophenols adsorbed to manganese oxide surfaces. *Environ. Sci. Technol.* **1989**, *23*, 421–428. [CrossRef]
18. Li, H.; Lee, L.S.; Jafvert, C.T.; Graveel, J.G. Effect of substitution on irreversible binding and transformation of aromatic amines with soils in aqueous systems. *Environ. Sci. Technol.* **2000**, *34*, 3674–3680. [CrossRef]
19. Pizzigallo, M.D.R.; Ruggiero, P.; Crecchio, C.; Mascolo, G. Oxidation of chloroanilines at metal oxide surfaces. *J. Agric. Food Chem.* **1998**, *46*, 2049–2054. [CrossRef]
20. Lin, K.; Liu, W.; Gan, J. Oxidative removal of bisphenol A by manganese dioxide: Efficacy, products, and pathways. *Environ. Sci. Technol.* **2009**, *43*, 3860–3864. [CrossRef]
21. Lu, Z.J.; Lin, K.D.; Gan, J. Oxidation of bisphenol F (BPF) by manganese dioxide. *Environ. Pollut.* **2011**, *159*, 2546–2551. [CrossRef] [PubMed]

22. Sarmah, P.; Dutta, D.K. Manganese mediated aqueous reduction of aromatic nitro compounds to amines. *J. Chem. Res.* **2003**, 236–237. [CrossRef]
23. Rubert, K.F.; Pedersen, J.A. Kinetics of oxytetracycline reaction with a hydrous manganese oxide. *Environ. Sci. Technol.* **2006**, *40*, 7216–7221. [CrossRef] [PubMed]
24. Sabirova, J.S.; Cloetens, L.F.F.; Vanhaecke, L.; Forrez, I.; Verstraete, W.; Boon, N. Manganese-oxidizing bacteria mediate the degradation of 17α-ethinylestradiol. *Microb. Biotechnol.* **2008**, *1*, 507–512. [CrossRef] [PubMed]
25. Xu, L.; Xu, C.; Zhao, M.; Qiu, Y.; Sheng, G.D. Oxidative removal of aqueous steroid estrogens by manganese oxides. *Water Res.* **2008**, *42*, 5038–5044. [CrossRef]
26. Huguet, M.; Deborde, M.; Papot, S.; Gallard, H. Oxidative decarboxylation of diclofenac by manganese oxide bed filter. *Water Res.* **2013**, *47*, 5400–5408. [CrossRef]
27. Forrez, I.; Carballa, M.; Verbeken, K.; Vanhaecke, L.; Schlusener, M.; Ternes, T.; Boon, N.; Verstraete, W. Diclofenac Oxidation by Biogenic Manganese Oxides. *Environ. Sci. Technol.* **2010**, *44*, 3449–3454. [CrossRef]
28. Nihemaiti, M.; Le Roux, J.; Hoppe-Jones, C.; Reckhow, D.A.; Croué, J.-P. Formation of haloacetonitriles, haloacetamides, and nitrogenous heterocyclic byproducts by chloramination of phenolic compounds. *Environ. Sci. Technol.* **2017**, *51*, 655–663. [CrossRef]
29. Chen, T.; Dou, H.; Li, X.; Tang, X.; Li, J.; Hao, J. Tunnel structure effect of manganese oxides in complete oxidation of formaldehyde. *Microporous Mesoporous Mater.* **2009**, *122*, 270–274. [CrossRef]
30. Luo, J.; Zhang, Q.; Garcia-Martinez, J.; Suib, S.L. Adsorptive and acidic properties, reversible lattice oxygen evolution, and catalytic mechanism of cryptomelane-type manganese oxides as oxidation catalysts. *J. Am. Chem. Soc.* **2008**, *130*, 3198–3207. [CrossRef]
31. Wu, C.; Xie, W.; Zhang, M.; Bai, L.; Yang, J.; Xie, Y. Environmentally friendly γ-MnO2 hexagon-based nanoarchitectures: Structural understanding and their energy-saving applications. *Chem. A Eur. J.* **2009**, *15*, 492–500. [CrossRef] [PubMed]
32. Zhang, J.; Li, Y.; Wang, L.; Zhang, C.; He, H. Catalytic oxidation of formaldehyde over manganese oxides with different crystal structures. *Catal. Sci. Technol.* **2015**, *5*, 2305–2313. [CrossRef]
33. Silberberg, M. *Chemistry: The Molecular Nature of Matter and Change with Advanced Topics*; McGraw-Hill: New York, NY, USA, 2018.
34. Stumm, W. *Chemistry of the Solid-Water Interface: Processes at the Mineral-Water and Particle-Water Interface in Natural Systems*; John Wiley & Son Inc.: New Jersey, NJ, USA, 1992.
35. Samuelsson, J.; Forssén, P.; Fornstedt, T. Sample conditions to avoid pH distortion in RP-LC. *J. Sep. Sci.* **2013**, *36*, 3769–3775. [CrossRef] [PubMed]
36. Fornstedt, T.; Forssén, P.; Westerlund, D. System peaks and their impact in liquid chromatography. *TrAC Trends Anal. Chem.* **2016**, *81*, 42–50. [CrossRef]
37. Xyla, A.G.; Sulzberger, B.; Luther III, G.W.; Hering, J.G.; van Cappellen, P.; Stumm, W. Reductive dissolution of manganese (III, IV)(hydr) oxides by oxalate: The effect of pH and light. *Langmuir* **1992**, *8*, 95–103. [CrossRef]
38. De Oliveira, T.; Guégan, R.; Thiebault, T.; Le Milbeau, C.; Muller, F.; Teixeira, V.; Giovanela, M.; Boussafir, M. Adsorption of diclofenac onto organoclays: Effects of surfactant and environmental (pH and temperature) conditions. *J. Hazard. Mater.* **2017**, *323*, 558–566. [CrossRef]
39. Das, R.N.; Roy, K. QSPR with extended topochemical atom (ETA) indices. 4. Modeling aqueous solubility of drug like molecules and agrochemicals following OECD guidelines. *Struct. Chem.* **2013**, *24*, 303–331. [CrossRef]
40. Loos, R. *Analytical Methods for Possible WFD 1st Watch List Substances*; European Commission, Joint Research Center, Onstitute fro Environment and Sustainability, Publications Office of the European Union: Luxembourg, 2015.
41. Zhao, J.; Wang, Q.; Fu, Y.; Peng, B.; Zhou, G. Kinetics and mechanism of diclofenac removal using ferrate (VI): Roles of Fe^{3+}, Fe^{2+}, and Mn^{2+}. *Environ. Sci. Pollut. Res.* **2018**, *25*, 22998–23008. [CrossRef]
42. Zhang, H.C.; Chen, W.R.; Huang, C.H. Kinetic modeling of oxidation of antibacterial agents by manganese oxide. *Environ. Sci. Technol.* **2008**, *42*, 5548–5554. [CrossRef]
43. Zhang, H.C.; Huang, C.H. Oxidative transformation of fluoroquinolone antibacterial agents and structurally related amines by manganese oxide. *Environ. Sci. Technol.* **2005**, *39*, 4474–4483. [CrossRef]
44. Bui, T.X.; Choi, H. Influence of ionic strength, anions, cations, and natural organic matter on the adsorption of pharmaceuticals to silica. *Chemosphere* **2010**, *80*, 681–686. [CrossRef] [PubMed]

45. Li, J.; Zhang, T.; Ye, M. Heterogeneous oxidation of diclofenac in the presence of α-MnO2 nanorods: Influence of operating factors and mechanism. *Water Sci. Technol.* **2015**, *71*, 1340–1346. [CrossRef] [PubMed]
46. Liu, W.; Sutton, N.B.; Rijnaarts, H.H.; Langenhoff, A.A.J.E.S.; Research, P. Anoxic conditions are beneficial for abiotic diclofenac removal from water with manganese oxide (MnO_2). *Environ. Sci. Pollut. Res. Int.* **2018**, *25*, 10141–10147. [CrossRef] [PubMed]
47. Su, D.; Ahn, H.-J.; Wang, G. Hydrothermal synthesis of α-MnO2 and β-MnO_2 nanorods as high capacity cathode materials for sodium ion batteries. *J. Mater. Chem. A* **2013**, *1*, 4845–4850. [CrossRef]
48. Wang, Y.; Feng, X.; Villalobos, M.; Tan, W.; Liu, F. Sorption behavior of heavy metals on birnessite: Relationship with its Mn average oxidation state and implications for types of sorption sites. *Chem. Geol.* **2012**, *292*, 25–34. [CrossRef]
49. Smith, P.F.; Deibert, B.J.; Kaushik, S.; Gardner, G.; Hwang, S.; Wang, H.; Al-Sharab, J.F.; Garfunkel, E.; Fabris, L.; Li, J.; et al. Coordination Geometry and Oxidation State Requirements of Corner-Sharing MnO6 Octahedra for Water Oxidation Catalysis: An Investigation of Manganite (γ-MnOOH). *ACS Catal.* **2016**, *6*, 2089–2099. [CrossRef]
50. Li, G.; Li, K.; Liu, A.; Yang, P.; Du, Y.; Zhu, M. 3D flower-like β-MnO_2/reduced graphene oxide nanocomposites for catalytic ozonation of dichloroacetic acid. *Sci. Rep.* **2017**, *7*, 43643. [CrossRef]
51. Kasprzyk-Hordern, B.; Ziółek, M.; Nawrocki, J. Catalytic ozonation and methods of enhancing molecular ozone reactions in water treatment. *Appl. Catal. B Environ.* **2003**, *46*, 639–669. [CrossRef]
52. Monteagudo, J.M.; El-taliawy, H.; Durán, A.; Caro, G.; Bester, K. Sono-activated persulfate oxidation of diclofenac: Degradation, kinetics, pathway and contribution of the different radicals involved. *J. Hazard. Mater.* **2018**, *357*, 457–465. [CrossRef]
53. Yu, H.; Nie, E.; Xu, J.; Yan, S.; Cooper, W.J.; Song, W. Degradation of diclofenac by advanced oxidation and reduction processes: Kinetic studies, degradation pathways and toxicity assessments. *Water Res.* **2013**, *47*, 1909–1918. [CrossRef]
54. Martínez, C.; Canle, L.M.; Fernández, M.I.; Santaballa, J.A.; Faria, J. Aqueous degradation of diclofenac by heterogeneous photocatalysis using nanostructured materials. *Appl. Catal. B Environ.* **2011**, *107*, 110–118. [CrossRef]
55. Soufan, M.; Deborde, M.; Legube, B. Aqueous chlorination of diclofenac: Kinetic study and transformation products identification. *Water Res.* **2012**, *46*, 3377–3386. [CrossRef] [PubMed]

Article

Kinetic and Prediction Modeling Studies of Organic Pollutants Removal from Municipal Wastewater using *Moringa oleifera* Biomass as a Coagulant

Bashir Adelodun [1,2], Matthew Segun Ogunshina [2], Fidelis Odedishemi Ajibade [3,4], Taofeeq Sholagberu Abdulkadir [5], Hashim Olalekan Bakare [6] and Kyung Sook Choi [1,7,*]

[1] Department of Agricultural Civil Engineering, Kyungpook National University, Daegu 41566, Korea; adelodun.b@unilorin.edu.ng
[2] Department of Agricultural and Biosystems Engineering, University of Ilorin, PMB 1515, Ilorin 240103, Nigeria; segunogunshina@gmail.com
[3] Department of Civil and Environmental Engineering, Federal University of Technology, PMB 704, Akure 340001, Nigeria; foajibade@futa.edu.ng
[4] Research Centre for Eco-Environmental Sciences, Chinese Academy of Sciences, Beijing 100085, China
[5] Department of Water Resources and Environmental Engineering, University of Ilorin, PMB 1515, Ilorin 240103, Nigeria; abdulkadir.ts@unilorin.edu.ng
[6] Department of Chemical Engineering, University of Ilorin, PMB 1515, Ilorin 240103, Nigeria; hashlek2016@gmail.com
[7] Institute of Agricultural Science & Technology, Kyungpook National University, Daegu 41566, Korea
* Correspondence: ks.choi@knu.ac.kr; Tel.: +82-53-950-5731; Fax: +82-53-950-6752

Received: 13 June 2020; Accepted: 16 July 2020; Published: 19 July 2020

Abstract: This study investigated the potential of *Moringa oleifera* (MO) seed biomass as a coagulant for the removal of turbidity, biochemical oxygen demand (BOD), and chemical oxygen demand (COD) of municipal wastewater. Triplicated laboratory experiments using MO coagulant added at varying treatment dosages of 50, 100, 150, 200 mg/L, and a control (0 mg/L) treatment were performed for a settling period of 250 min at room temperature. Kinetics and prediction variables of cumulative turbidity, BOD, and COD removal were estimated using simplified first order and modified Gompertz models. Results showed that the maximum removal of turbidity, BOD, and COD were 94.44%, 68.72%, and 57.61%, respectively, using an MO dose of 150 mg/L. Various kinetic parameters, such as rate constant (r), measured (RE_m) versus predicted (RE_p) cumulative removal, and specific pollutant removal rate (μ_m), were also maximum when an MO dose of 150 mg/L was added, the standard error being below 5%. The developed models were successfully validated over multiple observations. This study suggests low cost and sustainable removal of turbidity, BOD, and COD of municipal wastewater using MO seed biomass as a coagulant.

Keywords: kinetic studies; *Moringa oleifera*; plant seed biomass; prediction modeling; wastewater treatment

---

1. Introduction

An increasing human population and emerging lifestyles, among other factors, have greatly influenced the quality and quantity of municipal wastewater generation. The discharge of wastewater, which most times contains toxic organic and inorganic pollutants, without proper treatment into the ecosystem is widespread in developing countries and has been a great threat to human and aquatic life [1,2]. The growing concern of environmental degradation and water causing health-related issues including pollution of the water body and the scarcity of clean water has recently become the interest of researchers to achieve one of the Sustainable Development Goals (SDGs) of clean water and sanitation. It is estimated that about 1.1 billion people have no access to clean and safe potable water, and a significant number of them are in less developed countries of the world [3,4].

Owing to these, several techniques and processes have been developed and explored in the treatment of wastewater, among which are precipitation, ion exchange, filtration, electrodialysis, electrochemical process, coagulation, membrane process, and reverse osmosis [5]. However, the applicability of these technologies and techniques are limited in the developing world due to the cost and technical know-how constraints in the procurement of the raw materials and treatment processes required for these techniques [4–6]. Moreover, some of these treatment methods use excessive chemicals, generate secondary wastes which are often difficult to treat, and are found to be inefficient in the treatment of municipal wastewater due to the fact of varying forms of pollutants [7–9].

Recently, there has been a tremendous campaign on the use of sustainable approaches to address the environmental challenges of wastewater treatment and remediation [9–11]. In this regard, bio-coagulants such as *Moringa oleifera* (MO) biomass seeds have been recommended and extensively used as an effective alternative for the treatment of wastewaters, notably in Africa and Asia [2,12]. The MO biomass seeds, specifically, have gained remarkable attention in the research community due to its applicability as a coagulant and an antimicrobial agent in water and wastewater treatment [4,13]. The MO biomass offers a promising solution to wastewater treatment with less cost, availability of the raw materials, and higher efficiency in the removal of contaminants. In addition, the byproducts are ecofriendly as compared to other treatment processes or techniques including biological treatment methods [12,14].

The wastewater treatment using MO biomass is based on the coagulation-flocculation process in which the pollutant removal is achieved due to the presence of cationic protein in MO biomass seed [15,16], thereby forming small flocs with suspended particles and organic matters in the wastewater which are allowed to settle or sediment under varying contact times [10,17]. The coagulation mechanism of an MO biomass seed has been well discussed in the literature. Ndabigengesere et al. [17] described the coagulation mechanism of the MO biomass coagulant as adsorption and neutralization of charges, while Muyibi and Evison [18] attributed it to the bridging of destabilized particles. Okuda et al. [19] on the other hand, associated the variations in the coagulation mechanism of MO biomass coagulant in pollutant removal to the type of extractant used for the active component of the coagulant. Hoa and Hue [10] referred to the adsorption process as one of the main mechanisms of the coagulation process used by MO biomass. Despite the significant contributions on the coagulation mechanism of the MO biomass coagulant, the mechanism of pollutant removal by MO biomass through the settling ability of the flocs is not well discussed. Viotti et al. [20,21] re-emphasized the need for further studies on the removal of organic pollutants using the MO biomass. The inadequate details in this area create an important knowledge gap that needs to be considered. Appropriate models that describe the pollutant removal process could be valuable tools in this regard [22,23]. The studies that model the kinetic removal of pollutants in municipal wastewater using MO biomass as a coagulant are few, and to the best of our knowledge, there are no available publications on the applicability of a modified Gompertz model for the prediction of pollutant removal using MO biomass. The kinetic study offers valuable information on the mechanism of the reaction and pollutant removal process [14,24]. Also, MO biomass offers a cost-effective, easily accessible, and efficient pollutant removal from water and wastewater system as compared to other regular coagulants, such as alum and ferric chloride, making it a sustainable alternative to the developing countries [4,16].

Therefore, this study investigated the kinetic study of pollutants removal from municipal wastewater to further understand the MO biomass removal mechanism. The simplified form of the first-order kinetic model was applied to describe the pollutant removal process and the model parameters were subsequently estimated. The predictions of removal pollutants for turbidity, BOD, and COD were further conducted using the modified Gompertz model.

2. Materials and Methods

2.1. MO Biomass and Sample Preparation

The MO seed pods were collected from the Kwara State Ministry of Agriculture, Ilorin, Nigeria. The seeds were allowed to mature in the pod while still on the tree before plucking. This procedure was to ensure that the bioactive coagulants in the seed were effective [17]. The preparation of MO seeds involved two processes; the shelling of MO seeds from the dry pods, and the extraction of oil from the seeds to obtain MO powder as described by Farzadkia et al. [22] and Muyibi et al. [25], respectively. The MO pods were manually shelled using a knife to obtain the seeds after which the seeds were air-dried to maintain its uniform weight. The MO dried seeds were powdered using an electric grinder (Eurosonic Model ES-242) and sieved through a 0.5 mm mesh screen size to obtain MO biomass with uniform particle size.

Sixty grams of MO powder, mixed with 170 mL of hexane solvent, was fed into electro-thermal soxhlet extractor at 70 °C to extract the oil. The oil extraction process lasted for 4 h after which the solvent was left to condense and thereby separated from the oil. The resulting residue of MO after the oil extraction of 35% w/w was oven-dried at 50 °C overnight to obtain MO cake. Muyibi et al. [25] reported the enhanced coagulating potential of MO seeds after oil extraction. The obtained MO cake was, therefore, used in this study.

2.2. Sampling and Characterization of Wastewater

The samples of wastewater were taken from the main outlet of the sewer network system of the University of Ilorin main campus, Ilorin, Nigeria. The collected wastewater samples were transported to the laboratory within the University and then characterized about 1 h after the collection. The sample collection and analyses were carried out following the Standard Methods for Water and Wastewater Examination [26].

2.3. Experimental Design and Calculation of Turbidity, BOD and COD Removal Efficiency

The sampled wastewater of 1000 mL each, after the initial analyses for various parameters including turbidity, BOD, and COD, was filled into the five beakers in a Jar test set-up of Janke and Kunkel (Lovibond ET 730). The prepared samples of MO coagulants were then added at varying treatment dosages of 50, 100, 150, and 200 mg/L, with the last beaker used as control (0 mg/L). The samples of wastewater with the MO biomass coagulant were adjusted for pH levels (7.26 ± 0.09) using 0.5 M of NaOH or 0.5 M of HCl after which the samples were mixed rapidly at 100 rpm for 2 min, followed by slow mixing at 40 rpm for 20 min to aid sludge formation. Then the samples were gently transferred and allowed to settle in the sedimentation cones (Imhoff) at a different contact time of 50, 100, 150, 200, and 250 min. An aliquot of 250 mL was sampled from each sedimentation cone, and turbidity, BOD, COD were determined for each contact time, in triplicate. The removal efficiency of turbidity, BOD, and COD using varying MO dosage for each of the contact time was calculated using Equation (1)

$$\text{Removal efficiency (RE\%)} = \left(\frac{C_i - C_f}{C_i}\right) \times 100 \quad (1)$$

where C_i and C_f are the initial and final values of the analyzed parameters, respectively, expressed in NTU for turbidity, and mg/L for BOD and COD.

2.4. Kinetics of Turbidity, BOD and COD Removal

The pollutant removal rate of any wastewater treatment processes is one of the most important parameters used to define the effectiveness of reaction-based systems [23,27–29]. Therefore, the rate of solute sorption onto MO biomass was estimated. To evaluate the kinetics of turbidity, BOD, and COD removal process, a simplified form of the first-order kinetic model was used [28]. The form of the equation is given in Equation (2).

$$r = \log\left(\frac{C_i}{C_t}\right) \cdot \frac{2.303}{t_2 - t_1} \tag{2}$$

where r represents the rate constant of pollutant removal reaction (NTU min^{-1} and mg/L min^{-1}), while C_i and C_t are the initial and final parameter at the sampling interval ($t_2 - t_1$ min).

2.5. Modified Gompertz Model for Prediction of Turbidity, BOD and COD Removal

Recent studies have confirmed that the application of the Gompertz equation can be used for modeling the processes having a non-linear or exponential trend [29]. In this study, the cumulative removal of turbidity, BOD, and COD as a dependent variable to predict over settling time as an independent variable was considered. The cumulative parameter removal efficiency was calculated using Equation (3).

$$RE_m = \sum_{x=1}^{n} RE \tag{3}$$

where RE_m is the cumulative parameter removal efficiency (measured) in terms of the sum of n number of observations per timespan.

This model is a modified form of a modified Gompertz model which can be expressed as:

$$RE_p = P_{exp}\left\{-\exp\left[\frac{\mu_m}{P}(\lambda - t) + 1\right]\right\} \tag{4}$$

where RE_p is the cumulative pollutant removal (predicted), P is the maximum pollutant removal potential, μ_m is the maximum specific pollutant removal (%), λ and t are the lag phase and settling time (min), respectively.

2.6. Statistics

All the experiments were performed in the triplicated form. The analysis of variance (ANOVA) test was conducted with least significance difference (LSD) for the mean difference before and after treatment for the selected parameters. Data were processed using OriginPro Version 9 (Origin Corp., Northampton, MA, USA), IBM SPSS Version 23.0 (IBM Corp., Chicago, IL, USA). A model fitting tool, namely, Rank Models of OriginPro was used to simulate the prediction models. Finally, the model data were validated over multiple experimental trials.

3. Results and Discussion

3.1. Characteristics of Municipal Wastewater Used in This Study

The characteristics of municipal wastewater before and after treatment with MO biomass coagulants are presented in Table 1. The variables measured include pH, TSS, electrical conductivity, turbidity, BOD, COD, alkalinity, and total hardness, and each of which was found to be statistically significant after treatment with a varying dosage of MO (50, 100, 150, 200 mg/L) as indicated in their respective p-value ($p < 0.05$) in Table 1. In general, the sample wastewater before treatment was found to be neutral (7.26 ± 0.09) because of the anthropogenic activities in the University environment where the wastewater emanated. Nevertheless, the observed pH values, before and after treatment, were within the permissible limit for wastewater reuse in agriculture [30] and the national acceptable wastewater discharge range of 6–9 [31]. The cumulative variance (CV%) of wastewater parameters defined the less variation suitable for model building.

Table 1. Characteristics of municipal wastewater used for this study.

Parameter	MO Treatment	Before Treatment	CV (%)	After Treatment	USEPA Permissible Limit for Wastewater Reuse in Agriculture	NESREA Permissible Limit for Wastewater Discharge	Unit
	Control		0.01	7.37 ± 0.03 **			
	50 mg/L			6.93 ± 0.05 ***			
pH	100 mg/L	7.26 ± 0.09		6.88 ± 0.05 ***	6–8.4	6–9	-
	150 mg/L			7.02 ± 0.05 ***			
	200 mg/L			7.01 ± 0.05 ***			
	Control		13.50	91.88 ± 4.25 ***			
	50 mg/L			70.43 ± 2.05 ***			
Total suspended solids (TSS)	100 mg/L	111.93 ± 15.23		62.13 ± 2.36 ***	5	25	mg/L
	150 mg/L			41.70 ± 2.15 ***			
	200 mg/L			41.71 ± 2.00 ***			
	Control		4.80	10.88 ± 2.00 ***			
	50 mg/L			6.52 ± 1.15 ***			
Turbidity	100 mg/L	76.74 ± 3.69		5.35 ± 1.15 ***	2	5	NTU
	150 mg/L			4.27 ± 1.05 ***			
	200 mg/L			4.28 ± 1.25 ***			
	Control		8.72	144.40 ± 4.28 ns			
	50 mg/L			76.28 ± 2.75 ***			
Biochemical oxygen demand (BOD)	100 mg/L	147.45 ± 12.87		55.83 ± 2.55 ***	30	30	mg/L
	150 mg/L			46.12 ± 2.05 ***			
	200 mg/L			46.22 ± 3.16 ***			
	Control		2.12	448.18 ± 3.45 **			
	50 mg/L			237.28 ± 3.00 ***			
Chemical oxygen demand (COD)	100 mg/L	474.18 ± 10.09		204.30 ± 1.50 ***	120	60	mg/L
	150 mg/L			200.88 ± 2.52 ***			
	200 mg/L			201.00 ± 2.50 ***			
	Control		0.31	401.01 ± 0.03 ns			
	50 mg/L			429.09 ± 0.05 ***			
Electrical conductance (EC)	100 mg/L	402.44 ± 1.27		438.41 ± 0.05 ***	700	400	µS/cm
	150 mg/L			447.55 ± 0.01 ***			
	200 mg/L			447.76 ± 0.05 ***			
	Control		0.48	383 ± 1.26 ns			
	50 mg/L			311.85 ± 0.95 ***			
Alkalinity	100 mg/L	385.00 ± 1.85		327.25 ± 0.09 ***	50–150	-	mg/L
	150 mg/L			329.95 ± 1.06 ***			
	200 mg/L			330.05 ± 0.28 ***			
	Control		3.38	125.25 ± 1.25 ns			
	50 mg/L			111.36 ± 0.12 ***			
Total hardness (TH)	100 mg/L	128.50 ± 4.36		104.96 ± 1.00 ***	-	-	mg/L
	150 mg/L			101.12 ± 1.05 ***			
	200 mg/L			98.56 ± 1.28 ***			

ns: not significant; *, **, ***: Statistically significant at $p < F$ values of 0.05, 0.01, and 0.001, respectively. MO is *Moringa oleifera*; CV is Coefficient of variation; USEPA is United States Environmental Protection Agency; NESREA is National Environmental Standards and Regulations Enforcement Agency (Establishment) Act.

The turbidity and TSS measure the suspended particulate matters in the wastewater and can also describe the extent of wastewater pollution. Before the treatment, TSS and turbidity parameters (111.93 ± 15.23 and 76.74 ± 3.69) exceeded allowable national standard limits of 25 mg/L and 5 NTU [32], respectively, for wastewater discharge. However, after treatment with MO biomass coagulant, the TSS value was significantly reduced while the turbidity fell within the acceptable national limit of wastewater discharge. The initial high turbidity and TSS levels could be attributed to the presence of inorganic particulate matter in the wastewater.

The initial BOD and COD values before treatment with MO biomass (147.45 ± 12.87 and 474.18 ± 10.09 mg/L, respectively) were remarkably higher than the permissible discharge levels in Africa and Asia [32]. Although the values of both BOD and COD after treatment with 200 mg/L of MO dosage (46.22 ± 3.16 and 201.00 ± 2.50) were above the permissible limit for wastewater reuse in agriculture [30] and wastewater discharge [31]. However, there was a statistically significant difference ($p < 0.001$) between before treatment and after treatment with MO dosage. The BOD and COD indicate the extent of organic pollution in the wastewater and the high levels of these pollutants could be detrimental to the plant, animal, and human health when discharged into the environment, especially without proper treatment.

Similarly, the electrical conductivity of the wastewater, both before and after treatment, was found to be within the permissible limit of USEPA guidelines [30], and there was a significant difference between the before treatment and after treatment with MO biomass. Other measured parameters including alkalinity and total hardness exceeded the national standard value range of 50–100 mg/L for alkalinity whereas the standard limit for the hardness is yet to be established, revealing an indication of polluted wastewater. The three parameters, turbidity, BOD, and COD that represent both the particulate and organic pollution of the wastewater were therefore used to investigate the kinetics of the pollutant removal process.

3.2. *Effect of MO Dose on Turbidity, BOD and COD Removal*

The effect of MO biomass on turbidity, BOD, and COD removal is shown in Figure 1. The removal of all the three investigated pollutants increased with the increase in the MO biomass dosages under 250 min settling time. The result depicts that turbidity removal reached a maximum of 94.44% with the MO biomass dosage of 150 mg/L, which slightly decreased to 94.42% by increasing the dosage to 200 mg/L. Therefore, there was no significant difference for turbidity removal between 150 and 200 mg/L of MO.

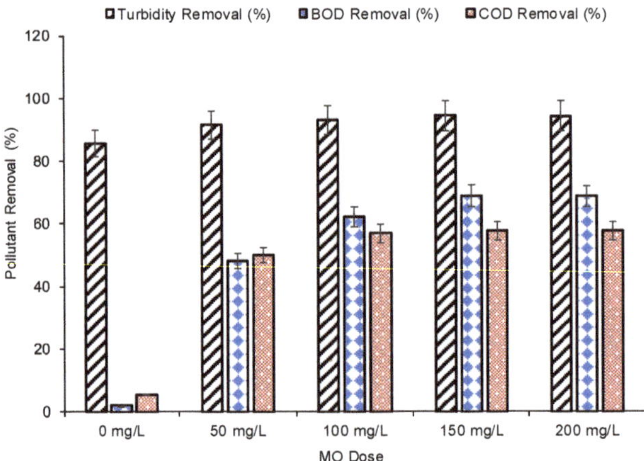

Figure 1. Effect of MO biomass coagulant dosage on turbidity, BOD, COD removal of municipal wastewater.

These results are similar to those obtained by Kayode et al. [32], Kane et al. [33], and Boulaadjoul et al. [34] for municipal wastewater and paper mill effluent, respectively. Dotto et al. [35] also confirmed an improved COD removal with increasing settling time using MO coagulant in textile wastewater. The further decrease in turbidity removal could be explained by the absence of opposite charged colloidal particles needed by the excess MO biomass to react with, which have been exhaustively neutralized and precipitated with the optimum biomass dosage. The maximum BOD and COD removal of 68.72% and 57.64%, respectively, was achieved at 150 mg/L of MO biomass dosage. The removal of the two pollutants is associated with the reduction of turbidity and organic suspended materials in wastewater [10] and as a result, the removal of both BOD and COD followed the same trend as that of turbidity removal. These results are in agreement with those obtained by Al-gheethi et al. [36] and Rosmawanie et al. [37].

3.3. Kinetics of Turbidity, BOD, and COD Removal

In this regard, the kinetics study of turbidity, BOD, and COD removal from municipal wastewater using MO biomass exhibited that the pollutant removal process in this study followed the reaction trend of the first-order kinetic model. The results of the first-order model for removal of turbidity, COD, and BOD are presented in Table 2.

Table 2. First-order kinetic variables for percent turbidity, BOD, and COD removal.

Parameter	Model Variable	MO Treatment Dose				
		Control	50 mg/L	100 mg/L	150 mg/L	200 mg/L
Turbidity	R^2	0.8432	0.8534	0.8386	0.8335	0.8399
	r	2.2216	2.8276	3.1608	3.6531	3.6472
	y	$-0.0042\,x + 1.8143$	$-0.0051\,x + 1.8256$	$-0.0056\,x + 1.82$	$-0.0062\,x + 1.8265$	$-0.0062\,x + 1.8266$
BOD	R^2	0.8064	0.9774	0.9623	0.9638	0.9697
	r	1.0664	1.2234	1.3184	1.3841	1.3833
	y	$-0.0005\,x + 2.1667$	$-0.0012\,x + 2.1557$	$-0.0018\,x + 2.1478$	$-0.0022\,x + 2.146$	$-0.0022\,x + 2.1462$
COD	R^2	0.9912	0.8755	0.8534	0.8675	0.8696
	r	0.8686	0.9696	0.9969	0.9999	0.9999
	y	$-0.0001\,x + 2.6760$	$-0.0013\,x + 2.6357$	$-0.0016\,x + 2.6262$	$-0.0016\,x + 2.6269$	$-0.0016\,x + 2.6261$

R^2: coefficient of determination; r: first-order rate constant; y: linear fitness equation of log(C) versus time t (x).

The data were best fitted in Equation (2) and gave satisfactory results for the applicability of the model in turbidity, BOD, and COD removal. The plot of logarithms of turbidity, BOD, and COD versus settling time (t) showed the good fitness of data points derived from experimental observations (Figure 2). It was evidenced that the coefficient of determination (R^2) values for time course turbidity, BOD, and COD removal ranged from 0.8335–0.8534, 0.8064–0.9774, and 0.8686–0.9999, respectively. Besides this, the rate constant (r) was encountered in control treatment for turbidity (2.2216 NTU min^{-1}), BOD (1.0664 mg/L·min^{-1}), and COD (0.8686 mg/L·min^{-1}) removal, correspondingly (Figure 3). Though, there was a variation in the k values of 150 mg/L and 200 mg/L MO treatments. However, the overall maximum removal of turbidity, BOD, and COD in a total of 250 min of settling allowance was achieved in 150 mg/L of MO dose along with a reasonable k value.

The pollutant removal rate in a reaction-based system depends on numerous internal and external factors such as temperature, dose, and particle of plant biomass, agitation, aeration, retention period, and characteristics of wastewater itself [38]. The plant biomass has both physical and chemical binding sites that capture the pollutant particles and forms a complex. As the reaction initiates, the pollutant particle starts attaching to these sites, and the net weight of plant biomass particles is increased [39]. The reaction lasts until the complete saturation of the free sites, and finally, the rapid settling of the particle occurs. The reaction is largely dependent on solute and solvent properties which determines the net pollutant removal efficiency of such systems. Previously, Viotti et al. [20] investigated the adsorption kinetic removal of diclofenac from wastewater using MO biomass. Their results indicated an adsorption capacity of 60.805 mg/g for MO biomass with a significant efficiency as compared to the activated carbon with a maximum adsorption capacity of 71.150 mg/g.

Adsorption of nonylphenol in wastewater using kinetics study was also investigated by Dai et al. [40], and the adsorption process fitted well into the Elovich kinetics. Similarly, Sacher et al. [41] compared three classical kinetic models including zero-, first-, and second-order kinetics to determine the removal of monochloride in Loire river water. The only first-order model fitted well into their kinetic data under the applied experimental conditions, while other forms of kinetic models showed significant random variations. Chen et al. [42] studied the kinetics of zinc removal from wastewater using the first-order model. They confirmed that zinc removal was significantly affected by the initial dose of zinc ions during the electrocoagulation process. Similarly, Kumar et al. [28] conducted lab-scale experiments of phyto-treatment of sugar mill effluent using water hyacinth plants in the CSTR system assisted with direct current (DC). They analyzed the kinetics of BOD and COD reduction in the CSTR system and found the maximum BOD and COD rates of 0.54 mg/L week^{-1} and 0.90 mg/L week^{-1}, respectively, using the first-order model.

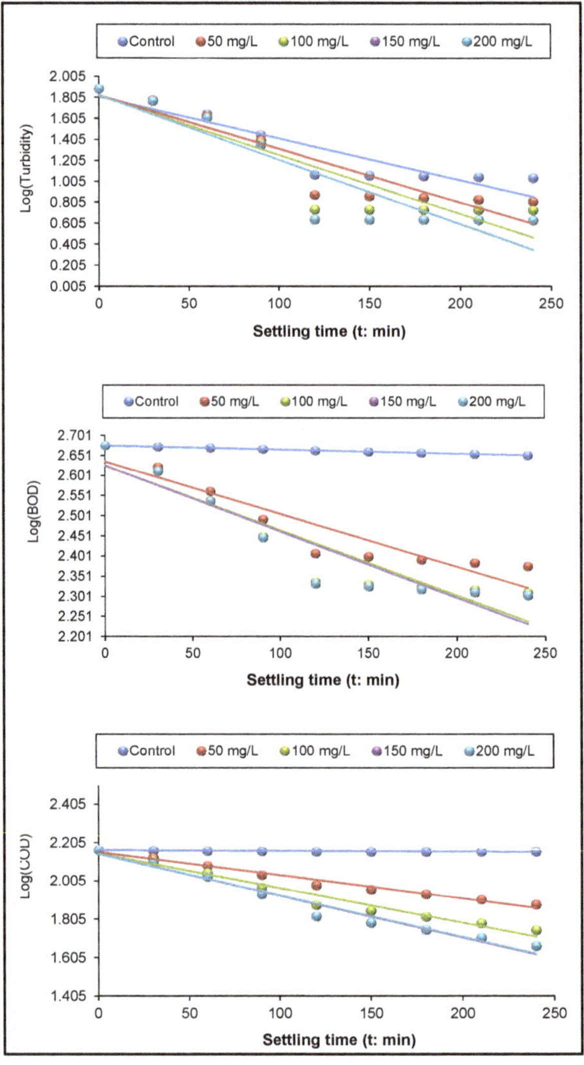

Figure 2. Log(C) versus time (t: min) plots of turbidity, BOD, COD removal of municipal wastewater.

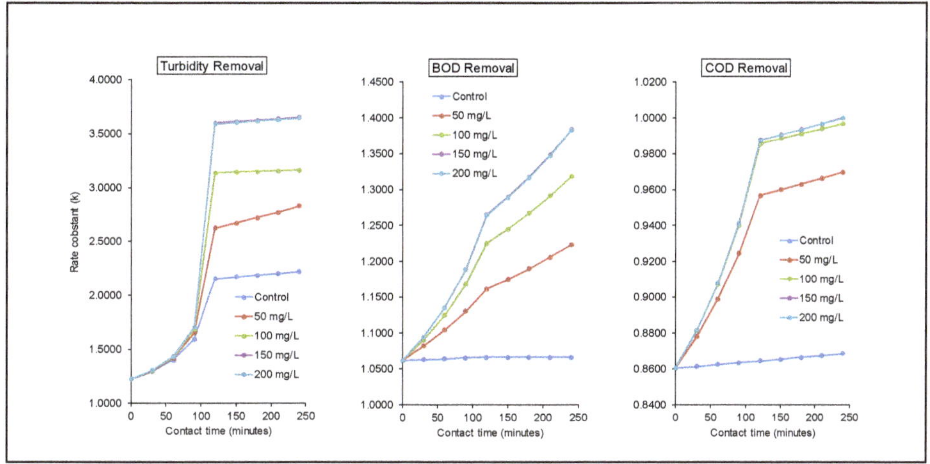

Figure 3. The first-order rate constant (*r*) of cumulative turbidity (NTU min^{-1}), BOD (mg/L·min^{-1}), and COD (mg/L·min^{-1}) removal of municipal wastewater using MO coagulant.

3.4. Prediction and Validation of Model Results

Prediction modeling results of cumulative turbidity, BOD, and COD removal using the modified Gompertz kinetic model are given in Table 3. From the final results, it was evidenced that the model can be successfully implemented for simulation of turbidity, BOD, and COD removal from municipal wastewater using MO biomass. The results showed that the coefficient of determination (R^2) had strong fitness of experimental data, which ranged from 0.98–0.99 for cumulative turbidity, BOD, and COD removal.

Table 3. Measured versus predicted percent (%) removal of turbidity, BOD, and COD with Gomphertz model variables.

Parameter	Variable	MO Treatment Dose				
		Control	50 mg/L	100 mg/L	150 mg/L	200 mg/L
Turbidity	RE_m	85.82	91.49	93.02	94.44	94.42
	RE_p	87.50	93.24	95.22	96.97	96.95
	S_E	2.24	2.37	2.51	2.55	2.53
	Λm	45.94	45.99	45.56	45.57	45.58
	μ	0.0291	0.0290	0.0293	0.0296	0.0294
	R^2	0.98	0.98	0.98	0.98	0.98
BOD	RE_m	2.07	48.26	62.14	68.72	68.65
	RE_p	2.10	46.58	61.71	68.15	68.08
	S_E	0.05	1.64	1.57	1.61	1.62
	Λm	45.99	63.32	56.82	55.16	55.24
	μ	0.0290	0.0177	0.0205	0.0214	0.0213
	R^2	0.99	0.99	0.99	0.99	0.99
COD	RE_m	5.48	49.96	56.92	57.64	57.61
	RE_p	5.37	50.24	57.38	58.05	58.02
	S_E	0.74	1.10	1.34	1.35	1.34
	Λm	117.89	48.31	47.15	47.28	47.29
	μ	0.0092	0.0265	0.0277	0.0276	0.0276
	R^2	0.99	0.99	0.99	0.99	0.99

RE_m: Measured removal (%); RE_p: predicted removal (%) using modified Gompertz model; S_E: standard error in prediction; λ_{max}: lag phase in min; μm: specific removal rate per min; and R^2: coefficient of determination.

Furthermore, the predicted removal in all MO treatments was intensely close to the measured removal. Figures 4 and 5 give graphical representations of comparison of predicted and measured turbidity, BOD, and COD removal. The maximum measured (R_m) versus predicted (R_p) removal of turbidity (RE_m: 94.44 and RE_p: 96.97), BOD (RE_m: 68.72 and RE_p: 68.15), and COD (RE_m: 57.64 and RE_p: 58.05) indicated that the models had an acceptable and small standard error in prediction (<5%). The maximum specific removal rate (μ_m) was maximum in 150 mg/L MO treatment for all the three selected parameters. However, all the model parameters in 200 mg/L MO treatment were slightly less than the parameter values of 150 mg/L MO treatment, therefore, increasing the MO dose after 150 mg/L was not recommended. Also, the medium values of the lag phase (λ_{max}) revealed that the initial settling time of 150 min was the most determining phase in which the maximum specific pollutant removal was achieved. However, the models were developed based on the specific municipal wastewater and the estimated variables may vary according to the water quality. Still, the methodology suggests the good fitness of the models in determining the major factors of the municipal wastewater treatment process.

In our findings, the modified Gompertz model-based variables increased from control treatment to 150 mg/L MO treatment and further decreased to 200 mg/L. The stability of these model variables may be associated with the kinetic factors, i.e., solute (MO biomass) concentration. A high dose of MO may not be kinetically feasible for the pollutant capture reaction, and, therefore, the reaction rate was stabilized. The good fitness of the non-linear curve, as demonstrated in Figure 4, explained the feasibility of the modified Gompertz model in the present study. Recently, Carvajal et al. [43] studied the pollutant removal kinetics and their relative impact of anoxic BTEX biodegradation using the modified Gompertz model. They precisely modeled the BTEX biodegradation and showed that the higher concentration mixture might cause an inhibitory effect on the degradation process. Báez et al. [29] investigated the pollutant load removal efficiency of whey using a modified Gompertz model with validation of the results by specific model verification tools like R^2, efficiency, and standard error in prediction. Hernández-Martínez et al. [44] performed batch mode experiments in a bubble column reactor for pollutant load reduction petroleum hydrocarbons using microbial biomass. They also modeled the hydrocarbon biosorption process using the modified Gompertz model and explained the model predicted removal was near to the experimental removal.

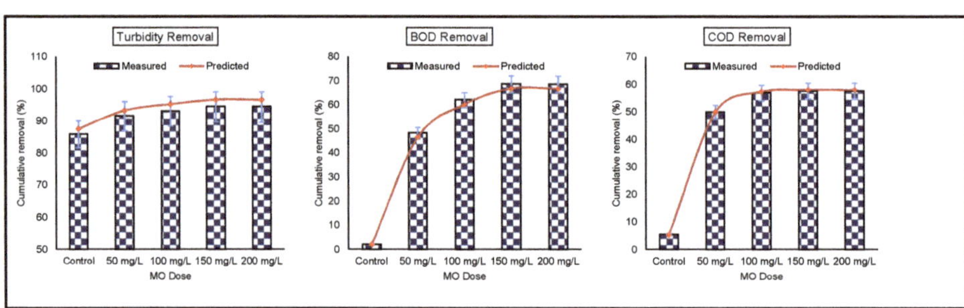

Figure 4. Effect of MO dose with measured versus predicted cumulative removal of turbidity, BOD, and COD of municipal wastewater using MO coagulant.

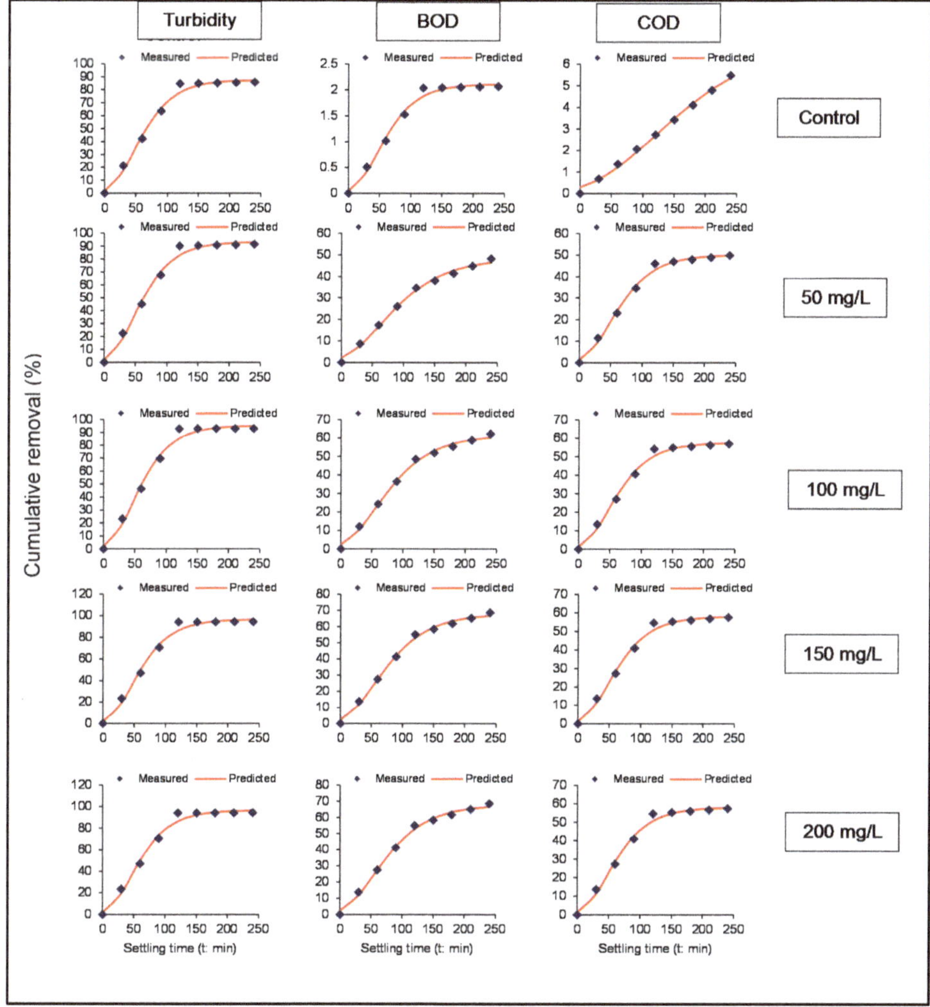

Figure 5. Experimental versus modified Gompertz model predicted cumulative removal (%) of turbidity, BOD, COD of municipal wastewater using MO coagulant.

4. Conclusions

In this study, we deduced that *Moringa oleifera* (MO) seed biomass was an effective coagulant to reduce the pollutant load of municipal wastewater. The results revealed that a significant reduction ($p < 0.05$) of the selected wastewater parameters, viz., turbidity, BOD, and COD was achieved. Also, the kinetic modeling of the reactor system utilizing the first-order and modified Gompertz model helped to enhance the turbidity, BOD, and COD removal process. We found that among the different MO treatments (0–200 mg/L), the maximum measured (R_m) and predicted (RE_p) removal of turbidity (RE_m: 94.44 and RE_p: 96.97), BOD (RE_m: 68.72 and RE_p: 68.15), and COD (RE_m: 57.64 and RE_p: 58.05) were achieved using a MO dose of 150 mg/L after 250 min of settling retention, respectively. However, the models should be tested for the treatment of other wastewaters as the variables may vary according to the water quality. The findings of this study will be essentially useful for low-cost primary treatment of municipal wastewater employing MO seed biomass and minimize its possible, disposable environmental risks.

Author Contributions: B.A. conceived the idea, conducted the experiment, analyzed the data, and wrote the manuscript. M.S.O. collected the data, performed sampling analysis, and wrote the manuscript. F.O.A. performed the data curation, analyzed the data, and revised the manuscript. T.S.A. analyzed the data and revised the manuscript. H.O.B. collected the data and revised the manuscript. K.S.C. resources and reviewed the manuscript. All authors have read and agreed to the published version of the manuscript.

Funding: This research received no external funding.

Acknowledgments: The authors would like to express their gratitude to the University of Ilorin, Ilorin for providing the necessary facilities for conducting this research. We gratefully acknowledge the assistance of Pankaj Kumar of Agro-ecology and Pollution Research Laboratory, Department of Zoology and Environmental Science, Gurukula Kangri Vishwavidyalala, Haridwar-249404 (Uttarakhand), India, who thoroughly read the manuscript and offered brilliant suggestions to improve the manuscript. In addition, the efforts of the three anonymous reviewers to further improve the manuscript are appreciated.

Conflicts of Interest: The authors declare no conflict of interest.

References

1. Freitas, J.H.E.S.; de Santana, K.V.; do Nascimento, A.C.C.; de Paiva, S.C.; de Moura, M.C.; Coelho, L.C.B.B.; de Oliveira, M.B.M.; Paiva, P.M.G.; do Nascimento, A.E.; Napoleão, T.H. Evaluation of using aluminum sulfate and water-soluble Moringa oleifera seed lectin to reduce turbidity and toxicity of polluted stream water. *Chemosphere* **2016**, *163*, 133–141. [CrossRef] [PubMed]
2. Vunain, E.; Masoamphambe, E.F.; Mpeketula, P.M.G.; Monjerezi, M.; Etale, A. Evaluation of coagulating efficiency and water borne pathogens reduction capacity of Moringa oleifera seed powder for treatment of domestic wastewater from Zomba, Malawi. *J. Environ. Chem. Eng.* **2019**, *7*, 103118. [CrossRef]
3. Abaliwano, J.K.; Ghebremichael, K.A.; Amy, G.L. *Application of the Purified Moringa Oleifera Coagulant for Surface Water Treatment*; United Nations Educational, Scientific and Cultural Organization, Institute of water Education (UNESCO-IHE): Delft, The Netherlands, 2008; Volume 5.
4. Kansal, S.K.; Kumari, A. Potential of M. oleifera for the treatment of water and wastewater. *Chem. Rev.* **2014**, *114*, 4993–5010. [CrossRef]
5. Crini, G.; Lichtfouse, E. Advantages and disadvantages of techniques used for wastewater treatment. *Environ. Chem. Lett.* **2019**, *17*, 145–155. [CrossRef]
6. Xiong, B.; Piechowicz, B.; Wang, Z.; Marinaro, R.; Clement, E.; Carlin, T.; Uliana, A.; Kumar, M.; Velegol, S.B. Moringa oleifera f-sand Filters for Sustainable Water Purification. *Environ. Sci. Technol. Lett.* **2018**, *5*, 38–42. [CrossRef]
7. Tripathi, A.; Mishra, A.K. Knowledge and passive adaptation to climate change: An example from Indian farmers. *Clim. Risk Manag.* **2017**, *16*, 195–207. [CrossRef]
8. Valverde, K.C.; Coldebella, P.F.; Silva, M.F.; Nishi, L.; Bongiovani, M.C.; Bergamasco, R. Moringa oleifera Lam. and Its Potential Association with Aluminium Sulphate in the Process of Coagulation/Flocculation and Sedimentation of Surface Water. *Int. J. Chem. Eng.* **2018**, *2018*, 12–15. [CrossRef]
9. Nwoba, E.G.; Vadiveloo, A.; Ogbonna, C.N.; Ubi, B.E.; Ogbonna, J.C.; Moheimani, N.R. Algal Cultivation for Treating Wastewater in African Developing Countries: A Review. *Clean Soil Air Water* **2020**, *48*, 1–14. [CrossRef]
10. Hoa, N.T.; Hue, C.T. Enhanced water treatment by Moringa oleifera seeds extract as the bio-coagulant: Role of the extraction method. *J. Water Supply Res. Technol.* **2018**, *67*, 634–647. [CrossRef]
11. Shirani, Z.; Santhosh, C.; Iqbal, J.; Bhatnagar, A. Waste Moringa oleifera seed pods as green sorbent for efficient removal of toxic aquatic pollutants. *J. Environ. Manag.* **2018**, *227*, 95–106. [CrossRef]
12. Rocha, V.V.F.; dos Santos, I.F.S.; Silva, A.M.L.; Sant'Anna, D.O.; Junho, A.L.; Barros, R.M. Clarification of high-turbidity waters: A comparison of Moringa oleifera and virgin and recovered aluminum sulfate-based coagulants. *Environ. Dev. Sustain.* **2019**, 1–12. [CrossRef]
13. Adelodun, B.; Odedishemi, F.; Segun, M.; Choi, K.-S. Dosage and settling time course optimization of Moringa oleifera in municipal wastewater treatment using response surface methodology. *Desalin. Water Treat.* **2019**, *167*, 45–56. [CrossRef]

14. Bello, O.S.; Lasisi, B.M.; Adigun, O.J.; Ephraim, V. Scavenging Rhodamine B dye using moringa oleifera seed pod. *Chem. Speciat. Bioavailab.* **2017**, *29*, 120–134. [CrossRef]
15. Camacho, F.P.; Sousa, V.S.; Bergamasco, R.; Ribau Teixeira, M. The use of Moringa oleifera as a natural coagulant in surface water treatment. *Chem. Eng. J.* **2017**, *313*, 226–237. [CrossRef]
16. Villaseñor-Basulto, D.L.; Astudillo-Sánchez, P.D.; del Real-Olvera, J.; Bandala, E.R. Wastewater treatment using Moringa oleifera Lam seeds: A review. *J. Water Process Eng.* **2018**, *23*, 151–164. [CrossRef]
17. Ndabigengesere, A.; Narasiah, K.S.; Talbot, B.G. Active agents and mechanism of coagulation of turbid waters using Moringa oleifera. *Water Res.* **1995**, *29*, 703–710. [CrossRef]
18. Muyibi, S.A.; Evison, L.M. Optimizing physical parameters affecting coagulation of turbid water with Morninga oleifera seeds. *Water Res.* **1995**, *29*, 2689–2695. [CrossRef]
19. Okuda, T.; Baes, A.U.; Nishijima, W.; Okada, M. Coagulation mechanism of salt solution-extracted active component in Moringa oleifera seeds. *Water Res.* **2001**, *35*, 830–834. [CrossRef]
20. Viotti, P.V.; Moreira, W.M.; dos Santos, O.A.A.; Bergamasco, R.; Vieira, A.M.S.; Vieira, M.F. Diclofenac removal from water by adsorption on Moringa oleifera pods and activated carbon: Mechanism, kinetic and equilibrium study. *J. Clean. Prod.* **2019**, *219*, 809–817. [CrossRef]
21. Natarajan, R.; Manivasagan, R. Effect of operating parameters on dye wastewater treatment using Prosopis cineraria and kinetic modeling. *Environ. Eng. Res.* **2019**, *25*, 788–793. [CrossRef]
22. Farzadkia, M.; Ehrampush, M.H.; Mehrizi, E.A.; Sadeghi, S.; Talebi, P.; Salehi, A.; Kermani, M. Investigating the efficiency and kinetic coefficients of nutrient removal in the subsurface artificial wetland of Yazd wastewater treatment plant. *Environ. Health Eng. Manag.* **2015**, *2*, 23–30.
23. Adelodun, B.; Ajibade, F.O.; Abdulkadir, T.S.; Bakare, H.O.; Choi, K.S. SWOT analysis of agro-waste based adsorbents for persistent dye pollutants removal from wastewaters. In *Environmental Degradation: Causes and Remediation Strategies*; Kumar, V., Singh, J., Kumar, P., Eds.; Agro Environ Media—Agriculture and Ennvironmental Science Academy: Haridwar, India, 2020; pp. 88–103.
24. Delelegn, A.; Sahile, S.; Husen, A. Water purification and antibacterial efficacy of Moringa oleifera Lam. *Agric. Food Secur.* **2018**, *7*, 1–10. [CrossRef]
25. Muyibi, S.A.; Noor, M.J.M.M.; Leong, T.K.; Loon, L.H. Effects of oil extraction from Moringa oleifera seeds on coagulation of turbid water. *Int. J. Environ. Stud.* **2002**, *59*, 243–254. [CrossRef]
26. Rice, E.W.; Baird, R.B.; Eaton, A.D.; Clesceri, L.S. *APHA Standard Methods for the Examination of Water and Wastewater*, 22nd ed.; APHA, AWWA, WEF: Washington, DC, USA, 2012. [CrossRef]
27. El Shahawy, A.; Heikal, G. Organic pollutants removal from oily wastewater using clean technology economically, friendly biosorbent (*Phragmites australis*). *Ecol. Eng.* **2018**, *122*, 207–218. [CrossRef]
28. Kumar, V.; Singh, J.; Kumar, P.; Kumar, P. Response surface methodology based electro-kinetic modeling of biological and chemical oxygen demand removal from sugar mill effluent by water hyacinth (*Eichhornia crassipes*) in a Continuous Stirred Tank Reactor (CSTR). *Environ. Technol. Innov.* **2019**, *14*, 100327. [CrossRef]
29. Báez, L.C.; Castro Rosas, J.; Villagómez Ibarra, J.R.; Palma Quiroz, I.; Páez Lerma, J.B.; Gómez Aldapa, C.A. Production of benzyl carbonyl (rose aroma) from whey and its effect on pollutant load removal. *Environ. Dev. Sustain.* **2019**, *21*, 609–619. [CrossRef]
30. United States Environmental Protection Agency (USEPA). *Guidelines for Water Reuse*; United States Environmental Protection Agency: Washington, DC, USA, 2012.
31. NESREA. *National Environmental Standards and Regulations Enforcement Agency (Establishment) Act*; NESREA: Abuja, Nigeria, 2009; pp. 1–21.
32. Kayode, O.; Luethi, C.; Rene, E. Management Recommendations for Improving Decentralized Wastewater Treatment by the Food and Beverage Industries in Nigeria. *Environments* **2018**, *5*, 41. [CrossRef]
33. Kane, C.; Bâ, A.; Mahamat, S.A.M.; Ayessou, N.; Mbacké, M.K.; Mar Diop, C.G. Combination of alum and extracted Moringa oleifera bioactive molecules powder for municipal wastewater treatment. *Int. J. Biol. Chem. Sci.* **2016**, *10*, 1918–1929. [CrossRef]
34. Boulaadjoul, S.; Zemmouri, H.; Bendjama, Z.; Drouiche, N. A novel use of Moringa oleifera seed powder in enhancing the primary treatment of paper mill effluent. *Chemosphere* **2018**, *206*, 142–149. [CrossRef]
35. Dotto, J.; Fagundes-Klen, M.R.; Veit, M.T.; Palácio, S.M.; Bergamasco, R. Performance of different coagulants in the coagulation/flocculation process of textile wastewater. *J. Clean. Prod.* **2019**, *208*, 656–665. [CrossRef]

36. Al-gheethi, A.A.; Mohamed, R.M.S.R.; Wurochekke, A.A.; Nurulainee, N.R.; Mas Rahayu, J.; Amir Hashim, M.K. Efficiency of Moringa oleifera Seeds for Treatment of Laundry Wastewater. *MATEC Web Conf.* **2017**, *103*, 1–8. [CrossRef]
37. Rosmawanie, M.; Mohamed, R.; Al-Gheethi, A.; Pahazri, F.; Amir-Hashim, M.K.; Nur-Shaylinda, M.Z. Sequestering of pollutants from public market wastewater using Moringa oleifera and Cicer arietinum flocculants. *J. Environ. Chem. Eng.* **2018**, *6*, 2417–2428. [CrossRef]
38. Gao, Q.; Xu, J.; Bu, X.H. Recent advances about metal–organic frameworks in the removal of pollutants from wastewater. *Coord. Chem. Rev.* **2019**, *378*, 17–31. [CrossRef]
39. Samal, K.; Dash, R.R.; Bhunia, P. Effect of hydraulic loading rate and pollutants degradation kinetics in two stage hybrid macrophyte assisted vermifiltration system. *Biochem. Eng. J.* **2018**, *132*, 47–59. [CrossRef]
40. Dai, Y.D.; Shah, K.J.; Huang, C.P.; Kim, H.; Chiang, P.C. Adsorption of nonylphenol to multi-walled carbon nanotubes: Kinetics and isotherm study. *Appl. Sci.* **2018**, *8*, 2295. [CrossRef]
41. Sacher, F.; Gerstner, P.; Merklinger, M.; Thoma, A.; Kinani, A.; Roumiguières, A.; Bouchonnet, S.; Richard-Tanaka, B.; Layousse, S.; Ata, R.; et al. Determination of monochloramine dissipation kinetics in various surface water qualities under relevant environmental conditions-Consequences regarding environmental risk assessment. *Sci. Total Environ.* **2019**, *685*, 542–554. [CrossRef] [PubMed]
42. Chen, X.; Ren, P.; Li, T.; Trembly, J.P.; Liu, X. Zinc removal from model wastewater by electrocoagulation: Processing, kinetics and mechanism. *Chem. Eng. J.* **2018**, *349*, 358–367. [CrossRef]
43. Carvajal, A.; Akmirza, I.; Navia, D.; Pérez, R.; Muñoz, R.; Lebrero, R. Anoxic denitrification of BTEX: Biodegradation kinetics and pollutant interactions. *J. Environ. Manag.* **2018**, *214*, 125–136. [CrossRef]
44. Hernández-Martínez, R.; Valdivia-Rivera, S.; Betto-Sagahon, J.; Coreño-Alonso, A.; Tzintzun-Camacho, O.; Lizardi-Jiménez, M.A. Solubilization and removal of petroleum hydrocarbons by a native microbial biomass in a bubble column reactor. *Rev. Mex. Ing. Quím.* **2019**, *18*, 181–189. [CrossRef]

Article

Evaluation of Groundwater and Grey Water Contamination with Heavy Metals and Their Adsorptive Remediation Using Renewable Carbon from a Mixed-Waste Source

Taghrid S. Alomar [1], Mohamed A. Habila [2,*], Zeid A. Alothman [2], Najla AlMasoud [1,*] and Saad Saeed Alqahtany [2]

1 Department of Chemistry, College of Science, Princess Nourah bint Abdulrahman University, Riyadh 11671, Saudi Arabia; tsalomar@pnu.edu.sa

2 Department of Chemistry, College of Science, King Saud University, P.O. Box 2455, Riyadh 11451, Saudi Arabia; zaothman@ksu.edu.sa (Z.A.A.); ksasa911@hotmail.com (S.S.A.)

* Correspondence: mhabila@ksu.edu.sa (M.A.H.); nsalmasoud@pnu.edu.sa (N.A.); Tel.: +96-614-674-198 (M.A.H.); +966-11-8236046 (N.A.); Fax: +96-614-675-992 (M.A.H.)

Received: 23 December 2019; Accepted: 15 June 2020; Published: 24 June 2020

Abstract: The contamination of water sources with heavy metals is a serious challenge that humanity is facing worldwide. The aim of this work was to evaluate and remediate the metal pollution in groundwater and greywater resources from Riyadh, Saudi Arabia. In addition, we investigated the application of ultrasonic power before adsorption to assess the dispersion of renewable carbon from mixed-waste sources (RC-MWS) as an adsorbent and enhance the water purification process. The renewable carbon adsorbent showed high ability to adsorb Pb(II), Zn(II), Cu(II), and Fe(II) from samples of the actual water under study. The conditions for the remediation of water polluted with heavy metals by adsorptive-separation were investigated, including the pH of the adsorption solution, the concentration of the heavy metal(s) under study, and the competition at the adsorption sites. The enhanced adsorption process exhibited the best performance at a pH of 6 and room temperature, and with a contact time of 60 min. Kinetic studies showed that the pseudo-second-order kinetic model was fitted with the adsorption of Pb(II), Zn(II), Cu(II), and Fe(II) onto the RC-MWS. The adsorption data were well fitted by Langmuir isotherms. The Freundlich isotherm was slightly fitted in the cases of Cu(II), Zn(II), and Fe(II), but not in the cases of Pb(II). The developed adsorption process was successfully applied to actual water samples, including water samples from Deria and Mozahemia and samples from clothes and car washing centers in Riyadh city.

Keywords: heavy metals determination; groundwater; greywater; adsorption; separation; inductively coupled plasma mass spectroscopy

1. Introduction

Determining heavy metals contamination in water systems, including groundwater and greywater systems, is a prerequisite for employing or treating them. Greywater includes water used by clothes and car washing machines and basins in kitchens and bathrooms in houses and mosques. However, the term 'black water' refers to the water used by toilets. The water from groundwater or greywater resources is expected to be more suitable for economic wastewater treatment than black water [1–5]. Groundwater may be contaminated with heavy metals; for example, a risk of human exposure to heavy metals through groundwater used as a source of drinking water has been reported [6,7]. Water resource shortages have forced countries to use different water supply sources, such as groundwater, seawater, rainwater, riverwater, and wastewater [8–10]. The availability and purity of water have a direct impact on public health. The desalination and reuse of wastewater have attracted interest in fields of applied science in some countries, including Saudi Arabia [8,11–15]. The most common permanent pollutants of water systems are heavy metals from natural or industrial sources [16,17]. Heavy metals are characterized by having an atomic weight in the range of 63.5–200.6. The major industries that discharge heavy metals are the mining and metal plating industries. The discharge from these industries has led to a large amount of heavy metals in the water system [18,19]. The most damaging impact of heavy metals pollution is their accumulation inside living beings, particularly fishes, in the marine environment. In addition, they accumulate in plants and animals to reach higher levels in the food chain and become very dangerous to humans. Moreover, it is very difficult to biodegrade heavy metals in the environment [20]. Many dangerous and chronic diseases result from contamination with heavy metal cations, such as chromium, copper, zinc, lead, iron, manganese, cadmium, and mercury cations. Municipal wastewater represents another prominent source of metal pollution, which may include cadmium, arsenic, selenium, zinc, and nickel [16,19,20]. The World Health Organization (WHO) reported that the serious toxicity of heavy metals may have damaging health impacts on many human organs, such as the nervous system and gastrointestinal (GI) systems. In addition to that, it may have harmful effects on the lungs, the renal system, and the liver, and in some cases may lead to more complicated diseases, including cancer [20]. The treatment of wastewater helps to protect the environment and solve water shortage problems [21–25]. This trend has attracted increasing interest, especially for greywater, which is produced by hand-washing basins and washing machines and is not allowed to come into contact with toilet water at all [1–3]. There are many technologies for treating wastewater containing heavy metals, such as membrane separation, chemical oxidation, and chemical reduction, carbon adsorption, liquid extraction, electrolytic treatment, ion exchange, electro precipitation, coagulation, evaporation, flotation, hydroxide and sulfide precipitation, ultrafiltration, crystallization, and electro-dialysis [26–28]. The flotation process has successfully been applied for the separation of ions with the assistance of surfactants or dispersant gases [29–34]. However, adsorption exhibits superior efficiency in the removal of heavy metals from wastewater [35,36]. The main goal of the adsorption process is to decrease the level of heavy metals pollution in wastewater; however, different adsorbents have different efficiencies and the cost is highly variable [26,36–40]. Improvements in the adsorptive treatment are usually based on the development of a highly porous or functionalized adsorbent [18,41]. Activated carbon is known to be a highly effective adsorbent for the removal of heavy metals from wastewater, and it readily dissolves in an extreme pH medium, which facilitates its application and makes the process more suitable for a large number of pollutants of organic and inorganic species [35,42–44]. The main advantage of the application of activated carbon for controlling pollution by adsorption is its low price compared with other materials as well as being easy to produce from various waste sources [45–47]. In addition, activated carbon from waste sources is considered to be a safe adsorbent for wastewater treatment [48–51]. Ultrasonic waves have recently been proposed and used to enhance ion transfer, penetration, and separation. In addition, ultrasonic waves are applied in liquid extraction to improve interactions and enhance the mass transfer [52,53]. This study aimed to develop an improved separation process for removing metal pollutants from groundwater and grey water, investigate the characteristics of adsorption with the help of ultrasonic waves, and evaluate the competition in the adsorption process in the case of a mixed heavy metal solution system. We also optimized the pH, contact time, and metal ion concentrations, in order to adjust the adsorption process. The developed adsorptive separation process was employed to remove Pb(II), Zn(II), Cu(II), and Fe(II) from groundwater and greywater to improve the overall water quality.

2. Experimental

2.1. Water Samples from Different Regions of Riyadh City and Surrounding Cities

Water samples were collected from Riyadh city as well as from cities surrounding it, including Deria and Mozahemia. The water samples included drinking water samples, groundwater samples, and samples from a clothes washing center and a car washing center. The samples were acidified with nitric acid after being put in polypropylene bottles, and then analyzed for the heavy metals Pb(II), Zn(II), Cu(II), and Fe(II). In addition, the collected samples were applied in an adsorption process for the evaluation of the heavy metal separation efficiency.

2.2. Determining Heavy Metals and Optimizing the Adsorptive Remediation Process

Nitric acid, hydrochloric acid, sodium hydroxide, lead nitrate, zinc nitrate, copper nitrate, and ferrous nitrate were purchased from Sigma-Aldrich, St. Louis, MO, USA. All reagents were of analytical grade. The renewable carbon from a mixed-waste source (RC-MWS) used in this study is prepared in our laboratory and it was previously characterized [44]. The adsorptive remediation process for adsorbing Pb(II), Zn(II), Cu(II), and Fe(II) from an aqueous solution onto renewable carbon from a mixed-waste source (RC-MWS) was assessed by a group technique in which the primary model metal solutions for Pb(II), Zn(II), Cu(II), and Fe(II) (500 parts per million, ppm) were made in the laboratory by employing nitrate salts, and metal solutions (400, 300, 200, 100, and 50 ppm) were prepared daily by dilution. A 50-mL metal cation solution was added to 0.03 g of RC-MWS. Blank experiments were carried out without the addition of an adsorbent. The mixtures were exposed to ultrasonic waves for 2 min, and then shacked for a certain period of time. Then, after filtration, the decrease in the amount of the heavy metal in the solution was measured throughout the treatment by inductively coupled plasma mass spectrometry (ICPMS) using Equation (1):

$$q_e = \frac{(C_0 - C_e) * V}{M} \quad (1)$$

where C_0 represents the initial concentration of metal ions in the solution, C_e is the equilibrium concentration of metal ions in the solution, V is volume of the solution (L), and M is the mass of the adsorbent (g).

The adsorption procedure was optimized for heavy metals by determining the impact of pH in the range from 2 to 7, contact time from 5–120 min, and metal cation concentration in the range 25–500 ppm on the treatment procedure.

Finally, the collected water samples, including drinking water and groundwater samples and samples from clothes and car washing centers, were used to evaluate the treatment's efficiency, which was calculated using Equation (2):

$$\text{The removal efficiency } \% = \left(\frac{(C_0 - C_e)}{C_0}\right) * 100 \quad (2)$$

where C_0 represents the initial concentration of metal ions in the solution, and C_e is the equilibrium concentration of metal ions in the solution.

3. Results and Discussion

3.1. Evaluation of the Adsorptive Remediation Process for Separation of Pb(II), Zn(II), Cu(II), and Fe(II) from an Aqueous Medium Using the Model Solutions

3.1.1. Effect of pH on the Metal Ion Solution Medium

Because the pH of an aqueous solution that contains metal ions is a crucial parameter that controls an adsorption process, the role of the hydrogen ion concentration was determined by employing solutions with pH values ranging from 2 to 7. As shown in Figure 1, the adsorption capacity, q_e (mg/g), of RC-MWS for Pb(II), Zn(II), Cu(II), and Fe(II) became larger by increasing the pH value from 2 to 6. Indeed, the maximum adsorption capacity of Pb(II), Zn(II), Cu(II), and Fe(II) was achieved at a pH value of 6. Similar results were reported in studies adsorbing heavy metals in a less acidic medium onto different biomass systems [39]. These results could be attributed to the chelation's lower stability in a highly acidic medium.

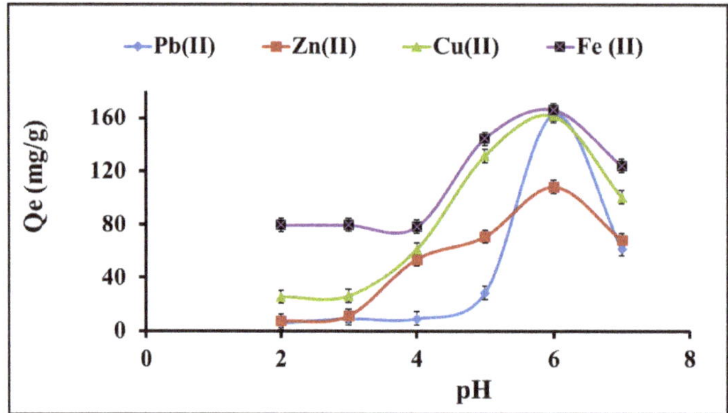

Figure 1. Effect of pH on the adsorption of Pb(II), Zn(II), Cu(II), and Fe(II) onto 0.03 g of renewable carbon from a mixed-waste source (RC-MWS) at room temperature and with a contact time of 60 min.

3.1.2. Competition for the Adsorption of Pb(II), Zn(II), Cu(II), and Fe(II) Onto RC-MWS

The application of RC-MWS for the adsorption of Pb(II), Zn(II), Cu(II), and Fe(II) was done and the adsorption capacities are shown in Figure 2. The highest adsorption capacity was reached with Fe(II), followed by Pb(II), Cu(II), and then Zn(II).

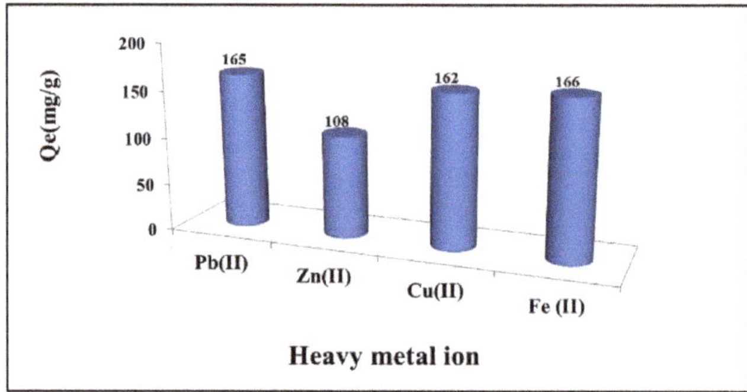

Figure 2. The competition for the adsorption of Pb(II), Zn(II), Cu(II), and Fe(II) onto RC-MWS.

3.1.3. Kinetic Studies on the Adsorption of Pb(II), Zn(II), Cu(II), and Fe(II) Onto RC-MWS

The rate of adsorption of Pb(II), Zn(II), Cu(II), and Fe(II) onto RC-MWS with the assistance of ultrasonic waves was determined by measuring the concentration of the remaining Pb(II), Zn(II), Cu(II), and Fe(II) in the aqueous solution at different times. By testing the effect of different contact times (from 5 to 120 min), it was discovered that the maximum adsorption capacity was achieved with a contact time of 60 min (Figure 3). The adsorption capacity did not significantly change with further increases in the contact time.

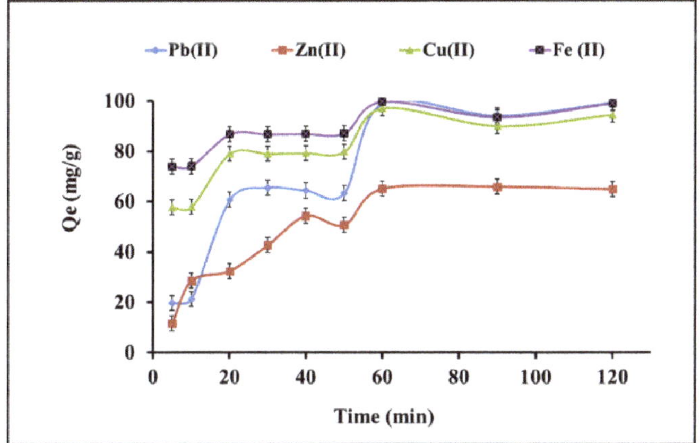

Figure 3. The effect of contact time on the adsorption of Pb(II), Zn(II), Cu(II), and Fe(II) onto 0.05 g of RC-MWS.

The pseudo-first-order equation, in its integrated form [54], is expressed as in Equation (3):

$$\log(q_e - q_t) = -\frac{k_1}{2.303}t + log(q_e) \quad (3)$$

where q_e and q_t are the amounts of adsorbate uptake per mass of adsorbent at equilibrium and at any time t (min), respectively, and k_1 (min^{-1}) is the rate constant of the pseudo-first-order kinetic model.

The pseudo-first-order rate constant, k_1, was obtained by plotting $\log(q_e - q_t)$ versus time t, (Figure 4). It can be seen that the correlation coefficient value, R_1^2, is weak. In addition, the calculated adsorption capacity (Table 1) was different and far from the experimental q_e, therefore, the adsorption data for Pb(II), Zn(II), Cu(II), and Fe(II) onto RC-MWS is not fitted well with the pseudo-first-order kinetic model. McKay et al. suggested that in some adsorption cases the pseudo-first-order kinetic model is not suitable, due to the boundary layer which may control the adsorption at the beginning [55].

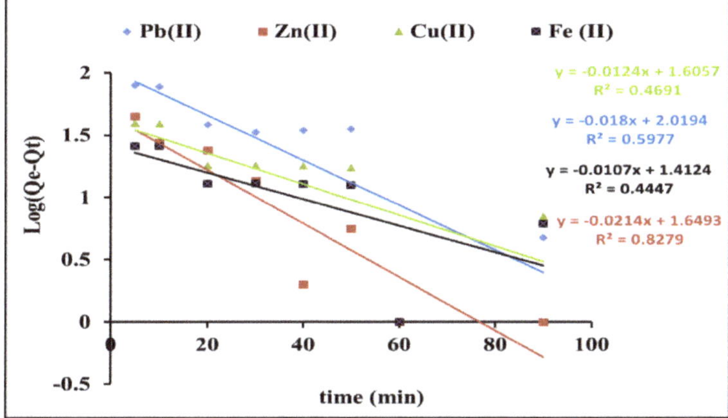

Figure 4. The pseudo-first-order model of the adsorption of Pb(II), Zn(II), Cu(II), and Fe(II) onto RC-MWS, (A) considering time from 0 to 40 min and (B) considering time from 0 to 90 min.

Table 1. Parameters for the kinetics of the adsorption of Pb(II), Zn(II), Cu(II), and Fe(II) onto RC-MWS.

		Pseudo First Order		R^2	Pseudo Second Order		R^2
	q_e, exp (mg/g)	k_1 (min^{-1})	q_e, cal (mg/g)		k_2 (g/mg·min)	q_e, cal (mg/g)	
Pb(II)	99	0.041	104.5	0.59	9.2×10^{-4}	66.6	0.92
Zn(II)	65	0.049	44.5	0.82	3.6×10^{-3}	97	0.99
Cu(II)	97	0.028	40.3	0.46	2.9×10^{-4}	116.2	0.85
Fe(II)	100	0.024	25.8	0.44	1.8×10^{-3}	96.1	0.98

The pseudo-second-order kinetic equation, in its integrated form, is expressed as in Equation (4) [56]:

$$\frac{t}{q_e} = \left(\frac{1}{q_e}\right) \cdot t + \frac{1}{K_2 q_e^2} \tag{4}$$

where q_e and q_t are the amounts of adsorbate uptake per mass of adsorbent at equilibrium and at any time t (min), respectively, and k_2 (g/mg.min) is the pseudo-second-order rate constant.

q_e and k_2 are calculated by plotting t/q_t versus t from the slope and intercept (Figure 5). From the data shown in Table 1, it can be seen that the data of the adsorption of Pb(II), Zn(II), Cu(II), and Fe(II) onto the RC-MWS is fitted with the pseudo-second-order kinetic model, which suggest that the reaction rate is primarily controlled by the movement of the metal ions from the solution to the surface of the adsorbent [27].

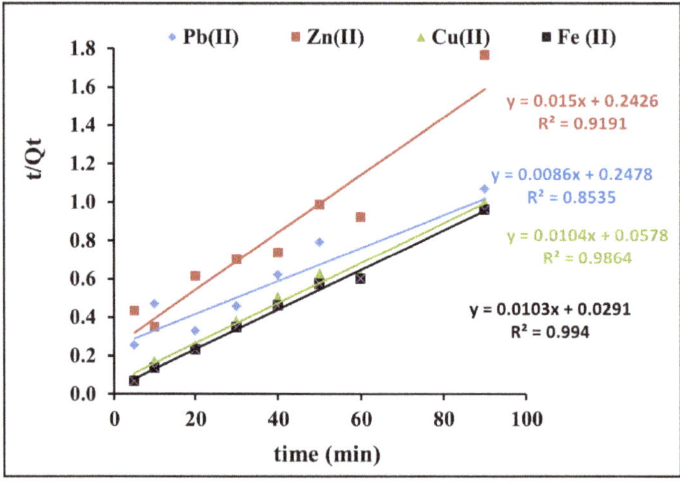

Figure 5. The pseudo-second-order model of the adsorption of Pb(II), Zn(II), Cu(II), and Fe(II) onto RC-MWS.

3.1.4. Isotherm Studies

The data for adsorption of Pb(II), Zn(II), Cu(II), and Fe(II) onto RC-MWS at equilibrium were analyzed by using the Langmuir model [57] as expressed in Equation (5):

$$\frac{C_e}{q_e} = \left(\frac{1}{Q_{max}^0}\right) C_e + \frac{1}{Q_{max}^0 K_L} \tag{5}$$

where Q_{max}^0 (mg/g) is the maximum saturated monolayer adsorption capacity of the RC-MWS, C_e (mg/L) is the adsorbate concentration at equilibrium, q_e (mg/g) is the amount of adsorbate uptake at equilibrium, and K_L (L/mg) is a constant related to the affinity between an adsorbent and adsorbate. The correlation coefficient, R^2, for the adsorption of Pb(II), Zn(II), Cu(II), and Fe(II) onto RC-MWS showed that the adsorption data was well fitted by the Langmuir isotherm (Figure 6) suggesting monolayer adsorption.

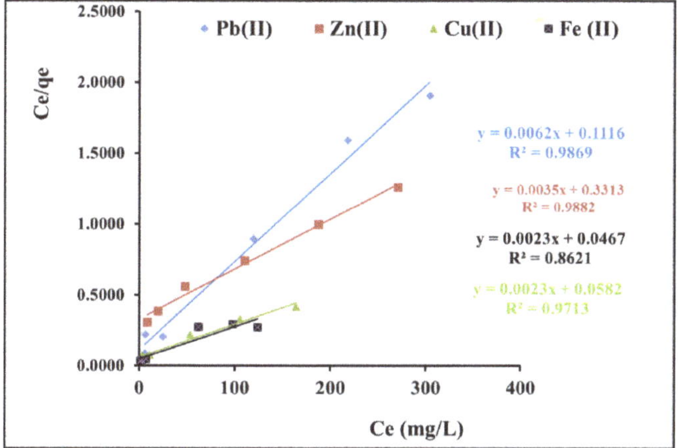

Figure 6. Langmuir adsorption isotherms for the adsorption of Pb(II), Zn(II), Cu(II), and Fe(II) onto RC-MWS.

The Freundlich isotherm [58] is expressed as in Equation (6):

$$\log q_e = n \log C_e + \log K_f \quad (6)$$

where q_e (mg/g) is the amount of adsorbate uptake at equilibrium, C_e (mg/L) is the adsorbate concentration at equilibrium, K_f (mg/g)/(mg/L)^n is the Freundlich constant, and n (dimensionless) is the Freundlich intensity parameter, which indicates the magnitude of the adsorption driving force or the surface heterogeneity.

The Freundlich isotherm model is commonly used to describe the adsorption data characteristic for heterogeneous surfaces at equilibrium [59–62]. The plots of log q_e and log C_e (Figure 7) gave a linear line, from which n and K_f were calculated (Table 2). The values of the Freundlich intensity parameter (n) lower than 1 indicate favorable adsorption; however, the value of correlation coefficient, R^2 (Table 2), indicate that the adsorption of Pb(II), Zn(II), Cu(II), and Fe(II) onto RC-MWS tends to be more comfortable with the Langmuir model rather than the Freundlich model.

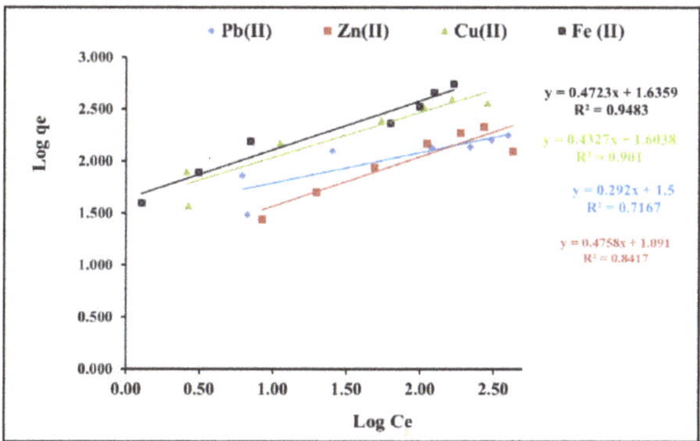

Figure 7. Freundlich adsorption isotherms for the adsorption of Pb(II), Zn(II), Cu(II), and Fe(II) onto RC-MWS.

Table 2. Langmuir and Freundlich constants for the adsorption of Pb(II), Zn(II), Cu(II), and Fe(II) onto RC-MWS.

	Langmuir Constants		Correlation Coefficients R^2	Freundlich Constants		Correlation Coefficients R^2
	K_L (L/mg)	Q^0_{max} (mg/g)		K_f (mg/g)/(mg/L)^n	n	
Pb(II)	0.05	161.2	0.98	31.6	0.29	0.71
Zn(II)	0.01	285.7	0.98	12.3	0.47	0.84
Cu(II)	0.039	434.7	0.97	40.1	0.43	0.90
Fe(II)	0.049	434.7	0.86	43.2	0.47	0.94

The following Equation (7) is applied for the Freundlich isotherm:

$$q_e = K_f C_e^{\frac{1}{n}} \tag{7}$$

where K_f and $1/n$ are Freundlich constants, related to the adsorption capacity and adsorption intensity (heterogeneity factor), respectively. In this case, the n with a value in the range of 0 to 10 indicates the favorable adsorption; however, this index has some applicable limitation practically due to the fact that the n value should not exceed 10 [63].

The Application of the Adsorption Process to Actual Water Samples

The adsorption process was applied to remove Pb(II), Zn(II), Cu(II), and Fe(II) from actual water samples. Water samples were collected from Deria, Mozahemia, and clothes and car washing centers in different regions. The initial Pb(II), Zn(II), Cu(II), and Fe(II) concentrations in the samples were determined by ICP-MS [64,65]. Then, the treatment by adsorption onto RC-MWS was applied under the optimal conditions (a pH of 6, room temperature, and a contact time of 60 min). After the treatment, the Pb(II), Zn(II), Cu(II), and Fe(II) concentrations were once again measured by ICP-MS. Then, we used Equation (2) to calculate the removal efficiency percentage (%). The results are presented in Tables 3 and 4. All of the tested water samples were found to contain a very small amount of Pb(II), Cu(II), and Fe(II), at values below the permitted concentrations, which demonstrates the high level of safety of the tested water samples. However, the detected concentrations were further reduced after the adsorption of Pb(II), Cu(II), and Fe(II) onto RC-MWS.

Table 3. Efficiency of the removal of Pb(II), Zn(II), Cu(II), and Fe(II) from different water samples from Deria and Muzahemia.

Water Sample	Pb(II) Investigations			Zn(II) Investigations			Cu(II) Investigations			Fe(II) Investigations		
	Initial Concentration of Pb(II) (ppm)	Final Concentration of Pb(II) (ppm)	Efficiency %	Initial Concentration of Zn(II) (ppm)	Final Concentration of Zn(II) (ppm)	Efficiency %	Initial Concentration of Cu(II) (ppm)	Final Concentration of Cu(II) (ppm)	Efficiency %	Initial Concentration of Fe(II) (ppm)	Final Concentration of Fe(II) (ppm)	Efficiency %
Deria-1	0.2151	0	100	1.3	0	100	1.26	0.00	100	3.54	1.17	97
Deria-2	0.1357	0	100	1.2	0.091	92	0.33	0.00	100	0.70	3.31	53
Deria-3	0.2313	0	100	2.1	0.064	97	0.11	0.00	100	2.42	8.71	64
Deria-4	0.0436	0	100	1.4	0.075	95	0.25	0.00	100	0.28	0.18	94
Deria-5	0.1305	0	100	1.6	0	100	0.16	0.00	100	0.22	0.66	70
Deria-6	0	0	-	1.8	0.019	99	0.46	0.00	100	0.36	0.05	99
Deria-7	0	0	-	1.4	0.063	96	0.00	0.00	-	1.23	1.90	85
Deria-8	0	0	-	1.4	0	100	1.10	0.00	100	5.42	2.89	95
Deria-9	0	0	-	1.2	0	100	0.22	0.00	100	4.47	2.23	95
Deria-10	0	0	-	1.1	0.08	93	0.24	0.29	88	1.94	2.31	88
Deria-11	0.0405	0	100	0.8	0.01	99	0.25	0.00	100	14.60	1.48	99
Mozahemia-1	0.0833	0	100	2.1	0.056	97	0.25	0.00	100	0.31	0.14	96
Mozahemia-2	0	0	-	1.2	0.038	97	0.00	0.00	-	0.17	0.00	100
Mozahemia-3	0	0	-	1.0	0.033	97	0.06	0.00	100	11.19	2.22	98

Table 4. Efficiency of the removal of Pb(II), Zn(II), Cu(II), and Fe(II) from water samples from different clothes and car washing centers in Riyadh city.

Water Samples	Pb(II) Investigations			Zn(II) Investigations			Cu(II) Investigations			Fe(II) Investigations		
	Initial Concentration of Pb(II) (ppm)	Final Concentration of Pb(II) ppm)	Efficiency %	Initial Concentration of Zn(II) (ppm)	Final Concentration of Zn(II) (ppm)	Efficiency %	Initial Concentration of Cu(II) (ppm)	Final Concentration of Cu(II) (ppm)	Efficiency %	Initial Concentration of Fe(II) (ppm)	Final Concentration of Fe(II) (ppm)	Efficiency %
clothes-1	0	0	-	905.21	12.59	99	11.18	1.92	98	5.31	3.87	93
clothes-2	0	0	-	997.08	1.69	100	0.16	0.00	100	2.20	1.09	95
clothes-3	6.1151	0.215386	96.5	862.14	0.85	100	1.84	0.87	95	1.32	3.17	76
clothes-4	0.4854	0.003688	99.2	4600.01	1.09	100	1.86	1.16	94	4.21	9.25	78
clothes-5	5.039	0.025297	99.5	1019.94	0.87	100	0.39	0.00	100	2.15	3.42	84
clothes-6	0	0	-	638.83	0.35	100	0.53	0.00	100	4.48	3.09	93
clothes-7	0.0024	0	100	977.57	0.00	100	0.24	0.00	100	9.42	3.08	97
clothes-8	0.2102	0	100	904.20	1.56	100	0.71	0.62	91	4.56	5.68	88
cars-1	0	0	-	923.95	3.78	100	15.09	28.41	81	47.54	0.59	100
cars-2	0	0	-	1217.16	7.21	99	21.11	7.31	97	1.95	0.18	99
cars-3	1.787	0	100	842.75	14.36	98	39.94	73.41	82	15.87	6.13	96
cars-4	0	0	-	865.13	8.00	99	92.67	270.09	71	373.65	3.71	100
cars-5	0	0	-	1056.29	12.91	99	50.99	79.33	84	7.86	1.00	99
cars-6	0	0	-	1228.33	0.00	100	0.30	0.00	100	41.20	19.86	95
cars-7	0	0	-	956.51	0.25	100	0.11	0.00	100	9.18	0.36	100
cars-8	0	0	-	1174.11	10.11	99	30.34	27.44	91	81.47	2.91	100

4. Conclusions and Recommendations

The most important parameters for optimizing the adsorption of Pb(II), Zn(II), Cu(II), and Fe(II) onto RC-MWS were investigated to enhance water purification. The optimized conditions were at a pH of 6, a contact time of 60 min, an adsorbent dose of 0.03 g RC-MWS, and at room temperature. Water samples, including groundwater and greywater samples, were collected from the Riyadh, Deria, and Mozahemia regions, and analyzed for contamination with heavy metals and subjected to treatment by adsorption of Pb(II), Zn(II), Cu(II), and Fe(II) onto RC-MWS. The trace amounts of Pb(II), Zn(II), Cu(II), and Fe(II) that were detected by ICP-MS in water samples indicated the existence of low concentrations of these heavy metals in some cases. However, by applying the adsorption procedure, the heavy metals were successfully removed from all tested samples. The process for adsorbing Pb(II), Zn(II), Cu(II), and Fe(II) onto RC-MWS is recommended for the optimization of water quality.

Author Contributions: Formal analysis, T.S.A. and Z.A.A.; investigation, M.A.H., N.A. and S.S.A.; methodology, M.A.H., Z.A.A., N.A. and S.S.A.; project administration, T.S.A.; Resources, Z.A.A. and N.A.; writing—original draft, M.A.H.; writing—review & editing, T.S.A. and M.A.H. All authors have read and agreed to the published version of the manuscript.

Funding: This work was funded by the deanship of scientific research at Princess Nourah Bint Abdulrahman University through the research group program grant number RGP-1440-0023.

Acknowledgments: The authors acknowledge the deanship of scientific research at Princess Nourah Bint Abdulrahman University for funding this research through the research group program grant number RGP-1440-0023.

Conflicts of Interest: The authors declare no conflict of interest.

References

1. Alharbi, S.K.; Shafiquzzaman, M.; Haider, H.; AlSaleem, S.S.; Ghumman, A.R. Treatment of Ablution Greywater for Recycling by Alum Coagulation and Activated Carbon Adsorption. *Arab. J. Sci. Eng.* **2019**, *44*, 8389–8399. [CrossRef]

2. Pradhan, S.; Al-Ghamdi, S.G.; Mackey, H.R. Greywater recycling in buildings using living walls and green roofs: A review of the applicability and challenges. *Sci. Total Environ.* **2019**, *652*, 330–344. [CrossRef]
3. Al-Jasser, A.O. Greywater reuse in Saudi Arabia: Current situation and future potential. *WIT Trans. Ecol. Environ.* **2011**, *153*, 159–169.
4. Abdula'aly, A.I.; Chammem, A.A. Groundwater treatment in the Central Region of Saudi Arabia. *Desalination* **1994**, *96*, 203–214. [CrossRef]
5. Villamar, C.A.; Vera-Puerto, I.; Rivera, D.; De la Hoz, F.D. Reuse and recycling of livestock and municipal wastewater in Chilean agriculture: A preliminary assessment. *Water (Switz.)* **2018**, *10*, 817. [CrossRef]
6. Amin, S.; Farjoud, M.R.; Shabani, A. Groundwater Contamination by Heavy Metals in Water Resources of Shiraz Area. *Iran Agric. Res.* **2011**, *30*, 21–32.
7. Vetrimurugan, E.; Brindha, K.; Elango, L.; Ndwandwe, O.M. Human exposure risk to heavy metals through groundwater used for drinking in an intensively irrigated river delta. *Appl. Water Sci.* **2017**, *7*, 3267–3280. [CrossRef]
8. Ouda, O.K.M. Towards Assessment of Saudi Arabia Public Awareness of Water Shortage Problem. *Resour. Environ.* **2013**, *3*, 10–13.
9. DeNicola, E.; Aburizaiza, O.S.; Siddique, A.; Khwaja, H.; Carpenter, D.O. Climate change and water scarcity: The case of Saudi Arabia. *Ann. Glob. Health* **2015**, *81*, 342–353. [CrossRef]
10. Drewes, J.E.; Garduño, C.P.R.; Amy, G.L. Water reuse in the kingdom of Saudi Arabia—Status, prospects and research needs. *Water Sci. Technol. Water Supply* **2012**, *12*, 926–936. [CrossRef]
11. Abdel-Satar, A.M.; Al-Khabbas, M.H.; Alahmad, W.R.; Yousef, W.M.; Alsomadi, R.H.; Iqbal, T. Quality assessment of groundwater and agricultural soil in Hail region, Saudi Arabia. *Egypt. J. Aquat. Res.* **2017**, *43*, 55–64. [CrossRef]
12. Balkhair, K.S.; Ashraf, M.A. Field accumulation risks of heavy metals in soil and vegetable crop irrigated with sewage water in western region of Saudi Arabia. *Saudi J. Biol. Sci.* **2016**, *23*, S32–S44. [CrossRef] [PubMed]
13. Al-Ahmadi, M.E. Groundwater quality assessment in Wadi Fayd, Western Saudi Arabia. *Arab. J. Geosci.* **2013**, *6*, 247–258. [CrossRef]
14. Alhababy, A.; Al-Rajab, A. Groundwater Quality Assessment in Jazan Region, Saudi Arabia. *Curr. World Environ.* **2015**, *10*, 22–28. [CrossRef]
15. Mohamaden, M.I.I.; Khalil, M.K.; Draz, S.E.O.; Hamoda, A.Z.M. Ecological risk assessment and spatial distribution of some heavy metals in surface sediments of New Valley, Western Desert, Egypt. *Egypt. J. Aquat. Res.* **2017**, *43*, 31–43. [CrossRef]
16. Gupta, V.K.; Gupta, M.; Sharma, S. Process development for the removal of lead and chromium from aqueous solutions using red mud—An aluminium industry waste. *Water Res.* **2001**, *35*, 1125–1134. [CrossRef]
17. Iqbal, M.; Edyvean, R.G.J. Loofa sponge immobilized fungal biosorbent: A robust system for cadmium and other dissolved metal removal from aqueous solution. *Chemosphere* **2005**, *61*, 510–518. [CrossRef]
18. Dubey, A.; Shiwani, S. Adsorption of lead using a new green material obtained from Portulaca plant. *Int. J. Environ. Sci. Technol.* **2012**, *9*, 15–20. [CrossRef]
19. Tyler, G.; Påhlsson, A.M.B.; Bengtsson, G.; Bååth, E.; Tranvik, L. Heavy-metal ecology of terrestrial plants, microorganisms and invertebrates—A review. *Water Air Soil Pollut.* **1989**, *47*, 189–215. [CrossRef]
20. Järup, L. Hazards of heavy metal contamination. *Br. Med. Bull.* **2003**, *68*, 167–182. [CrossRef]
21. Shukia, S.; Mostaghimi, S.; Shanholtz, V.O.; Collins, M.C. A gis-based modeling approach for evaluating groundwater vulnerability to pesticides. *J. Am. Water Resour. Assoc.* **1998**, *34*, 1275–1293. [CrossRef]
22. Al-Jassim, N.; Ansari, M.I.; Harb, M.; Hong, P.-Y. Removal of bacterial contaminants and antibiotic resistance genes by conventional wastewater treatment processes in Saudi Arabia: Is the treated wastewater safe to reuse for agricultural irrigation? *Water Res.* **2015**, *73*, 277–290. [CrossRef] [PubMed]
23. Zouboulis, A.I.; Matis, K.A.; Lanara, B.G.; Loos-Neskovic, C. Removal of Cadmium from Dilute Solutions by Hydroxyapatite. II. Flotation Studies. *Sep. Sci. Technol.* **1997**, *32*, 1755–1767. [CrossRef]
24. Esalah, J.O.; Weber, M.E.; Vera, J.H. Removal of lead, cadmium and zinc from aqueous solutions by precipitation with sodium Di-(n-octyl) phosphinate. *Can. J. Chem. Eng.* **2000**, *78*, 948–954. [CrossRef]
25. Canet, L.; Ilpide, M.; Seta, P. Efficient facilitated transport of lead, cadmium, zinc, and silver across a flat-sheet-supported liquid membrane mediated by lasalocid A. *Sep. Sci. Technol.* **2002**, *37*, 1851–1860. [CrossRef]
26. Senthilkumar, R.; Vijayaraghavan, K.; Thilakavathi, M.; Iyer, P.V.R.; Velan, M. Application of seaweeds for the removal of lead from aqueous solution. *Biochem. Eng. J.* **2007**, *33*, 211–216. [CrossRef]

27. El-Toni, A.M.; Habila, M.A.; Ibrahim, M.A.; Labis, J.P.; ALOthman, Z.A. Simple and facile synthesis of amino functionalized hollow core-mesoporous shell silica spheres using anionic surfactant for Pb(II), Cd(II), and Zn(II) adsorption and recovery. *Chem. Eng. J.* **2014**, *251*, 441–451. [CrossRef]
28. Symposium on Waste Water—Treatment and Reuse | British Council Saudi Arabia. Available online: https://www.britishcouncil.sa/en/events/symposium-waste-water-treatment-and-reuse (accessed on 11 October 2019).
29. Polat, H.; Erdogan, D. Heavy metal removal from waste waters by ion flotation. *J. Hazard. Mater.* **2007**, *148*, 267–273. [CrossRef] [PubMed]
30. Wang, H.; Wen, S.; Han, G.; Xu, L.; Feng, Q. Activation mechanism of lead ions in the flotation of sphalerite depressed with zinc sulfate. *Miner. Eng.* **2020**, *146*, 106132. [CrossRef]
31. Hassanzadeh, A.; Azizi, A.; Kouachi, S.; Karimi, M.; Celik, M.S. Estimation of flotation rate constant and particle-bubble interactions considering key hydrodynamic parameters and their interrelations. *Miner. Eng.* **2019**, *141*, 105836. [CrossRef]
32. Feng, Q.; Wen, S.; Bai, X.; Chang, W.; Cui, C.; Zhao, W. Surface modification of smithsonite with ammonia to enhance the formation of sulfidization products and its response to flotation. *Miner. Eng.* **2019**, *137*, 1–9. [CrossRef]
33. Kyzas, G.; Matis, K. Flotation in Water and Wastewater Treatment. *Processes* **2018**, *6*, 116. [CrossRef]
34. Kitchener, J.A.; Gochin, R.J. The mechanism of dissolved air flotation for potable water: Basic analysis and a proposal. *Water Res.* **1981**, *15*, 585–590. [CrossRef]
35. Özer, A. Removal of Pb (II) ions from aqueous solutions by sulphuric acid-treated wheat bran. *J. Hazard. Mater.* **2007**, *141*, 753–761. [CrossRef] [PubMed]
36. Badmus, M.O.A.; Audu, T.O.K.; Anyata, B. Removal of copper from industrial wastewaters by activated carbon prepared from periwinkle shells. *Korean J. Chem. Eng.* **2007**, *24*, 246–252. [CrossRef]
37. Amarasinghe, B.M.W.P.K.; Williams, R.A. Tea waste as a low cost adsorbent for the removal of Cu and Pb from wastewater. *Chem. Eng. J.* **2007**, *132*, 299–309. [CrossRef]
38. Qaiser, S.; Saleemi, A.R.; Umar, M. Biosorption of lead from aqueous solution by Ficus religiosa leaves: Batch and column study. *J. Hazard. Mater.* **2009**, *166*, 998–1005. [CrossRef] [PubMed]
39. Othman, Z.A.A.; Hashem, A.; Habila, M.A. Kinetic, equilibrium and thermodynamic studies of cadmium (II) adsorption by modified agricultural wastes. *Molecules* **2011**, *16*, 10443–10456. [CrossRef] [PubMed]
40. (PDF) Adsorption of Chromium Ion from Industrial Effluent Using Activated Carbon Derived from Plantain. Available online: https://www.researchgate.net/publication/322677407_Adsorption_of_Chromium_Ion_from_Industrial_Effluent_Using_Activated_Carbon_Derived_from_Plantain (accessed on 11 October 2019).
41. Bulut, Y.; Tez, Z. Adsorption studies on ground shells of hazelnut and almond. *J. Hazard. Mater.* **2007**, *149*, 35–41. [CrossRef] [PubMed]
42. Amuda, O.S.; Giwa, A.A.; Bello, I.A. Removal of heavy metal from industrial wastewater using modified activated coconut shell carbon. *Biochem. Eng. J.* **2007**, *36*, 174–181. [CrossRef]
43. Gong, J.; Liu, J.; Chen, X.; Jiang, Z.; Wen, X.; Mijowska, E.; Tang, T. Converting real-world mixed waste plastics into porous carbon nanosheets with excellent performance in the adsorption of an organic dye from wastewater. *J. Mater. Chem. A* **2015**, *3*, 341–351. [CrossRef]
44. Habila, M.A.; ALOthman, Z.A.; Al-Tamrah, S.A.; Ghafar, A.A.; Soylak, M. Activated carbon from waste as an efficient adsorbent for malathion for detection and removal purposes. *J. Ind. Eng. Chem.* **2015**, *32*, 336–344. [CrossRef]
45. Saleem, J.; Shahid, U.B.; Hijab, M.; Mackey, H.; McKay, G. Production and applications of activated carbons as adsorbents from olive stones. *Biomass Convers. Biorefin.* **2019**, *9*, 775–802. [CrossRef]
46. Balcerek, M.; Pielech-Przybylska, K.; Patelski, P.; Dziekońska-Kubczak, U.; Jusel, T. Treatment with activated carbon and other adsorbents as an effective method for the removal of volatile compounds in agricultural distillates. *Food Addit. Contam. Part A Chem. Anal. Control Expo. Risk Assess.* **2017**, *34*, 714–727. [CrossRef]
47. Ghaedi, M.; Nasab, A.G.; Khodadoust, S.; Rajabi, M.; Azizian, S. Application of activated carbon as adsorbents for efficient removal of methylene blue: Kinetics and equilibrium study. *J. Ind. Eng. Chem.* **2014**, *20*, 2317–2324. [CrossRef]
48. Dias, J.M.; Alvim-Ferraz, M.C.M.; Almeida, M.F.; Rivera-Utrilla, J.; Sánchez-Polo, M. Waste materials for activated carbon preparation and its use in aqueous-phase treatment: A review. *J. Environ. Manag.* **2007**, *85*, 833–846. [CrossRef]
49. Wong, S.; Ngadi, N.; Inuwa, I.M.; Hassan, O. Recent advances in applications of activated carbon from biowaste for wastewater treatment: A short review. *J. Clean. Prod.* **2018**, *175*, 361–375. [CrossRef]

50. Wang, A.; Zheng, Z.; Li, R.; Hu, D.; Lu, Y.; Luo, H.; Yan, K. Biomass-derived porous carbon highly efficient for removal of Pb (II) and Cd (II). *Green Energy Environ.* **2019**, *4*, 414–423. [CrossRef]
51. Deng, J.; Li, M.; Wang, Y. Biomass-derived carbon: Synthesis and applications in energy storage and conversion. *Green Chem.* **2016**, *18*, 4824–4854. [CrossRef]
52. Yang, S.; Fang, X.; Duan, L.; Yang, S.; Lei, Z.; Wen, X. Comparison of ultrasound-assisted cloud point extraction and ultrasound-assisted dispersive liquid liquid microextraction for copper coupled with spectrophotometric determination. *Spectrochim. Acta Part A Mol. Biomol. Spectrosc.* **2015**, *148*, 72–77. [CrossRef]
53. Alothman, Z.A.; Habila, M.A.; Yilmaz, E.; Soylak, M.; Alfadul, S.M. Ultrasonic supramolecular microextration of nickel (II) as N,N′-Dihydroxy-1,2-cyclohexanediimine chelates from water, tobacco and fertilizer samples before FAAS determination. *J. Mol. Liq.* **2016**, *221*, 773–777. [CrossRef]
54. Lagergren, S. About the Theory of So-Called Adsorption of Soluble Substances. *Sven. Vetensk.* **1898**, *24*, 1–39. Available online: https://www.oalib.com/references/5821801 (accessed on 12 June 2020).
55. McKay, G.; Ho, Y.S.; Ng, J.C.Y. Biosorption of copper from waste waters: A review. *Sep. Purif. Methods* **1999**, *28*, 87–125. [CrossRef]
56. Blanchard, G.; Maunaye, M.; Martin, G. Removal of heavy metals from waters by means of natural zeolites. *Water Res.* **1984**, *18*, 1501–1507. [CrossRef]
57. Langmuir, I. The adsorption of gases on plane surfaces of glass, mica and platinum. *J. Am. Chem. Soc.* **1918**, *40*, 1361–1403. [CrossRef]
58. Freundlich, H.M.F. Over the Adsorption in Solution. *J. Phys. Chem. A* **1906**, *57*, 385–470.
59. Mohanty, K.; Jha, M.; Meikap, B.C.; Biswas, M.N. Removal of chromium (VI) from dilute aqueous solutions by activated carbon developed from Terminalia arjuna nuts activated with zinc chloride. *Chem. Eng. Sci.* **2005**, *60*, 3049–3059. [CrossRef]
60. Arshadi, M.; Amiri, M.J.; Mousavi, S. Kinetic, equilibrium and thermodynamic investigations of Ni(II), Cd(II), Cu(II) and Co(II) adsorption on barley straw ash. *Water Resour. Ind.* **2014**, *6*, 1–17. [CrossRef]
61. Abdel Salam, O.E.; Reiad, N.A.; ElShafei, M.M. A study of the removal characteristics of heavy metals from wastewater by low-cost adsorbents. *J. Adv. Res.* **2011**, *2*, 297–303. [CrossRef]
62. Wang, L.; Zhang, J.; Zhao, R.; Li, Y.; Li, C.; Zhang, C. Adsorption of Pb(II) on activated carbon prepared from Polygonum orientale Linn.: Kinetics, isotherms, pH, and ionic strength studies. *Bioresour. Technol.* **2010**, *101*, 5808–5814. [CrossRef]
63. Tran, H.N.; You, S.J.; Hosseini-Bandegharaei, A.; Chao, H.P. Mistakes and inconsistencies regarding adsorption of contaminants from aqueous solutions: A critical review. *Water Res.* **2017**, *120*, 88–116. [CrossRef] [PubMed]
64. Albratty, M.; Arbab, I.; Alhazmi, H.; Attafi, I.; Al-Rajab, A. Icp-Ms Determination of Trace Metals in Drinking Water Sources in Jazan Area, Saudi Arabia. *Curr. World Environ.* **2017**, *12*, 06–17. [CrossRef]
65. Kausar, R.; Ahmad, Z. Determination of toxic inorganic elements pollution in ground waters of Kahuta Industrial Triangle Islamabad, Pakistan using inductively coupled plasma mass spectrometry. *Environ. Monit. Assess.* **2009**, *157*, 347–354. [CrossRef] [PubMed]

Article

Reasons of Acceptance and Barriers of House Onsite Greywater Treatment and Reuse in Palestinian Rural Areas

Rehab A. Thaher [1], Nidal Mahmoud [2], Issam A. Al-Khatib [2,*] and Yung-Tse Hung [3]

1 Faculty of Graduate Studies, Birzeit University, P.O. Box 14, Birzeit, West Bank, Palestine; rehabthaher@gmail.com

2 Institute of Environmental and Water Studies, Birzeit University, P.O. Box 14, Birzeit, West Bank, Palestine; nmahmoud@birzeit.edu

3 Department of Civil and Environmental Engineering, Cleveland State University, Cleveland, OH 44115, USA; y.hung@csuohio.edu

* Correspondence: ikhatib@birzeit.edu

Received: 18 May 2020; Accepted: 9 June 2020; Published: 11 June 2020

Abstract: In the last twenty years, house onsite wastewater management systems have been increasing in the West Bank's rural areas. The aim of this research was to reveal, in the context of providing onsite Grey Water Treatment Plants (GWTPs) for wastewater management in the rural communities in Palestine, the local population's perceptions, in the sense of acceptance of and barriers towards such a type of wastewater management, so as to figure out successes, failures and lessons. The data collection tool was a questionnaire that targeted the households served with GWTPs. The findings show that 13% of the total constructed treatment plants were not operative. The most important barrier as mentioned by 66.5% is odor emission and insect infestation. Then, 25.1% of the implementing agencies never monitor or check the treatment plants, and 59.3% of them monitor and check the plants only during the first 2–3 months. The next barrier is inadequate beneficiary experience in operation and maintenance. Health concerns regarding quality of crops irrigated by treated grey water were another barrier. The results revealed that the reuse of treated grey water in irrigation was the main incentive for GWTPs as stated by 88.0% of beneficiaries. The second incentive was the saving of cesspit discharge frequency and its financial consequences, as stated by 71.3%. Finally, 72.5% of the beneficiaries stated that they had a water shortage before implementing GWTPs, and the GWTPs contributed to solving it. The highest percentage (82.6%) of beneficiaries accepted the treatment units because of their willingness to reuse treated water for irrigation and agricultural purposes. Education level has an impact on GWTP acceptance, with 73% of not educated beneficiaries being satisfied and 58.8% of educated people being satisfied. Islamic religion is considered a driver for accepting reuse of treated grey water in irrigation, according to the majority of people (70%). Women play a major role on GWTP management; 68.9% of the treatment systems are run by men side-by-side with women (fathers and mothers), and 24% are run completely by women. The majority of GWTP beneficiaries (70.4%) are satisfied with GWTPs. Little effort is required for operation and maintenance, with only an average of 0.4 working hours per week. Therefore, house onsite grey water management systems are acceptable in rural communities, but attention should be given to the reasons of acceptance and barriers highlighted in this research.

Keywords: greywater treatment; house onsite; reuse; irrigation; acceptance; barriers

1. Introduction

1.1. General Information

Palestinian territories face significant and growing shortfalls in the water supply available for domestic use. The World Health Organization (WHO) considers 100 liters per capita per day (L/c.d) as the benchmark minimum for domestic consumption to achieve full health and hygiene benefits. In contrast, available water resources for domestic consumption in the West Bank are only 62 L/c.d [1].

The results of the Palestinian Central Bureau of Statistics [2] showed that 93% of the households in the Palestinian Territory live in housing units connected to a water network [2]. During 2015, data indicated that 53.9% of households in Palestine used wastewater networks to dispose their wastewater, while 31.8% of households used porous cesspits [2]. The estimated quantity of wastewater generated in the West Bank for the year 2015 was estimated at 65.82 million cubic meters (MCM) [3].

In the last twenty years, house onsite grey water management systems have been implemented in the rural communities of the West Bank, justified by lack of adequate wastewater services and driven by business opportunities for the implementing Non-Governmental Organizations (NGOs) funded by donors. Some of those projects were not successful, but some others are still operating very successfully. The reasons for acceptance and barriers of providing onsite grey water treatment plants from the beneficiaries' points of view have not yet been investigated.

A decentralized wastewater management system may consist of individual onsite systems and/or cluster systems, either singly or in combination with more highly collectivized facilities [4]. The degree of collectivization at any stage of the treatment and reuse or dispersal processes will be determined by a variety of local circumstances, including topography, site and soil characteristics, development density, type of development, community desires with regard to land use issues and sites of potential reuse and/or sites where discharge would be allowable. Decentralized wastewater systems in particularly arid regions promote wastewater uptake by plants [5].

The composition of grey water varies greatly and reflects the lifestyle of the residents and the choice of household chemicals for washing up, laundry etc. [6]. Characteristic of grey water is that it often contains high concentrations of easily degradable organic material, i.e., fat, oil and other organic substances from cooking, as well as residues from soap and detergents [7–9].

The generated amount of grey water greatly varies as a function of the dynamics of the household. It is influenced by factors such as existing water supply systems, number of household members, age distribution, life style characteristics, typical water usage patterns, etc. Reuse of treated grey water in irrigation can significantly contribute to reducing water bills and increasing food security [10–13], at the same time leading to saving of drinking water for domestic consumption.

Grey water, in contrast to common perception, may be quite polluted, and thus may pose health risks and negative aesthetics (i.e., offensive odor and color) and environmental effects [14–16].

Grey water contains pollutants and microorganisms stemming from household and personal cleaning activities. Laundry and shower water are slightly polluted, but kitchen water is highly polluted with organic matter from food wastes, and so requires special attention [17]. Indeed, grey water contains by far fewer pathogens than total sewage [18]. Therefore, grey water should not be considered as a waste, but a beneficial resource. It is increasingly agreed that grey water can alleviate water shortages [19,20]. Grey water is a valuable water resource that can be utilized for irrigating home gardens or agricultural land [10].

The willingness of households to adopt a grey water treatment and reuse system depends on many factors such as sociocultural acceptance, public awareness, economic situation and institutional capacity in the field of the onsite treatment [21]. The public perception of wastewater reuse is still suspicious although generally grey wastewater reuse is more acceptable than black water reuse [22].

The aim of this research was to investigate the reasons of GWTPs' acceptance by the beneficiaries and the barriers of implementing these systems in the Palestinian rural communities.

1.2. Grey Water Practices in Palestine

Great efforts have been undertaken by Palestinian institutions, governmental and non-governmental, to advance the wastewater infrastructure centralized and onsite systems. Nevertheless, low population densities in rural and suburban areas and limited funding are major obstacles for the development of wastewater services. The Palestinian institutions promote implementation of house and community onsite treatments and agricultural reuse of treated effluents. However, sociocultural acceptance and public awareness should be addressed, as well as the institutional capacity to administer the decentralized wastewater management systems. Figure 1 shows an example of onsite grey water treatment plant, and Figure 2 shows a reuse scheme by treated grey water in Palestine.

Figure 1. Onsite grey water treatment plant, Duara Al Qare'-Ramallah.

Figure 2. Reuse scheme by treated grey water in Palestine, Al Qubeba-Jerusalem.

1.3. Description of House Onsite Grey Water Treatment Plant

The house wastewater piping systems are modified to separate the grey and black wastewaters. The (black) toilet wastewater is disposed into an available cesspit. The grey wastewater (wastewater sources except toilet wastewater) is transported to the household grey water treatment plant (GWTP).

The onsite GWTP is comprised of a septic tank (first compartment) ahead of two up-flow gravel filters (second and third compartments) as presented in Figure 1. Grease is tapped in the septic tank using an outlet T-shaped pipe as shown in Figure 3. The fourth compartment is a pumping wet well tank where the anaerobically pre-treated wastewater is lifted to a multi-layer coal–sand filter. Afterwards, the treated wastewater is stored in an irrigation tank connected to the garden irrigation network. More details about the system can be found in Burnat and Shtayye [23].

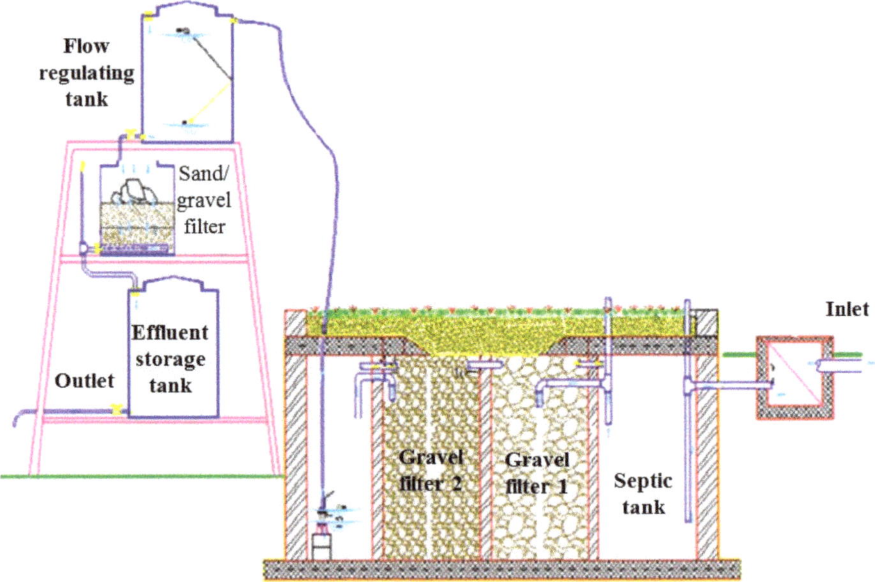

Figure 3. Onsite grey water treatment plant [23].

2. Methodology

The study area included 18 different rural communities in eight governorates of the West Bank: Ramallah, Jerusalem, Bethlehem, Hebron, Jenin, Tubas, Tulkarem, Nablus. The study area was selected according to the availability of onsite GWTPs distributed mostly in all governorates of the West Bank as illustrated in Figure 4.

In this study, a survey by questionnaire was conducted. A sample of 185 owner "beneficiaries" of onsite GWTPs at the household level was randomly selected and the questionnaire was distributed at the household level. The recovery rate was 89.2%, as 166 questionnaires were filled. The questionnaire was finalized after consulting several experts from different institutions and key people who work in the water and sanitation fields. The questionnaire was written in Arabic and included questions about family size, job, income, general information regarding the treatment plant, monitoring of the treatment plant, satisfaction regarding the GWTP, current status of the sanitation system, aesthetic concerns and the treatment plant's impact, the impact of the sanitation system on health, reasons of acceptance and barriers, social and managerial aspects, financial aspects, confidence in the applied systems, etc.

Figure 4. Study area in eight governorates of the West Bank.

The surveyors were educated about the GWTP, its operation and maintenance methods and the environmental, economic and agriculture values that grey water provides. The questionnaire also included inquiries regarding inspection of the treatment systems and testing of the quality of treated water. The obtained data were analyzed using the Statistical Science Software Program (SPSS) software version 20 [24]. Acceptance of providing onsite GWTPs was statistically tested using logistic regression analysis.

3. Results and Discussion

3.1. General Information on Households

The survey results revealed that the average family size in the study area was 9.3 people, which is considered a large family size as the average family size in the West Bank is 5 persons [25]. Out of the total implemented GWTPs, 76.5% served one household, 14.2% served two households and 9.2% served three to four households. The average monthly income of the onsite GWTP owners ranged from 280 up to 830 US$ as illustrated in Figure 5; the latest official Palestinian statistics reveal that 13.9% of the West Bank population is below the national poorness standard, as their average income is less than 580 US$ [26].

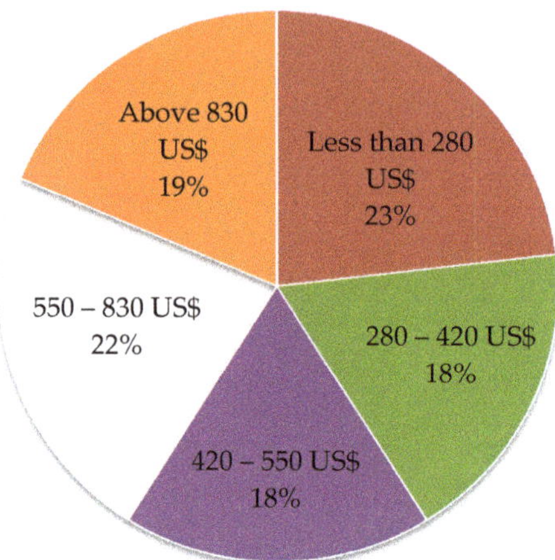

Figure 5. Percentage distribution of families based on family monthly income in US$.

3.2. General Information on Onsite GWTPs

The treatment plants which are distributed in the rural communities have been constructed over the last twenty years, with 99.3% of them constructed over the last fifteen years. All of them were constructed by local or international NGOs supported by external donors.

The findings of this study showed that 13% of the total implemented GWTPs do not currently operate. The reasons include: (1) production of strong odors and their impact on the owners and neighbors; (2) not being effective in the treatment process, as stated by beneficiaries; (3) changing of the function of the plant to a rainwater harvesting cistern by some of the beneficiaries since construction; and (4) not being adequately trained on operation and maintenance.

The data revealed that 25.1% of the executing agencies had not ever monitored or checked the treatment plants. Furthermore, 59.3% of these agencies had followed the plants only during the early phase (2–3 months as per beneficiaries) after completion of construction; only 11.4% of them had monitored and operated the plants through regular visits and giving support to ensure the performance of the plants. Similarly, Ahmad et al. [22] reported that most onsite GWTPs did not operate after the funded projects had been terminated, a sequence of no identifications for ownership. No monitoring systems were available for the treatment plants, although those systems were used for irrigation. Moreover, Sandec [10] found that the main system malfunctions resulted from improper operation and maintenance attributable to the owners' poor understanding of the systems. Consequently, beneficiaries should be trained in proper management of the system, and as such their involvement during the planning and implementation stages is decisive.

The results showed that 61.7% of the implementing agencies had never inspected or monitored the quality of treated water, 26.9% of them had monitored the quality and process performance of the plants during the first period after implementation and only 7.2% had monitored the plants on a regular basis ranging between 1–2 times per month. This reveals that there was no reliable or continuous monitoring system of the plants by the implementing agencies. In fact, the monitoring tasks were shifted directly to the owners without adequate knowledge and experience of the system's monitoring and evaluation. The results showed that 48.8% of the system owners were not satisfied with the implementing agencies' behavior upon completion of the construction phase of the project. This high percentage of dissatisfaction shows the limited role and responsibility of the implementing agency which negatively affected the sustainability of these onsite wastewater treatment systems.

The beneficiaries stated that the implementing agencies made many mistakes throughout the planning and construction phases of the projects. During the planning phase, mistakes included inappropriate site selection of the treatment plant, improper technical design and capacity, lack of consultation with community representatives such as community-based organizations (CBOs) and not conducting feasibility studies for the projects. During the construction phase, mistakes included leakage from the treatment plant, low quality of the main construction works and poor finish due to poor monitoring and supervision of construction.

3.3. Water and Sanitation Household Conditions

Treatment plants require available space surrounding the home; 95% of the household respondents had gardens. The average area of the gardens was between 100–500 m^2. Furthermore, 79% of the houses had rainwater harvesting systems. Treatment plants affect irrigation and saving of fresh water, as 51.5% of the interviewers used the fresh water from water network in irrigation before construction of GWTP. However, this percent considerably decreased after construction of the treatment plants: 15% of the beneficiaries still used a network water source in irrigation after construction of an onsite GWTP and 30% of them used a water network from time to time. Indeed, most of the investigated rural communities face chronic water shortages, with 72.5% of the beneficiaries reporting that before implementation of the treatment plants, they had a water shortage and the onsite GWTPs helped in alleviating this problem. A total of 35.3% of beneficiaries stated that the GWTPs contributed to solving the water shortage and 44.3% stated that GWTPs contribute partially to solving water shortages, since as they used treated water for irrigation, consequently they save fresh water. The average household planted area before and after establishment of a GWTP was 153 m^2, while the average planted area after establishing a GWTP was 156 m^2. Though the difference in the planted area is not significant, the agriculture practices became more efficient and productive. Likewise, Sandec [10] stated that treated grey water reuse in irrigation might considerably influence in lessening water bills and contributing to food security.

Findings revealed that there are two types of agriculture. The majority of the interviewees (77.8%) stated that they use treated water in open agriculture and 15.6% of them use treated water in a green house. The percentages of beneficiaries who use treated grey water in irrigating fruit trees, vegetables, flowers and fodder are 71.9, 44.3, 4.8 and 1.2%, respectively. The fruits are mostly consumed by the system owners' families (77.4%); around 10% are gifted to relatives, neighbors and friends and 7.5% are usually sold in the market. Therefore, the availability of a house onsite GWTP leads to utilizing treated effluent in irrigation, contributing to food security. Acceptance of implementing GWTPs for the purpose of treated effluent reuse in irrigation is varied according to the many reasons mentioned in Table 1.

- Governorate: From the results of Table 2, it appears that the percentages of acceptance have close values between all governorates, which shows that onsite GWTPs in the West Bank are acceptable for the purpose of reuse in irrigation. However, acceptance was not at the same level in all governorates.
- Family Size: From Figure 6 it is noticeable that the acceptance of a GWTP for reuse in irrigation is influenced by the number of family members, where the percentage increases with increasing family size.
- Job: Acceptance of GWTPs was different for people with different jobs as per Table 3, where a high percentage was found for workers and farmers (who have less income), while employees or wholesalers have relatively less interest in GWTPs.
- Education Level: 87.7% of less educated people accept GWTPs for reuse in agriculture, but a lower percent (81.2%) of educated people accept GWTPs. This emphasizes that educated people have more concerns regarding the quality of treated water.
- Suffering from Water Shortage before Construction of GWTPs: 85.6% of people who were suffering from water shortages accept construction of GWTPs for reuse in irrigation, while a lower percent (75.6%) is found for people who had no problems with water shortages.

- Garden Availability: 86.1% of people who have a home garden would be willing to reuse treated grey water in irrigation, however 22% of those who do not have a home garden were not able to reuse for irrigation.
- Frequency of Cesspit's Emptying before Providing GWTP: Acceptance of reuse in irrigation depends on discharge of cesspits per year: 76.2% of people who empty cesspits 1–3 times per year accept reuse in agriculture, while 88.3% of people who discharge their cesspits more than 4 times per year accept reuse in agriculture.
- Owner's Satisfaction of Cesspits: 73.9% of people who are satisfied in applying cesspits accept reuse in irrigation, while a larger percent (87.4%) of people who are not satisfied accept providing onsite GWTPs for the purpose of reuse in irrigation.

Table 1. Acceptance of providing GWTPs (grey water treatment plants) for reuse in irrigation.

Independent Value	Acceptance of GWTPs Asymp. Sig. (2-sided) Value *	Status
Age	0.526	Not significant
Governorate	0.002	Significant
Number of households	0.433	Not significant
Family size	0.0135	Significant
Job	0.00	Significant
Age of responsible person for managing GWTP	0.501	Not significant
Education level of those responsible of GWTP	0.00	Significant
Suffering of water shortage before construction of GWTPs	0.003	Significant
frequency of cesspit's emptying before providing GWTP	0.002	Significant
Level of noise	0.32	Not significant
Garden availability	0.00	Significant
Owner's satisfaction of cesspit's	0.001	Significant

*: Significant value, if Asymp. Sig. (2-sided) value is less than or equal 0.05.

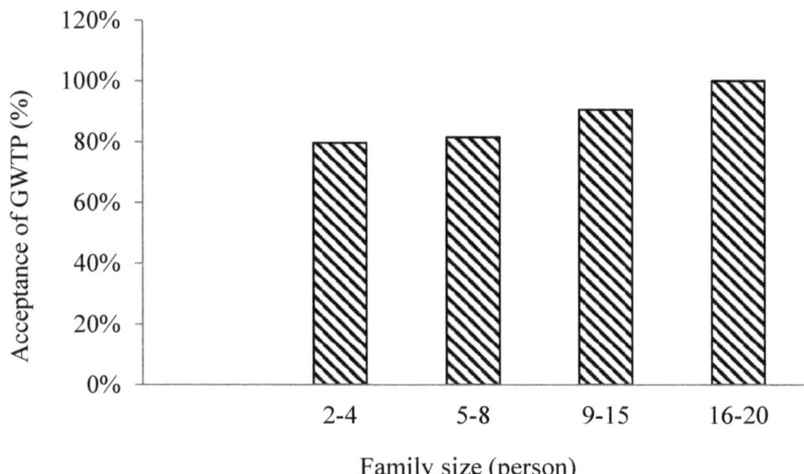

Figure 6. Acceptance of GWTPs for reuse in irrigation vs. family size.

Table 2. Acceptance of GWTPs for reuse in irrigation per governorate.

Governorate	Number of Respondents	Acceptance of GWTPs (%)
Bethlehem	4	83.3
Ramallah	55	68.1
Jerusalem	7	100
Hebron	38	85.7
Nablus	6	100
Tulkarem	7	85.7
Jenin	39	82.1
Tubas	10	100

Table 3. Acceptance of GWTPs for reuse in irrigation versus job.

Job	Acceptance of GWTPs (%)
Worker	85
Employee	78.6
Farmer	90
Wholesaler	80%

3.4. Reasons for Acceptance GWTPs

The reasons for acceptance of GWTPs to replace the previous sanitation system "cesspits" were different across many aspects. The highest percentage (82.6%) of beneficiaries who accepted having treatment units was due to their willingness to reuse treated water in irrigation and agricultural purposes, and the lowest percentage was in regards to saving on the water bill, as illustrated in Figure 7.

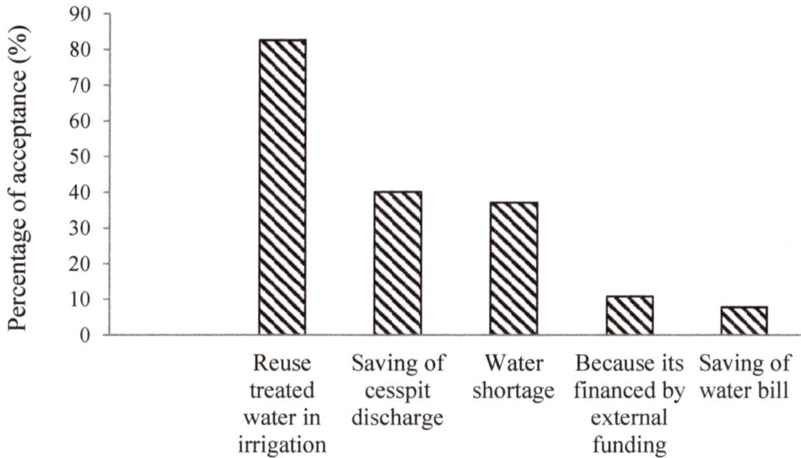

Figure 7. Percentage of beneficiaries' acceptance of GWTPs based on influencing factors.

Similarly, Adilah [27] reported that treated effluent reuse for irrigating fruit trees and flowers in the home garden has the highest potential to be accepted. Saving of cesspit discharge is another important reason for accepting GWTP, as only black wastewater goes to the cesspit. Water shortage is also a reason for accepting GWTP, as the majority experience water shortages, especially in the summer. The lowest percentage is for savings on the water bill because providing a GWTP does not have much effect on utilization of fresh water, as they were not used for irrigation before construction of the GWTPs.

3.5. The Barriers for Application of Onsite GWTPS

Many barriers were raised by interviewees for the application of GWTPs; the following barriers are arranged by priority, as illustrated in Figure 8. The first and biggest barriers are odor emission and insect infestation. This emphasizes the importance of further developing the systems to improve their performance. The second barrier is the lack of implementing agency (NGOs) follow-up, especially after the end of implementation. The NGOs do not implement evaluation and monitoring of system performance after the projects have ended. Accordingly, the beneficiaries do not have the required experience in operation and maintenance. Health risks and worries about water quality are other barriers since people are not confident about the quality of treated grey water. A lower percentage of beneficiaries stated other barriers such as operation and maintenance burden on the householder, lack of experience in operation and maintenance and the financial burden of operation and maintenance. Likewise, Ahmad et al. [22] reported that no monitoring systems were available for the treatment plants, although those systems were used for irrigation. This emphasizes the importance of considering the follow-up process and practical training in operation and maintenance as a part of project implementation.

- Replacement of GWTPs in Case of Providing Sewerage Networks: 52.1% of GWTPs owners would not replace the treatment plant in the case of providing sewerage networks, while 37.7% of them stated that they would replace the treatment plant in the case of providing sewerage networks. The mentioned results refer to many aspects that interfere with the replacement of GWTPs in the case of providing sewerage networks, as discussed below.
- Water shortage: 56.6% of GWTP beneficiaries who accepted GWTPs because of water shortages were not willing to replace the onsite GWTP in the case of providing sewerage network, while 43% of GWTP beneficiaries who did not face water shortages preferred replacing the onsite GWTP in the case of providing sewerage network. This result indicates that water shortage is a significant reason to maintain the onsite GWTP.
- Availability of Fund by External Donor: 66.6% of GWTP beneficiaries who accept GWTPs because they are supported by external funds were not willing to replace the onsite GWTPs in the case of providing sewerage network, while 52.1% of GWTP beneficiaries who accept GWTPs when they are not supported by external funds were not willing to replace the onsite GWTPs in case of providing sewerage network, which means that fund availability was not a significant reason for replacing the onsite GWTPs.
- Reduction of Cesspit Discharge Frequency: 53.9% of GWTP beneficiaries who accepted a GWTP for the reduction of cesspit discharge frequency were not willing to replace the onsite GWTP in the case of providing a sewerage network, while 37.4% of GWTP beneficiaries who accepted a GWTP for not saving cesspit discharge were willing to replace the onsite GWTP in case of providing sewerage network. From the mentioned results it is concluded that reduction of cesspit discharge frequency is a major reason for preference of GWTPs.
- Reuse in Irrigation: 54.4% of GWTP beneficiaries who accept a GWTP for the purpose of reuse in irrigation were not willing to replace the onsite GWTP in the case of providing a sewerage network, while 47.4% of GWTP beneficiaries who accept a GWTP not for reuse in irrigation were willing to replace the onsite GWTP in the case of providing sewerage network. From the mentioned results it is concluded that reuse in irrigation is an important reason for preference of GWTPs.
- Saving in Water Bill: 61.5% of GWTP beneficiaries who accept GWTPs for saving on their water bill were not willing to replace the onsite GWTP in the case of providing a sewerage network, while 40.0% of GWTP beneficiaries who accepted a GWTP not for saving on their water bill were willing to replace the onsite GWTP. This means that saving on their water bill is a very important reason for preference of GWTPs.

- Satisfaction of Applied System: 68.1% of GWTP beneficiaries who were satisfied with the unit's performance were not willing to replace the onsite GWTP in the case of providing a sewerage network, while 71.3% of GWTP beneficiaries who were not satisfied with the unit's performance were willing to replace the onsite GWTP. This indicates that the satisfaction of the existing sanitation system is a significant issue regarding the replacement of it with another one.
- Contribution of GWTPs to Solve the Water Shortage: 60.3% of GWTP beneficiaries who benefited from the treatment units by their contribution in solving water shortages were not willing to replace the onsite GWTP in the case of providing sewerage network, while 72.0% of GWTP beneficiaries who did not get benefits from the treatment units regarding their water shortages were willing to replace the onsite GWTP.
- Monitoring, Operation and Maintenance of the GWTPs: GWTPs are basically managed by women: the GWTPs are operated by men (fathers) and women (mothers) side-by-side (68.9%), while 24% are operated solely by women. Therefore, more focus should be placed on o women in terms of training and managing onsite sanitation systems, since they are more involved in household water management. The majority of the interviewees (73.1%) completed high school only or less, 20.4% had a university degree and higher education. Little effort is required for running the GWTPs, since the average yearly working time is 19.7 hours, corresponding to 0.4 hours per week. Operation and maintenance work include cleaning and checking the inlet manhole, removing scum from the first compartment (septic tank), pipe cleaning and cleaning of the whole treatment plant.

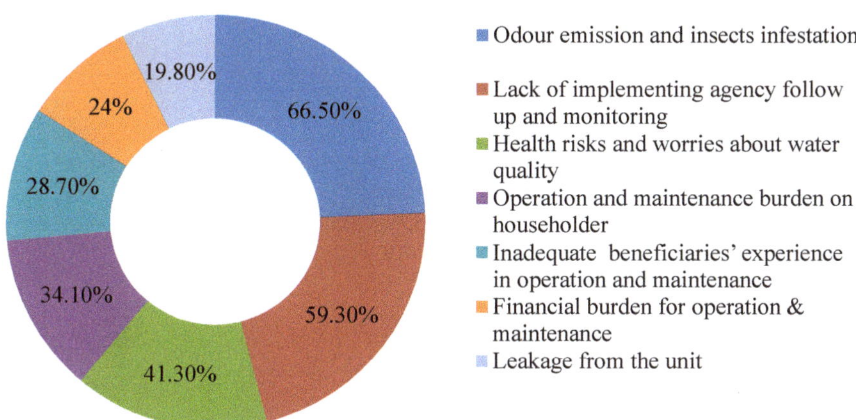

Figure 8. Barriers of providing onsite GWTPs.

4. Conclusions and Recommendations

4.1. Conclusions

Reasons for acceptance of GWTPs. Reuse of treated grey water in irrigation was the main incentive for a GWTP as stated by 88.0% of respondents, followed by reduction in cesspit discharge frequency and its financial consequences as stated by 71.3%. A total of 35.3% of respondents mentioned water shortages, reduction of potential risks of ground water pollution, reduction of water bill and enhanced hygiene. Availability of funds was an important driver for the construction of GWTPs as stated by 70.7%. Islamic religion was considered a driver; the majority of people (70%) accepted reuse of treated grey water in irrigation. Women play a major role in GWTP management since they are more involved in household water and sanitation management; 68.9% of the treatment systems are run by men side-by-side with women (fathers and mothers), and 24% is run completely by women. The aesthetic impact of the system is very positive, as mentioned by 74.9% of beneficiaries. The majority of GWTP beneficiaries (70.4%) are satisfied. Little effort is required for operation and maintenance, with only an average 0.4 working hours per week.

Barriers for Onsite GWTPs. The first barrier as mentioned by 66.5% is odor emission and insect infestation. A total of 59.3% stated that the systems lack follow-up and monitoring from the implementing agency's side. System failures were also caused by inadequate beneficiary experience in operation and maintenance and lack of system understanding was stated by 34.1% of beneficiaries. Health concerns and doubt surrounding the quality of crops irrigated by treated grey water was another barrier raised by beneficiaries.

Success and Failure Lessons. Water shortage is a main driver for providing an onsite grey water system, and farmers with experience in agriculture are more capable of managing the grey water system and reuse schemes than others. Failure of GWTPs happened as a result of inappropriate operation and maintenance and lack of system understanding, as well as lack of technical support from the implementing agency. Sometimes failures happened as a result of improper utilization of treated water and seepage of water into the surrounding area, lack of reuse schemes and agricultural plans and finally beneficiaries' limited experience in agricultural practices.

House onsite grey water management systems are acceptable in rural communities; therefore, a more proper system is required to handle the wastewater and replace cesspits and their harmful implications on environment, ground water and public health.

4.2. Recommendations

- There is an essential need to improve the performance of the treatment plants, to increase treatment efficiency and to introduce well-operated wastewater treatment facilities.
- Ensure treated water quality complies with applied local and international standards and its suitability for reuse purposes.
- At the policy level, the government should encourage and be more aware of the potential application of onsite GWTPs in rural communities, so the government should be more involved in wastewater management in rural areas to replace cesspits.
- The government should encourage the use of non-conventional water resources in agriculture, especially treated grey water.
- Implementing agencies should implement regular monitoring and maintenance of the onsite GWTPs, especially after the end of implementation and consider this phase as a part of the project implementation.
- Implementation of GWTPs should be applied according to social and technical feasibility studies, including involvement of people in the planning and implementation process to ensure understanding of the whole system.
- GWTP beneficiaries require the necessary training in operation and maintenance of the system management to maintain sustainability and to handle the system successfully.
- Development of public awareness programs, to better understand and improve public knowledge of wastewater systems and perception toward reuse schemes, in parallel with field visits of local people to other wastewater treatment and reuse facilities for sharing of knowledge and ideas.
- A more proper system is required to handle the wastewater to replace cesspits and their implications on the environment, ground water and health in rural communities.

Author Contributions: Conceptualization, R.A.T.; Methodology, N.M. and R.A.T.; Validation, N.M. and R.A.T.; Formal Analysis, R.A.T.; Investigation, I.A.A.-K.; Data Curation, R.A.T.; Writing—Original Draft Preparation, R.A.T.; Writing—Review & Editing, I.A.A.-K.; Visualization, Y.-T.H.; Supervision, N.M.; Project Administration, R.A.T.; Funding Acquisition, N.M. All authors have read and agreed to the published version of the manuscript.

Funding: This research was supported by the Palestinian Water Authority through the project of "Wastewater Need Assessment in the Palestinian Rural Communities" funded by the Austrian Development Agency (ADA)/Australian Government.

Conflicts of Interest: The authors declare no conflict of interest.

References

1. WBG. *Securing Water for Development in West Bank and Gaza*; International Bank for Reconstruction and Development: Washington, DC, USA, 2018.
2. PCBS. *Results Household Environmental Survey*; Palestinian Central Bureau of Statistics: Ramallah, Palestine, 2015.
3. ARIJ. *Status of the environment in the state of Palestine*; Applied Research Institute—Jerusalem: Bethlehem, Palestine, 2015.
4. Lambert, L.A.; Jordan Lee, J. Nudging greywater acceptability in a Muslim country: Comparisons of different greywater reuse framings in Qatar. *Environ. Sci. Policy.* **2018**, *89*, 93–99. [CrossRef]
5. Chirisa, I.; Bandauko, E.; Matamanda, A.; Mandisvika, G. Decentralized domestic wastewater systems in developing countries: The case study of Harare (Zimbabwe). *Appl. Water Sci.* **2017**, *7*, 1069–1078. [CrossRef]
6. Boano, F.; Caruso, A.; Costamagna, E.; Ridolfi, L.; Masi, F.; Demichelis, F.; Galvão, A.; Pisoeiro, J.; Rizzo, A.; Masi, F. A review of nature-based solutions for greywater treatment: Applications, hydraulic design, and environmental benefits. *Sci. Total Environ.* **2020**, *711*, 134731. [CrossRef] [PubMed]
7. Ridderstolpe, P. *Introduction to Grey Water Management*; EcoSanRes Programme, Stockholm Environment Institute: Stockholm, Sweden, 2004.
8. Maimon, A.; Gross, A. Greywater: Limitations and perspective. *Curr. Opin. Environ. Sci. Health* **2018**, *2*, 1–6. [CrossRef]
9. Oh, K.S.; Leong, J.Y.C.; Poh, P.E.; Chong, M.N.; Lau, E.V. A review of greywater recycling related issues: Challenges and future prospects in Malaysia. *J. Clean. Prod.* **2018**, *171*, 17–29. [CrossRef]
10. Morel, A.; Diener, S. *Greywater Management in Low and Middle-Income Countries, Review of Different Treatment Systems for Households or Neighbourhoods*; Sandec Report No. 14/06; Sandec (Water and Sanitation in Developing Countries) at Eawag (Swiss Federal Institute of Aquatic Science and Technology): Dübendorf, Switzerland, 2006.
11. Zraunig, A.; Estelrich, M.; Gattringer, H.; Kisser, J.; Langergraber, G.; Radke, M.; Rodriguez-Roda, I.; Buttiglieri, G. Long term decentralized greywater treatment for water reuse purposes in a tourist facility by vertical ecosystem. *Ecol. Eng.* **2019**, *138*, 138–147. [CrossRef]
12. Leong, J.Y.C.; Balan, P.; Chong, M.N.; Poh, P.E. Life-cycle assessment and life-cycle cost analysis of decentralised rainwater harvesting, greywater recycling and hybrid rainwater-greywater systems. *J. Clean. Prod.* **2019**, *229*, 1211–1224. [CrossRef]
13. Gorgich, M.; Mata, T.M.; Martins, A.; Caetano, N.S.; Formigo, N. Application of domestic greywater for irrigating agricultural products: A brief study. *Energy Rep.* **2020**, *6*, 811–817. [CrossRef]
14. Diaper, C.; Dixon, A.; Butler, D.; Fewkes, A.; Persons, S.A.; Stephenson, T.; Strathern, M.; Strutt, J. Small scale water recycling systems—Risk assessment and modelling. *Water Sci. Technol.* **2001**, *43*, 83–90. [CrossRef] [PubMed]
15. Siggins, A.; Burton, V.; Ross, C.; Lowe, H.; Horswell, J. Effects of Long-Term Greywater Disposal on Soil: A Case Study. *Sci. Total Environ.* **2016**, *557–558*, 627–635. [CrossRef] [PubMed]
16. Oteng-Peprah, M.; de Vries, N.; Acheampong, M.A. Households' willingness to adopt greywater treatment technologies in a developing country—Exploring a modified theory of planned behaviour (TPB) model including personal norm. *J. Environ. Manage.* **2020**, *254*, 109807. [CrossRef] [PubMed]
17. Redwood, M. Greywater use in the Middle East and North Africa region. In Proceedings of the Greywater Stock-taking Meeting, IDRC–CSBE, Aqaba, Jordan, 12–15 February 2007.
18. WHO. *Guidelines for the Safe Use of Waste-Water, Excreta and Grey Water—Volume 1–4—Excreta and Grey Water Use in Agriculture*; World Health Organization: Geneva, Switzerland, 2006.

19. Ashok, S.S.; Kumar, T.; Bhalla, K. Integrated Greywater Management Systems: A Design Proposal for Efficient and Decentralised Greywater Sewage Treatment. *Procedia CIRP* **2018**, *69*, 609–614. [CrossRef]
20. Oteng-Peprah, M.; Acheampong, M.A.; deVries, N.K. Greywater Characteristics, Treatment Systems, Reuse Strategies and User Perception—a Review. *Water Air Soil Pollut.* **2018**, *229*, 255. [CrossRef] [PubMed]
21. Radingoana, M.P.; Dube, T.; Mollel, M.H.; Letsoalo, J.M. Perceptions on greywater reuse for home gardening activities in two rural villages of Fetakgomo Local Municipality, South Africa. *Phys. Chem. Earth Parts A B C* **2019**, *112*, 21–27. [CrossRef]
22. Ahmad, A.; Lubbad, I.; Shaheen, A.; Mogheir, Y. *Small Scale Wastewater Treatment Plants in Palestinian Rural Areas: An Environmentally Sound Option*; Environment Quality Authority: Gaza, Palestine, 2009.
23. Burnat, J.; Shtayye, I. On-site grey water treatment in Qebia Village, Palestine, Greywater Use in the Middle East. Technical, Social, Economic and Policy Issues. 2009. Available online: https://www.idrc.ca/sites/default/files/openebooks/466-6/index.html (accessed on 14 December 2019).
24. *SPSS Statistics for Windows*; Version 20.0; IBM Corp.: Armonk, NY, USA, 2011.
25. Palestinian Central Bureau of Statistics (PCBS). On the Occasion of the International Population Day 11/7/2019. Ramallah, Palestine. 2019. Available online: http://www.pcbs.gov.ps/post.aspx?lang=en&ItemID=3503# (accessed on 20 February 2020).
26. PCBS. *Poverty Profile in Palestine*; Palestinian Central Bureau of Statistics: Ramallah, Palestine, 2017.
27. Adilah, O. Assessment of Wastewater Reuse Potential in Palestinian Rural Areas. Master's Thesis, Faculty of Graduate Studies, Program in Water and Environmental Engineering, Birzeit University, Palestine, Birzeit, 2011.

Article

Mercury, Arsenic and Lead Removal by Air Gap Membrane Distillation: Experimental Study

Abdullah Alkhudhiri [1,*], Mohammed Hakami [2], Myrto-Panagiota Zacharof [3], Hosam Abu Homod [1] and Ahmed Alsadun [1]

1 King Abdulaziz City for Science and Technology (KACST), National Centre for Desalination & Water Treatment Technology, Riyadh 11442, Saudi Arabia; habuhomod@kacst.edu.sa (H.A.H.); aalsadun@kacst.edu.sa (A.A.)
2 Chemical Engineering Technology Department, Jubail Industrial College, Jubail Industrial City 31961, Saudi Arabia; hakami_mw@jic.edu.sa
3 Sustainable Environment Research Centre (SERC), Faculty of Engineering, Computing and Science, University of South Wales, Pontypridd CF37 1DL, UK; myrto-panagiota.zacharof@southwales.ac.uk
* Correspondence: khudhiri@kacst.edu.sa

Received: 31 March 2020; Accepted: 29 May 2020; Published: 31 May 2020

Abstract: Synthetic industrial wastewater samples containing mercury (Hg), arsenic (As), and lead (Pb) ions in various concentrations were prepared and treated by air gap membrane distillation (AGMD), a promising method for heavy metals removal. Three different membrane pore sizes (0.2, 0.45, and 1 μm) which are commercially available (TF200, TF450, and TF1000) were tested to assess their effectiveness in combination with various heavy metal concentrations and operating parameters (flow rate 1–5 L/min, feed temperature 40–70 °C, and pH 2–11). The results indicated that a high removal efficiency of the heavy metals was achieved by AGMD. TF200 and TF450 showed excellent membrane removal efficiency, which was above 96% for heavy metal ions in a wide range of concentrations. In addition, there was no significant influence of the pH value on the metal removal efficiency. Energy consumption was monitored at different membrane pore sizes and was found to be almost independent of membrane pore size and metal type.

Keywords: air gap membrane distillation; wastewater treatment; heavy metal removal; industrial wastewater

1. Introduction

Industrial wastewater is one of the most serious pollutants, contributing significantly to the current load constraints of conventional wastewater treatment plants. Several industrial sectors such as petroleum, petrochemicals, tanning and electroplating are generating large amounts of toxic heavy metal wastewater, which needs to be extensively treated prior to its release to the ecosystem. The type of contaminants and physicochemical properties of industrial wastewater effluent such as temperature, viscosity, salinity, and turbidity vary with each stream. Nevertheless, most industrial wastewater streams contain heavy metals such as zinc, copper, mercury, lead, and arsenic in amounts that if left untreated will exceed the limit allowable by the national public health and safety regulations for their safe disposal [1].

Heavy metals have different properties based on their atomic number and chemical structure that contribute to their effects and toxicity to the environment and human health. There are 17 elements that are considered to be very toxic, including mercury (Hg), lead (Pb), and arsenic (As). Toxicity levels depend on the type of metal and the type of organism that is exposed to it [2].

1.1. Membrane Technology

Membrane technologies can be used to effectively treat wastewater that includes heavy metals due to their flexibility, scalability, and easiness to operate and maintain. Heavy metal concentrations, type of contaminants, and the level of filtration required determines the treatment process type. Moreover, the membrane performance (rejection) can be affected by several parameters such as the material used, the membrane pore size, and the membrane composition; for example, metallic, ceramic, and composite materials are utilized to remove heavy metals. In addition, the influence of heavy metal concentration and type of contaminants on the rejection factor were reported.

Kurniawan et al. [3] conducted a comparative study of heavy metal removal (Cd, Cr, Cu, Ni, and Zn) via ultrafiltration (UF), nanofiltration (NF), and reverse osmosis (RO). They concluded that UF could remove about 90% of heavy metal concentration at a pH of 5 to 9.5. In addition, they noticed that the heavy metal concentration is essential in the case of UF and NF; however, the rejection factor for RO was not affected by the metal concentration. Moreover, Seidel et al. [4] investigated the rejection factor of As(III) and As(V) using NF and RO. They found that As(V) can be effectively removed by RO and NF. However, RO can be used to achieve a high removal rate in As(III) only. In addition, they observed that the removal of As(III) declined from 28% to 5% as the metal concentration increased from 0.01 to 0.316 ppm.

A hybrid process has been implemented to treat wastewater that contains heavy metals. For example, Blocher et al. [5] developed a flotation/microfiltration process to remove copper, nickel, and zinc from aqueous solutions. Moreover, synthetic waste water was treated by electrocoagulation followed by a microfiltration process [6].

1.2. Membrane Distillation (MD)

Membrane distillation (MD) is a promising technology for treating saline water and wastewater with high rejection rates, which cannot be accomplished by conventional technologies. MD is a thermally driven separation process in which only the vapor molecules pass through a microporous hydrophobic membrane. The vapor pressure difference which is caused by the temperature difference across the membrane surface is considered to be the driving force for MD. In order to avoid a wetting incidence of the membrane pores, the membrane pore size must be as small as possible. However, MD permeate flux will decline. Therefore, an optimum membrane pore size needs to be determined for each MD application and the feed type to be treated [7,8].

Liquid entry pressure (LEP) is an important membrane characteristic. LEP is defined as the minimum transmembrane pressure that is required for a feed solution to penetrate a large pore size ($r_{\max}$). Therefore, the hydrostatic pressure should be lower than LEP to avoid membrane wetting [9,10]. LEP can be estimated from the following equation:

$$\Delta P = P_F - P_P = \frac{-2B\gamma_l \cos\theta}{r_{max}} \quad (1)$$

where P_F and P_P are the hydraulic pressures on the feed and permeate side, B is the geometric pore coefficient (equal to 1 for cylindrical pores), γ_l is the liquid surface tension, θ is the contact angle, and r_{max} is the maximum pore size.

Operating conditions such as the temperature, flow rate, membrane type, as well as membrane pore structure play a significant role in the system's efficiency [10].

The mass flux (J) in MD is assumed to be proportional to the vapor pressure difference across the membrane, and is given by [10,11]

$$J = C\left(P_f - P_p\right) \quad (2)$$

where C is the membrane coefficient and P_f and P_p is the difference of vapor pressure at the membrane feed and permeate surfaces.

Membrane distillation (MD) has not been widely applied for heavy metal removal; for example, contaminated groundwater with arsenic (40–2000 ppm) has been treated by direct contact MD (DCMD) to 10 ppm [12]. Almost 100% arsenic removal was achieved without wetting the membrane pore. Moreover, arsenite (As(III)) and arsenate (As(V)) removals were examined by DCMD. It was pointed out that the rejection of As was independent of the solution pH and the temperature [13]. Treatment of heavy metal wastewater by vacuum membrane distillation (VMD) was achieved by Zhongguang Ji [14]. The effect of pH on VMD performance was studied, and the author specified that the VMD process showed good acid resistance. A modified PVDF/TiO_2 electrospun membrane was prepared to remove heavy metal traces from water via VMD [15]. Improvements in the rejection factor and permeate flux were noticed.

To the authors' current knowledge, there are no studies available which deal with the removal of heavy metal from industrial wastewater via AGMD. Therefore, this study investigates the application of suitable membrane technologies over a wide range of industrial heavily polluted wastewater. Air gap membrane distillation (AGMD) is a promising method for heavy metals removal (Pb, As and Hg) from industrial wastewaters. AGMD was chosen to treat industrial wastewater due to the high quality of permeate flux and low risk of membrane wetting. Three different membrane pore sizes (0.2, 0.45 and 1 μm) which are commercially available (TF200, TF450 and TF1000) were tested to assess their effectiveness in combination with various heavy metal concentrations and operating parameters (flow rate −5 L/min, feed temperature 40–70 °C, pH 2–11). The AGMD process could be used for localized, low-cost deployment treatment in the industry.

2. Experimental Procedure and Material

The impact of a wide range of concentrations for Pb, As, and Hg on permeate flux and on the rejection factor was examined as shown in Table 1. Metal ion solutions of Pb^{2+}, As^{3+}, and Hg^{2+} were prepared from $Pb(NO_3)_2$, AsO_3, and $HgCl_2$ compounds, respectively. The influence of pH on the rejection factor at room temperature was explored at three different pH values: 2, 7, and 11. In addition the influence of pore size was studied using three commercial membranes (0.2, 0.45 and 1.0 μm). The experiments were conducted over a wide range of heavy metal concentrations at a constant feed flow rate (1.5 L/min), feed temperature (50 °C) and coolant temperature (10 °C). Moreover, a mixed metal ion solution of Pb^{2+}, As^{3+}, Hg^{2+}, and NaCl (synthetic industrial wastewater) was prepared and treated by different membrane pore sizes for 5 h. The experimental tests were achieved by the AGMD module in a horizontal position with a membrane effective area of 0.006 m^2, as shown in Figure 1. The air gap width was about 5 mm. Three types of flat sheet polytetrafluoroethylenes (PTFE) microporous hydrophobic membranes were employed in this study, as shown in Table 2.

Table 1. Range of concentration of heavy metals used in the filtration process.

Heavy Metal	Concentration (ppm)				
Lead Pb^{2+}	50	100	200	1000	1500
Arsenic As^{3+}	2	5	10	25	100
Mercury Hg^{2+}	5	10	20	-	-
Synthetic wastewater	Pb^{2+} (1000 ppm); As^{3+} (5 ppm); Hg^{2+} (5 ppm) and NaCl (2000 ppm).				

The effect of the flow rate, feed temperature, and condensing temperature on the rejection factor was examined. The influence of contaminated water flow rate at 1, 3 and 5 L/min was explored. The flow rate could be manipulated by adjusting the pump speed to achieve the desired flow rate. The contaminated water was heated and pumped to the top part of the membrane cell at a constant temperature (50 °C). Moreover, cooling fluid was also pumped at a constant flow rate (2.5 L/min) to the bottom cell compartment at a constant temperature of 10 °C.

With regard to the feed temperature influence, contaminated water temperature at 40 °C, 50 °C, 60 °C and 70 °C was analysed. The temperature was adjusted to the desired point and controlled as well throughout the process. The contaminated water was heated and pumped at 1.5 L/min to the top part of the membrane cell. The cooling fluid temperature was kept at a constant temperature of 10 °C and pumped to the bottom part of membrane cell.

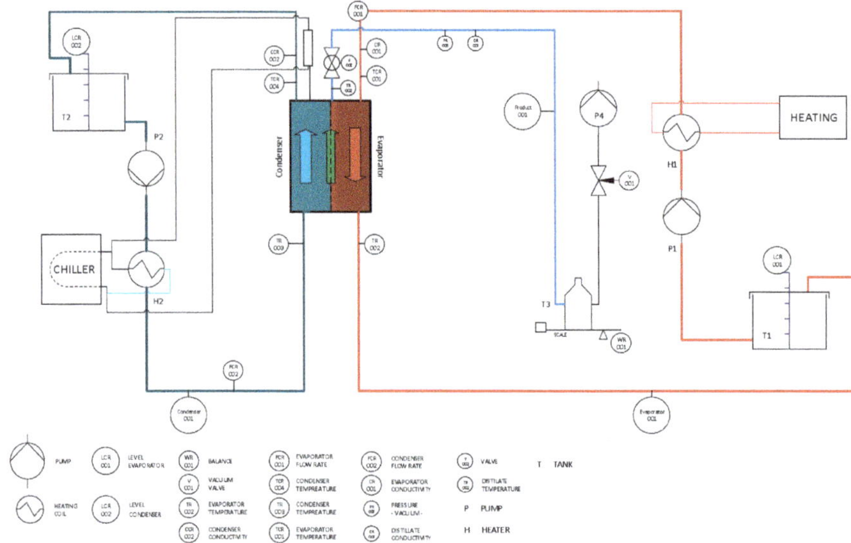

Figure 1. Schematic diagram of the air gap membrane distillation (AGMD) experimental apparatus used in this study.

Table 2. Properties of membranes that were used in this work as specified by the manufacturer. PTFE: polytetrafluoroethylenes.

Specification	Description
Trade name	TF200 TF450 TF1000
Manufacturer Material Membrane support Thickness	Sterlitech corporation PTFE Polypropylene 175 µm
Mean pore size and liquid entry pressure (LEP)	0.2 µm (2.55 bar) 0.45 µm (0.76 bar) 1.00 µm (0.28 bar)

The permeate flux (J) was measured by weighing the obtained permeate for a predetermined time using an electronic balance which was connected to a computer:

$$J = \frac{W}{A\ \Delta t} \tag{3}$$

where W is the obtained permeate weight and A is the effective membrane area.

Furthermore, the concentration of heavy metals in the feed and permeate was measured by atomic absorption spectrometry (AAS) and inductive couple plasma (ICP). Standard solutions for Pb(II),

As(III) and Hg(II) were used to ensure the readings of AAS and ICP were correct. Therefore, the rejection factor can be estimated as

$$\text{Rejection Factor} = (1 - \frac{C_p}{C_f}) \times 100 \quad (4)$$

where (C_f) and (C_p) are the feed and permeate concentrations (in ppm), respectively.

3. Results and Discussion

The effect of various metal ion concentrations of Pb^{2+}, As^{3+}, and Hg^{2+} in different operating conditions on the membrane flux and the rejection factor were tested. Furthermore, the impact of the pH solution on the rejection factor was investigated.

3.1. Membrane Pore Size and Metal Concentration Effect

The impact of the membrane pore size on the heavy metal rejection by AGMD was investigated. The experiments were conducted over a wide range of heavy metal concentrations at a constant feed flow rate, feed temperature, and coolant temperature.

The effect of membrane pore size on the heavy metal rejection is shown in Figure 2. It should be noted that the TF200 membrane has a pore size of 0.2 µm, the TF450 membrane has a pore size of 0.45 µm and the TF1000 membrane has a pore size of 1.0 µm. TF200 shows excellent membrane removal for heavy metal ions Pb(II), As(III), and Hg(II) with a wide range of concentrations. The rejection was almost 100% due to the small membrane pore size. For instance, the permeate flux of Pb (1000 ppm), As (25 ppm), and Hg (10 ppm) was 0.271, 0.293, and 0.302 $g/m^2{\cdot}s$, respectively. As is evident from Figure 2, the efficiency of metal ion removal by the TF200 membrane was not affected by the metal concentration. For TF450, the rejection factor for Hg(II) was almost 100% and was not affected by the metal concentration due to its ionic size. On the other hand, the rejection factor for Pb(II) at 1000 ppm and 1500 ppm was 98% and 96%, respectively. In contrast to TF200, the TF1000 membrane showed less membrane efficiency for metal removal due to its increased membrane pore size. This result can be attributed to the decrease of the membrane hydrophobicity due to the wetting incidence in the membrane pores. Few hydrophobic membrane pores were wetted, which allowed the feed solution to penetrate through the permeate side due to the lower value of the liquid entry pressure (LEP) for TF1000. It is worth noting that the LEP value depends on several factors such as the maximum membrane pore size and the membrane hydrophobicity. Nonetheless, it was noticed that the permeate flux increased compared to TF200 and TF450. As a consequence, TF1000 was not further used.

Moreover, the rejection factor of Pb(II), As(III) and Hg(II) for high concentrations was lower than the rejection factor for low concentrations, which implies that a concentration polarization phenomenon occurred. Concentration polarization is defined as an increase of feed concentration near the membrane surface. Therefore, permeate flux slightly increased and was accompanied with a decrease in the permeate quality due to the wetting incidence of the PTFE membrane. Eykens et al [16] indicated that the membrane defects and the maximum pore size strongly affects LEP and is a possible cause of membrane wetting. For instance, the rejection factor for Pb(II) 1500 ppm, As(III) 100 ppm and Hg(II) 20 ppm was 93%, 97% and 98%, respectively. As a result, the TF200 and TF450 were selected to study the impact of feed flow rate, feed temperature, and pH.

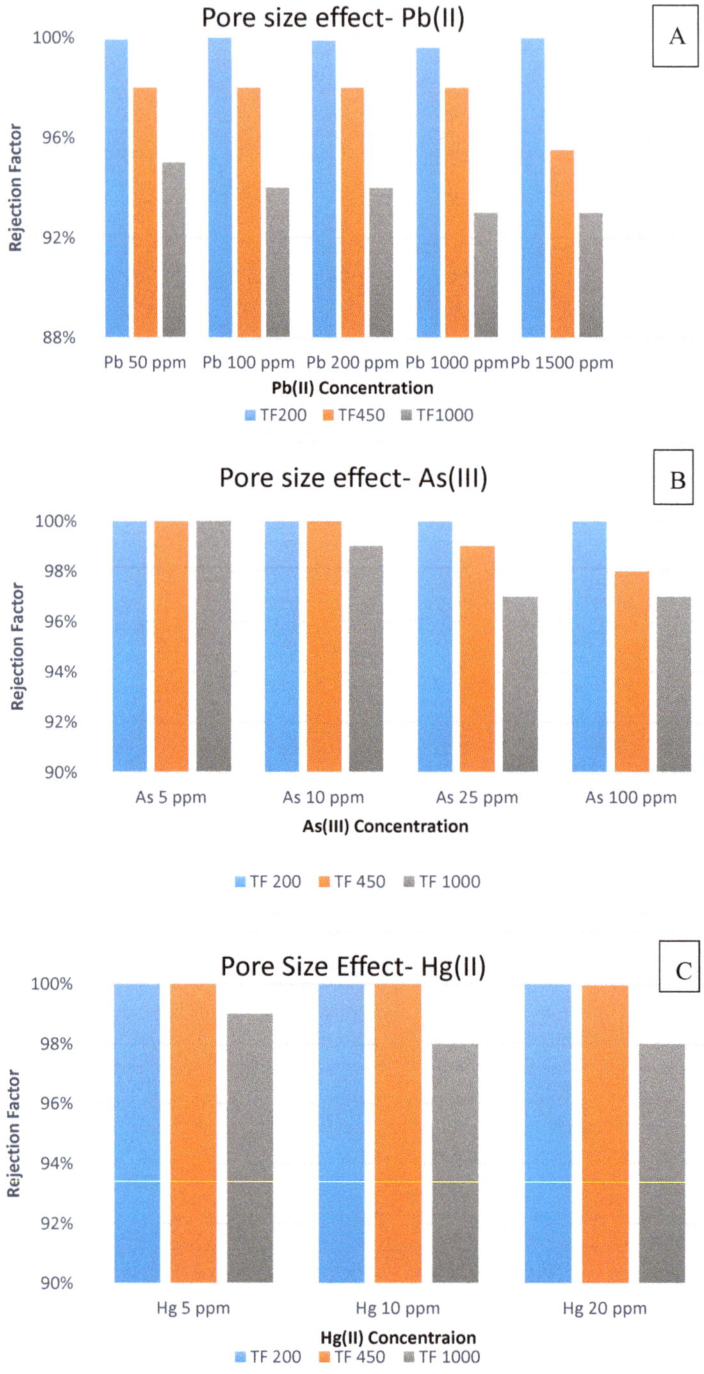

Figure 2. Pore size effect for different heavy metals using TF200, TF450 and TF1000: (**A**) Pb(II), (**B**) As(III), and (**C**) Hg(II).

3.2. Effect of Feed Flow Rate

In order to study the impact of feed flow rate on the rejection of heavy metal removal, several experiments were conducted by changing the initial feed concentration and feed flow rate from 1 to 5 L/min at a constant feed and cooling temperatures.

Figures 3–5 reveal that Pb(II), As(III) and Hg(II) were almost rejected completely (99–100%) by the TF200 membrane due to the high LEP value. Similarly, the TF450 membrane as shown in Figure 5 could reject Hg(II) completely at different flow rates. The high removal of Hg(II) by the TF450 membrane might be due to the membrane hydrophobicity and the size of the mercury ions. Olatunji and Camacho [17] pointed out that the mass transfer coefficient at the feed boundary layer increases due to the increase in feed flow rate which reduces the concentration polarization impact and can lead to membrane wetting.

It was also noticed that Pb(II) rejection increases when the feed flow rate increases regardless of the concentration (Figure 6). For example, the rejection factor of the membrane TF450 for 1000 ppm Pb(II) was 97%, 98% and 100% at 1, 3 and 5 L/min, respectively. This can be explained by the decrease of the concentration polarization phenomenon resulting in the increase of the feed flow rate [11,18].

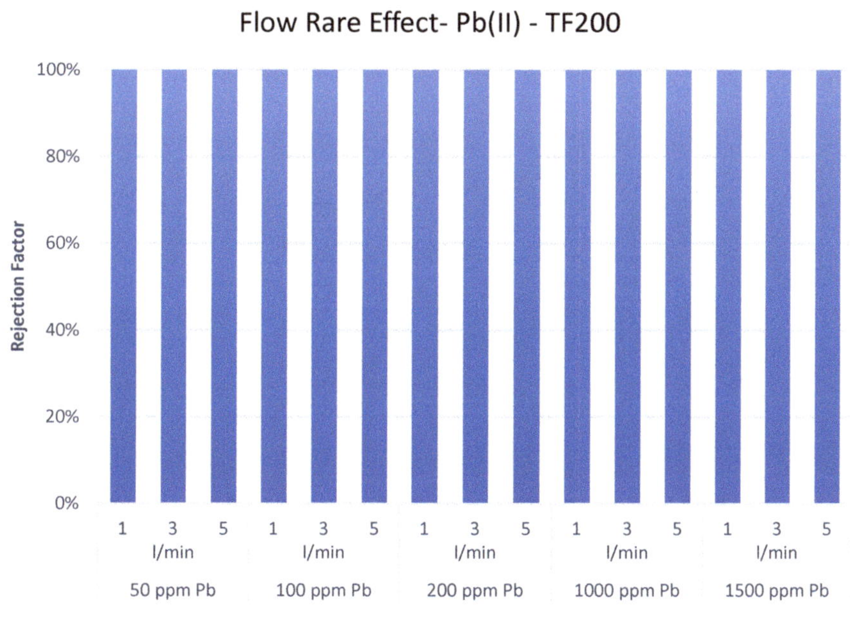

Figure 3. Effect of feed flow rate on the rejection factor using the TF200 membrane for Pb(II) (feed temperature = 50 °C, condensing temperature = 10 °C).

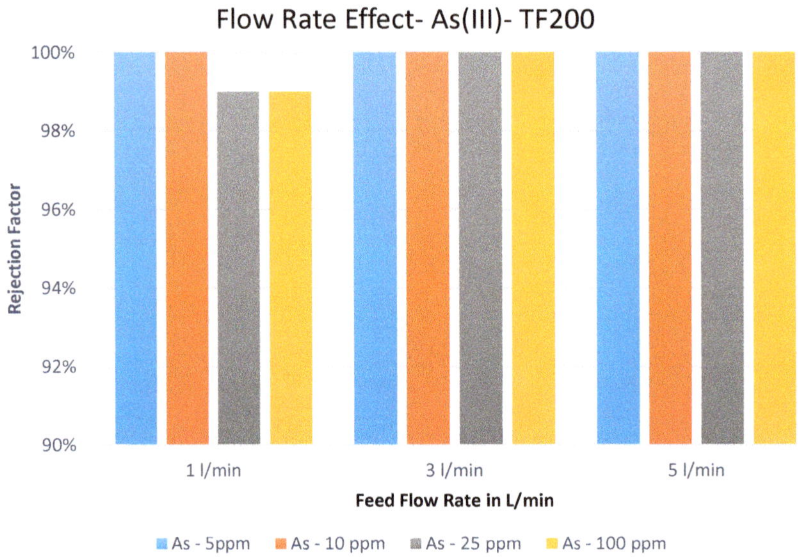

Figure 4. Effect of feed flow rate on the rejection factor using TF200 membrane for As(III) (feed temperature = 50 °C, condensing temperature = 10 °C).

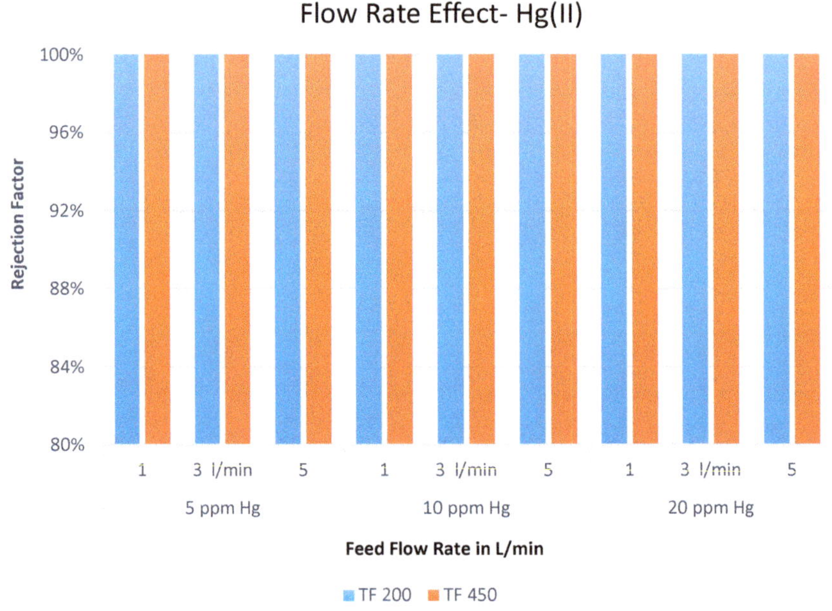

Figure 5. Effect of feed flow rate on the rejection factor using TF200 and TF450 membranes for Hg(II) (feed temperature = 50 °C, condensing temperature = 10 °C).

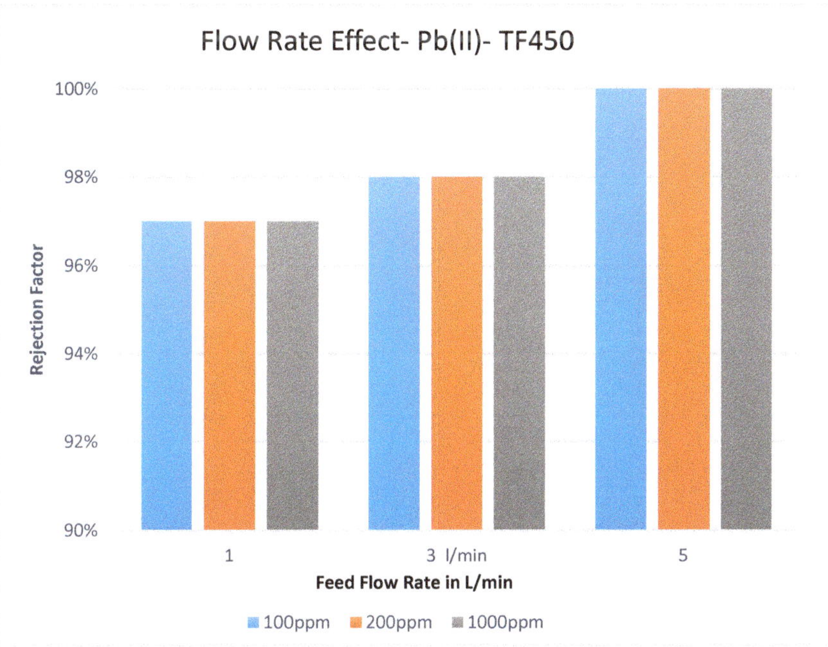

Figure 6. Effect of feed flow rate on the rejection factor using TF450 membrane for Pb(II) (feed temperature = 50 °C, condensing temperature = 10 °C).

3.3. *Effect of Feed Temperatures*

The impact of feed temperature on heavy metal rejection by the AGMD was investigated. The experiments were performed with a wide feed temperature range—i.e., 40–70 °C—and a constant feed flow rate and coolant temperature.

The impact of feed temperature on heavy metal removal by the AGMD process with a wide temperature range is shown in Figures 7 and 8. The heavy metal removal efficiency for TF200 membrane was perfect and stable over a wide concentration range of heavy metals and temperature. The removal was almost 100%, indicating the minimum influence of feed temperature on the membrane's performance. Likewise, the TF450 membrane showed excellent membrane efficiency for the metal removal of Hg(II) and As(III), which might be attributed to the membrane hydrophobicity. Additionally, a higher flux occurs at higher feed temperatures, which improves the rejection factor due to a "dilution" of the leakage, if it occurred. For example, the Hg(II) removal at 5, 10 and 20 ppm was almost 100% over a wide temperature range. However, feed temperature plays a role in Pb(II) removal, as shown in Figure 8C. This decrease in the Pb(II) removal at high concentration (1500 ppm) might be attributed to the decrease of the feed surface tension with the increase of the feed concentration and temperature [19,20] which negatively affects the LEP value.

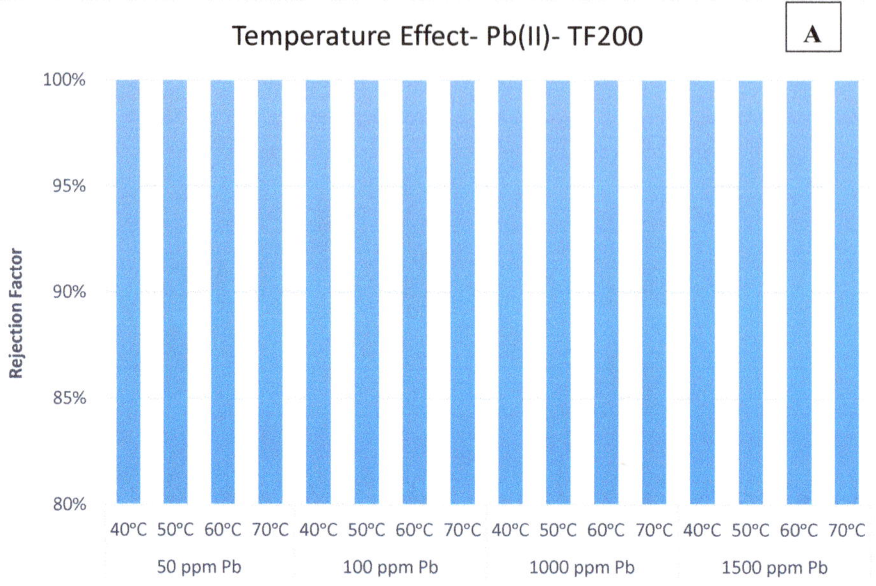

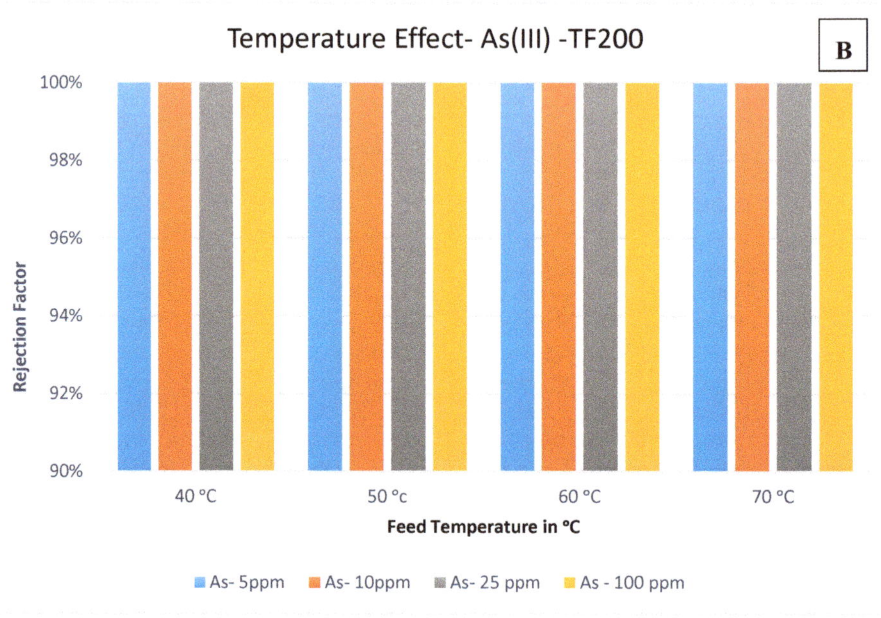

Figure 7. *Cont.*

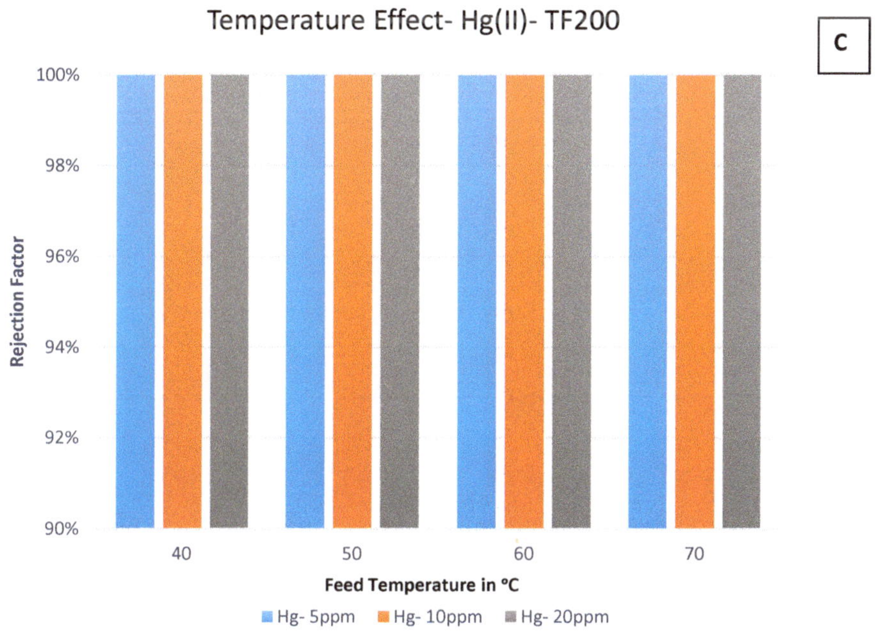

Figure 7. Effect of feed temperature on the rejection factor for TF200 membrane: (**A**) Pb(II), (**B**) As(III) and (**C**) Hg(II) (coolant temperature = 10 °C, flow rate = 3 L/min).

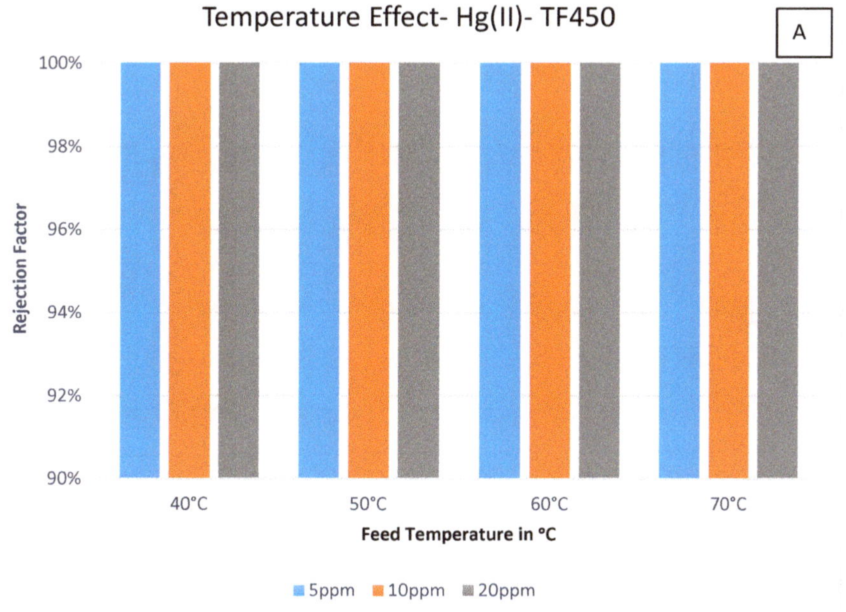

Figure 8. *Cont.*

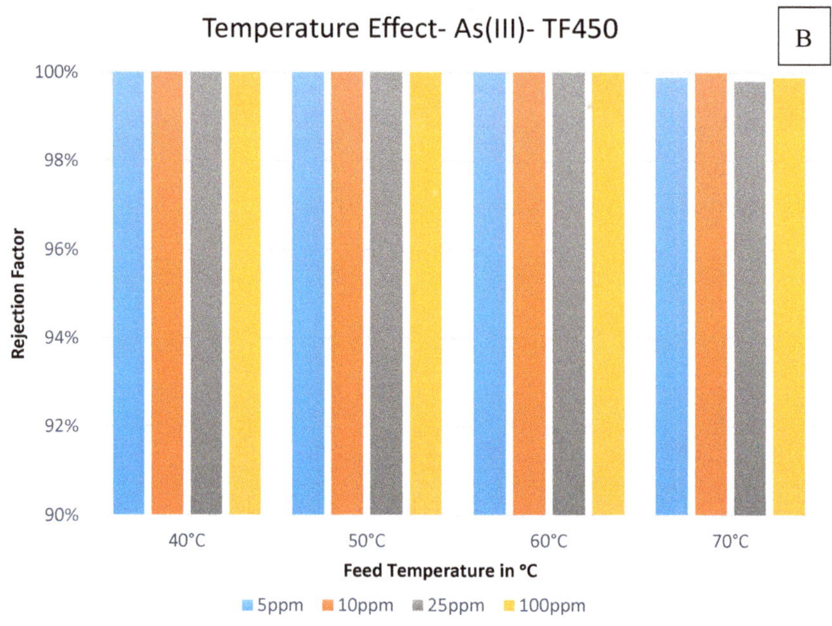

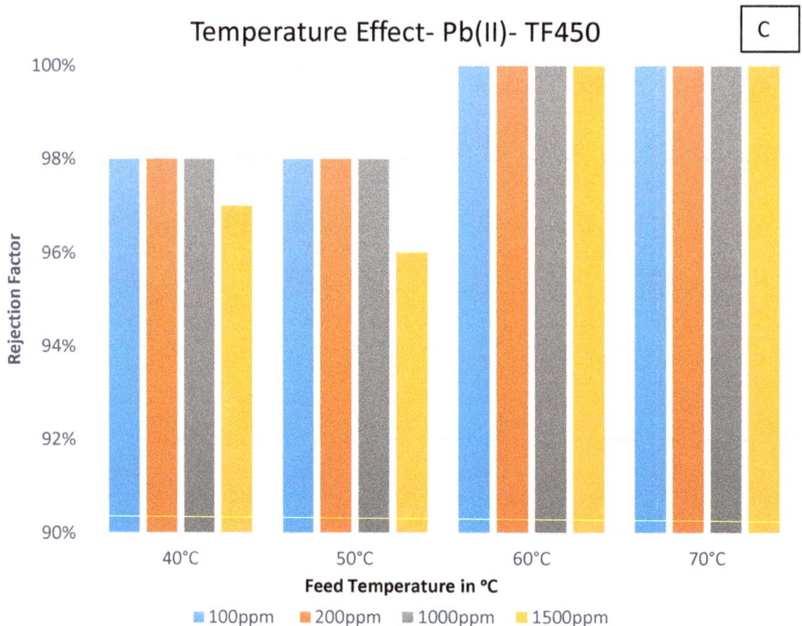

Figure 8. Effect of feed temperature on the rejection factor for TF450 membrane: (**A**) Hg(II), (**B**) As(III) and (**C**) Pb(II) (coolant temperature = 10 °C, flow rate = 3 L/min).

3.4. Effect of pH

Conventional treatment methods—especially membrane processes—are highly dependent on the pH value of the feed solution. The effect of the pH on heavy metal retention was studied at different pH values and flow rates to assess the effectiveness of MD.

The influence of pH on the heavy metal removal from wastewater by AGMD was investigated using TF200 and TF450. Experiments were performed with pH values ranging from 2–11. Metal ion removal was measured to study the impact of solution acidity. As evident from Figure 9, the removal of heavy metals is stable and stayed at an excellent level for the entire treatment process, which indicates that high acidity has no significant impact on the membrane's performance. A similar finding was reported by Zhongguang [14] and Qu et al. [13]. It is important to point out that the rejection factor for TF450 slightly decreased with a low pH value and flow rate, resulting in a rejection factor varying from 96% to 100%. For instance, the concentrations of the heavy metals in the permeate side at a pH of 2 and (3 L/min) for Pb (1500 ppm initial concentration), As (100 ppm initial concentration) and Hg (20 ppm initial concentration) were 36 ppm, 2 ppm, and 0.6 ppm, respectively. It is important to note that the PTFE membrane is an excellent chemical resistor.

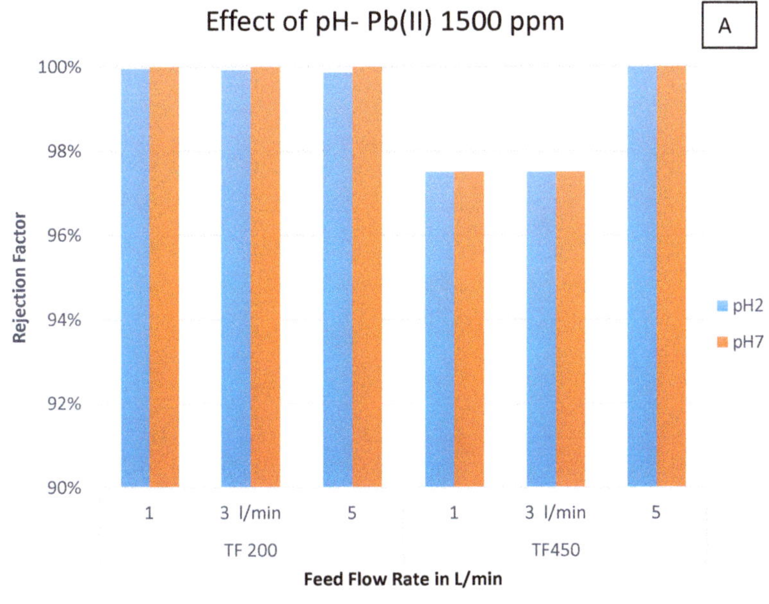

Figure 9. *Cont.*

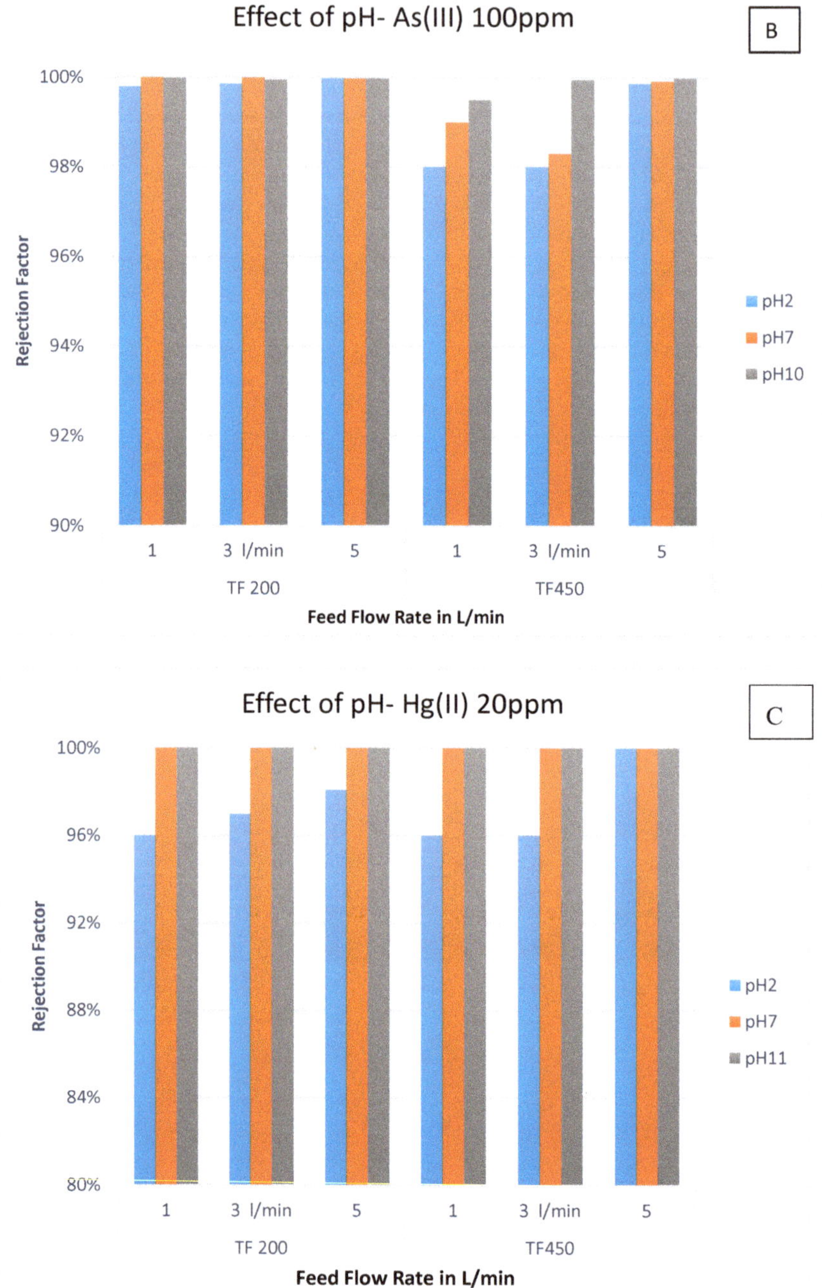

Figure 9. Effect of pH on the rejection factor using TF200/TF450 membranes for (**A**) Pb(II), (**B**) As(III) and (**C**) Hg(II) (feed temperature = 50 °C, flow rate = 3 L/min).

3.5. Energy Consumption

It was observed in this work that the operating parameters of a 5 L/min feed flow rate and a 60 °C feed temperature are the optimum operating parameters due to an excellent rejection factor and a high permeate flux for both TF200 and TF450. Therefore, synthetic wastewater was treated via TF200 and TF450. The permeate flux, energy consumption, and rejection factor were measured. The energy consumption in this study referred to the amount of electrical energy consumed (in kWh) for the heating, cooling, and pump systems. It was inferred that the energy consumption was almost independent of the membrane pore size and metal type. This result can be attributed to the fact that the stagnant air gap represents the predominant resistance to the mass and heat transfer. The energy consumption per gram of treated water was about 0.02 kWh/g for TF200 and about 0.016 for TF450. It is worth noting that the permeate flux for TF200 and TF450 was 0.242, and 0.323 g/m^2 s, respectively. The results of the permeate flux, energy consumption, and the rejection factor for both membranes are shown in Table 3. It was observed that the rejection factor of the TF200 membrane was almost stable; however, the rejection factor slightly decreased in the case of the TF450 membrane for Pb(II) and As(III).

Table 3. Permeate flux, energy consumption and rejection factor for synthetic wastewater via TF200 and TF450.

Membrane	Permeate Flux (g/m^2 s)	Energy Consumption (kWh/g)	Heavy Metal Removal Percentage
TF200	0.242	0.022	Pb(II) 100% As(III) 100% Hg(II) 100%
TF450	0.323	0.016	Pb(II) 99.4% As(III) 99.7% Hg(II) 100%

4. Conclusions

Air gap membrane distillation (AGMD) has been implemented at different operating parameters to treat industrial wastewater. Three different membrane pore sizes of 0.2, 0.45, and 1 μm which are commercially available (TF200, TF450 and TF1000) were tested to assess their effectiveness. Moreover, AGMD was employed to treat synthetic industrial wastewater and assess the effectiveness of the heavy metal removal. The major findings can be summarized as follows:

- AGMD is a promising technology for wastewater treatment and is expected to be a cost-effective process.
- TF200 and TF450 membranes presented a better metal removal efficiency over a wide concentration range, temperature and flow rate.
- The heavy metal rejection increases when the feed flow rate increases, regardless of its concentration.
- The TF1000 membrane showed less membrane efficiency for metal removal due to the increase in the membrane pore size.
- A high acidity of metal solution (PH value) has no significant impact on the membrane's performance.
- The energy consumption is almost independent of the membrane pore size and metal type.

Author Contributions: Conceptualization, A.A. (Abdullah Alkhudhiri), M.H. and M.-P.Z.; methodology, A.A. (Abdullah Alkhudhiri), H.A.H. and A.A. (Ahmed Alsadun); validation, A.A. (Abdullah Alkhudhiri), M.H. and M.-P.Z.; formal analysis, A.A. (Abdullah Alkhudhiri); investigation, A.A. (Abdullah Alkhudhiri), H.A.H. and A.A. (Ahmed Alsadun); writing—original draft preparation, A.A. (Abdullah Alkhudhiri); writing—review and editing, A.A. (Abdullah Alkhudhiri), M.H. and M.-P.Z. All authors have read and agreed to the published version of the manuscript.

Funding: This work was financially supported by the Energy and Water Research Institute / KACST, Saudi Arabia

Acknowledgments: The authors would like to thank Nora Alkhudhiri, for proof reading of the manuscript.

Conflicts of Interest: The authors declare no conflict of interest.

References

1. Alkhudhiri, A.; Darwish, N.B.; Hilal, N. Analytical and forecasting study for wastewater treatment and water resources in Saudi Arabia. *J. Water Process. Eng.* **2019**, *32*, 100915. [CrossRef]
2. Rashdi, B.A. Removal of Heavy Metals from a Concentrated Aqueous Solution: Adsorption and Nanofiltration Techniques. Ph.D. Thesis, Swansea University, Wales, UK, 2012.
3. Kurniawan, T.A.; Chan, G.Y.S. Physico–chemical treatment techniques for wastewater laden with heavy metals. *Chem. Eng. J.* **2006**, *118*, 83–98. [CrossRef]
4. Seidel, A.; Waypa, J.J.; Elimelech, M. Role of Charge (Donnan) Exclusion in Removal of Arsenic from Water by a Negatively Charged Porous Nanofiltration Membrane. *Environ. Eng. Sci.* **2001**, *18*, 105–113. [CrossRef]
5. Blöcher, C.; Dorda, J.; Matis, K.A. Hybrid flotation—Membrane filtration process for the removal of heavy metal ions from wastewater. *Water Res.* **2003**, *37*, 4018–4026. [CrossRef]
6. Keerthi, V.V.; Balasubramanian, N. Removal of heavy metals by hybrid electrocoagulation and microfiltration processes. *Environ. Technol.* **2013**, *34*, 2897–2902. [CrossRef] [PubMed]
7. Alkhudhiri, A.; Hilal, N. 3-Membrane distillation—Principles, applications, configurations, design, and implementation. In *Emerging Technologies for Sustainable Desalination Handbook*; Gude, V.G., Ed.; Butterworth-Heinemann: Oxford, UK, 2018; pp. 55–106.
8. El-Bourawi, M.S.; Ding, Z.; Ma, R.; Khayet, M. A framework for better understanding membrane distillation separation process. *J. Membr. Sci.* **2006**, *285*, 4–29. [CrossRef]
9. Franken, A.C.M.; Nolten, J.A.M.; Mulder, M.H.V.; Bargeman, D.; Smolders, C.A. Wetting criteria for the applicability of membrane distillation. *J. Membr. Sci.* **1987**, *33*, 315–328. [CrossRef]
10. Alkhudhiri, A.; Darwish, N.; Hilal, N. Membrane distillation: A comprehensive review. *Desalination* **2012**, *287*, 2–18. [CrossRef]
11. Alkhudhiri, A.; Darwish, N.; Hilal, N. Treatment of saline solutions using Air Gap Membrane Distillation: Experimental study. *Desalination* **2013**, *323*, 2–7. [CrossRef]
12. Pal, P.; Manna, A.K. Removal of arsenic from contaminated groundwater by solar-driven membrane distillation using three different commercial membranes. *Water Res.* **2010**, *44*, 5750–5760. [CrossRef] [PubMed]
13. Qu, D.; Wang, J.; Hou, D.; Luan, Z.; Fan, B.; Zhao, C. Experimental study of arsenic removal by direct contact membrane distillation. *J. Hazard. Mater.* **2009**, *163*, 874–879. [CrossRef] [PubMed]
14. Ji, Z. Treatment of heavy-metal wastewater by vacuum membrane distillation: Effect of wastewater properties. *IOP Conf. Ser.: Earth Environ. Sci.* **2018**, *108*, 042019. [CrossRef]
15. Moradi, R.; Monfared, S.M.; Amini, Y.; Dastbaz, A. Vacuum enhanced membrane distillation for trace contaminant removal of heavy metals from water by electrospun PVDF/TiO_2 hybrid membranes. *Korean J. Chem. Eng.* **2016**, *33*, 2160–2168. [CrossRef]
16. Eykens, L.; De Sitter, K.; Dotremont, C.; De Schepper, W.; Pinoy, L.; Van Der Bruggen, B. Wetting Resistance of Commercial Membrane Distillation Membranes in Waste Streams Containing Surfactants and Oil. *Appl. Sci.* **2017**, *7*, 118. [CrossRef]
17. Olatunji, S.O.; Camacho, L.M. Heat and Mass Transport in Modeling Membrane Distillation Configurations: A Review. *Front. Energy Res.* **2018**, *6*. [CrossRef]
18. Hwang, H.J.; He, K.; Gray, S. Direct contact membrane distillation (DCMD): Experimental study on the commercial PTFE membrane and modeling. *J. Membr. Sci.* **2011**, *371*, 90–98. [CrossRef]
19. Staszak, K.; Wieczorek, D.; Michocka, K. Effect of Sodium Chloride on the Surface and Wetting Properties of Aqueous Solutions of Cocamidopropyl Betaine. *J. Surfactants Deterg.* **2015**, *18*, 321–328. [CrossRef] [PubMed]
20. Park, H.; Thompson, R.B.; Lanson, N.; Tzoganakis, C.; Park, C.B.; Chen, P. Effect of Temperature and Pressure on Surface Tension of Polystyrene in Supercritical Carbon Dioxide. *J. Phys. Chem. B* **2007**, *111*, 3859–3868. [CrossRef] [PubMed]

Article

Purification Effect of Sequential Constructed Wetland for the Polluted Water in Urban River

Xueyuan Bai [1], Xianfang Zhu [2], Haibo Jiang [1], Zhongqiang Wang [1], Chunguang He [1,*], Lianxi Sheng [1,*] and Jie Zhuang [3]

1 State Environmental Protection Key Laboratory of Wetland Ecology and Vegetation Restoration, Key Laboratory for Vegetation Ecology, Ministry of Education, Jilin Provincial Key Laboratory of Ecological Restoration and Ecosystem Management, Northeast Normal University, Changchun 130117, China; baixy324@nenu.edu.cn (X.B.); jianghb625@nenu.edu.cn (H.J.); wangzq027@nenu.edu.cn (Z.W.)

2 Key Laboratory of Water and Sediment Sciences, Ministry of Education, Department of Environmental Engineering, Peking University, Beijing 100871, China; xianfangzhu@pku.edu.cn

3 Department of Biosystems Engineering and Soil Science, University Tennessee, Knoxville, TN 37996, USA; jzhuang@utk.edu

* Correspondence: he-cg@nenu.edu.cn (C.H.); shenglx@nenu.edu.cn (L.S.); Tel.: +86-431-891-65611

Received: 19 February 2020; Accepted: 3 April 2020; Published: 8 April 2020

Abstract: Constructed wetlands can play an active role in improving the water quality of urban rivers. In this study, a sequential series system of the floating-bed constructed wetland (FBCW), horizontal subsurface flow constructed wetland (HSFCW), and surface flow constructed wetland (SFCW) were constructed for the urban river treatment in the cold regions of North China, which gave full play to the combined advantages. In the Yitong River, the designed capacity and the hydraulic loading of the system was 100 m^3/d and 0.10 m^3/m^2d, respectively. The hydraulic retention time was approximately 72 h. The monitoring results, from April to October in 2016, showed the multiple wetland ecosystem could effectively remove chemical oxygen demand (COD), ammonia nitrogen (NH_4^+-N), total nitrogen (TN), total phosphate (TP), and suspended solids (SS) at average removal rates of 74.79%, 80.90%, 71.12%, 78.44%, and 91.90%, respectively. The removal rate of SS in floating-bed wetland was the largest among all the indicators (80.24%), which could prevent the block of sub-surface flow wetland effectively. The sub-surface flow wetland could remove the NH_4-N, TN, and TP effectively, and the contribution rates were 79.20%, 64.64%, and 81.71%, respectively. The surface flow wetland could further purify the TN and the removal rate of TN could reach 23%. The total investment of this ecological engineering was $12,000. The construction cost and the operation cost were $120 and $0.02 per ton of polluted water, which was about 1/3 to 1/5 and 1/6 to 1/3 of the conventional sewage treatment, respectively. The results of this study provide a technical demonstration of the restoration of polluted water in urban rivers in northern China.

Keywords: polluted urban river; sequential constructed wetlands; purification effect; water restoration; Yitong River

---

1. Introduction

In recent decades, with the rapid development of urban economies and urbanization, urban rivers all over the world have been facing the threat of pollution. Japan, the United States, and some European countries began to implement water purification of urban rivers in the 1950s and 1960s [1]. However, the research and application of further technology was only carried out in the past 20 to 30 years [2,3]. At present, river water purification technologies can be categorized into physical, chemical, biological, and ecological methods. Physical methods include aeration [4] and sediment dredging [5–7]. Chemical methods include chemical precipitation [8] and the application of chemical algaecide [9]. Biological methods include bioremediation [10], biofilms [11,12], contact oxidation [13–15], and membrane bioreactor technology [16,17]. Ecological methods include ecological ponds [18,19], plant purification treatment [20,21], ecological floating beds [22–24], and constructed wetlands [25–27]. Among these methods, aeration is the easiest, fastest, most effective, and most widely applied. It has been applied to the restoration of the Emscher River in Germany, the Homewood Canal in the United States, and the Thames River in the United Kingdom [28]. However, the operation and maintenance costs are high. Sediment dredging can remove endogenous pollution sources and is also a widely applied method to improve the water environment. There are problems associated with this method, such as the large amount of construction work, challenges to processing the sludge, and the additional pollution created by improper dredging [29,30]. Physical and chemical methods are often used only as secondary approaches to mediate pollution or to treat emergent water pollution. Although biological methods effectively clean up pollution and have low energy consumption requirements and small environmental impacts, they also have issues such as time-consuming processes to cultivate microorganisms and purification that is subject to restrictions from external conditions such as the temperature and water flow. Biofilm technology requires large-scale construction and involves land use issues. Constructed wetlands, on the other hand, use physical, chemical, and biological synergies in an ecosystem to efficiently remove pollutants by simulating the natural environment [31]. The operating costs are low, and they are easy to maintain and manage without secondary pollution. They also offer ecological benefits: they have been demonstrated to be an economical and efficient sewage treatment and management method [32], and they have become a preferred ecological method to improve the water quality of rivers in cities around the world. Therefore, the development of optimized functions of constructed wetlands that suit different goals and requirements has been a focal area of research. However, the ecological ponds occupy a large area of land, which is easily affected by environmental conditions, and the removal effect of nitrogen and phosphorus is not stable. The plant purification treatment and ecological floating beds only rely on aquatic plants to absorb nitrogen, phosphorus, and organic matter; thus, the capacity is limited. The constructed wetlands have disadvantages for purifying water because of the large area and low hydraulic load for the surface flow constructed wetland (SFCW); moreover, the horizontal subsurface flow constructed wetland (HSFCW) has a poor resistance to shock loads and is easily blocked. Therefore, the river water purification is a complex system engineering, and needs the joint application of various ecological technologies. The Yitong River is the mother river of Changchun, a capital city in northeast China. To create a water landscape for the city, multiple dams have been constructed to reserve water on the river section within Changchun. As a result, the flow rate has become slow, the dissolved oxygen (DO) level has decreased sharply, the river's self-purification capacity has been lost, and the water quality has begun to deteriorate. Eutrophication is especially severe in seasons with high temperatures. This situation has brought hidden risks to the water environment and the ecological security of the areas of Changchun near the lower reaches of the river, which severely affects the surrounding ecological environment and the daily life of nearby residents and restricts the sustainable development of the urban waters.

In this paper, we built a sequential combination system of the floating-bed constructed wetland (FBCW) + horizontal subsurface flow constructed wetland (HSFCW) + surface flow constructed wetland (SFCW) to purify the polluted water of Yitong River, using the favorable terrain of the park on the shore. This system was continuously monitored for one year and the results were analyzed. They can provide necessary data support for the popularization and application of the technology.

2. Method

2.1. Study Area

The study area is along the Yitong River in the city of Changchun, Jilin Province, northeast China (Figure 1, 43°50′32″ N, 125°21′32″ E), which has a temperate continental monsoon climate zone.

The annual rainfall in the study area is 600 to 700 mm. The average annual temperature is 4.6 °C. The maximum temperature in the summer is 40 °C. The annual freezing period is 5 months.

Figure 1. Location and overview of the study area.

The water quality was monitored in the study area from April to October 2014. Water quality data are shown as averages ± standard deviations (Table 1). The average chemical oxygen demand (COD) and ammonia nitrogen (NH_4^+-N) concentrations were 57.65 mg/L and 5.60 mg/L, respectively, which were both beyond the lower limits of Class V water in the Environmental Quality Standards for Surface Water (GB3838-2002) [33]. Therefore, the water quality was classified as worse than Class V.

Table 1. Results of water quality monitoring in the study area of the Yitong River.

Water Quality Indicators	Range (mg/L)	Mean
pH	6.89–7.28	7.08 ± 0.13
COD	45.48–63.46	57.65 ± 6.28
NH_4^+-N	4.65–6.42	5.60 ± 0.68
TN	5.54–7.39	6.57 ± 0.71
TP	1.43–2.24	1.81 ± 0.31
SS	40.26–62.37	54.04 ± 8.04
DO	1.32–2.16	1.69 ± 0.28

COD—chemical oxygen demand; NH_4^+-N—ammonia nitrogen; TN—total nitrogen; TP—total phosphate; SS—suspended solids; DO—dissolved oxygen.

2.2. Ecological Engineering Design

The engineering process is shown in Figure 2. The wetland system was composed of FBCW, HSFCW, and SFCW in sequence. Firstly, the water of the Yitong River was introduced into FBCW by gravity from the upstream rubber. The FBCW mainly realized the pre-treatment function, introducing the adsorption carrier and suspended biological filler, which attached a large number of microbial flora, and formed a collaborative purification system of aquatic plants, porous substrate, and biofilm. This is to improve the degradation efficiency of organics, nitrogen, and phosphorus, reduce the pollution load for follow-up HSFCW, and improve the impact resistance of the system. Additionally, the FBCW can remove suspended solids (SS) effectively and reduce the risk of clogging of the HSFCW. The FBCW used the terrain elevation difference to design a multi-level water drop for reaeration, and created favorable conditions of the denitrification for HSFCW. The HSFCW was the core of the system, and formed the aerobic-anoxic-anaerobic environment, which contributed to the smooth progress of nitrification and denitrification, and thus, could effectively promote the removal of nitrogen in sewage. Falling water and reoxygenation by FBCW improved the concentration of DO, which was conducive to the transformation of $NH_4{}^+$-N and improved the nitrification reaction capacity in the front end of HSFCW. However, in the middle and back end of HSFCW, the anoxic and anaerobic environment, which for the DO was lower, could promote the denitrification process. Then, the effluent from the HSFCW entered the SFCW. The SFCW was rich in microtopography, which could improve the DO, further removed the pollutants of nitrogen and phosphorus, and provided a safety guarantee for the effluent reaching the standard. The effluent of SFCW flowed back to Yitong River. The system has played an effective collaborative role of FBCW, HSFCW, and SFCW, which reduced the risk of wetland blocking and improved the antipollution load capacity and operation stability. At last, the system improved the efficiency of pollutant purification.

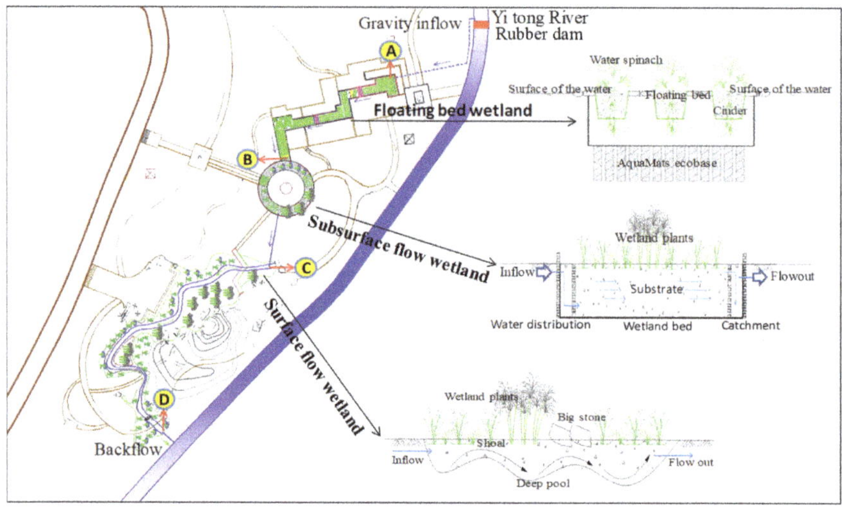

Figure 2. Processing flow and configuration.

The designed water volume was 100 m^3/d. According to the water quality monitoring results in 2014, the design of influent water was COD ≤ 70 mg/L, $NH_4{}^+$-N ≤ 8.0 mg/L, SS ≤ 60 mg/L. Therefore, the design of effluent water was COD ≤ 40 mg/L, $NH_4{}^+$-N ≤ 2.0 mg/L, SS ≤ 6.0 mg/L. The total surface hydraulic loading was approximately 0.10 m^3/m^2d. The total area of the wetlands was 1000 m^2. The hydraulic retention time was 3 d. Among them, the surface hydraulic loading of FBCW, HSFCW, and SFCW was 0.42 m^3/m^2d, 0.50 m^3/m^2d, and 0.18 m^3/m^2d, respectively. The area of FBCW, HSFCW, and SFCW was 240 m^2, 200 m^2, and 560 m^2, respectively. The hydraulic retention time of FBCW, HSFCW, and SFCW was 0.83 d, 1.17 d, and 1.0 d, respectively.

The FBCW comprised 25 floating beds, with 3.00 × 2.00 m per floating bed. The total area of floating beds was 150 m^2, and a coverage rate was 62.5% in FBCW. The top of the floating beds was set up with 3000 planters with diameters of 10cm and heights of 30 cm (20 planters/m^2). They were filled with absorbent volcanic rocks as the substrate to a height of 25 cm. Two water spinach plants were planted in every planter. AquaMats (Hangzhou Zijing Envrionmental Engineering CO.,LTD., Hangzhou, China), a biofiltration media, were hung 20 cm apart from one another under the floating beds. The height of the media was 30 cm.

The depth of the HSFCW was 1.2 m. The SSFCW was filled with natural volcanic rocks with a particle size of 10–20 cm and a filling height of 1.0 m. *Lythrum salicaria L.*, a wetland plant, was planted in a density of 25 clumps/m^2 and three plants/clump. The distribution zone and the catchment zone were established front and after end in the wetland to ensure the even distribution of water in the system and to maximize the wetland efficiency.

The designed depth of SFCW was 0.1 m to 0.8 m. The SFCW was comprised of shallow waters, deep pools, sandbars, and ecological islands, with *Lythrum salicaria L.*, *Iris pseudacorus L.*, *Salix integra*, and other wetland plants. It was also supplemented by ecological bag revetment and slope bank with vegetation.

2.3. Data Collection and Analysis

The water quality was continuously monitored from April to October 2016. Samples were taken from 9:00–11:00 on the 17th of each month. Sampling staff used clean plastic samplers to take water samples at the inflow and outflow of each wetland along the direction of the flow. The sampling points are shown in Figure 2: sampling point A (system inflow), sampling point B (outflow of the FBCW), sampling point C (outflow of the HSFCW), and sampling point D (outflow of SFCW). Three samples of 500 mL water were taken in clean plastic sampling bottles at each sampling point as duplicate samples. A water quality analysis of the samples was conducted within 24 h. The main monitoring indicators included pH, SS, DO, COD, NH_4^+-N, total nitrogen (TN) and total phosphate (TP). The methods were based on the Water and Wastewater Monitoring and Analysis Method (Fourth Edition) [34].

3. Results

The system was operated from April to June in 2016. The effectiveness of the removal of various pollutants gradually stabilized, and the water quality of the system improved significantly. Monitoring data showed that the SS concentration of inflow fluctuated between 46.38 and 89.26 mg/L (the mean was 64.84 mg/L) while the system was operating. The SS concentration of the outflow was 4.35–5.68 mg/L (the mean was 4.90 mg/L). The removal rate of SS was 87.86%–94.80% (the mean was 91.90%). From June to September, the root system of the plants in the floating-bed wetland continued to grow. After the formation of a stable biofilm on the hanging biofiltration media, the filtration by the plant roots and the interception and adsorption of the biofiltration media became prominent. The maximum removal rate of SS was 94.80%. As seen in Figure 3, the average removal rate of SS by the FBCW was 80.24%. The removal of SS in the HSFCW increased by 9.45% on average compared to the outflow of the FBCW. The removal of SS in the SFCW increased by 2.21% on average compared to the outflow of HSFCW. These results indicated that the removal of SS was mainly completed in the FBCW. The HSFCW and the SFCW enhanced the removal of SS.

During the operation of the system, the DO concentration of the inflow was basically maintained at 1.43–1.86 mg/L. The average DO concentration of the system inflow was 1.59 mg/L. The DO concentration of the outflow was 3.86–5.58 mg/L. The average DO concentration of the outflow was 4.79 mg/L. The DO concentration of the outflow met the water quality standard of Class IV water in the Environmental Quality Standards for Surface Water (GB3838-2002) [33]. As seen in Figure 4, oxygenation by the waterfalls in the FBCW significantly increased the DO concentration of the outflow. The consumption of the HSFCW brought the DO concentration back down to a lower level at the outflow of the HSFCW. Then, the effect of the SFCW gradually improved the DO concentration of the system outflow. The DO concentration in general showed a pattern of first increasing, then decreasing, and increasing again. The DO concentration of the outflow increased by an average of 2.03 times. The DO concentration of the outflow reached a maximum of 5.58 mg/L. The increase in DO was mainly attributed to the waterfalls in the FBCW. The terrain changes in the SFCW, such as deep ponds, shallow waters, jumps, and waterfalls, adjusted the form and speed of the flow.

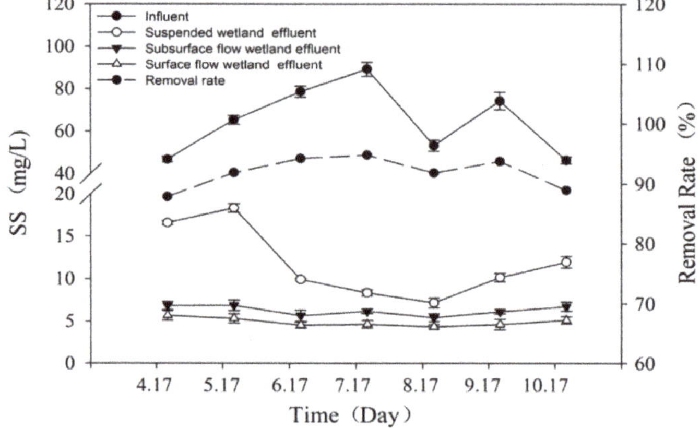

Figure 3. Concentration of SS in the inflow and outflow and the removal rate.

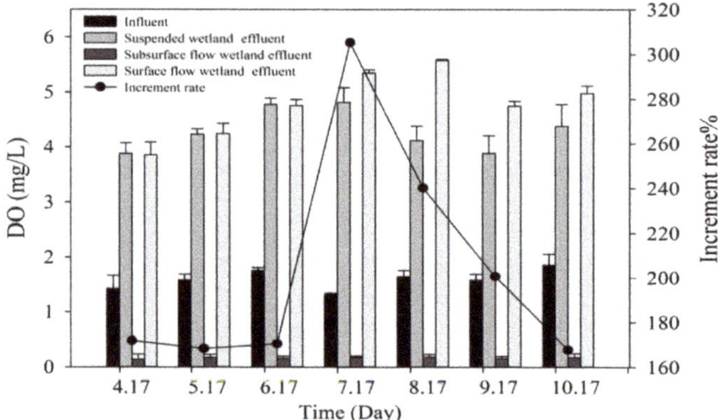

Figure 4. DO concentration of the inflow and the increase rate.

During the operation of the system, the COD concentration of the inflow fluctuated between 62.35 and 83.48 mg/L. The average COD concentration of the system inflow was 71.57 mg/L. The average COD concentration of the outflow was 20.57 mg/L. The COD removal rate of the system was 38.14%–82.05%, and the average removal rate was 74.79%. As seen in Figure 5, the COD removal rate was the lowest in April at 38.14%. It rose to 61.29% in May. With the exception of April, May, and October, when the system just started and reached the end of operation, the COD removal rate could reach more than 80%. The COD concentration of the outflow in June to September met the water quality standard of Class III water in the Environmental Quality Standards for Surface Water (GB3838-2002) [33]. In this system, the contribution of the HSFCW to the COD removal was close to 54.42%. The floating-bed wetland contributed approximately 34.56% to the COD removal. The contribution of the SFCW was only approximately 11.02% because the COD removal by the HSFCW was mainly achieved through microbial degradation and adsorption by plants and filtration media. Compared to the COD removal rate in the HSFCW, the COD removal rate in the FBCW was lower due to the shorter retention time. Additionally, the SFCW was at the downstream end of the system where the COD concentration of the inflow was lower. In addition, the SFCW had a limited ability to remove pollutants thus, its COD removal rate was not high.

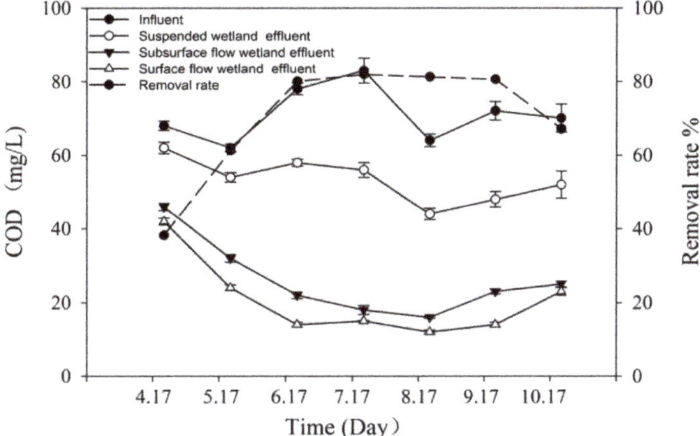

Figure 5. COD concentration of the inflow and outflow and the removal rate.

During the operation of the system, the NH_4^+-N concentration of the inflow was 5.65–8.26 mg/L. The average NH_4^+-N concentration of the inflow was 6.98 mg/L. The average NH_4^+-N concentration of the outflow was 1.25–4.66 mg/L, and the average removal rate was 80.90%. As seen in Figure 6, excluding the initial stage in April and May and the late operation season in October, the average NH_4^+-N concentration of the outflow from June to September was 1.35 mg/L. The removal rate of NH_4^+-N was higher than 78%. The NH_4^+-N concentration of the outflow met the water quality standard of Class IV water in the Environmental Quality Standards for Surface Water (GB3838-2002) [33]. The removal of NH_4^+-N in the entire wetland system was largely achieved in the HSFCW, which contributed to 79.20% in the system purification.

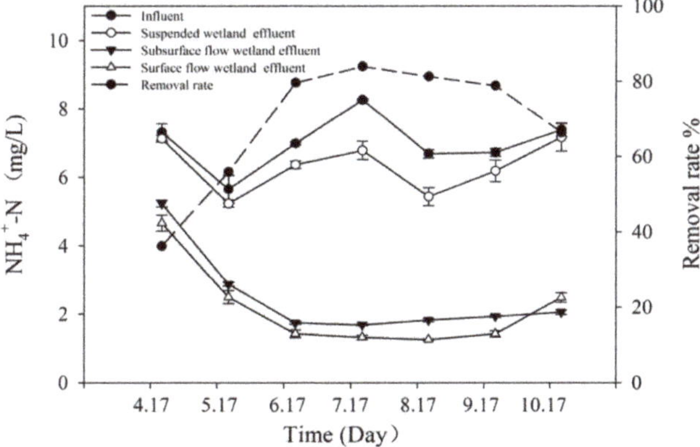

Figure 6. Concentration of NH_4^+-Nin the inflow and outflow and the removal rate.

During the operation of the system, the concentration of TN in the inflow was 6.43–8.89 mg/L. The average TN concentration of the system inflow was 7.73 mg/L. The TN concentration in the outflow was 1.75–6.28 mg/L, and the average removal rate was 71.12%. As seen in Figure 7, in April, the TN concentration in the outflow was 6.28 mg/L, and the removal rate was only 19.00%. In May, the TN concentration in the outflow was 4.16 mg/L, and the removal rate was approximately 35%. From June to September, the average TN concentration of the outflow was 1.83 mg/L, and the removal rate reached approximately 76.75% and was stable. In October, the TN concentration in the outflow was 2.64 mg/L, and the removal rate decreased. The TN removal rate by the FBCW, the HSFCW, and the SFCW was 12.35%, 64.66%, and 23.00%, respectively

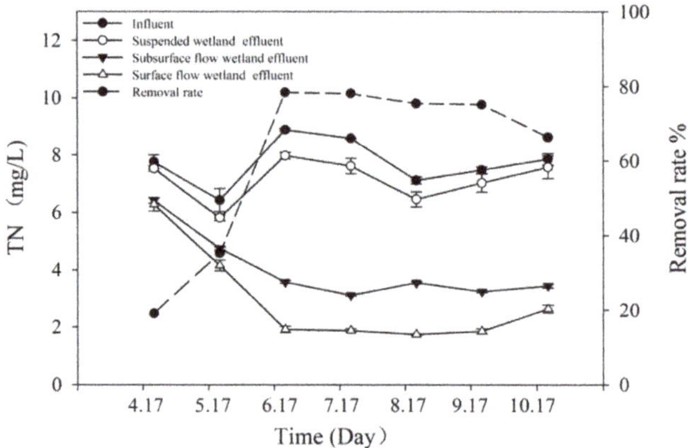

Figure 7. Concentration of TN in the inflow and outflow and the removal rate.

During the operation of the system, the concentration of TP in the inflow was 1.48–1.69 mg/L. The average TP concentration in the inflow was 1.56 mg/L. The TP concentration in the outflow was 0.32–0.38 mg/L. The average TP concentration in the outflow was 0.35 mg/L. As seen in Figure 8, the removal rate of TP reached an average of 78.39%. The TP concentration in the outflow met the water quality standard of Class V water in the Environmental Quality Standards for Surface Water (GB3838-2002) [33]. The system had a strong removal capacity for TP, which was mainly attributed to the HSFCW. The removal of TP by the FBCW and the SFCW was not significant, with an average removal rate of 12.86% and 5.43%, respectively. This might be related to the form of phosphorus and the mechanism and method of removal. It is generally believed that the removal of phosphorus in artificial wetlands is mainly achieved by adsorption to the substrates. Therefore, it is possible that the substrate structures of the HSFCW played a major role in the adsorption and retention of phosphorus. The FBCW and SFCW only removed a small portion of the suspended phosphorus.

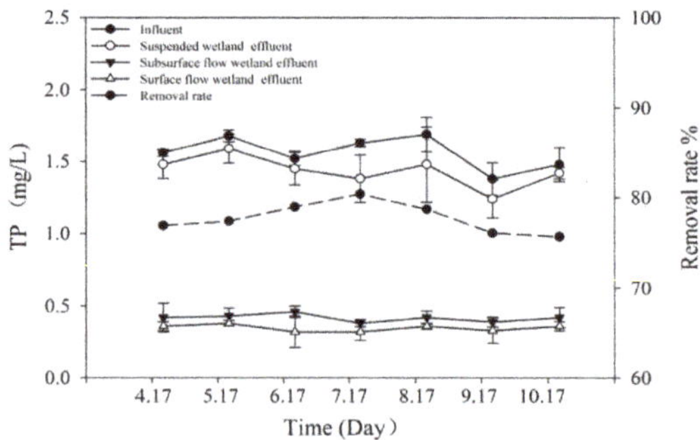

Figure 8. Concentration of TP in the inflow and outflow and the removal rate.

4. Discussion

Due to the cold winter in northern China, artificial wetlands cannot function properly in the winter. As a result, the wetland system stops operating in November every year. It restarts in March to April the next year when the weather becomes warmer again. When the system first started operating, the low temperature inhibited microbial activities, and the plants were not growing yet. The system mainly relied on substrate adsorption and interception to purify the water, and the efficiency of pollutant removal was low. As shown in Figures 5–7, in April, with an average temperature of 3–12 °C, the removal rates of COD and $NH_4{}^+$-N were less than 40%, and the removal rate of TN was only 19.00%. In May, the removal rates of COD, $NH_4{}^+$-N and TN increased to 61.29%, 55.92%, and 35.30%, respectively, when the average temperature was 12–23 °C and the wetland plants started to grow. Between June to September, the temperature gradually rose and the average temperature was 16–29 °C, when the plants grew up. As the operating time increased, the operation of the system gradually became stable and the microbial activities increased significantly. The integrated effect of the wetland plants, substrate, and microorganisms was fully functional, thus, the purification efficiency was best. The both removal rates of COD and $NH_4{}^+$-N were over 80%, when the TN could exceed 75%. After September, the temperature gradually decreased, and the plants stopped growing and started to wither. When the average temperature was 2–4 °C in October, the removal rates of COD, $NH_4{}^+$-N, and TN decreased to 67.14%, 66.39%, and 66.41%, respectively. The average temperature was −7–2 °C in November, and the wetland system progressively stopped running. Existing studies have shown that the nitrogen removal in a wetland is achieved by microbial activities, substrate adsorption, and plant absorption [35,36]. Low temperatures seriously affect the efficiency of water purification in artificial wetlands [37,38], because denitrifying microorganisms in artificial wetlands are active only when the temperature is above 15 °C [39,40]. In addition, during the operation of the system from June to September, despite the poor water quality of the inflow, the wetland system still operated stably. The removal rates of COD and NH4+-N still exceeded 75%. The COD concentration of the outflow was ≤20 mg/L, and the NH4+-N concentration of the outflow was ≤1.5 mg/L. The system showed a good effectiveness for pollutant removal.

The purpose of having a combination of different types of wetlands is to balance the advantages and disadvantages of each type of wetland [41]. Such a design can break through the restrictions imposed by the technical characteristics and environmental conditions of a single wetland type, enhance the system's resistance to loading fluctuations, and improve the efficiency of the water treatment [42]. According to a survey of more than 100 constructed wetlands in the United States, nearly half of them became clogged within five years of operation [43]. It is believed that the accumulation of SS is the main cause of blockage in the constructed wetlands [44]. Studies have also shown that horizontal and vertical flow wetlands can remove more than 90% of organic matter and almost all N and P if effective pretreatment facilities are set up prior to the constructed wetlands or if the wetland area is large enough [45]. Compared to a settling basin or an oxidation pond, using a FBCW as a pretreatment unit for the HSFCW can greatly improve the removal rate of SS and the purification efficiency of COD, nitrogen, and phosphorus. It can reduce the pollution load for the subsequent wetland process, improve the resistance of the loading capacity of the entire wetland system, and reduce the risk of clogging the HSFCW. The HSFCW is the core of the multiple wetlands and the key to effective denitrification in the system. The nitrification and denitrification by the microorganisms are the important processes for denitrification in constructed wetlands [46]. Golterman found that there must be both aerobic and anaerobic conditions in a wetland to achieve the nitrification and denitrification processes and the effective removal of nitrogen [47]. The nitrification process is aerobic. $NH_4{}^+$-N is mainly removed by nitrification. The DO concentration should normally be maintained above 2.0 mg/L for nitrification. Although the FBCW was not efficient in removing $NH_4{}^+$-N, the waterfalls designed by using the height differences in the terrain enabled oxygenation. This maintained a higher concentration of DO before water went into the HSFCW and promoted nitrification in the upper stream of the HSFCW and the conversion and removal of $NH_4{}^+$-N in the HSFCW. Denitrification requires anaerobic environments.

The DO level in the middle and lower streams of the HSFCW decreased rapidly. A hypoxic or anaerobic environment was formed, which promoted the process of denitrification and achieved an ideal removal of TN from the system. The removal of TN by a single type of wetland is not effective, because it is difficult to form both aerobic and hypoxic environments [31]. Multiple wetlands can effectively improve the level of DO in the outflow through the FBCW and form an anaerobic environment in the middle and lower streams in the HSFCW, which are conducive to denitrification in the wetland and critical to the effective denitrification of the system. Compared to the HSFCW, the average removal rate of NH_4^+-N was much lower in the SFCW in the lower stream of the system because the concentration of NH_4^+-N was lower in the inflow. However, the removal rate of TN in the SFCW was significant, accounting for 23% of the total removal of TN in the wetland system. This might be because the SFCW was usually in a saturated state during the operation and had considerable oxygen consumption, which was conducive to denitrification. In addition, the DO concentration going into the SFCW was very low. The multiple changes in the terrain between deep pools and shallow waters helped form a local environment with aerobic-hypoxic-anaerobic microcirculations, which promoted denitrification and the removal of TN. In addition, the seasonal temperature differences are large in northern China, and seasonal changes have a great impact on the effectiveness of constructed wetlands [25]. Multiple constructed wetlands can effectively make up for the disadvantage of a single wetland type [41], prolong the operation time of wetlands, and improve the efficiency of wetland purification.

For improved sewage treatment methods such as cyclic-activated sludge technology, activated charcoal adsorption, biofilms, and contact oxidation, the construction cost is usually $300–600 per ton of sewage, and the operating cost is approximately $0.07–0.15 per ton of sewage. Not only are the construction and operation costs high, but the management is also complicated. The construction cost of a constructed wetland system is approximately $100–180 per ton of sewage. The operating cost is approximately $0.014–0.056 per ton of sewage. Compared with the traditional biochemical treatment, constructed wetlands have the advantages of low investment, low operating cost, simple operation and management, and high ecological efficiency [48]. Therefore, these systems have great advantages in the restoration of polluted rivers. In this study, the total investment in construction was only $0.12 million, which was $120 per ton of water, and was about 1/5–1/3 of the investment in the construction of traditional sewage treatment. The system used gravity flow, no power was needed for operation, and the operating cost only involved regular maintenance. The maintenance costs were covered by the park maintenance budget, and the park created a part-time maintenance job for the wetlands. The part-time staff's wage was 70 dollar/month, and the system maintenance costs were only 0.023 dollar per ton of water, equal to 1/6–1/3 of the traditional sewage treatment. Compared with generic constructed wetland systems, the system described in this paper has a lower operating cost. Currently, the economic factor has become an important parameter for the selection of sewage treatment methods. Compared with operating constructed wetlands and other river water purification technologies in China and other countries, the multiple wetland technology in this study has obvious advantages and practicality.

5. Conclusions

(1) The one-year continuous monitored results of the sequential wetland system of FBCW, HSFCW, and SFCW showed that the system had a good effect on pollution removal. The system was mature and stable between June and September, and the average removal rates of COD, NH_4^+-N, TN, TP, and SS were 74.79%, 80.09%, 71.12%, 78.39%, and 91.90%, respectively. The FBCW had the highest removal rate of SS (80.24%), which could effectively prevent the blocking of HSFCW. The HSFCW contributed 79.20%, 64.64%, and 81.71% to the total removal of NH_4^+-N, TN, and TP, respectively. However, the SFCW could further remove TN, which contributed 23.00% to the total TN. The construction cost of the system was only $120 per ton of water, and the operation cost was only 0.023 dollar per ton of water. The system was an economic and efficient river water purification technology.

(2) The temperature directly affected the operation and treatment of the sequential series system of the constructed wetlands. The system was in a stage of operation and growth in April and May,

while the removal efficiency of pollutants was relatively low, and then gradually increases over time. The system was in a stable stage from June to September; thus, the removal efficiency of pollutants was the highest, and both removal rates of COD and NH_4^+-N were over 80%, while the TN exceeded 75%. The removal efficiency of pollutants decreased with the temperatures in the October. Then, the system stopped running when the water froze. In a temperate continental monsoon climate zone, the system cannot run stably all year; thus, we should take the measures of heat preservation and regulation of cold-adapted microorganism to prolong the operation time and improve purification efficiency.

(3) The sequential wetland system of FBCW, HSFCW, and SFCW by using the nearshore land on both sides of the river is an effective measure to treatment of polluted river water. The system can give full play to the advantages of different wetland and realize the maximization of the overall function. The system has high pollutant purification efficiency, good operation stability, strong anti-pollution load capacity, lower construction investment and operation cost, and simple management and maintenance. The technology can provide a reference for the restoration of polluted water bodies in urban rivers in temperate continental monsoon climate zone.

(4) The system integrates water purification, ecological restoration, landscape beautification, and entertainment. Based on the concept of "close to nature, multi-functional and sustainable" in the ecological restoration, the research gives a great theory and practice to promote the health of the river environment and the sustainable development of regional ecology economy society.

Author Contributions: Conceptualization, L.S.; methodology, L.S.; software, H.J.; validation, X.B., X.Z. and C.H.; formal analysis, X.B.; investigation, H.J.; resources, X.B.; data curation, X.B.; writing—original draft preparation, X.B.; writing—review and editing, X.B. and X.Z.; visualization, L.S. and J.Z.; supervision, L.S. and C.H.; project administration, X.B. and Z.W.; funding acquisition, C.H. and H.J. All authors have read and agreed to the published version of the manuscript.

Funding: This research was founded by the Foundation of Jilin Scientific and Technological Development Project, grant numbers (20190103137JH; 20190701048GH); the National Natural Science Foundation of China, grant numbers: 41901116.

Conflicts of Interest: The authors declare no conflicts of interest.

References

1. Magdaleno, A.; Puig, A.; Cabo, L.D.; Salinas, C. Water Pollution in an Urban Argentine River. *B. Environ. Contam. Tox.* **2001**, *67*, 0408–0415. [CrossRef] [PubMed]
2. Cao, J.X.; Sun, Q.; Zhao, D.H.; Xu, M.Y.; Shen, Q.S.; Wang, D.; Wang, Y.; Ding, S.M. A critical review of the appearance of black-odorous water bodies in China and treatment methods. *J. Hazard. Mater.* **2020**, *385*, 121511. [CrossRef] [PubMed]
3. Zhong, F.; Wu, J.; Dai, Y.; Xiang, D.; Deng, Z.; Cheng, S. Responses of water quality and phytoplankton assemblages to remediation projects in two hypereutrophic tributaries of Chaohu Lake. *J. Environ. Manag.* **2019**, *248*, 109276. [CrossRef] [PubMed]
4. Zhu, W.-B.; Wang, H.-X.; Liu, C.; Zhang, J.; Liang, S. Improvement of River Water Quality by Aeration: WASP Model Study. *Environ. Sci.* **2015**, *36*, 1326–1331.
5. Fathollahzadeh, H.; Kaczala, F.; Bhatnagar, A.; Hogland, W. Significance of environmental dredging on metal mobility from contaminated sediments in the Oskarshamn Harbor, Sweden. *Chemosphere* **2015**, *119*, 445–451. [CrossRef] [PubMed]
6. Chen, M.; Cui, J.; Lin, J.; Ding, S.; Gong, M.; Ren, M.; Tsang, D.C. Successful control of internal phosphorus loading after sediment dredging for 6, years: A field assessment using high-resolution sampling techniques. *Sci. Total Environ.* **2018**, *616*, 927–936. [CrossRef]
7. Jing, L.; Bai, S.; Li, Y.; Peng, Y.; Wu, C.; Liu, J.; Liu, G.; Xie, Z.; Yu, G. Dredging project caused short-term positive effects on lake ecosystem health: A five-year follow-up study at the integrated lake ecosystem level. *Sci. Total Environ.* **2019**, *686*, 753–763. [CrossRef]
8. Kasprzak, P.; Gonsiorczyk, T.; Grossart, H.P.; Hupfer, M.; Koschel, R.; Petzoldt, T.; Wauer, G. Restoration of aeutrophic hard-water lake by applying an optimised dosage of poly-aluminium chloride (PAC). *Limnologica* **2018**, *70*, 33–48. [CrossRef]

9. Jin, X.; Bi, L.; Lyu, T.; Chen, J.; Zhang, H.; Pan, G. Amphoteric starch-based bicomponent modified soil for mitigation of harmful algal blooms (HABs) with broad salinity tolerance: Flocculation, algal regrowth, and ecological safety. *Water Res.* **2019**, *165*, 115005. [CrossRef]
10. Vezzulli, L.; Pruzzo, C.; Fabiano, M. Response of the bacterial community to in situ bioremediation of organic-rich sediments. *Mar. Pollut. Bull.* **2004**, *49*, 740–751. [CrossRef]
11. Jarvie, H.P.; Neal, C.; Warwick, A.; White, J.; Neal, M.; Wickham, H.D.; Hill, L.K.; Andrews, M.C. Phosphorus uptake into algal biofilms in a lowland chalk river. *Sci. Total. Environ.* **2002**, *282*, 353–373. [CrossRef]
12. Cai, X.; Yao, L.; Sheng, Q.; Jiang, L.; Wang, T.; Dahlgren, R.A.; Deng, H. Influence of a biofilm bioreactor on water quality and microbial communities in a hypereutrophic urban river. *Environ. Technol.* **2019**, *1*, 1–34. [CrossRef] [PubMed]
13. Juang, D.F.; Tsai, W.P.; Liu, W.K. Treatment of polluted river water by a gravel contact oxidation system constructed under riverbed. *Int. J. Environ. Sci. Technol.* **2008**, *5*, 305–314. [CrossRef]
14. Li, L.; Xie, S.; Zhang, H.; Wen, D. Field experiment on biological contact oxidation process to treat polluted river water in the Dianchi Lake watershed. *Front. Environ. Sci. Eng. China* **2009**, *3*, 38–47. [CrossRef]
15. Xu, J.; Lo, S.L. A case study on gravel-contact-aeration-oxidation system to treat the combined sewage and rainwater flowing into Keelung River, Taiwan. *Desalin. Water Treat.* **2019**, *156*, 32–37. [CrossRef]
16. Li, L.; Visvanathan, C. Membrane technology for surface water treatment: advancement from microfiltration to membrane bioreactor. *Rev. Environ. Sci. Bio/Technology* **2017**, *16*, 737–760. [CrossRef]
17. Sun, W.; Li, C.; Dong, B.; Chu, H. A Pilot-scale diatomite membrane bioreactor for slightly polluted surface water treatment. *CLEAN-Soil, Air, Water* **2018**, *46*, 1700117. [CrossRef]
18. Rushton, B.T.; Bahk, B.M. Treatment of stormwater runoff from row crop farming in Ruskin, Florida. *Water Sci. Technol.* **2001**, *44*, 531–538. [CrossRef]
19. Jurczak, T.; Wojtal-Frankiewicz, A.; Kaczkowski, Z.; Oleksińska, Z.; Bednarek, A.; Zalewski, M. Restoration of a shady urban pond-The pros and cons. *J. Environ. Manag.* **2018**, *217*, 919–928. [CrossRef]
20. Lu, X.M.; Huang, M.S. Nitrogen and phosphorus removal and physiological response in aquatic plants under aeration conditions. International journal of Environ. *Int. J. Environ. Sci. Technol.* **2010**, *7*, 665–674. [CrossRef]
21. Yu, L.; Zhang, Y.; Liu, C.; Xue, Y.; Shimizu, H.; Wang, C.; Zou, C. Ecological responses of three emergent aquatic plants to eutrophic water in Shanghai, P. R. China. *Ecol. Eng.* **2019**, *136*, 134–140. [CrossRef]
22. Zhou, X.; Wang, G. Nutrient concentration variations during Oenan the javanica growth and decay in the ecological floating bed system. *J. Environ. Sci.* **2010**, *22*, 1710–1717. [CrossRef]
23. Zheng, L.; Yang, R.; Wang, H.; Song, J. Study on water remediation and uptake ability of nitrogen and phosphorus by plants using combined ecological floating bed. *Chinese J. Environ. Eng.* **2013**, *7*, 2153–2159.
24. Samal, K.; Kar, S.; Trivedi, S. Ecological floating bed (EFB) for decontamination of polluted water bodies: Design, mechanism and performance. *J. Environ. Manag.* **2019**, *251*, 109550. [CrossRef]
25. Jing, S.R.; Lin, Y.F. Seasonal Effects on Ammonium Nitrogen Removal by Constructed Wetlands Treating Polluted River Water in Southern Taiwan. *Environ. Pollut.* **2004**, *127*, 291–301. [CrossRef]
26. Jia, H.; Sun, Z.; Li, G. A four-stage constructed wetland system for treating polluted water from an urban river. *Ecol. Eng.* **2014**, *71*, 48–55. [CrossRef]
27. Liu, G.; He, T.; Liu, Y.; Chen, Z.; Li, L.; Huang, Q.; Xie, Z.; Xie, Y.; Wu, L.; Liu, J. Study on the purification effect of aeration-enhanced horizontal subsurface-flow constructed wetland on polluted urban river water. *Environ. Sci. Pollut. Res.* **2019**, *26*, 12867–12880. [CrossRef]
28. Spencer, K.L.; Dewhurst, R.E.; Penna, P. Potential impacts of water injection dredging on water quality and ecotoxicity in Limehouse Basin, River Thames, SE England, UK. *Chemosphere* **2006**, *63*, 509–521. [CrossRef]
29. Whalen, P.; Toth, L.; Koebel, J.; Strayer, P. Kissimmee River restoration: a case study. *Water Sci. Technol.* **2002**, *45*, 55–62. [CrossRef]
30. Kondratyev, S. A system for ecological and economic assessment of the use, preservation and restoration of urban water bodies: St Petersburg as a case study. *Intern. Assoc. Hydrol. Sci.* **2003**, *281*, 327–333.
31. Vymazal, J. Removal of nutrients in various types of constructed wetlands. *Sci. Total. Environ.* **2007**, *380*, 48–65. [CrossRef] [PubMed]
32. Konnerup, D.; Koottatep, T.; Brix, H. Treatment of domestic wastewater in tropical, subsurface flow constructed wetlands planted with Canna and Heliconia. *Ecol. Eng.* **2009**, *35*, 248–257. [CrossRef]

33. *Environmental Quality Standards for Surface Water*; GB 3838-2002; China Environmental Science Press: Beijing, China, 2002.
34. Editorial board of water and wastewater monitoring and analysis methods, State Environmental Protection Administration. *Water and Wastewater Monitoring and Analysis Method*, 4th ed.; China Environmental Science Press: Beijing, China, 2002.
35. Tanner, C.C.; Kadlec, R.H.; Gibbs, M.M.; Sukias, J.P.; Nguyen, M. Nitrogen processing gradients in subsurface-flow treatment wetlands—influence of wastewater characteristics. *Ecol. Eng.* **2002**, *18*, 499–520. [CrossRef]
36. Pang, Y.; Zhang, Y.; Yan, X.; Ji, G. Cold Temperature Effects on Long-Term Nitrogen Transformation Pathway in a Tidal Flow Constructed Wetland. *Environ. Sci. Technol.* **2015**, *49*, 13550–13557. [CrossRef] [PubMed]
37. Steer, D.; Fraser, L.; Boddy, J.; Seibert, B. Efficiency of small constructed wetlands for subsurface treatment of single-family domestic effluent. *Ecol. Eng.* **2002**, *18*, 429–440. [CrossRef]
38. Zhi, W.; Yuan, L.; Ji, G.; He, C. Enhanced Long-Term Nitrogen Removal and Its Quantitative Molecular Mechanism in Tidal Flow Constructed Wetlands. *Environ. Sci. Technol.* **2015**, *49*, 4575–4583. [CrossRef]
39. Kuschk, P.; Wiesner, A.; Kappelmeyer, U.; Weißbrodt, E.; Kastner, M.; Stottmeister, U. Annual cycle of nitrogen removal by a pilot-scale subsurface horizontal flow in a constructed wetland under moderate climate. *Water Res.* **2003**, *37*, 4236–4242. [CrossRef]
40. Zhang, Y.; Cheng, Y.; Yang, C.; Luo, W.; Zeng, G.; Lu, L. Performance of system consisting of vertical flow trickling filter and horizontal flow multi-soil-layering reactor for treatment of rural wastewater. *Bioresour. Technol.* **2015**, *193*, 424–432. [CrossRef]
41. Vymazal, J. Horizontal sub-surface flow and hybrid constructed wetlands systems for wastewater treatment. *Ecol. Eng.* **2005**, *25*, 478–490. [CrossRef]
42. Fang, T.; Bao, S.; Sima, X.; Jiang, H.; Zhu, W.; Tang, W. Study on the application of integrated eco-engineering in purifying eutrophic river waters. *Ecol. Eng.* **2016**, *94*, 320–328. [CrossRef]
43. Perfler, R.; Laber, J. Constructed wetlands for rehabilitation and reuse of surfacewater in tropical and subtropical areas-first results from small scale Plots using vertical flow beds. *Water Sci.Technol.* **1999**, *40*, 155–162. [CrossRef]
44. Langergraber, G.; Haberl, R.; Laber, J.; Pressl, A. Evaluation of substrate clogging processes in vertical flow constructed wetlands. *Water Sci. Technol.* **2004**, *48*, 25–34.
45. Luederitz, V.; Eckert, E.; Lange-Weber, M.; Lange, A.; Gersberg, R.M. Nutrient removal efficiency and resource economics of vertical flow and horizontal flow constructed wetlands. *Ecol. Eng.* **2001**, *18*, 157–171. [CrossRef]
46. Lee, C.; Fletcher, T.D.; Sun, G. Review: nitrogen removal in constructed wetland systems. *Eng. Life Sci.* **2009**, *9*, 11–22. [CrossRef]
47. Golterman, H.L. *The Chemistry of Phosphate and Nitrogen Compounds in Sediments*; Springer: Dordrecht, The Netherlands, 2004.
48. Kurzbaum, E.; Kirzhner, F.; Armon, R. Improvement of water quality using constructed wetland systems. *Rev. Environ. Health* **2012**, *27*, 59–64. [CrossRef] [PubMed]

Article

The Effect of Submergence and Eutrophication on the Trait's Performance of *Wedelia Trilobata* over Its Congener Native *Wedelia Chinensis*

Ahmad Azeem [1], Jianfan Sun [1,*], Qaiser Javed [1], Khawar Jabran [2] and Daolin Du [1,*]

[1] School of the Environment and Safety Engineering, Jiangsu University, Zhenjiang 212013, China; ahmadazeem631@yahoo.com (A.A.); qjaved_uaf@yahoo.com (Q.J.)

[2] Department of Plant Production and Technologies, Faculty of Agricultural Sciences and Technologies, Nigde Omer Halisdemir University, Nigde 51200, Turkey; khawarjabran@gmail.com

* Correspondence: zxsjf@ujs.edu.cn (J.S.); ddl@ujs.edu.cn (D.D.); Tel.: +86-178-5100-3075 (J.S.); +86-187-9600-7971 (D.D.)

Received: 20 January 2020; Accepted: 23 March 2020; Published: 26 March 2020

Abstract: Climate change and artificial disturbance may lead to increased submergence and eutrophication near a riparian zone and the shift of terrestrial plants into a riparian zone. In this study, the responses of terrestrial invasive *Wedelia trilobata* (WT) and congener native *Wedelia chinensis* (WC) plants were examined under submergence and eutrophication. A greenhouse experiment was conducted in which ramets of WT and WC were investigated under two levels of submergence (S1 and S2) and three levels of nutrients (N1, N2 and N3) along with two cultures (mono and mixed). Submergence (S) did not affect the morphological traits of both the species but nutrients (N), culture (C) and their interaction, along with submergence, had a significant effect on the morphological traits of both the species. The growth of WC under high submergence and high nutrients was decreased compared with low nutrients (N1, N2) but WT maintained its growth in monoculture. In mixed culture, low submergence (S1) and low nutrients (N1, N2) made WC more dominant but high submergence (S2) and high nutrients (N3) made WT more successful than WC due to its high phenotypic plasticity and negative effect of competition intensity. It was concluded that both species survive and grow well under submergence and eutrophication, but high submergence and eutrophication provide better conditions for WT to grow well.

Keywords: submergence; eutrophication; invasive-native competition; growth rate; morphological traits

---

1. Introduction

Globalization causes many invasive plant species to develop their wild population, where they were not introduced before [1,2]. Many of these species have successfully spread in the introduced ranges and now they occupy large areas because of decreased environment suitability for native plant species [3,4] and increased resource availability due to global change that helps invasive plant species [5,6]. Native diversity is under threat due to increases in precipitation and nutrients enrichment [7], and disturbance in environmental conditions makes invasive plant species more dominant, especially in wetland or a riparian zone [8]. Subsequently, better traits' performance under different environmental conditions makes invasive plant species successful and allows them to spread in the introduced habitat [9–12].

High phenotypic plasticity seems to benefit plants to cope with environmental changes and get the advantage of increases in resources, and invasive plant species have been better on it compared to native plant species [13,14]. The success of invasive plant species in a new region was faster than native plant species because of higher phenotypic plasticity, and invasive plant species could outperform native plant species under interspecific competition [15,16]. It was found in a meta-analysis that with increasing resources, invasive plant species were more dominant than native plant species because of higher phenotypic plasticity and a more negative response under interspecific competition [17–19]. Several studies were conducted to test this hypothesis based on phenotypic plasticity and relative competition intensity against the environmental factors, such as water, temperature, nutrients, soil types, light and CO_2 [2,20]. While environmental variability is affected by environmental factors, such as water fluctuation, nutrient variations and light availability as well [21]. Therefore, it is important to test these responses of invasive and native plant species to variability, other than changes in mean environmental factors.

There are many global change environmental factors, i.e., temperature, light, water, CO_2 and nutrients, that are affecting the growth of plant species, but in riparian zones or wetland, the major factors are submergence and eutrophication [22]. Submergence is the modern form of flood, in which shoots of the plant are under water. A discharge from riversides, canals and dams will lead to submergence near these areas and plants will face a submergence period. The strategies in which plants cope with submergence have been reported in many studies [23–25]. Mostly, plants use two main techniques, "escape or quiescence", to cope with submergence [26,27]. In the escape technique, plants re-establish a relationship with the atmosphere to increase shoot elongation and the number of leaves [28,29]. Under the quiescence technique, plants limit or cease their growth under water to tolerate submergence [26,30]. Submergence imposes considerable stress or negative effects on all plant characteristics because of a decrease in energy and carbohydrates [31] and also severely influences the photosynthesis [32]. Although, adaptive growth of invasive plant species assists in submergence and complete or incomplete submergence significantly inhibits their growth [25,27].

Eutrophication on the water surface is another major global environmental factor, which has negative effects on aquatic ecosystems. Eutrophication is the major threat in the 21st century because it reduces the diversity of native plant species in the aquatic ecosystems and boosts invasive plant species [16,33]. Eutrophication is the syndrome associated with an excess of macronutrients that boost invasive plant growth to outcompete the native plant species [34]. Eutrophication has been increasing in the aquatic ecosystems because of agriculture and urban activities (i.e., point and nonpoint source pollution). Eutrophication creates excessive nutrient enrichment on the water surface that promotes algae biomass accumulation on the water surface, which enhances invasive plant growth [35]. The native plant species diversity has declined in the aquatic ecosystems due to the occurrence and expansion of certain macrophytes [20,36]. Invasive plant species require light, water and nutrients for growth development, and CO_2 for photosynthesis and oxygen for respiration [34]. Eutrophication reduces the stress of submergence by increasing shoots of invasive plant species that ultimately get more light, CO_2 and oxygen for photosynthesis and also increase belowground biomass under submergence to outcompete the native competitor [27].

Nutrients travel from terrestrial to aquatic ecosystems through the primary source, i.e., surface runoff [35]. Flooding and excess rain boost the surface runoff, collecting more nutrients into aquatic ecosystems. By considering the negative effect of submergence and the positive effect of a high amount of nutrients on plant growth, we hypothesized that increasing nutrients under a submergence condition might alleviate submergence stress, that is, plants would grow well under high nutrients compared to low nutrients when the plants were submerged. We also hypothesized that invasive plant species under submergence and high nutrient conditions would be growing faster than native plant species when growing together, even they were introduced from terrestrial regions to aquatic regions.

To test this hypothesis, a greenhouse experiment has been conducted. The species chosen for this study belong to the *Asteraceae* family. *Wedelia trilobata* (WT), an annual invasive plant species, and *Wedelia chinensis* (WC), which is a congener native species in China. In China, initially, WT was introduced as a groundcover species, but it spread rapidly amongst gardens, roadsides, agricultural fields and near riversides [37,38]. Furthermore, WT was mostly found in the arid and semi-arid regions and also in the wetland area of Hainan province of China, which indicated that water fluctuation and nutrient enrichment made WT successful in these habitats [39]. WC, the native congener of WT, is mostly used as a medicinal plant, but the growth rate of WC is very slow compared to WT [39,40]. The introduction of WT in the riparian or wetland areas is not well understood, so studying the effects of submergence and eutrophication on WT is a crucial step toward understanding the mechanism of its invasion and predicting its development trends in the riparian and wetland regions. For this purpose, plants of WT and WC were tested with two levels of submergence combined with three levels of nutrient concentrations to attain a eutrophication effect, although biomass allocation and morphological traits were measured.

2. Materials and Methods

This study was conducted in the school of Environmental and Safety Engineering at Jiangsu University, Zhenjiang China, from September to mid-November 2019 (32.20° N, 119.45° E). Ramets of *Wedelia trilobata* (WT) and *Wedelia chinensis* (WC) were collected from the vicinity of the study site. These collected ramets of WT and WC were prepared in the seedling tray with sand as the growing medium. These trays were placed in a greenhouse that had a 25 ± 5 °C temperature with 60% relative humidity. The distilled water was supplied every day and Hoagland solution was supplied weekly. When these ramets had two fully expanded leaves, then the ramets were transformed into a plastic pot (12 cm diameter and 7 cm height) containing sand as a growing medium, and these pots were placed in a bin (80 × 40 × 20 cm) for the mesocosm experiment under greenhouse. There were two cultures used: monoculture and mixed culture of both species, with six replicates of each treatment. These ramets were allowed to be settled in these new habitats for one week, and then divided equally into two submergence and three nutrients groups, in order to stimulate the naturally occurring submergence and eutrophication in a wetland ecosystem. Submergence levels differed from water levels in the bins (7 and 14 cm, labeled as S1 and S2, respectively) so that the actual water level at which WT and WC were growing was 0 and 7 cm. Eutrophication levels were made according to the nutrient concentration in the Yangzi river [41]. There were three nutrient concentrations with Ck, 10- and 100-fold dilution of modified concentration respectively, as N1, N2 and N3. Submergence, nutrient concentrations and the plant culture method have been explained in Figure 1. There was a total of 108 pots. After applying treatments to each bin, tap water was added every day to keep the submergence level constant and renewal of the solution was done once a week. After four weeks of treatments, the plants were harvested.

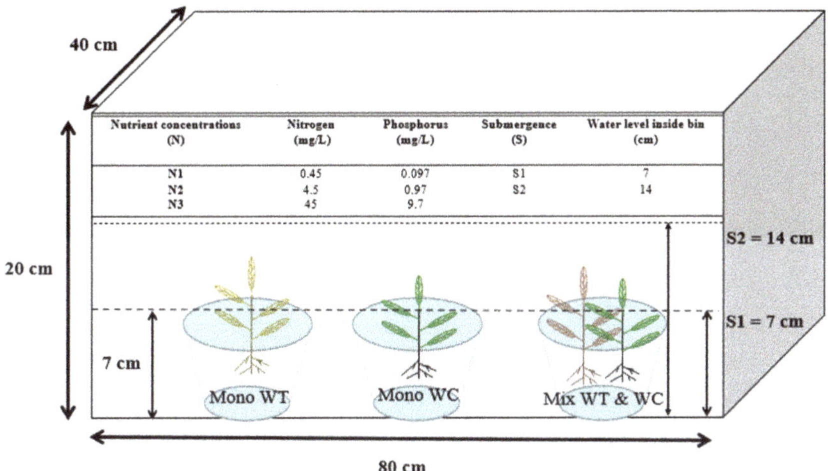

Figure 1. Diagram illustrating experiment treatments and culture methods.

2.1. Morphological Traits Measurement

Plant height and root length were measured with the help of a ruler at the end of the experiment. The number of leaves and nodes per plant were counted carefully. The leaf area of each treatment plant with six replicates was measured with the help of ImageJ software. The plants were oven-dried at ≤80 °C for 48 h to measure total biomass of each treatment plant; afterwards, plants were divided into leaves, stems and roots. We also calculated leaves mass ratio, stem mass ratio and root-to-shoot ratio of each treatment. Specific leaf area (SLA) was measured with leaf area to dry mass.

The plasticity index (P_I) of plant height and root length of WT and WC under different submergence and eutrophication treatments were calculated by using the following equation [42,43]:

$$P_I = \frac{\text{Maximum Value} - \text{Minimum Value}}{\text{Maximum Value}}$$

where maximum and minimum value represent the maximum and minimum values of plant height and root length of WT and WC under all treatments to determine plasticity indices of plant height and root length for different species and for different treatments, respectively. The value of P_I ranged from 0 to 1, where 0 represents zero and 1 represents a higher P_I.

2.2. Relative Interaction Index

The relative interaction index (RII) between invasive and native competitor was measured according to Reference [44]:

$$RII = \frac{(M_c - M_i)}{(M_c + M_i)}$$

where Mc is the mass of WT or WC under mixed culture and M_i is the mass of WT or WC in the one plant pots (monoculture). RII represents the interspecific interaction between invasive and native plant species. RII ranges from −1 to 1. If RII < 0, then interaction intensity has a negative impact, if RII > 0, then interaction intensity shows a positive impact on plant, but if RII = 0, then there is no impact of a competitor.

2.3. Statistical Analyses

Before analyses of the data, the normality and homogeneity of the variances of all data were checked with the help of the Shapiro–Wilk Normality test and Levene's test. A three-way analysis of variance (ANOVA) along with submergence, nutrient and plant culture as the main factors, was performed to determine the main effects and interaction effects on plant morphological traits, total biomass accumulation and total biomass allocation. Furthermore, a post hoc Student–Newman–Keuls test, $p < 0.05$, was used for multiple comparison to measure significant differences between treatments. A one-way ANOVA was used to assess the interactive effect of submergence and nutrient levels on morphological traits, total biomass accumulation and total biomass allocation. All analyses were conducted in SPSS (Version 22.0, IBM, USA) and graphs were made in the software Origin pro9.

3. Results

3.1. Biomass Accumulation and Allocation

Plants of both species survived under submergence and nutrient concentrations. Submergence (S), nutrient (N) and culture (C) had different effects on different traits of both plant species. S had non-significant results on total biomass (Table 1, F = 1.797, $p > 0.05$) but N and C and the interactions S × N × C were significant (Table 1, F = 16.347, $p < 0.01$). In monoculture, total biomass of WT was significantly higher than WC in all submergence and nutrient treatments (Figure 2a), but under low submergence and low nutrients (S1.N1 and S1.N2), total biomass of WC was higher than WT, while under high submergence and all nutrients levels (S2.N1, S2.N2 and S2.N3), total biomass of WT was higher than WC. This indicated that increasing submergence along with nutrients made WT more dominant (Figure 2b).

Table 1. Analysis of variance (ANOVA) results of biomass accumulation and allocation under submergence and nutrient concentrations.

Sources	Total Biomass		Root-to-Shoot Ratio		Stem Mass Ratio		Leaf Mass Ratio	
	F	*p*	F	*p*	F	*p*	F	*p*
S	1.797	0.186	4.403	<0.05	1.585	0.214	38.525	<0.01
N	10.303	<0.01	17.383	<0.01	17.051	<0.01	16.281	<0.01
C	10.830	<0.01	18.101	<0.01	16.588	<0.01	23.454	<0.01
S * N	61.212	<0.01	2.901	0.065	0.683	0.510	21.778	<0.01
S * C	62.137	<0.01	0.252	0.859	1.110	0.354	11.904	<0.01
N * C	24.333	<0.01	5.347	<0.01	4.288	<0.01	21.006	<0.01
S * N * C	16.347	<0.01	6.120	<0.01	5.003	<0.01	16.088	<0.01

Significance level: $p < 0.05$, $p < 0.01$. Note: S represented submergence, N represented nutrient, C represented culture.

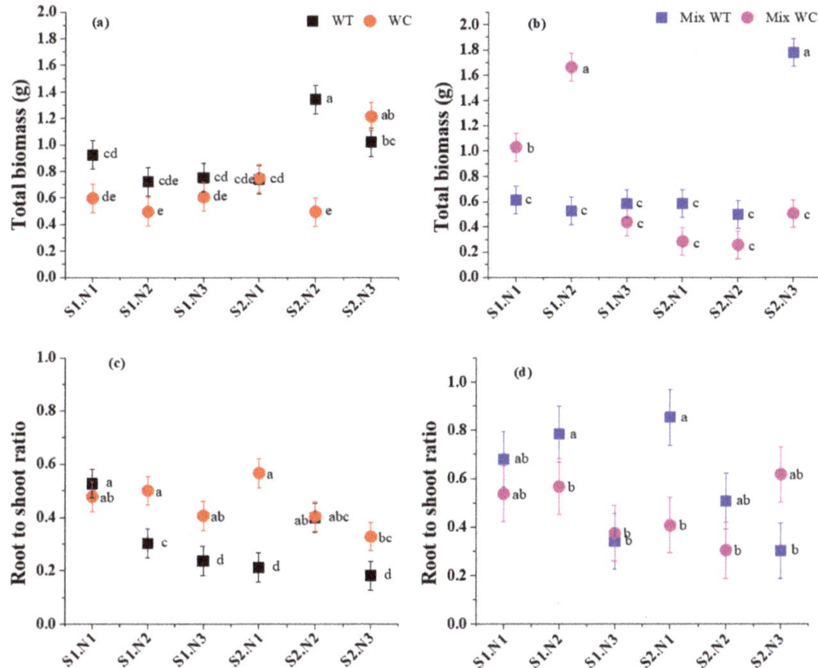

Figure 2. Total biomass and root-to-shoot ratio of *Wedelia trilobata* (WT) and *Wedelia chinensis* (WC) under mono and mixed culture with six replicates, (**a**) total biomass of both species under monoculture, (**b**) total biomass of both species under mixed culture, (**c**) root-to-shoot ratio of both species under monoculture, (**d**) root to shoot ratio of both species under mixed culture. Different letters indicate the significant difference between different treatments of submergence and nutrient concentrations according to ANOVA and Student–Newman–Keuls, $p < 0.05$.

For biomass allocation, the effects of S, N and C on the root-to-shoot ratio were independent (Table 1, $p < 0.05$). Their interaction, S × N × C, was significant (Table 1, F = 6.120, $p < 0.01$) but the interaction S × N was non-significant (Table 1, $p > 0.05$). The root-to-shoot ratio of WC was higher than WT in monoculture in every treatment (Figure 2c), while in mixed culture, higher submergence with low nutrients (S2.N2, S2.N1) and lower submergence with higher nutrients concentration (S1.N3), the root-to-shoot ratio of WT was higher than WC (Figure 2d).

However, S and S × N had a non-significant effect on stem mass ratio (Table 1, $p > 0.05$) but N, C and their interaction, N × C, had significant effects on stem mass ratio (Table 1, $p < 0.01$). However, the combined effect of S × N × C was significant on stem mass ratio (Table 1, F = 5.003, $p < 0.01$). The stem mass ratio of WT was higher than WC in monoculture because of the fast growth of WT below water. In mixed culture, the stem mass ratio of WT was higher in both submergence levels along with high nutrient concentration (S1.N3, S2.N3). However, WC had a higher stem mass ratio in low submergence along with low nutrients (S1.N1) (Figure 3b), illustrating that low submergence and nutrients made WC more dominant than WT in competition.

S, N, C and their interaction, S × N × C, had significant effects on leaves mass ratio (Table 1, $p < 0.01$). The leaves mass ratio of WC was higher at both submergence and low nutrient concentration levels (S1.N2 and S2.N2) but lower at higher nutrient concentration along with both submergence levels (S1.N3 and S2.N3). This was because increasing nutrients and submergence reduced the leaves mass ratio of WC, but opposite results were found under the same treatment for WT in monoculture (Figure 3c). Under competition, WT had a higher leaves mass ratio in all treatments of submergence and nutrients compared to WC (Figure 3d). According to these results, mass allocation of WT was higher than WC in higher submergence and higher nutrients levels.

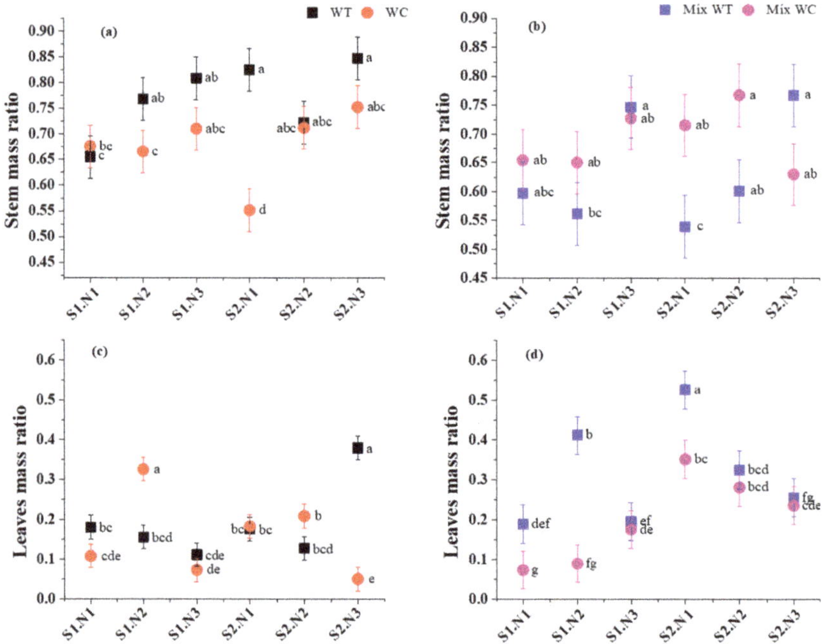

Figure 3. Stem mass ratio and leaves mass ratio of *Wedelia trilobata* and *Wedelia chinensis* under mono and mixed culture with six replicates. (**a**) stem mass ratio under monoculture of both species, (**b**) stem mass ratio under mixed culture of both species, (**c**) leaves mass ratio under monoculture of both species, (**d**) leaves mass ratio under mixed culture of both species. Different letters indicate the significant difference between different treatments of submergence and nutrient concentrations according to ANOVA and Student–Newman–Keuls, $p < 0.05$.

3.2. Morphological Traits

S, N, C and their interactions, S × N, S × C, N × C and S × N × C, were significant (Table 2, $p < 0.05$). Plant height of WC was more than WT in monoculture under high submergence and high nutrient (S1.N3). While under low submergence and low nutrients conditions (S1.N2) plant height of WT was higher than WC (Figure 4a). Plant height of WC in mixed culture was more than WT in low submergence along with low nutrients (S1.N1, S1.N2). Under low submergence with high nutrients (S1.N3) and high submergence with all nutrient concentrations (S2.N1, S2.N2 and S2.N3), WT has more height than WC because increasing submergence and high nutrients make WT more destructive than WC under competition (Figure 4b).

The root length of both species was not affected by S but was effected by N and C (Table 2, $p > 0.05$, $p < 0.05$), while their interactions, S × N and S × N × C, had a significant effect on both plant species (Table 2, $p < 0.01$). Root length of WC under competition was higher than WT in low submergence along with low nutrients (S1.N1 and S1.N2), while for the rest of all other levels, WT root length was taller than WC due to higher nutrients along with submergence, which facilitate WT (Figure 4d). In monoculture, root length of WT was higher, indicating more biomass allocation below-ground to make it more successful in submergence and high nutrient conditions (Figure 4c).

Table 2. ANOVA results of morphological traits.

Sources	Plant Height		Root Length		Number of Nodes		Number of Leaves		Specific Leaf Area	
	F	*p*	F	*P*	F	*p*	F	*p*	F	*p*
S	4.590	<0.05	3.291	0.076	0.053	0.818	4.900	<0.05	3.617	0.063
N	52.098	<0.01	26.067	<0.01	53.560	<0.01	3.878	<0.05	16.504	<0.01
C	14.013	<0.01	26.582	<0.01	10.329	<0.01	5.196	<0.01	5.450	<0.01
S * N	17.396	<0.01	7.899	<0.01	17.613	<0.01	9.233	<0.01	11.578	<0.01
S * C	3.491	<0.05	5.923	<0.01	3.076	<0.05	5.463	<0.01	1.694	0.181
N * C	9.070	<0.01	2.490	<0.05	4.209	<0.01	1.685	0.145	1.093	0.380
S * N * C	8.875	<0.01	9.458	<0.01	3.676	<0.01	1.619	0.163	5.693	<0.01

Significance level: $p < 0.05$, $p < 0.01$. Note: S represented submergence, N represented nutrient, C represented culture.

Submergence had no effect on the number of nodes but nutrient and culture had significant results (Table 2, $p > 0.05$, $p < 0.01$). Their interactions, S × N and S × N × C, also had significant findings for number of nodes (Table 2, F = 3.676, $p < 0.01$). WC had a greater number of nodes within monoculture, under high submergence and high nutrient (S1.N3). While under low submergence and low nutrients conditions (S1.N2) number of nodes of WC was not higher) (Figure 5a).

In mixed culture, WT had more number of nodes than WC in all treatment but only under both submergence and low nutrients (S1.N1, S2.N2), WC had more nodes than WT (Figure 5b).

S, N and C had independent effects on number of leaves per plant but their interaction, S × N × C, had no effect (Table 2, F = 1.619, $p > 0.05$). Monoculture WC had a greater number of leaves per plant compared to WT (Figure 5c). Mixed culture WT had a greater number of leaves per plant except in both submergence and low nutrients (S1.N1 and S2.N2) (Figure 5d). According to the results, submergence and nutrient richness helped both species to grow fast, but under competition, WT gets more benefit than WC.

The interactions of S × N and S × N × C had significant effects on specific leaf area (SLA) of both plant species (Table 2, $p < 0.01$) but independently, S has no effect on SLA (Table 2, F = 3.617, $p > 0.05$). SLA of both plant species enhanced with increasing submergence and amount of nutrients. For both monoculture and mixed culture, mostly, SLA of WC was higher than WT but under high submergence along with high nutrients (S2.N3), WT had a higher SLA (Figure 5e,f).

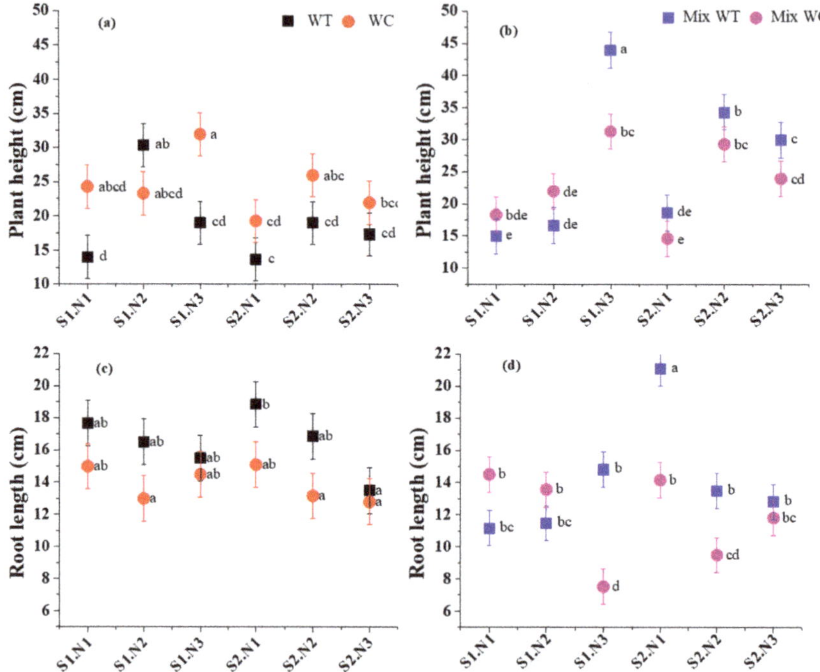

Figure 4. Plant height and root length of *Wedelia trilobata* and *Wedelia chinensis* under mono and mixed culture with six replicates, (**a**) plant height of both species under monoculture, (**b**) plant height of both species under mixed culture, (**c**) root length of both species under monoculture, (**d**) root length of both species under mixed culture. Different letters indicate the significant difference between different treatments of submergence and nutrient concentrations according to ANOVA and Student–Newman–Keuls, $p < 0.05$.

3.3. Plasticity Index

Plasticity index of plant height was significantly affected by N ($F = 9.489$, $p < 0.01$) and the interaction of S × N ($F = 8.454$, $p < 0.01$) but S ($F = 0.033$, $p > 0.05$) and C ($F = 1.705$, $p > 0.05$) had no impact, while root length plasticity index was significantly affected by S, N, C and their interaction ($p < 0.01$). Plasticity indices of plant height and root length of both plant species were significantly different in every treatment (Figure 6). Under monoculture, the plasticity index of plant height was higher for WC compared to WT, but for the plasticity index of root length, WT was higher than WC due to more below-ground biomass allocation. For treatments with high nutrients and both submergence levels like S1.N3, S2.N2 and S2.N3, WT showed more plastic than WC and this made WT more successful under competition in nutrient-rich and submergence conditions (Figure 6).

3.4. Relative Interaction Index

The relative competition interaction index of both species was significantly affected by S ($F = 8.290$, $p < 0.01$) and the interaction of S × N ($F = 3.876$, $p < 0.05$) but not significantly by N ($F = 0.013$, $p > 0.05$). Competition interactions were mostly negative for both of the species (Figure 7). The competition intensity of both species was significantly increased with enhancing submergence and nutrients. The competition intensity of WT was higher than WC in high submergence and high nutrients.

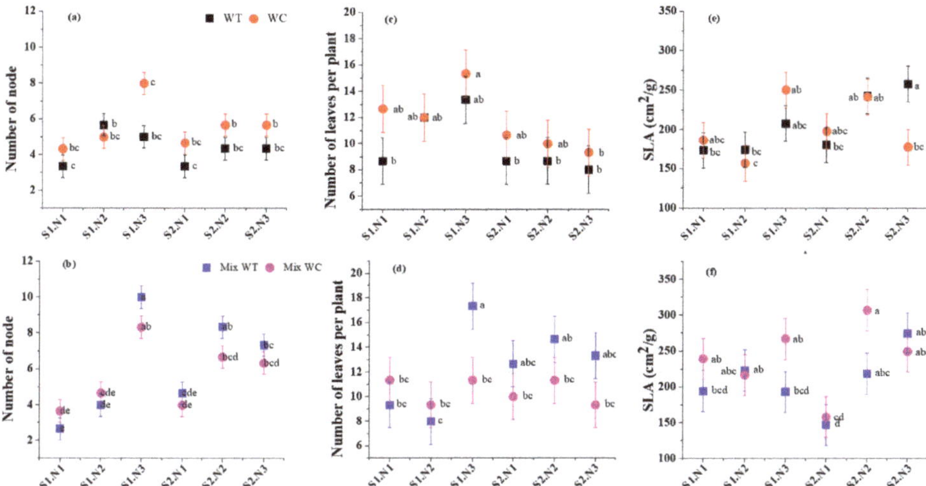

Figure 5. Number of node, number of leaves per plant and specific leaf area (SLA) of *Wedelia trilobata* and *Wedelia chinensis* under mono and mixed culture with six replicates, (**a**) number of nodes of both species under monoculture, (**b**) number of nodes of both species under mixed culture, (**c**) number of leaves per plant of both species under monoculture, (**d**) number of leaves per plant of both species under mixed culture, (**e**) SLA of both species under monoculture, (**f**) SLA of both species under mixed culture. Different letters indicate the significant difference between different treatments of submergence and nutrient concentrations according to ANOVA and Student–Newman–Keuls, $p < 0.05$.

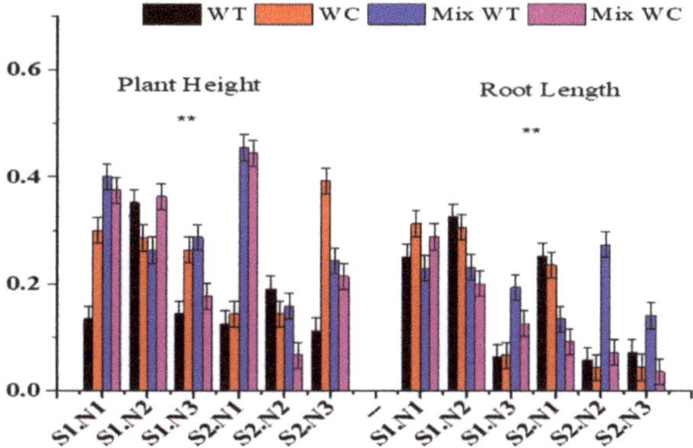

Figure 6. Plasticity index of plant height and root length of *Wedelia trilobata* and *Wedelia chinensis* under mono and mixed culture with six replicates. Submergence and nutrient treatments were significantly different according to ANOVA, ** $p < 0.05$.

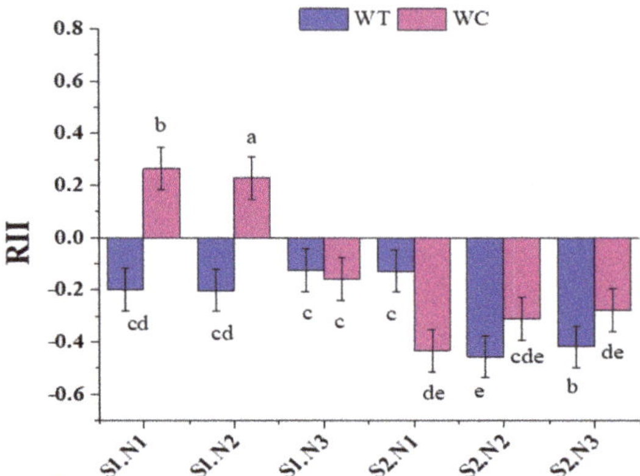

Figure 7. Relative interaction index of *Wedelia trilobata* and *Wedelia chinensis* under competition. Different letters indicate the significant difference between different treatments of submergence and nutrient concentrations according to ANOVA and Student–Newman–Keuls, $p < 0.05$.

4. Discussion

4.1. Biomass Response under Submergence and Eutrophication

The results of this study confirmed our hypothesis, that increasing nutrient concentrations can promote the growth of both the plant species; however, this occurred only when the amount of nutrient concentrations attained high eutrophic status [27]. Submergence (S) along with low nutrients concentration (S1.N1 and S1.N2) decreased the growth of WT under mixed and mono culture compared with WC; however, its inhibitory effects were much greater than the elevation effects of enhancing nutrient concentrations along with submergence (Figure 2). Similar results were found in previous studies [23,24,30]. While it can be realized that little eutrophication and a low submergence level would not promote the growth of WT, because invasive plant species, especially clonal plants, like to grow in nutrient-rich habitats [45], that is why high amounts of eutrophication and submergence would increase the growth of WT under mono and mixed culture (Figure 2).

Biomasses of many plants was increased or maintained under submergence [46] (Table 1, Figure 2) because plants used the quiescence technique to tolerate submergence [26]. WC maintained a higher root-to-shoot ratio in all treatments under monoculture, which was the plants strategy to cope with submergence and eutrophication [36], while WT maintained a better stem mass ratio under all treatments in monoculture, because adventitious roots were a tolerance stratagem of WT to survive these conditions and also provided structural stability to maintain sexual reproduction [7,24]. Total biomass of both species was increased under both submergence and three nutrient levels' interaction because more than 65% of plants of both species were above the water surface due to plant height elongation [47]. Submerged plants have less biomass compared to plants above the water surface, which agreed with our finding [48]. Submergence along with nutrient concentrations increased biomass allocation, plant height and compensatory growth of invasive plant species, when nutrient concentrations reached eutrophication [24,49,50]. WT showed its destructive behavior under competition, as biomass, root-to-shoot ratio and stem mass ratio of WT was higher than WC under higher levels of submergence and nutrients (S2.N3), while with low submergence along with low nutrients (S1.N1, S1.N2), WC had higher growth (Figures 2 and 3). This demonstrated that WC has the ability to bear with a low shortage of oxygen that was created due to submergence and eutrophication under competition [51]. It was also indicated that low nutrient concentration does not alleviate the effect of submergence particularly on invasive plant species [30], while increasing nutrient concentrations and submergence gave more success to WT with a boost in its invasion because growth rate, photosynthetic capacity and phenotypic plasticity of invasive plant species were usually increased more intensively than those of native hydrophytes under nutrient-rich water [52,53].

4.2. Morphological Traits under Submergence and Eutrophication

Submergence (S) did not have significant effects on most of the morphological traits like root length, number of leaves per plant and SLA (Table 2, Figures 4 and 5), because submergence created a stressful environment due to which both native and invasive plant species suffered [24]. However, nutrients (N) and their interactions, S × N, S × N × C, had significant effects on the morphological traits of both species. This result agrees with our hypothesis, and several researchers have found that increasing nutrient concentrations overcomes the stress of submergence and encourages plants to grow under these conditions, especially for invasive plant species [7,23,27]. Plant heights of both the species were increasing under submergence and nutrients treatments (Table 2, Figure 4a,b). Plant height increment was the plants' response to submergence and high nutrients, to restore contact between stem, leaves and air above the water surface, which was the plants' approach to deal with submergence and eutrophication [24,27]. If the plant fails to reconnect with air, that certainly gives rise to carbohydrate depletion [54] and finally, causes the death of the plant [55]. Therefore, plant height increment under submergence and high nutrients conditions was seen to be a favorable trait of both species under these environmental conditions. In monoculture, WC had a higher number of nodes, leaves per plant and higher SLA compared with WT, but within treatments, WC had higher morphological traits at S1.N2, representing zero submergence (water surface equal to pot height) along with high nutrients that make WC successful, but an increasing submergence level along with nutrients decreased the growth of WC (Figures 4 and 5). This finding indicates that WC could bear oxygen deficiency and maintain its growth when it was growing alone but within competition, WC growth reduced because WT increased its below-ground biomass to capture the resources that cause two effects on WC, with one being oxygen deficiency and the other competition [27]. Number of nodes and leaves per plant were decreased with increasing submergence and nutrient concentrations of both species (Figure 5). It was consistent with the results that submergence and eutrophication reduced the number of nodes and number of leaves per plant [24,27]. Under interspecific competition, number of nodes and leaves per plant of WT was higher than WC at high submergence and high nutrients levels because of the allelochemicals' effect that reduced the growth of the neighboring plant [21].

Plant height of WC was higher under monoculture because S × N gives benefit to its fast growth, but in mixed culture, WT was taller than WC (Figure 4) due to higher phenotypic plasticity and the negative effect under the competition interaction index (Figures 6 and 7). High nutrients and submergence level gave the advantage to WT under competition because of higher resource acquisition ability and increased below-ground biomass along with the allelochemicals' effect [56]. Taller plants and longer roots made WT more dominant under higher submergence and nutrient levels (S2.N2, S2.N3) because WT creates oxygen deficiency [57]. Shorter height of WC under competition did not allow for easily regaining contact with air. This situation caused photosynthesis limitation [48], and also serious carbohydrate depletion [54]. The number of nodes, leaves per plant and SLA of WT under competition were higher in high submergence and high nutrients (S2.N2 and S2.N3) (Figure 5) because of the high plasticity index [27]. However, faster growth rate, photosynthetic capacity, metabolism enzyme activity, nutrients assimilation and higher phenotypic plasticity of many invasive plant species give a benefit in nutrient-rich waters over those of native hydrophytes [53,58]. High nutrients enhance biomass accumulation, compensatory growth, SLA [49,59] and increase total biomass, shoot length and total number of leaves per plant that increase the intensity of interspecific competition of invasive plants over native plants [59] that was found in this study (Figure 7).

5. Conclusions

Our results exhibited that biomass and morphological traits of WT and WC were increased to cope with submergence and eutrophication. Under low submergence and low nutrient levels (S1.N2), WC illustrated more growth; meanwhile, compared with high submergence and high nutrients, the growth of WC was decreased. Furthermore, under low submergence and low nutrient conditions (S1.N1, S2.N2), WT was not dominant but had more growth in high submergence and high nutrient levels (S2.N2 and S2.N3). WC has better morphological traits under monoculture compared to mixed culture, while WT was prominently successful under high submergence and high nutrients (S2.N2 and S2.N3) due to high phenotypic plasticity and better competition intensity. It was concluded that both species survived under submergence and eutrophication conditions. Environmental modeling suggested that artificial disturbance and change in climate will ensure submergence and eutrophication. The findings of this research contribute to the understanding of terrestrial plant response, when subjected to a riparian zone, because the increased below-ground and above-ground biomass under submergence and eutrophication helped them to capture the resources and outcompete their native competitor.

Author Contributions: Conceptualization, A.A. and Q.J.; methodology, A.A.; software, Q.J.; validation, J.S., K.J. and D.D.; formal analysis, A.A.; investigation, A.A.; resources, J.S.; data curation, Q.J.; writing—original draft preparation, A.A.; writing—review and editing, K.J.; visualization, Q.J; supervision, J.S.; project administration, D.D.; funding acquisition, D.D. All authors have read and agreed to the published version of the manuscript.

Funding: This research was funded by the State Key Research Development Program of China, grant number "2017YFC1200100" and the National Natural Science Foundation of China "31971427, 31770446 and 31570414".

Acknowledgments: This work was supported by the State Key Research Development Program of China (2017YFC1200100), the National Natural Science Foundation of China (31971427, 31770446 and 31570414), the Priority Academic Program Development of Jiangsu Higher Education Institutions (PAPD) and Jiangsu Collaborative Innovation Center of Technology and Material of Water Treatment.

Conflicts of Interest: The authors declare no conflict of interest

References

1. Seebens, H.; Essl, F.; Dawson, W.; Fuentes, N.; Moser, D.; Pergl, J.; Pyšek, P.; van Kleunen, M.; Weber, E.; Winter, M. Global trade will accelerate plant invasions in emerging economies under climate change. *Glob. Chang. Biol.* **2015**, *21*, 4128–4140. [CrossRef] [PubMed]
2. Van Kleunen, M.; Dawson, W.; Essl, F.; Pergl, J.; Winter, M.; Weber, E.; Kreft, H.; Weigelt, P.; Kartesz, J.; Nishino, M.; et al. Global exchange and accumulation of non-native plants. *Nature* **2015**, *525*, 100. [CrossRef] [PubMed]
3. Alpert, P. The advantages and disadvantages of being introduced. *Biol. Invasions* **2006**, *8*, 1523–1534. [CrossRef]
4. Guan, B.; Yu, J.; Cao, D.; Li, Y.; Han, G.; Mao, P. The ecological restoration of heavily degraded saline wetland in the Yellow River Delta. *CLEAN Soil Air Water* **2013**, *41*, 690–696. [CrossRef]
5. Davis, M.A.; Grime, J.P.; Thompson, K. Fluctuating resources in plant communities: A general theory of invasibility. *J. Ecol.* **2000**, *88*, 528–534. [CrossRef]
6. Blumenthal, D. Interrelated causes of plant invasion. *Science* **2005**, *310*, 243–244. [CrossRef]
7. Zhao, H.; Yang, W.; Xia, L.; Qiao, Y.; Xiao, Y.; Cheng, X.; An, S. Nitrogen-enriched eutrophication promotes the invasion of Spartina Alterniflora in coastal China. *CLEAN Soil Air Water* **2015**, *43*, 244–250. [CrossRef]
8. Vilà, M.; Espinar, J.L.; Hejda, M.; Hulme, P.E.; Jarošík, V.; Maron, J.L.; Pergl, J.; Schaffner, U.; Sun, Y.; Pyšek, P. Ecological impacts of invasive alien plants: A meta-analysis of their effects on species, communities and ecosystems. *Ecol. Lett.* **2011**, *14*, 702–708. [CrossRef]
9. Funk, J.L.; Standish, R.J.; Stock, W.D.; Valladares, F. Plant functional traits of dominant native and invasive species in mediterranean-climate ecosystems. *Ecology* **2016**, *97*, 75–83. [CrossRef]
10. Heberling, J.M.; Fridley, J.D. Resource-use strategies of native and invasive plants in Eastern North American forests. *New Phytol.* **2013**, *200*, 523–533. [CrossRef]
11. Jia, J.; Dai, Z.; Li, F.; Liu, Y. How will global environmental changes affect the growth of alien plants? *Front. Plant Sci.* **2016**, *7*, 1623. [CrossRef]

12. Van Kleunen, M.; Schlaepfer, D.R.; Glaettli, M.; Fischer, M. Preadapted for invasiveness: Do species traits or their plastic response to shading differ between invasive and non-invasive plant species in their native range? *J. Biogeogr.* **2011**, *38*, 1294–1304. [CrossRef]
13. Gratani, L. Plant phenotypic plasticity in response to environmental factors. *Adv. Bot.* **2014**. [CrossRef]
14. Nicotra, A.B.; Atkin, O.K.; Bonser, S.P.; Davidson, A.M.; Finnegan, E.J.; Mathesius, U.; Poot, P.; Purugganan, M.D.; Richards, C.L.; Valladares, F. Plant phenotypic plasticity in a changing climate. *Trends Plant Sci.* **2010**, *15*, 684–692. [CrossRef] [PubMed]
15. Richards, C.L.; Bossdorf, O.; Muth, N.Z.; Gurevitch, J.; Pigliucci, M. Jack of all trades, master of some? On the role of phenotypic plasticity in plant invasions. *Ecol. Lett.* **2006**, *9*, 981–993. [PubMed]
16. Liu, Y.; van Kleunen, M. Responses of common and rare aliens and natives to nutrient availability and fluctuations. *J. Ecol.* **2017**, *105*, 1111–1122. [CrossRef]
17. Davidson, A.M.; Jennions, M.; Nicotra, A.B. Do invasive species show higher phenotypic plasticity than native species and, if so, is it adaptive? A meta-analysis. *Ecol. Lett.* **2011**, *14*, 419–431. [CrossRef]
18. Palacio-López, K.; Gianoli, E. Invasive plants do not display greater phenotypic plasticity than their native or non-invasive counterparts: A meta-analysis. *Oikos* **2011**, *120*, 1393–1401. [CrossRef]
19. Liu, Y.; Oduor, A.M.O.; Zhang, Z.; Manea, A.; Tooth, I.M.; Leishman, M.R.; Xu, X.; van Kleunen, M. Do invasive alien plants benefit more from global environmental change than native plants? *Glob. Chang. Biol.* **2017**, *23*, 3363–3370. [CrossRef]
20. Gołdyn, H. Changes in plant species diversity of aquatic ecosystems in the agricultural landscape in West Poland in the last 30 years. *Biodivers. Conserv.* **2010**, *19*, 61–80. [CrossRef]
21. Parepa, M.; Fischer, M.; Bossdorf, O. Environmental variability promotes plant invasion. *Nat. Commun.* **2013**, *4*, 1604. [CrossRef] [PubMed]
22. Evtimova, V.V.; Donohue, I. Quantifying ecological responses to amplified water level fluctuations in standing waters: An experimental approach. *J. Appl. Ecol.* **2014**, *51*, 1282–1291. [CrossRef]
23. Chen, Y.; Zhou, Y.; Yin, T.-F.; Liu, C.-X.; Luo, F.-L. The invasive wetland plant Alternanthera philoxeroides shows a higher tolerance to waterlogging than its native congener Alternanthera sessilis. *PLoS ONE* **2013**, *8*, 81456. [CrossRef] [PubMed]
24. Fan, S.; Yu, H.; Liu, C.; Yu, D.; Han, Y.; Wang, L. The effects of complete submergence on the morphological and biomass allocation response of the invasive plant Alternanthera philoxeroides. *Hydrobiologia* **2015**, *746*, 159–169. [CrossRef]
25. Short, F.T.; Kosten, S.; Morgan, P.A.; Malone, S.; Moore, G.E. Impacts of climate change on submerged and emergent wetland plants. *Aquat. Bot.* **2016**, *135*, 3–17. [CrossRef]
26. Strange, E.; Hill, J.; Coetzee, J. Evidence for a new regime shift between floating and submerged invasive plant dominance in South Africa. *Hydrobiologia* **2018**, *817*, 349–362. [CrossRef]
27. Zhang, H.; Liu, J.; Chen, X.; Du, Y.; Wang, Y.; Wang, R. Effects of submergence and eutrophication on the morphological traits and biomass allocation of the invasive plant Alternanthera philoxeroides. *J. Freshw. Ecol.* **2016**, *31*, 341–349. [CrossRef]
28. Colmer, T.; Voesenek, L. Flooding tolerance: Suites of plant traits in variable environments. *Funct. Plant Biol.* **2009**, *36*, 665–681. [CrossRef]
29. Dalmolin, Â.C.; Dalmagro, H.J.; Lobo, F.D.A.; Junior, M.Z.A.; Ortíz, C.E.R.; Vourlitis, G.L. Effects of flooding and shading on growth and gas exchange of Vochysia divergens Pohl (Vochysiaceae) of invasive species in the Brazilian Pantanal. *Braz. J. Plant Physiol.* **2012**, *24*, 75–84. [CrossRef]
30. Luo, F.-L.; Nagel, K.A.; Scharr, H.; Zeng, B.; Schurr, U.; Matsubara, S. Recovery dynamics of growth, photosynthesis and carbohydrate accumulation after de-submergence: A comparison between two wetland plants showing escape and quiescence strategies. *Ann. Bot.* **2010**, *107*, 49–63. [CrossRef]
31. Webb, J.A.; Wallis, E.M.; Stewardson, M.J. A systematic review of published evidence linking wetland plants to water regime components. *Aquat. Bot.* **2012**, *103*, 1–14. [CrossRef]
32. Jackson, M.; Colmer, T. Response and adaptation by plants to flooding stress. *Ann. Bot.* **2005**, *96*, 501–505. [CrossRef]
33. Schindler, D.W. Recent advances in the understanding and management of eutrophication. *Limnol. Oceanogr.* **2006**, *51*, 356–363. [CrossRef]

34. O'Hare, M.T.; Baattrup-Pedersen, A.; Baumgarte, I.; Freeman, A.; Gunn, I.D.M.; Lázár, A.N.; Sinclair, R.; Wade, A.J.; Bowes, M.J. Responses of aquatic plants to eutrophication in rivers: A revised conceptual model. *Front. Plant Sci.* **2018**, *9*, 451. [CrossRef] [PubMed]
35. Smith, V.H.; Tilman, G.D.; Nekola, J.C. Eutrophication: Impacts of excess nutrient inputs on freshwater, marine, and terrestrial ecosystems. *Environ. Pollut.* **1999**, *100*, 179–196. [CrossRef]
36. Mäemets, H.; Palmik, K.; Haldna, M.; Sudnitsyna, D.; Melnik, M. Eutrophication and macrophyte species richness in the large shallow North-European Lake Peipsi. *Aquat. Bot.* **2010**, *92*, 273–280. [CrossRef]
37. Song, L.; Chow, W.S.; Sun, L.; Li, C.; Peng, C. Acclimation of photosystem II to high temperature in two Wedelia species from different geographical origins: Implications for biological invasions upon global warming. *J. Exp. Bot.* **2010**, *61*, 4087–4096. [CrossRef]
38. Talukdar, T.; Talukdar, D. Response of antioxidative enzymes to arsenic-induced phytotoxicity in leaves of a medicinal daisy, Wedelia chinensis Merrill. *J. Nat. Sci. Biol. Med.* **2013**, *4*, 383. [CrossRef]
39. Dai, Z.-C.; Fu, W.; Qi, S.-S.; Zhai, D.-L.; Chen, S.-C.; Wan, L.-Y.; Huang, P.; Du, D.-L. Different responses of an invasive clonal plant Wedelia trilobata and its native congener to gibberellin: Implications for biological invasion. *J. Chem. Ecol.* **2016**, *42*, 85–94. [CrossRef]
40. Talukdar, T.; Mukherjee, S.K. Comparative study of cypselas in three common species of Asteraceae. *Pleione* **2008**, *2*, 147–149.
41. Chen, Y.; Liu, R.; Sun, C.; Zhang, P.; Feng, C.; Shen, Z. Spatial and temporal variations in nitrogen and phosphorous nutrients in the Yangtze River Estuary. *Mar. Pollut. Bull.* **2012**, *64*, 2083–2089. [CrossRef] [PubMed]
42. Funk, J.L. Differences in plasticity between invasive and native plants from a low resource environment. *J. Ecol.* **2008**, *96*, 1162–1173. [CrossRef]
43. Lamarque, L.J.; Porte, A.J.; Eymeric, C.; Lasnier, J.-B.; Lortie, C.J.; Delzon, S. A test for pre-adapted phenotypic plasticity in the invasive tree Acer negundo L. *PLoS ONE* **2013**, *8*, 74239. [CrossRef] [PubMed]
44. Armas, C.; Ordiales, R.; Pugnaire, F.I. Measuring plant interactions: A new comparative index. *Ecology* **2004**, *85*, 2682–2686. [CrossRef]
45. Wilson, J.R.; Yeates, A.; Schooler, S.; Julien, M.H. Rapid response to shoot removal by the invasive wetland plant, alligator weed (Alternanthera philoxeroides). *Environ. Exp. Bot.* **2007**, *60*, 20–25. [CrossRef]
46. Mommer, L.; Visser, E.J. Underwater photosynthesis in flooded terrestrial plants: A matter of leaf plasticity. *Ann. Bot.* **2005**, *96*, 581–589. [CrossRef]
47. Wang, T.; Hu, J.; Gao, Y.; Yu, D.; Liu, C. Disturbance, trait similarities, and trait advantages facilitate the invasion success of Alternanthera Philoxeroides (Mart) Griseb. *CLEAN Soil Air Water* **2017**, *45*, 1600378. [CrossRef]
48. Pierik, R.V.; van Aken, J.; Voesenek, L. Is elongation-induced leaf emergence beneficial for submerged Rumex species? *Ann. Bot.* **2008**, *103*, 353–357. [CrossRef]
49. Ding, W.; Zhang, H.; Zhang, F.; Wang, L.; Cui, S. Morphology of the invasive amphiphyte Alternanthera philoxeroides under different water levels and nitrogen concentrations. *Acta Biol. Crac. Bot.* **2014**, *56*, 136–147. [CrossRef]
50. Zuo, S.; Ma, Y.; Shinobu, I. Differences in ecological and allelopathic traits among Alternanthera philoxeroides populations. *Weed Biol. Manag.* **2012**, *12*, 123–130. [CrossRef]
51. Sauter, M. Root responses to flooding. *Curr. Opin. Plant Biol.* **2013**, *16*, 282–286. [CrossRef] [PubMed]
52. Riis, T.; Lambertini, C.; Olesen, B.; Clayton, J.S.; Brix, H.; Sorrell, B.K. Invasion strategies in clonal aquatic plants: Are phenotypic differences caused by phenotypic plasticity or local adaptation? *Ann. Bot.* **2010**, *106*, 813–822. [CrossRef]
53. Yu, H.; Wang, L.; Liu, C.; Fan, S. Coverage of native plants is key factor influencing the invasibility of freshwater ecosystems by exotic plants in China. *Front. Plant Sci.* **2018**, *9*, 250. [CrossRef] [PubMed]
54. Kawano, N.; Ito, O.; Sakagami, J.-I. Morphological and physiological responses of rice seedlings to complete submergence (flash flooding). *Ann. Bot.* **2008**, *103*, 161–169. [CrossRef] [PubMed]
55. Bailey-Serres, J.; Voesenek, L. Flooding stress: Acclimations and genetic diversity. *Annu. Rev. Plant Biol.* **2008**, *59*, 313–339. [CrossRef] [PubMed]
56. Liu, G.; Yang, Y.-B.; Zhu, Z.-H. Elevated nitrogen allows the weak invasive plant Galinsoga quadriradiata to become more vigorous with respect to inter-specific competition. *Sci. Rep.* **2018**, *8*, 3136. [CrossRef] [PubMed]

57. Liu, D.; Wang, R.; Gordon, D.R.; Sun, X.; Chen, L.; Wang, Y. Predicting plant invasions following China's water diversion project. *Environ. Sci. Technol.* **2017**, *51*, 1450–1457. [CrossRef]
58. You, W.; Yu, D.; Xie, D.; Yu, L.; Xiong, W.; Han, C. Responses of the invasive aquatic plant water hyacinth to altered nutrient levels under experimental warming in China. *Aquat. Bot.* **2014**, *119*, 51–56. [CrossRef]
59. Wu, H.; Ding, J. Global change sharpens the double-edged sword effect of aquatic alien plants in China and beyond. *Front. Plant Sci.* **2019**, *10*, 787. [CrossRef]

Article

Optimization and Analysis of Zeolite Augmented Electrocoagulation Process in the Reduction of High-Strength Ammonia in Saline Landfill Leachate

Mohd Azhar Abd Hamid [1], Hamidi Abdul Aziz [1,2,*], Mohd Suffian Yusoff [1,2] and Sheikh Abdul Rezan [3]

1 School of Civil Engineering, Engineering Campus, Universiti Sains Malaysia, Nibong Tebal 14300, Pulau Pinang, Malaysia; mohdazhar@student.usm.my (M.A.A.H.); suffian@usm.my (M.S.Y.)
2 Solid Waste Management Cluster, Engineering Campus, Universiti Sains Malaysia, Nibong Tebal 14300, Pulau Pinang, Malaysia
3 School of Materials and Mineral Resources Engineering, Universiti Sains Malaysia, Nibong Tebal 14300, Pulau Pinang, Malaysia; srsheikh@usm.my
* Correspondence: cehamidi@usm.my; Tel.: +60-45996215; Fax: +60-45996906

Received: 15 December 2019; Accepted: 11 January 2020; Published: 16 January 2020

Abstract: This work examined the behavior of a novel zeolite augmented on the electrocoagulation process (ZAEP) using an aluminum electrode in the removal of high-strength concentration ammonia (3471 mg/L) from landfill leachate which was saline (15.36 ppt) in nature. For this, a response surfaces methodology (RSM) through central composite designs (CCD) was used to optimize the capability of the treatment process. Design-Expert software (version 11.0.3) was used to evaluate the influences of significant variables such as zeolite dosage (100–120 g), current density (540–660 A/m^2), electrolysis duration (55–65 min), and initial pH (8–10) as well as the percentage removal of ammonia. It is noted that the maximum reduction of ammonia was up to 71%, which estimated the optimum working conditions for the treatment process as follows: zeolite dosage of 105 g/L, the current density of 600 A/m^2, electrolysis duration of 60 min, and pH 8.20. Furthermore, the regression model indicated a strong relationship between the predicted values and the actual experimental results with a high R^2 of 0.9871. These results provide evidence of the ability of the ZAEP treatment as a viable alternative in removing high-strength landfill leachate of adequate salinity without the use of any supporting electrolyte.

Keywords: ammonia; zeolite; electrocoagulation; response surface methodology; stabilized; leachate

1. Introduction

In Malaysia, the rise in inhabitants, active urbanization development, and revolution of industrialization over the years has been causing fast economic growth. As a result, generation of municipal waste is increasing annually (6.2 million ton/year) and is linearly proportionate to the increase in population and migration of urbanization. Generally, in most developed countries, organic waste is the main contributor of about 40% to 60% of the total weight of municipal solid waste (MSW). In this sense, solid waste disposal due to rapid solid waste generation is one of the primary environmental concerns that must be addressed [1]. Presently, landfilling is the main option for waste disposal because of its low operation cost and simpler operating management [2]. Heterogeneous waste (municipal solid waste) deposited in landfills undergoes several physicochemical reactions, which result in an extremely polluted dark liquid with bad odor known as leachate [3]. Furthermore, as water passes over a landfill, most of the organic and inorganic contents such as ammoniacal nitrogen (NH_3-N) and heavy metals are transported into the leachates [4]. Lack of efficient leachate management may lead to pollution of the water bodies resources and groundwater, which could harm human wellness and aquatic habitats [5].

A landfill is typically categorized into three major phases—acidogenic, intermediate, and methanogenic—according to their respective biodegradability ratios (BOD_5/COD); more than 0.5 (young, <5 years), between 0.1 and 0.5 (intermediate, 5 to 10 years), and less than 0.1 (stabilized, >10 years) [6,7]. Landfill leachate can be classified as a concentrated wastewater pollutant, which is very hard to be treated. Organic compounds such as ammonia, chemical oxygen demand (COD), and color are the most challenging soluble contaminants to be completely removed from landfill leachate [8]. Regrettably, conventional treatment techniques (single process) in normal operation are incapable to fully comply with the acceptable effluent limit as described accordingly in the Environmental Quality Act 1974. In contrast, an affiliation of multiple biological, chemical, and physical treatment techniques seems to be more effective for stabilized and high-strength landfill leachate [9].

A previous review by [10] had explained that a biological treatment includes aerobic and anaerobic processes as well as an efficient treatment of young leachate; meanwhile, physical and chemical treatments that involve several techniques such as ion exchange, adsorption, coagulation–flocculation, and oxidation are further effective in treating stabilized landfill leachate. Among advanced oxidation processes in physico-chemical treatment, the electrocoagulation (EC) technique offers an effective alternative method for treating various pollutants in water and wastewater [11]. In terms of simplicity of equipment, including reaction time, ease of operation, electro-generation of flocs, level of maintenance, supporting electrolyte requirement, and quantity of sludge, the EC method offers better advantages in comparison with other current processes [12]. Generally, an EC process consists of three essential mechanisms of coagulation, flotation, and electrochemistry [13]. In a landfill leachate treatment, electrocoagulation also includes the processes of ionization, electrolysis, hydrolysis, and free hydroxyl radicals, which lead to significant changes in the liquid properties as well as enhancement in the overall contaminant removal [14].

A prior review by [15] reported that the effectiveness of an EC process is subjected to several basic working parameters such as electrode type, current density, electrolyte concentration, pH, electrode distance, reaction time, and arrangement of the electrode. However, other researchers [16] with their study on a similar method claimed that the most vital factor affecting the overall efficiency of an electrocoagulation process is the type of electrode material used. Normally for an EC electrode, aluminum (Al) or iron (Fe) materials are mostly preferred as they are low priced, readily accessible, and effective in pollutant removal [17]. Several studies by [18–20] had proven that an aluminum electrode has a better performance in eliminating various contaminants in the treatments of landfill leachate. Moreover, in the removal of dissolved organic pollutants, for example, color, turbidity, chemical oxygen demand (COD), and so forth, aluminum electrodes prove to be more effective compared to other types of electrodes [1,21,22].

Despite the proven advantages and effectiveness of the EC method in landfill leachate treatment, current reviews have indicated that the removal of certain contaminants, particularly ammoniacal nitrogen (NH_3-N), is strictly ineffective and needs further attempts, and very limited studies were conducted in this case. There are various existing technologies focusing on the leachate treatment process that involve the removal of NH_3-N pollutants such as biological [23], air stripping [24], adsorption [25], chemical precipitation [26], and electro-oxidation [27] processes. In the physical treatment process, adsorption is a typical organic pollutant removal method used to transfer substances from catalysts to the solid surface. Since an adsorption process relates to the surface mechanism, the adsorbent surface area is the main affecting factor, where a high surface area provides more porosity for adsorbent interactions. A previous study by [28] revealed the effective application of activated carbon to mitigate COD and NH_3-N in landfill leachate by applying a fixed bed method.

Nevertheless, the preparation of activated carbon is costly and takes a longer time. As a comparison to this type of adsorbent for the reduction of NH_3-N, the mechanisms of adsorption and ion exchange displayed by zeolites are rather interesting, owing to their cost-effective and uncomplicated approaches [29]. Zeolite adsorbent is also used in the aeration tank of a sludge process as an augmented medium. As a result, the removal of NH_3-N, in particular, is significantly improved [30]. In addition, the natural zeolite mineral also offers a strong ion exchange and good adsorption bonding in a mixed suspension of contaminants [31].

Hence, in this work, a novel integrated treatment method was established to remove high-strength NH_3-H in saline landfill leachate using an electrocoagulation process with augmenting zeolite to enhance the efficiency of its performance. A preliminary experiment on the batch study of zeolite augmented on the electrocoagulation process (ZAEP) was rigorously conducted to evaluate its removal mechanism and the best fit of optimum conditions, where four vital variables affecting treatment capability were identified. The mechanisms of ammonia contaminant elimination from saline landfill leachate can be more complicated in ZAEP. According to [32], it was stated that the removal process mechanism for ammonia could be respectively distributed into four categories, namely direct oxidation by the anode, indirect oxidation of hydroxyl radicals, active chlorine sites, and adsorption by zeolite. The dissolution of aluminum particles (Al^{3+}) was directly reacted at the anode electrode when the electrical current was introduced. The bubbles produced indicate that the Al particles are being polarized. The reduced ammonia-forming nitrates are absorbed by zeolite through an ion exchange process with the aid of electrolysis and with synergistic effects of absorption. In addition, dissolved ammonia in electrolytes and zeolites can be oxidized via electro-generated $HOCl/OCl^-$. The electrical field generated by current facilitates the transition of ammonium towards the anode, speeding up the adsorption process of ammonium ions. The anode-generated hydroxyl radicals play a part in indirect ammonia oxidation [33] for the formation of flocs in the flotation process. Through the removal of ammonia using the EC process, organic pollutants are also removed by direct oxidation [16,27]. Considering this improvement mechanism, the main aim of this study was to achieve high removal contaminant efficiency of HN_3-H with the application of ZEAP in treating saline landfill leachate. Therefore, optimization of the ZAEP experiment using the four influential operating factor variables mentioned earlier, including current density, zeolite dosage, electrolysis time, and pH, on the elimination of NH_3-N was robustly investigated. A response surfaces methodology (RSM) based on central composite designs (CCDs) was applied to evaluate the interactions between the factors (variables) and response (removal). Experimental design by an analysis of variance (ANOVA) was performed to develop a statistical model of the second-order polynomial. Finally, a numerical optimization experiment based on the high desirability suggested by RSM was conducted to verify the correlation between the optimum conditions of operating variables and the predicted response values.

2. Materials and Methods

2.1. Leachate Sampling

The electrolytic landfill leachate, which was saline in nature, was taken from the Pulau Burung landfill site (PBLS) near the Byram forest reserve in Penang, Malaysia at the GPS location of 5°12′12.1″ N latitude and 100°25′30.2″ E longitude, which is situated next to the coastline. PBLS has an ordinary oceanic clay liner and receives around 2000 tons of municipal solid waste daily. On-site and laboratory testing were conducted using a multi-parameter instrument (YSI Pro Plus; YSI Incorporated, Yellow Springs, OH, USA) to determine the salinity value of the leachate. Approximately 15 sets of leachate sampling were done between January and September 2019 using a 30 L polyethylene container. The acquired samples were stored and immediately transferred into a refrigerator of 4 °C in accordance with the standard methods of the Examination of Water and Wastewater, American Public Health Association [34].

2.2. Characterization of Natural Zeolite

The granular of natural zeolite was purchased from the supplier by YM Multi trading company, Selangor, Malaysia. The mineral adsorbent size was provided in the range of 3–6 mm with a pH value between 6.5 and 7.5. Before any experiment, the zeolites were washed thoroughly with deionized water four times to remove any impurities from their surface area altogether and were then dehydrated overnight in a furnace (Memmert, Schwabach, Germany) at 105 °C to eradicate water and moisture from the pores. The dried granular zeolites were crushed using crusher machines (Retsch, Haan, Germany) and sieved to obtain an average sample size of 75 to 150 μm. Next, the natural zeolite was characterized by BET, XRF and XRD analyses.

2.2.1. Brunner–Emmet–Teller Analysis

The particular area of the adsorbent surface (SA), size (PS), and volume (PV) of zeolite pores were investigated by using nitrogen gas involving a process of adsorption and followed by desorption to obtain isotherm data at 77 K. The zeolite was degassed at 200 °C and 145 min earlier than the N_2 adsorption measurement. The total pore volume was obtained by data from N_2 adsorption under saturation conditions. The specific surface area of zeolite was analyzed using Brunner–Emmet–Teller (BET). The SA, PV, and PS of zeolite were recorded as 49.3371 m^2/g, 0.0927 cm^3/g, and 7.5141 nm, respectively. According to [35], explained that the adsorbent media can be categorized based on pore diameter such as micropores (less than 2 nm), mesopores (2 nm to 50 nm) and macropores (more than 50 nm). Thus, zeolite used in this study can be classified as mesopores because of its PS between the range of 2.0 nm to 50 nm.

2.2.2. X-ray Fluorescence (XRF) Analysis

The chemical composition of natural zeolite was determined using XRF instrument and is depicted in Table 1. The composition and structure of zeolite was identified as crystalline aluminosilicate with pores of molecular dimensions. The common formula of zeolite is ($Mx/n\ [(AlO_2) \times (SiO_2)\ y].\ mH_2O$), where M is the metal or hydrogen cations of valence (n) occupying the transferrable cationic sites on the zeolite structure. AlO_2 and SiO_2 are the important compounds sharing oxygen ions to form tetrahedral AlO_4 and SiO_4 building blocks for zeolite unit cell since the silicon ion has +4 and the aluminum ion has +3 charges. The cationic charge of the metals or hydrogen ions balance the negative charge on the aluminosilicate framework. The XRF results demonstrated a silicon oxide content of 74.17%, whereas that of aluminum oxide was only 13.42% by weight. The Si/Al ratio of the natural zeolite structure was found to be 5.68. According to [36], if the ratio of natural zeolite is more than 4, indicating that the zeolite is at a higher temperature, the structure would not be easily broken.

Table 1. Chemical composition of natural zeolite.

Component	SiO_2	TiO_2	Al_2O_3	Fe_2O_3	MnO	MgO	CaO	Na_2O	K_2O	P_2O_5
Weight (%)	71.8	0.13	12.63	1.43	0.03	0.61	2.11	0.93	2.56	0.02

2.2.3. X-ray Diffraction Analysis

The analysis results of XRD showed a raw clinoptilolite pattern, as illustrated in Figure 1. This identification of natural zeolite was done using X-pert High score Plus software (PANalytical, Almelo, The Netherlands). The sample exhibited crystalline structure elements that were confirmed to match with clinoptilolite characteristic peaks. The analyzed sample complied with the reference code 98-000-2606, and the density of zeolite was recorded at 2.151 g/cm^3. The finding also described that the chemical formula of clinoptilolite was recorded as $[Na_{1.66}\ K_{2.56}\ Ca_{1.9}\ (Al_{5.48}\ Si_{30.52}\ O_{72}).19.16\ H_2O]$. Although differing in practice, a similar clinoptilolite formula structure was also reported by [37] and revealed the Si/Al ratio was 5.569. Furthermore, this subfield of inorganic natural mineral also indicated a monoclinic structure type. The results were well in line with the literature [38], and the approximate overall raw mineral contents of clinoptilolite were 88% to 90%.

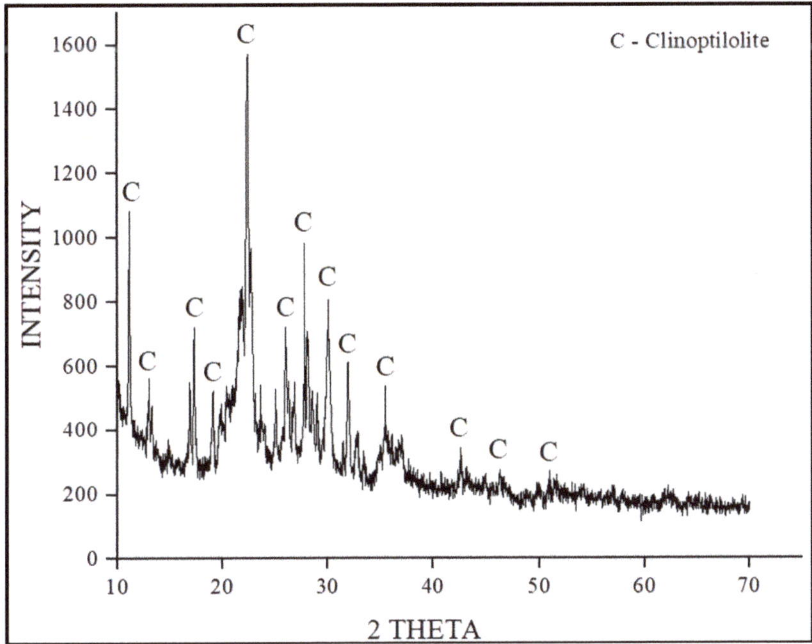

Figure 1. X-ray diffraction pattern of the natural zeolite (clinoptilolite) material.

2.3. Chemical Reagents and Instrument

The chemical reagents and instrument involved in this study were sulfuric acid, sodium hydroxide, pH meter (Eutech Instruments, Singapore), DR 2800 spectrometer (Hach Company, Loveland, CO, USA), measuring cylinder (NICE, London, UK), beaker (Fisher Scientific, Hampton, NH, USA), wired crocodile clip, laboratory film (Parafilm M; Bemis Company, Neenah, WI, USA), aluminum plate (grade 6061), DC power supply (OJE Model PS6005, 60 V/5 A, Suzhou, China), and stirrer (Heidolph MR Hei-Tec 220 V; Heidolph Instruments, Schwabach, Germany).

2.4. Analytical Methods

All experiments were conducted based on the standard methods 4500C (ammoniacal nitrogen) of the Examinations of Water and Wastewater [34]. The removal of pollutants of NH_3-N (mg/L) was tested via the Nessler Method 8038 using a Hach (Loveland, CO, USA) DR 2800 spectrometer. Meanwhile, salinity (ppt) was measured using a YSI Professional Plus multi-parameter probe (YSI Incorporated, Yellow Springs, OH, USA). All tests were done in triplicate.

2.5. Experimental Set-Up

The batch experiment of the zeolite augmented electrocoagulation process was conducted in the reduction of pollutants of NH_3-N, where the set-up is as shown in Figure 2. Landfill leachate electrolyte (750 mL) was firstly poured into a 1000 mL glass beaker. A DC power supply (OJE 60 V/5 A; Suzhou, China) was applied by connecting the positive and negative terminals to a pair of aluminum electrodes, which respectively acted as the anode and cathode.

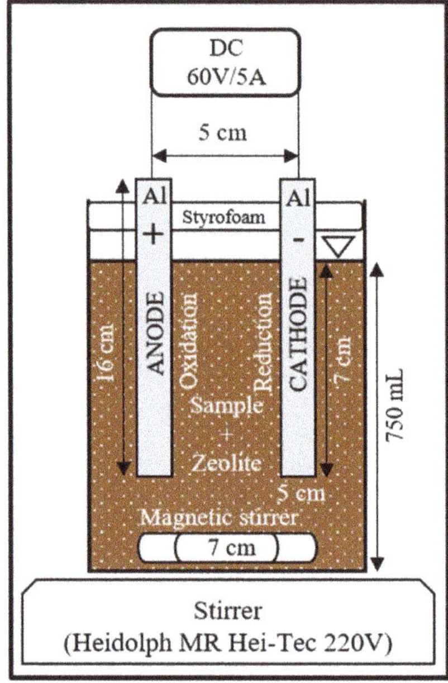

Figure 2. Experimental set-up for the zeolite augmented electrocoagulation treatment process.

Zeolite was introduced into the electrolyte, and the mixture was mixed for a 1 min duration for homogeneous mixing of the electrolyte. At the end of the stir, the electrodes were submerged into the electrolyte, and the power supply of DC was promptly switched on. The concentration of NH_3-N in the electrolyte solution was examined before and after the EC process took place. The total active area of the single electrode was 35 cm^2 when being dipped 70 mm into the electrolyte solution, while the electrode size was 16 × 5 × 1 cm. A stirrer (Heidolph MR Hei-Tec 220V; Heidolph Instruments, Schwabach, Germany) was set to produce a constant speed of 200 rpm, and a magnetic stirrer of 70 mm length was used for the stirring purpose. The spacing distance between both electrodes was set to 3 cm. At the beginning and end of each experiment, the electrodes were washed thoroughly with distilled water, dipped in 0.1 M HCl solution to remove impurities for 15 min, and again bathed with distilled water. At the end of the treatment process, the samples were undisturbed for 60 min [9]. The removal efficiencies of NH_3-N were determined based on Equation (1).

2.6. Concentration of Removed Pollutant

At designated intervals, samples were collected for analysis of the NH_3-N pollutant. The pollutant removal efficiency (Y%) was calculated based on Equation (1) below:

$$\mathrm{Y}(\%) = \left(\frac{C_o - C}{C_o}\right) \times 100 \tag{1}$$

where, C_o is the initial concentration of the removed pollutant before the electrocoagulation process, and C is the concentration of the removed pollutant after t min of the electrocoagulation process.

2.7. Response Surface Methodology Design

Design of the ZAEP experiment and statistical modelling were performed using the Design-Expert software version 11.0.3 (StatEase, Minneapolis, MN, USA) for analysis purposes. The central composite designs (CCDs) through response surfaces methodology (RSM) were used to optimize variables and investigate the correlation between the response and variable factors of ZAEP in the removal of NH_3-N contaminant. Four significant factorial variables were evaluated, including the zeolite dosage, current density, electrolysis duration, and pH. The coded variables and the second-order polynomial for response were determined, as shown in Equations (2) and (3). Table 2 tabulated the coded level of variables and their actual values (optimum values obtained from our previous experiment) involved in these studies

$$\text{Coded of variable } (X_i) = \frac{xi - Xcp}{\Delta x} \quad (2)$$

$$\text{Response } (Y\%) = \beta_0 + \sum_{i=1}^{k} \beta_i x_i + \sum_{i=1}^{k} \beta_{ii} x_i^2 + \sum_{i \neq 1 j=1}^{k} \beta_{ij} x_i x_j + \varepsilon \quad (3)$$

where, xi is the un-coded value, Xcp is the un-coded value at the center point, Δx is the value change between levels,β_0 is the constant-coefficient, β_i, β_{ii} and β_{ij} are the coefficients for the linear, quadratic and interaction effect respectively, x_i and x_j are the factors, and ε is the error.

Table 2. The coded and actual values of variables for the zeolite augmented on the electrocoagulation process (ZAEP) treatment.

Code	Factor	Unit	Coded Level of Variables		
			Low (−1)	Central (0)	High (+1)
A	Zeolite dosage	g	100	110	120
B	Current density	A/m^2	540	600	660
C	Electrolysis duration	Min	55	60	65
D	pH	-	8	9	10

3. Results and Discussion

3.1. Leachate Characteristics

The results of the compositions of the raw landfill leachate from PBLS are presented in Table 3. It is important to note that the biodegradability ratio (BOD_5/COD) of this landfill leachate was 0.051 < 0.1; thus, it can be considered as stabilized landfill leachate because of the high concentrated COD (4928 mg/L) and low concentration BOD_5 (254 mg/L). Furthermore, the raw leachate can be classified as alkaline since its average value of pH was 8.16, which was further categorized under the methanogenic phase because of a higher value than pH 7 [39]. A similar finding was also revealed by [40–42], which found alkaline conditions of landfill leachate in their studies on PBLS.

Table 3. Characteristics of Pulau Burung landfill leachate.

No.	Parameters	Value (January to June 2019)	Average	Standard Discharge Value from the Environmental Quality (Control of Pollution from Solid Waste Transfer Station and Landfill) Regulation 2009
1.	BOD_5 (mg/L)	207–283	254	20
2.	COD (mg/L)	4266–6648	4928	400
3.	Ammonia (mg/L)	3125–3782	3471	5
4.	Color (Pt-Co)	4930–18,380	8240	100 ADMI
5.	pH	7.52–8.21	8.16	6–9
6.	Salinity (ppt)	15.01–17.2	15.36	-
7.	BOD_5/COD	0.043–0.049	0.051	-

According to [7], the increase in the pH value of leachate is ascribed to the higher NH_3-N concentration as well as the lower concentration of nitrate in the landfill leachate. Thus, these results indicated that high-strength compounds of ammonia were present and recorded to be 3471 mg/L, which are beyond the standard discharge value of 5 mg/L. The concentration of NH_3-N in this sample was much higher than that reported by [43], which was only 2050 mg/L. The differentiation in the salinity value and concentration of ammonia may be the most vital indicator for microorganism survivability [6]. The development of bacteria has a strong correlation with the concentration of salinity and ammonia. The growth of microorganisms may be disturbed due to the high concentration levels of ammonia produced during the methanogenic phase [44]. Moreover, the landfill leachate produced from PBLS is aged and stabilized. Hence, biological treatment processes can hardly be fully employed.

3.2. Predicted Against Actual Values

Figure 3 illustrates a diagnostic plot of the comparison between the results from actual experiments and the predicted values from a linear regression model by the indication of color point as minimum and maximum removal rates. This figure clearly shows that both data points in this plot matched to each other and were close to the diagonal line (R^2), which illustrates strong adequacy to satisfy both the actual experimental and prediction values by the established model [45]. Furthermore, if the value of R^2 is over 0.9 for a regression linear equation model, it indicates an excellent correlation between the laboratory findings and the values of prediction [46]. Thus, as can be noted in Table 4, the regression coefficient, the R^2 value, was equal to 0.9871, confirming the fact that a strong correlation response existed in this model.

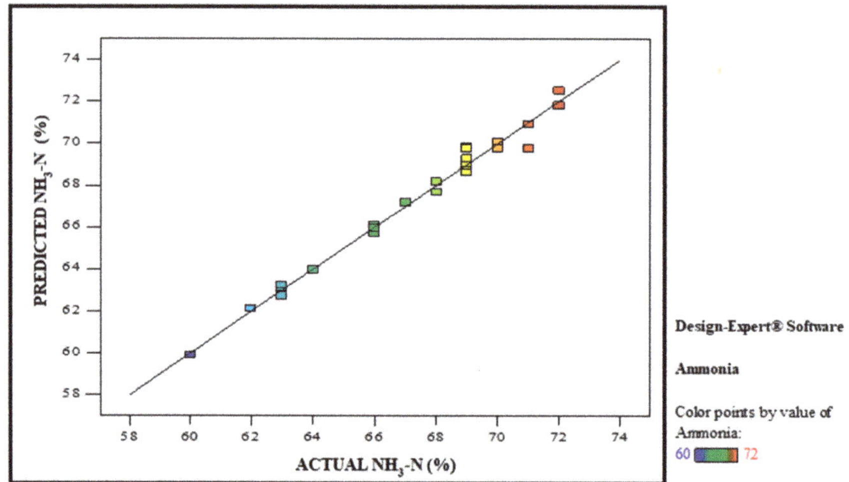

Figure 3. The plot of actual versus predicted values.

Table 4. Analysis of variance (ANOVA) and quadratic model validation on NH_3-N removal.

Final Equation of Actual Factor on Percentage Removal of NH_3-N					
% removal = 69.76 − 0.61A + 1.06B − 0.56C − 3.28D − 0.53A^2 − 1.53B^2 − 1.03C^2 − 0.53D^2 − 0.063AB − 0.69AC − 0.44AD + 0.31BC − 0.19BD + 0.44CD					
Analysis of Variance					
Source	**Sum^2**	**DF**	**Mean^2**	**F Value**	**Prob > F**
Model	317.06	14	22.65	82.06	<0.0001 (significant)
Lack of fit	2.14	10	0.21	0.53	0.8130 (not significant)
Model Validation					
R^2	**Adjusted R^2**		**Predicted R^2**		**Adequate Precision**
0.9871	0.9751		0.9633		33.938

3.3. Analysis of the Design of Experiments

Experimental validation was performed at an optimum condition of variables to study the interaction between the factors and response. The design matrix of independent variables as tabulated in Table 5 shows that a total of 30 runs of the experiment, with 16 factorial, 8 axial, and 6 centers, were suggested for ZAEP treatment in the reduction of NH_3-N.

Table 5. Design matrix and experimental results for optimized variables of ZAEP.

Run	Point Type	Variable Factor				Removal
		Zeolite Dosage (g)	Current Density (A/m^2)	Electrolysis Duration (min)	pH	NH_3-N (%)
1	Center	110	2.10	60	9	69
2	Center	110	2.10	60	9	70
3	Center	110	2.10	60	9	72
4	Center	110	2.10	60	9	71
5	Center	110	2.10	60	9	71
6	Center	110	2.10	60	9	70
7	Axial	100	2.10	60	9	70
8	Axial	110	2.10	55	9	68
9	Axial	110	2.10	65	9	71
10	Axial	110	2.31	60	9	69
11	Axial	110	2.10	60	8	72
12	Axial	110	1.89	60	8	67
13	Axial	110	2.10	60	10	65
14	Axial	120	2.10	60	9	68
15	Fact	100	1.89	65	8	68
16	Fact	100	1.89	55	8	69
17	Fact	100	2.31	55	8	71
18	Fact	100	2.31	65	10	66
19	Fact	100	2.31	65	8	71
20	Fact	100	1.89	55	10	63
21	Fact	100	2.31	55	10	64
22	Fact	100	1.89	65	10	63
23	Fact	120	2.31	65	8	68
24	Fact	120	1.89	55	8	70
25	Fact	120	1.89	65	8	69
26	Fact	120	1.89	55	10	62
27	Fact	120	2.31	55	10	63
28	Fact	120	2.31	65	10	62
29	Fact	120	2.31	55	8	72
30	Fact	120	1.89	65	10	60

As shown in Table 5, the results from the present research revealed that the reduction of NH_3-N was in the spectrum between 60% (min) and 72% (max). From this experimental finding, statistical Equation in Table 4 based on the second-order polynomial modelling and the analysis of variance (ANOVA) of the respective variables A (zeolite dosage), B (current density), C (electrolysis duration), and D (pH) were established to examine the percentage removal efficiency of the treatment process (Table 4). In order to validate this empirical model, several parameters were considered, such as prob < 0.05 (significant), Lack of Fit (LOF) > 0.05 (not significant), adequate precision ≥ 4, $R^2 \geq 0.8$, and $R^2 \approx R^2_{pre.}$ [46,47]. Thus, according to the results shown in Table 4, justification is to be drawn that the mathematical equation of the second-order polynomial model has satisfied all the vital requirements and may be used to evaluate the effectiveness of a ZEAP treatment on the elimination of NH_3-N pollutant.

3.4. The Effects of Factor Variables on Pollutant Removal

Three-dimensional (3D) response surface plots for all factor variables were generated using CCD to examine the influence of each variable combination and the interaction between variables on the removal of ammonia pollutant after ZAEP treatment on natural saline landfill leachate. Figures 4–6 show higher removal percentages of NH_3-N yielded at lower pH and among the reactions of zeolite dosage, current density, and electrolysis duration variables. The surface charge of the particles may be affected by pH variability. A progressive response would change the pH of an electrolyte that, in turn, affects organic matter reduction [48]. As can be observed in Figure 4, the maximum reduction of NH_3-N was found at a lower pH range, between 8.0 and 8.5, and zeolite dosage range of 105–110 g. This can be explained by the transformation of NH_3-N into uncharged ammonia at higher pH (pH > 9) [49]. As a result, it would reduce the removal of NH_3-N. As reviewed by [47], they mentioned that, at a lower pH, neutralization of charges is encouraged as higher metal amounts are dissolved owing to the greater current density, which leads to floc formation and increases in the effectiveness of removal. High removal percentages of NH_3-N were observed at the current density peak of 600 A/m^2 and pH between 8.0 and 8.5 (Figure 5).

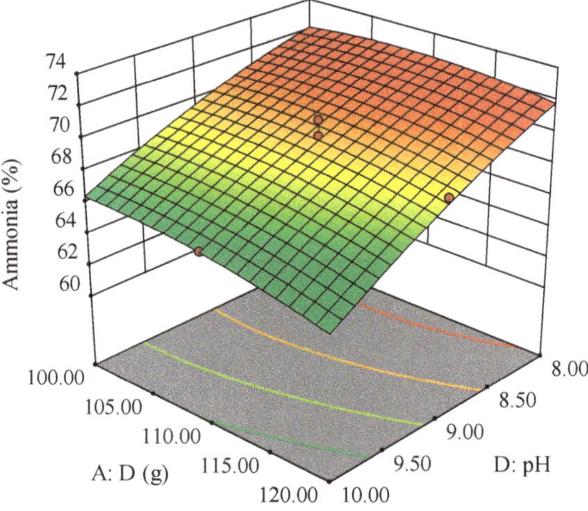

Figure 4. 3D surface plots for removal of ammonia on influences of dosage and pH.

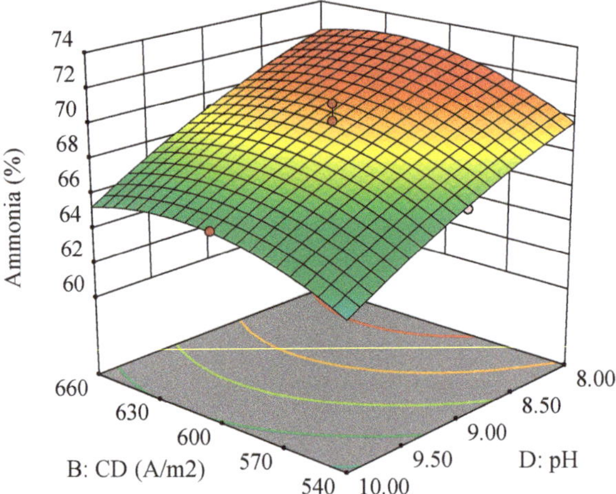

Figure 5. 3D surface plots for removal of ammonia on influences of current density and pH.

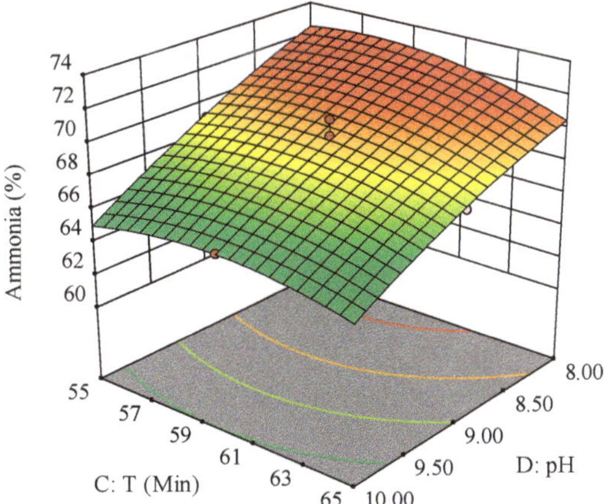

Figure 6. 3D surface plots for removal of ammonia on influences of electrolysis time and pH.

A substantial effect on the reduction of NH_3-N from the interaction between electrolysis duration and electrolytic pH was observed, which is as shown in Figure 6. This finding indicates that the removal percentage increased when the process was operated at an electrolysis time of 60 min and at the pH range mentioned earlier. The improvement in the removal efficiency is due to the duration of electrolysis, which allows a longer catalytic reaction time resulting in an increase in contaminant degradation [50] and a higher number of hydroxyl radicals, which results in metal-polymer species formation for further increase in the efficiency of removal [51]. Meanwhile, Figures 7 and 8 portray a curve shape, which indicates an interaction between zeolite dosage and other factor variables of electrolysis time and current density. It is noted that the percentage reduction of NH_3-N in saline landfill leachate increased with increasing current density and zeolite dosage (Figure 7). The current density of the produced electrical field may enhance the rate of NH_3-N ion exchange with lateral movement to the anode and, thus, favor the NH_3-N adsorption on zeolite [52]. Other research by [53] had also reported that molecular structures, as well as high-capacity cation exchanges of zeolite, accommodate free surface ion exchange and absorption. However, at a current density of more than 630 A/m^2, a slight insignificant drop was seen from the maximum removal.

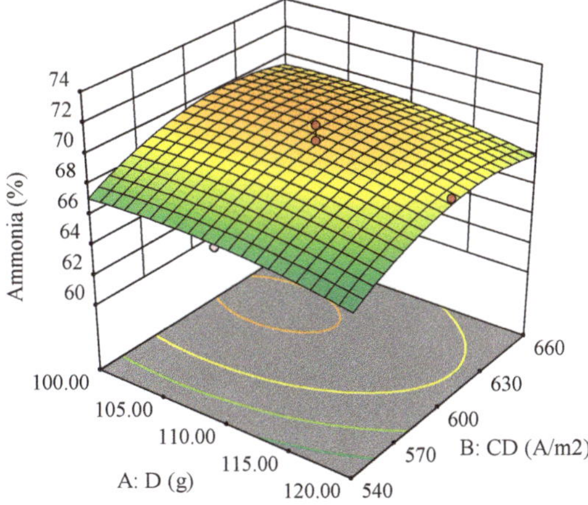

Figure 7. 3D surface plots for removal of ammonia on influences of dosage and current density.

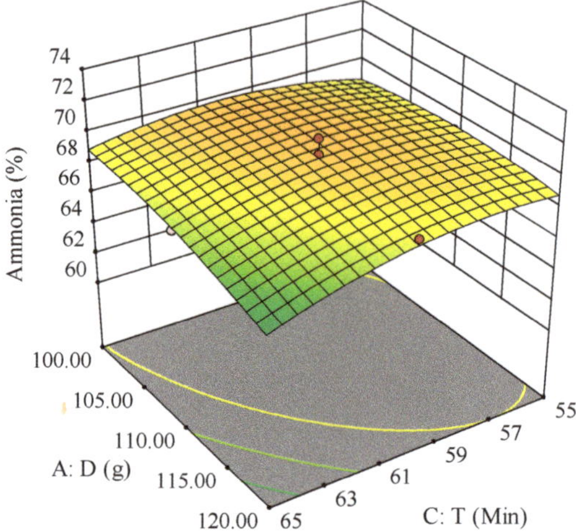

Figure 8. 3D surface plots for removal of ammonia on influences of dosage and electrolysis time.

According to [54,55], the removal efficiency of adsorbents increase when increasing the adsorbent dosage, leading to the accessibility of the active surface area. This fact is in agreement with the results from this study, which exhibited that higher percentage removal of NH_3-N was achieved with increasing zeolite dosage in the range of 105–115 g versus electrolysis time of 57–61 min, as illustrated in Figure 8. Furthermore, Figure 9 clearly shows that the maximum reduction of NH_3-N occurred in the range of 600–630 A/m^2 current density and at electrolysis duration between 59 and 63 min. This phenomenon can be described by a rise in the Al^{3+} charged cations and thus releasing $Al(OH)_3$ particles by the anode when increasing the current density [56]. Therefore, this analysis leads to the conclusion that the maximum reduction of NH_3-N from saline landfill leachate was revealed to be 72%, obtained from the range of zeolite dosage of 105–115 g, current density of 600–630 A/m^2, electrolysis time of 59–61 min, and pH of 8–8.5.

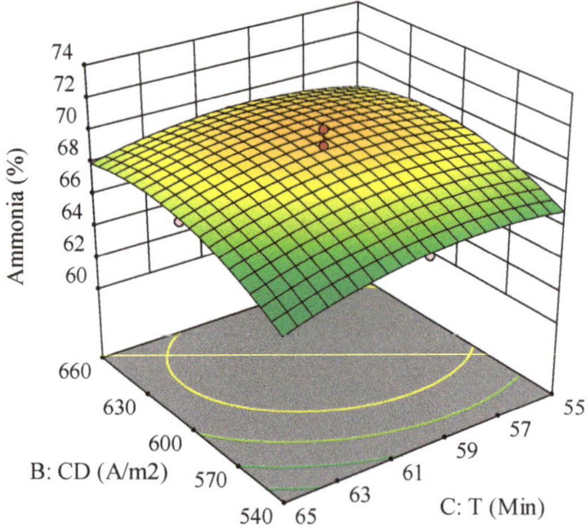

Figure 9. 3D surface plots for removal of ammonia on influences of current density and electrolysis time.

3.5. Analysis of Variables Optimization

Optimizing operational variables is essential in the process development of a model in order to validate an experiment and maximize the removal of ammonia pollutant from high-strength landfill leachate. Simultaneous RSM optimization for the removal response based on the operating variables was carried out in an attempt to investigate the optimum working variables for this novel zeolite augmented electrocoagulation process. All the input variables, including the zeolite dosage (A), current density (B), electrolysis duration (C), and pH (D), were set between ranges during optimization, whereas the output response of variables was collected as the percentage ammonia removal. The effects of the chosen operating parameters were configured to achieve optimum removal performance based on the desirability suggested by CCD [57,58]. The scale of desirability ranged from 0 to 1, with 0 indicating an undesirable response and 1 being the most desirable reaction [59]. A high desirability of 0.9987 was obtained and chosen as it was the closest to 1. Experiments were conducted to verify the optimal conditions. The predicted results through an optimization process, which were computed by the RSM system and the actual experimental results on pollutant removal, are expressed in Table 6.

Table 6. Optimization between predicted and experimental results on NH_3-N removal.

Zeolite Dosage (g)	Current Density (A/m^2)	Electrolysis Duration (min)	pH	NH_3-N Removal (%)	
				Predicted	Experimental
105	600	60	8.20	72.51	71.01

The recommended operating variables are a zeolite dosage of 105 g, current density of 600 A/m^2, electrolysis duration of 60 min, and pH 8.20. It is noted from the verification experiment that the actual removal result of NH_3-N was very close to that of the predicted result, which were 71.01% and 72.51%, respectively. This proves a good correlation between both parameters and confirms the eligibility of the model on ZAEP treatment using the optimized conditions.

3.6. Comparison of Treatment Performance

Numerous past studies reported the ability and effectiveness of the electrocoagulation system in the reduction of various organic and inorganic pollutants from landfill leachate. Although various literature exists on the treatment of landfill leachate in batch mode or continuous operation of the EC process, only a few studies reported on the reduction of ammonia pollutants. Current literature revealed that there are limitations in the conventional EC process on the treatment performances of specific parameters in leachate treatment, especially ammoniacal nitrogen [60,61]. In most cases, previous literature, as presented in Table 7, was in agreement with the mentioned fact of the ineffective ammonia pollutant removal performance in the treatment of landfill leachate.

As noted in Table 6, most of the works dealt with low concentrations of ammonia (50–2240 mg/L) as compared to the current study with a much higher concentration of 3471 mg/L. Ammonia removal efficiencies of 14% to 37% were reported in most studies, except by [65] where 80 removal efficiency of ammonia was reported at longer electrolysis time (2 hours) by using dual-electrode anode and much lower ammonia concentration (50–110 mg/L). Whereas, this current research uses more economical operating parameters (Al anode) and less treatment duration (60 min).

There have been a few studies on the removal of organic and inorganic pollutants using zeolite and other combination/coupling treatment processes in previous literature, as revealed by [33] and [32]. Nonetheless, they focused on the synthetic solution and soluble dye in pure water, in addition to municipal wastewater of concentrations between 30 and 50 mg/L. This is in contrast with the present study, which used natural high-strength ammonia through the novel ZAEP treatment. In addition, water matrix landfill leachate was also utilized as the electrolyte to evaluate the effectiveness of this process, which is known as the real wastewater that is difficult to be treated due to the various contents of complex organic compounds, heavy metals, and other contaminants [66,67]. Therefore, this present research produced a high-performance ammonia removal of up to 71%, without any required ancillary processes for further efficiency enhancement, owing to its ability and effectiveness in contaminant removal.

Table 7. The summary of studies on the treatment of landfill leachate using the electrocoagulation process.

No.	Water Matrix	Pollutant	Concentration	Experimental	Performance	References
1	Leachate	COD Ammonia	12,860 mg/L 2240 mg/L	Current density: 631 A/m^2, Time: 45 min, electrode: Al	59% 14%	[62]
2	Leachate	COD Ammonia	2566 mg/L 386 mg/L	Current density: 29.8 A/m^2, Time 30 min, pH 6	21% 20%	[63]
3	Leachate	COD Ammonia Color Turbidity Suspended solids	1992 mg/L 982 mg/L 3500 Pt–Co 181 NTU 330 mg/L	Current density: 200 A/m^2, pH: 4, Time: 20 min	60% 37% 94% 88% 89%	[61]
4	Leachate	Color Ammonia	2660 mg/L 577.04 mg/L	Current density: 150 A/m^2, Time: 60 min, Electrode: Al/Fe, Coagulant: 0.3 g/L, pH: 5	88% 25%	[64]
5	Leachate	COD Ammonia	167–180 mg/L 50–110 mg/L	Current: 1.6, Time: 120 min, Electrode: Dual anode, Type: TiO_2/IrO_2	75% 80%	[65]

3.7. The Economic Aspect of the ZAEP Treatment

Despite an enormous number of publications on coupling/combination with the EC method, very few studies have taken the optimization of processes and their cost efficiency into consideration. According to [68], they mentioned that the total cost estimation typically includes the chemical reagents, cost of electricity, sludge handlings, operators, and maintenance and equipment used in the EC system. However, in laboratory research, the electrode, consumption of electrical energy, and chemicals are the main parameters of the operating expenditure [69,70]. Also, the consumption of zeolite is also revealed in this analysis using Equation (4) as follows:

$$\text{Overall operating cost } (\$/m^3) = aEEC + bEMC + cCC + dNZ \tag{4}$$

where EEC is the energy of electrical consumption (kWh/m^3), EMC is the electrode material consumption (kg/m^3), CC is the chemical consumption (kg/m^3), and NZ is the natural zeolite consumption (kg/m^3). Meanwhile, the coefficients of a, b, c, and d respectively represent the ratios for electrical energy, Al electrode, chemical, and natural zeolites. For the Malaysian commercial market in 2019, the prices were 0.05 $/kWh, 0.96 $/kg for Al, 0.10 $/m^3, and 0.25 $/kg for NZ. These rates are similar to those of the international market. Based on the optimization of the ZAEP, the operation cost is computed to be approximately 36.46 $/m^3. Currently, no literature provides the operational cost calculation for this type of treatment process. However, it was found that in other wastewater treatments and combined/coupled EC processes, the operational cost study is much more significant, especially based on the studies by [71–73]. Although these studies had portrayed excellent results, which can provide assessments on the combined processes as cost-effective and more efficient than the process of electrocoagulation alone, further studies are strongly recommended to investigate the cost-saving aspect in this combined treatment method.

4. Conclusions

In this work, the treatment efficiency of the zeolite augmented electrocoagulation process (ZAEP) was evaluated for the ammonia reduction from high-strength landfill leachate, which was saline in nature. The process of condition optimization investigated the influences of interactions between the significant variables, including current density, zeolite dosage, electrolysis duration, and pH, using the experimental design of RSM. The results revealed that the most vital operating parameter affecting ammonia removal through the ZAEP treatment was pH. Based on the analysis of CCD, the actual experiment yielded a removal of ammonia of up to 71% at the optimal operating variables of zeolite dosage of 140 g/L, current density of 600 A/m^2, electrolysis duration of 60 min, and pH 8.20. The second-order polynomial model developed in this research was used to evaluate the percentage removal of ammonia with a strong correlation between the data points of experimental and prediction. Under these optimization conditions, the overall reasonable operating cost of the current treatment process was recorded as $36.46/m^2. The outcomes of various experiments lead to the conclusion that the treatment of concentrated landfill leachate using a ZAEP reactor is an effective process in the elimination of soluble organic pollutants.

Author Contributions: H.A.A. and M.S.Y. conceived and designed the experiments; M.A.A.H. performed the experiments and drafted the paper; H.A.A. and M.S.Y. supervised the work, conducted the data analysis, and revised the paper; S.A.R. proofread the paper. All authors have read and agreed to the published version of the manuscript.

Funding: This research was supported by the Universiti Sains Malaysia under RUI-grant (1001/PAWAM/8014081) and FRGS-grant (203/PAWAM/6071412) on research related to Solid Waste Management Cluster (SWAM), Engineering Campus, Universiti Sains Malaysia.

Conflicts of Interest: The authors declare no conflicts of interest.

References

1. De Pauli, A.R.; Espinoza-Quiñones, F.R.; Dall'Oglio, I.C.; Trigueros, D.E.G.; Módenes, A.N.; Ribeiro, C.; Borba, F.H.; Kroumov, A.D. New insights on abatement of organic matter and reduction of toxicity from landfill leachate treated by the electrocoagulation process. *J. Environ. Chem. Eng.* **2017**, *5*, 5448–5459. [CrossRef]
2. Aziz, H.A.; Rahim, N.A.; Ramli, S.F.; Alazaiza, M.Y.D.; Omar, F.M.; Hung, Y.T. Potential use of Dimocarpus longan seeds as a flocculant in landfill leachate treatment. *Water* **2018**, *10*, 1672. [CrossRef]
3. Peng, Y. Perspectives on technology for landfill leachate treatment. *Arab. J. Chem.* **2017**, *10*, S2567–S2574. [CrossRef]
4. Aziz, H.A.; Bashir, M.J.K.; Abu Amr, S.S. Trend of municipal landfill leachate treatment via a combination of ozone with various physic-chemical techniques. *Int. J. Environ. Eng.* **2016**, *8*, 95. [CrossRef]
5. Lestari, D.A.; David, J.P.; Park, J.E. Comparison study of bod & cod of leachate quality. *J. Environ. Eng. Waste Manag.* **2019**, *4*, 37–42.
6. Umar, M.; Aziz, H.A.; Yusoff, M.S. Variability in the concentration of indicator bacteria in landfill leachate—A comparative study. *Water Environ. Res.* **2015**, *87*, 223–226. [CrossRef] [PubMed]
7. Verma, M.; Kumar, R.N. Coagulation and electrocoagulation for co-treatment of stabilized landfill leachate and municipal wastewater. *J. Water Reuse Desalin.* **2018**, *8*, 234–243. [CrossRef]
8. Aziz, H.A.; Ramli, S.F. Recent development in sanitary landfilling and landfill leachate treatment in Malaysia. *Int. J. Environ. Eng.* **2019**, *9*, 201. [CrossRef]
9. Aziz, H.A.; Foul, A.A.; Isa, M.H.; Hung, Y.T. Physico-chemical treatment of anaerobic landfill leachate using activated carbon and zeolite: Batch and column studies. *Int. J. Environ. Waste Manag.* **2010**, *5*, 269. [CrossRef]
10. Aziz, S.Q.; Aziz, H.A.; Bashir, M.J.K.; Mojiri, A. Municipal landfill leachate treatment techniques: An Overview. *Wastewater Eng. Adv. Wastewater Treat. Syst.* **2014**, *3*, 1–18.
11. Gautam, P.; Kumar, S.; Lokhandwala, S. Advanced oxidation processes for treatment of leachate from hazardous waste landfill: A critical review. *J. Clean. Prod.* **2019**, *237*, 117639. [CrossRef]
12. Al-Raad, A.; Hanafiah, M.M.; Naje, A.S.; Ajeel, M.A.; O Basheer, A.; Ali Aljayashi, T.; Ekhwan Toriman, M. Treatment of saline water using electrocoagulation with combined electrical connection of electrodes. *Processes* **2019**, *7*, 242. [CrossRef]
13. Garg, K.K.; Prasad, B. Treatment of multicomponent aqueous solution of purified terephthalic acid wastewater by electrocoagulation process: Optimization of process and analysis of sludge. *J. Taiwan Inst. Chem. Eng.* **2016**, *60*, 383–393. [CrossRef]
14. Hassani, G.; Alinejad, A.; Sadat, A.; Esmaeili, A. Optimization of Landfill Leachate Treatment Process by Electrocoagulation, Electroflotation and Sedimentation Sequential Method. *Int. J. Electrochem. Sci.* **2016**, *11*, 6705–6706. [CrossRef]

15. Anjali, A.; Vivek, B.; Jayakumar, P.; Mithran, A. Electrochemical technologies for leachate treatment: A review. *Int. J. Sci. Eng. Res.* **2018**, *9*, 38–42.
16. Song, P.; Yang, Z.; Zeng, G.; Yang, X.; Xu, H.; Wang, L.; Xu, R.; Xiong, W.; Ahmad, K. Electrocoagulation treatment of arsenic in wastewaters: A comprehensive review. *Chem. Eng. J.* **2017**, *317*, 707–725. [CrossRef]
17. Cheballah, K.; Sahmoune, A.; Messaoudi, K.; Drouiche, N.; Lounici, H. Simultaneous removal of hexavalent chromium and COD from industrial wastewater by bipolar electrocoagulation. *Chem. Eng. Process. Process Intensif.* **2015**, *96*, 94–99. [CrossRef]
18. Tchamango, S.; Kamdoum, O.; Donfack, D.; Babale, D. Comparison of electrocoagulation and chemical coagulation processes in the treatment of an effluent of a textile factory. *J. Appl. Sci. Environ. Manag.* **2017**, *21*, 1317–1322. [CrossRef]
19. Zailani, L.W.M.; Zin, N.S.M. Application of Electrocoagulation in wastewater treatment: A general review. *IOP Conf. Ser. Earth Environ. Sci.* **2018**, *140*, 1–9. [CrossRef]
20. Mothil, S.; Devi, V.C.; Raam, R.S.; Senthilkumar, K. Electro-Coagulation of synthetic acid black 210 and acid red 1dye bath effluent using Fe and Al Electrodes in a recirculation cell. Am. *Int. J. Res. Sci. Technol. Eng. Math.* **2019**, *20*, 138–143.
21. Mahmad, M.K.N.; Rozainy, M.A.Z.M.R.; Abustan, I.; Baharun, N. Electrocoagulation process by using aluminium and stainless steel electrodes to treat total chromium, colour and turbidity. *Procedia Chem.* **2016**, *19*, 681–686. [CrossRef]
22. Asaithambi, P.; Beyene, D.; Aziz, A.R.A.; Alemayehu, E. Removal of pollutants with determination of power consumption from landfill leachate wastewater using an electrocoagulation process: Optimization using response surface methodology (RSM). *Appl. Water Sci.* **2018**, *8*, 11–12. [CrossRef]
23. Gao, J.; Chys, M.; Audenaert, W.; He, Y.; Van Hulle, S.W.H. Performance and kinetic process analysis of an Anammox reactor in view of application for landfill leachate treatment. *Environ. Technol.* **2014**, *10*, 1226–1233. [CrossRef]
24. Liu, Z.; Wu, W.; Shi, P.; Guo, J.; Cheng, J. Characterization of dissolved organic matter in landfill leachate during the combined treatment process of air stripping, Fenton, SBR and coagulation. *Waste Manag.* **2015**, *41*, 111–118. [CrossRef]
25. Bashir, M.J.K.; Aziz, H.A.; Amr, S.S.A.; Sethupathi, S.; Ng, C.A.; Lim, J.W. The competency of various applied strategies in treating tropical municipal landfill leachate. *Desalin. Water Treat.* **2015**, *54*, 2382–2395. [CrossRef]
26. Naveen, B.P.; Mahapatra, D.M.; Sitharam, T.G.; Sivapullaiah, P.V.; Ramachandra, T.V. Physico-chemical and biological characterization of urban municipal landfill leachate. *Environ. Pollut.* **2017**, *220*, 1–12. [CrossRef]
27. Ding, J.; Wei, L.; Huang, H.; Zhao, Q.; Hou, W.; Kabutey, F.T.; Yuan, Y.; Dionysiou, D.D. Tertiary treatment of landfill leachate by an integrated electro-oxidation/electro-coagulation/electro-reduction process: Performance and mechanism. *J. Hazard. Mater.* **2018**, *351*, 90–97. [CrossRef]
28. Abdul Halim, A.; Abu Sidi, S.F.; Hanafiah, M.M. Ammonia removal using organic acid modified activated carbon from landfill leachate. *Environ. Ecosyst. Sci.* **2017**, *1*, 28–30. [CrossRef]
29. Karadag, D.; Tok, S.; Akgul, E.; Turan, M.; Ozturk, M.; Demir, A. Ammonium removal from sanitary landfill leachate using natural Gördes clinoptilolite. *J. Hazard. Mater.* **2008**, *153*, 60–66. [CrossRef]
30. Bakar, A.A. Integrated Leachate Treatment by Sequencing Batch Reactor (Sbr) and Micro-Zeolite (MZ). Doctoral Dissertation, Universiti Tun Hussein Onn Malaysia, Johor, Malaysia, 2015.
31. Lim, C.K.; Seow, T.W.; Neoh, C.H.; Md Nor, M.H.; Ibrahim, Z.; Ware, I.; Mat Sarip, S.H. Treatment of landfill leachate using ASBR combined with zeolite adsorption technology. *3 Biotech* **2016**, *6*, 195. [CrossRef]
32. Song, L.; Li, C.; Huang, Y.; Zhou, Y. Enhanced electrolytic removal of ammonia from the aqueous phase with a zeolite-packed electrolysis reactor under a continuous mode. *J. Environ. Eng.* **2015**, *141*, 4014056.
33. Ding, J.; Zhao, Q.L.; Wang, K.; Hu, W.; Li, W.; Li, A.; Lee, D.J. Ammonia abatement for low-salinity domestic secondary effluent with a hybrid electrooxidation and adsorption reactor. *Ind. Eng. Chem. Res.* **2015**, *53*, 9999–10006. [CrossRef]
34. APHA. *Standard Methods for the Examination of Water and Wastewater*, 23rd ed.; American Public Health Association (APHA): Washington, DC, USA, 2017.
35. Thommes, M.; Kaneko, K.; Neimark, A.V.; Olivier, J.P.; Rodriguez-Reinoso, F.; Rouquerol, J.; Sing, K.S. Physisorption of gases, with special reference to the evaluation of surface area and pore size distribution (IUPAC Technical Report). *Pure Appl. Chem.* **2015**, *87*, 1051–1069. [CrossRef]

36. Alshameri, A.; Yan, C.; Al-Ani, Y.; Dawood, A.S.; Ibrahim, A.; Zhou, C.; Wang, H. An investigation into the adsorption removal of ammonium by salt activated Chinese (Hulaodu) natural zeolite: Kinetics, isotherms, and thermodynamics. *J. Taiwan Inst. Chem. Eng.* **2014**, *45*, 554–564. [CrossRef]
37. Yang, S.; Lach-hab, M.; Blaisten-Barojas, E.; Li, X.; Karen, V.L. Microporous and Mesoporous Materials Machine learning study of the heulandite family of zeolites. *Microporous Mesoporous Mater.* **2010**, *130*, 309–313. [CrossRef]
38. Favvas, E.P.; Tsanaktsidis, C.G.; Sapalidis, A.A.; Tzilantonis, G.T.; Papageorgiou, S.K.; Mitropoulos, A.C. Clinoptilolite, a natural zeolite material: Structural characterization and performance evaluation on its dehydration properties of hydrocarbon-based fuels. *Microporous Mesoporous Mater.* **2016**, *225*, 385–391. [CrossRef]
39. Bhalla, B.; Saini, M.S.; Jha, M.K. Characterization of Leachate from Municipal Solid Waste (MSW) Landfilling Sites of Ludhiana, India: A Comparative Study. *Int. J. Eng. Res. Appl.* **2012**, *2*, 732–745.
40. Bashir, M.J.K.; Aziz, H.A.; Yusoff, M.S.; Aziz, S.Q.; Mohajeri, S. Stabilized sanitary landfill leachate treatment using anionic resin: Treatment optimization by response surface methodology. *J. Hazard. Mater.* **2010**, *182*, 115–122. [CrossRef]
41. Zakaria, S.N.F.; Abdul Aziz, H.; Abu Amr, S.S. Performance of Ozone/$ZrCl_4$ Oxidation in stabilized landfill leachate treatment. *Appl. Mech. Mater.* **2015**, *802*, 501–506. [CrossRef]
42. Mohajeri, S.; Aziz, H.A.; Isa, M.H.; Zahed, M.A. Treatment of landfill leachate by electrochemicals using aluminium electrodes. *J. Appl. Res. Water Wastewater* **2018**, *10*, 435–440.
43. Zakaria, S.N.; Aziz, A.H. Characteristic of leachate at Alor Pongsu Landfill Site, Perak, Malaysia: A comparative study. *IOP Conf. Ser. Earth Environ. Sci.* **2018**, *140*, 12013. [CrossRef]
44. Aziz, H.A.; Aziz, S.Q.; Yusoff, M.S.; Bashir, M.J.K.; Umar, M. Leachate characterization in semi-aerobic and anaerobic sanitary landfills: A comparative study. *J. Environ. Manag.* **2010**, *91*, 2608–2614. [CrossRef] [PubMed]
45. Thirugnanasambandham, K.; Sivakumar, V. Treatment of egg processing industry effluent using chitosan as an adsorbent. *J. Serb. Chem. Soc.* **2014**, *79*, 743–757. [CrossRef]
46. Guo, J.; Du, J.; Chen, P.; Huang, X.; Chen, Q. Enhanced efficiency of swine wastewater treatment by the composite of modified zeolite and a bioflocculant enriched from biological sludge. *Environ. Technol.* **2018**, *39*, 3096–3103. [CrossRef] [PubMed]
47. Damaraju, M.; Bhattacharyya, D.; Panda, T.; Kurilla, K.K. Application of a continuous bipolar mode electrocoagulation (CBME) system for polishing distillery wastewater. *E3S Web Conf.* **2019**, *93*, 2005. [CrossRef]
48. Drogui, P.; Asselin, M.; Brar, S.K.; Benmoussa, H.; Blais, J.F. Electrochemical removal of pollutants from agro-industry wastewaters. *Purif. Technol.* **2008**, *61*, 301–310. [CrossRef]
49. Ye, Z.; Wang, J.; Sun, L.; Zhang, D.; Zhang, H. Removal of ammonium from municipal landfill leachate using natural zeolites. *Environ. Technol.* **2015**, *36*, 2919–2923. [CrossRef]
50. Eskandari, P.; Farhadian, M.; Solaimany Nazar, A.R.; Jeon, B.H. Adsorption and photodegradation efficiency of TiO_2/Fe_2O_3/PAC /TiO_2/Fe_2O_3/Zeolite nanophotocatalysts for the removal of cyanide. *Ind. Eng. Chem. Res.* **2019**, *58*, 2099–2112. [CrossRef]
51. Khan, S.U.; Islam, D.T.; Farooqi, I.H.; Ayub, S.; Basheer, F. Hexavalent chromium removal in an electrocoagulation column reactor: Process optimization using CCD, adsorption kinetics and pH modulated sludge formation. *Process Saf. Environ. Prot.* **2019**, *122*, 118–130. [CrossRef]
52. Kuang, P.; Chen, N.; Feng, C.; Li, M.; Dong, S.; Lv, L.; Zhang, J.; Hu, Z.; Deng, Y. Construction and optimization of an iron particle–zeolite packing electrochemical–adsorption system for the simultaneous removal of nitrate and by-products. *J. Taiwan Inst. Chem. Eng.* **2018**, *86*, 101–112. [CrossRef]
53. Chan, M.T.; Selvam, A.; Wong, J.W.C. Reducing nitrogen loss and salinity during "struvite" food waste composting by zeolite amendment. *Bioresour. Technol.* **2016**, *200*, 838–844. [CrossRef] [PubMed]
54. Khodam, F.; Rezvani, Z.; Amani-Ghadim, A.R. Enhanced adsorption of Acid Red 14 by co-assembled LDH/MWCNTs nanohybrid: Optimization, kinetic and isotherm. *J. Ind. Eng. Chem.* **2015**, *21*, 1286–1294. [CrossRef]

55. Ahmadi Azqhandi, M.H.; Ghaedi, M.; Yousefi, F.; Jamshidi, M. Application of random forest, radial basis function neural networks and central composite design for modeling and/or optimization of the ultrasonic assisted adsorption of brilliant green on ZnS-NP-AC. *J. Colloid Interface Sci.* **2017**, *505*, 278–292. [CrossRef] [PubMed]
56. Deghles, A.; Kurt, U. Treatment of tannery wastewater by a hybrid electrocoagulation/electrodialysis process. *Chem. Eng. Process. Process Intensif.* **2016**, *104*, 43–50. [CrossRef]
57. Lingamdinne, L.P.; Koduru, J.R.; Chang, Y.Y.; Karri, R.R. Process optimization and adsorption modeling of Pb(II) on nickel ferrite-reduced graphene oxide nano-composite. *J. Mol. Liq.* **2018**, *250*, 202–211. [CrossRef]
58. Tanyol, M.; Kavak, N.; Torlut, G. Synthesis of poly(AN-co-VP)/zeolite composite and its application for the removal of brilliant green by adsorption process: Kinetics, isotherms, and experimental design. *Adv. Polym. Technol.* **2019**, *2019*, 8482975. [CrossRef]
59. Bezerra, M.A.; Santelli, R.E.; Oliveira, E.P.; Villar, L.S.; Escaleira, L.A. Response surface methodology (RSM) as a tool for optimization in analytical chemistry. *Talanta* **2008**, *76*, 965–977. [CrossRef]
60. Abu Amr, S.S.; Aziz, H.A.; Hossain, M.S.; Bashir, M.J.K. Simultaneous removal of COD and color from municipal landfill leachate using Ozone/Zinc sulphate oxidation process. *Glob. Nest J.* **2017**, *19*, 498–504.
61. Mohamad Zailani, L.W.; Mohd Amdan, N.S.; Zin, N.S.M. Removal Efficiency of Electrocoagulation Treatment Using Aluminium Electrode for Stabilized Leachate. *IOP Conf. Ser. Earth Environ. Sci.* **2018**, *140*, 12049. [CrossRef]
62. Ilhan, F.; Kurt, U.; Apaydin, O.; Gonullu, M.T. Treatment of leachate by electrocoagulation using aluminium and iron electrodes. *J. Hazard. Mater.* **2008**, *154*, 381–389. [CrossRef]
63. Li, X.; Song, J.; Guo, J.; Wang, Z.; Feng, Q. Landfill leachate treatment using electrocoagulation. *Procedia Environ. Sci.* **2011**, *10*, 1159–1164. [CrossRef]
64. Mohd Amdan, N.S.; Mohd Zin, N.S.; Mohd Salleh, S.N.A.; Mohamad Zailani, L.W. Addition of composite coagulant (polyaluminium chloride and tapioca flour) into electrocoagulation (aluminium and ferum electrodes) for treatment of stabilized leachate. *MATEC Web Conf.* **2018**, *250*, 7. [CrossRef]
65. Ding, J.; Jiang, J.; Wei, L.; Geng, Y.; Zhao, Q.; Yuan, Y.; Dionysiou, D.D. Organic and nitrogen load removal from bio-treated landfill leachates by a dual-anode system. *Environ. Sci. Water Res. Technol.* **2018**, *4*, 2104–2112. [CrossRef]
66. Aziz, H.A.; Sobri, N.I.M. Extraction and application of starch-based coagulants from sago trunk for semi-aerobic landfill leachate treatment. *Environ. Sci. Pollut. Res.* **2015**, *22*, 16943–16950. [CrossRef]
67. Dia, O.; Drogui, P.; Buelna, G.; Dubé, R. Hybrid process, electrocoagulation-biofiltration for landfill leachate treatment. *Waste Manag.* **2018**, *75*, 391–399. [CrossRef]
68. Moussa, D.T.; El-Naas, M.H.; Nasser, M.; Al-Marri, M.J. A comprehensive review of electrocoagulation for water treatment: Potentials and challenges. *J. Environ. Manag.* **2017**, *186*, 24–41. [CrossRef]
69. Khaled, B.; Wided, B.; Béchir, H.; Elimame, E.; Mouna, L.; Zied, T. Investigation of electrocoagulation reactor design parameters effect on the removal of cadmium from synthetic and phosphate industrial wastewater. *Arab. J. Chem.* **2015**, *12*, 1848–1859. [CrossRef]
70. Kobya, M.; Gengec, E.; Demirbas, E. Operating parameters and costs assessments of a real dyehouse wastewater effluent treated by a continuous electrocoagulation process. *Chem. Eng. Process. Process Intensif.* **2016**, *101*, 87–100. [CrossRef]
71. Nippatla, N.; Philip, L. Electrocoagulation-floatation assisted pulsed power plasma technology for the complete mineralization of potentially toxic dyes and real textile wastewater. *Process Saf. Environ. Prot.* **2019**, *125*, 143–156. [CrossRef]
72. Karamati-Niaragh, E.; Alavi Moghaddam, M.R.; Emamjomeh, M.M.; Nazlabadi, E. Evaluation of direct and alternating current on nitrate removal using a continuous electrocoagulation process: Economical and environmental approaches through RSM. *J. Environ. Manag.* **2019**, *230*, 245–254. [CrossRef]
73. Ghazouani, M.; Akrout, H.; Jellali, S.; Bousselmi, L. Comparative study of electrochemical hybrid systems for the treatment of real wastewaters from agri-food activities. *Sci. Total Environ.* **2019**, *647*, 1651–1664. [CrossRef] [PubMed]

Article

Development of a Combined Aerobic–Anoxic and Methane Oxidation Bioreactor System Using Mixed Methanotrophs and Biogas for Wastewater Denitrification

I-Tae Kim [1,*], Ye-Eun Lee [1,2], Yeong-Seok Yoo [1,2], Wonsik Jeong [1], Young-Han Yoon [1], Dong-Chul Shin [1] and Yoonah Jeong [1]

1 Department of Land, Water and Environment Research, Korea Institute of Civil Engineering and Building Technology, 283, Goyang-daero, Ilsanseo-gu, Goyang-si, Gyeonggi-do 10223, Korea

2 Department of Construction Environment Engineering, University of Science and Technology, 217, Gajeong-ro, Yuseong-gu, Daejeon 34113, Korea

* Correspondence: itkim@kict.re.kr; Tel.: +82-31-910-0301

Received: 28 May 2019; Accepted: 2 July 2019; Published: 4 July 2019

Abstract: We developed a lab-scale aerobic–methane oxidation bioreactor (MOB)–anoxic system, combining a MOB and the aerobic–anoxic denitrification process, and evaluated its potential for advanced nitrogen treatment in wastewater treatment plants (WWTPs). The MOB used biogas generated from a WWTP and secondary-treated wastewater to support mixed methanotroph cultures, which mediated the simultaneous direct denitrification by methanotrophs and methanol production necessary for denitrifying bacteria in the anoxic chamber for denitrification. Compared to the aerobic–anoxic process, the aerobic–MOB–anoxic system with an influent concentration of 4.8 L·day^{-1} showed a marked increase in the reduction efficiency for total nitrogen (41.9% vs. 85.9%) and PO_4^{-3}-P (41.1% vs. 69.5%). However, the integrated actions of high nitrogen and phosphorus consumption are required for methanotroph growth, as well as the production and supply of methanol as a carbon source for denitrification and methane monooxygenase-mediated oxidation of NH_3 into N_2O by methanotrophs. After three months of continuous operation using actual wastewater, the total nitrogen removal rate was 76.3%, equivalent to the rate observed in a tertiary-advanced WWTP, while the total phosphorus removal rate reached 83.7%.

Keywords: aerobic–MOB–anoxic process; biogas; denitrification; mixed methanotroph culture; WWTP

1. Introduction

Methanotrophic bacteria are capable of removing nitrogen and phosphorus while consuming methane as the substrate in an oxidation reaction to produce methanol. To exploit this characteristic, Mechsner and Hamer [1] used methane as the carbon source for denitrification under aerobic conditions and observed an increase in the nitrate (NO_3^-) removal efficiency. Similarly, Werner and Kayser [2] obtained a high denitrification rate by introducing landfill-derived biogas into various types of wastewater treatment reactors, including a complete mixing reactor, trickling filter, and fluidized bed reactor, and subsequently studied the denitrification properties under aerobic conditions. In addition, the potential for methane-dependent denitrification has been verified based on the findings of laboratory-based reactor experiments [3–5]. When ammonia (NH_3) is present, methanotrophs oxidize NH_3 into NH_3OH, and then into N_2O as mediated by methane monooxygenase, a methane-oxidizing enzyme that uses NH_3 as a cometabolite; this reaction resembles the NH_3 oxidation pathway mediated by ammonia-oxidizing bacteria [6,7].

With the aim of improving the standard activated sludge process in wastewater treatment plants (WWTPs), Wuhrmann was the first to develop the nitrogen removal process [8]; Ludzack and Ettinger [9] subsequently developed the advanced nitrogen removal process. The difference between these two processes lies in the carbon source for denitrification and the location of the denitrification tank. In Wuhrmann's process (oxic–anoxic), the denitrification tank is positioned after the aeration tank; as a result, wastewater undergoes removal of organic substances in the aeration tank, and then maintains a long residence time in the denitrification tank for denitrification, which uses organic substances produced upon cell lysis as the carbon source. In the Ludzack–Ettinger process (anoxic–aerobic), the denitrification tank is positioned before the aeration tank and uses the organic substances in the influent wastewater as the carbon source for denitrification; then, the wastewater that has undergone nitrification in the aeration tank is transferred to the denitrification tank as return sludge. For nitrogen removal, the Wuhrmann process requires a retention time of at least 8 h in the denitrification tank; this led to commercialization failure due to the generation of NH_3-N and increase in the turbidity of the effluent. In the case of the Ludzack–Ettinger process, the nitrogen removal efficiency is very low, using only the return sludge sent from the secondary settling tank to the denitrification tank.

To date, most studies on the use of methanotrophs for methanol production have used NO_3^- mineral salts (NMS) medium [10], allowing the study of single species of type II methanotrophs [11,12]. Thus, previous studies of methanol production were conducted under limited, controlled laboratory conditions using only single methanotrophic species and pure methane. To ensure the economic feasibility of using methanotrophs for methanol production, the following issues must be resolved: (i) the need for a source of culture that can replace high-cost sources, such as those cultured in NMS medium; (ii) the practical difficulty of maintaining single-species cultures; and (iii) the cost of methane purification. A previous study [13] verified the potential for producing biomethanol using mixed methanotroph cultures in lieu of single methanotrophic species, as well as for using biogas instead of pure methane. Based on those results, in the present study, we developed and evaluated a process that directly produces methanol as the carbon source for denitrification in WWTPs. We used biogas generated from a WWTP and secondary-treated wastewater as the direct sources of the mixed methanotroph culture and designed a methane oxidation bioreactor (MOB) that could mediate direct denitrification by methanotrophs simultaneously with the production of methanol required for denitrifying bacteria and the anoxic process for denitrification. To combine the MOB with the denitrification process, a novel treatment system (aerobic–MOB–anoxic process) was developed, and its potential as an advanced wastewater treatment technology was verified in a lab-scale reactor.

2. Materials and Methods

2.1. Methanotrophs and Activated Sludge

The methane-oxidizing bacteria used in this study comprised a mixture of methanotrophic species cultured in a previous study [13], in which *Methylomonas methanica*, *Methylococcus capsulatus*, *Methylococcus bovis*, and *Methylobacter marinus* were the dominant species. The activated sludge used in the water treatment process was collected from the aeration tank of the Terminal Wastewater Treatment Plant in Ilsan, Goyang-si, Korea.

2.2. Methanotroph Culture Solution

The mixed methanotroph culture in the developed system was maintained using secondary-treated wastewater collected from the Terminal Wastewater Treatment Plant (Ilsan, Korea). The Terminal Wastewater Treatment Plant in Ilsan has a treatment capacity of 270,000 $m^3 \cdot day^{-1}$ and runs a tertiary advanced treatment facility for nitrogen and phosphorus removal. Table 1 shows the characteristics of the source water for the wastewater used in the unit batch test and the secondary-treated water used as the effluent of the secondary settling tank. All samples were stored in a refrigerator at 4 °C for subsequent use.

Table 1. Characteristics of raw and secondary-treated wastewater.

Parameter	Concentration (mg·L^{-1})	
	Raw Sewage	Treated Wastewater
COD	94.9	12.4
NH_3-N	44.8	16.7
NO_3^--N	4.6	15.3
PO_4^{3-}-P	5.8	1.6

2.3. Biogas Source

Biogas discharged from the anaerobic sludge digestor in the Terminal Wastewater Treatment Plant (Ilsan, Korea) was used in the experiment. The biogas was composed of CH_4 (67.7%), CO_2 (30.3%), N_2 (1.4%), and O_2 (0.6%). It was collected and stored using a 4.9 L high-pressure gas tank (GlobalGastec, Ltd., Buchun, Korea).

2.4. Analysis and Measurements

For the biogas (CH_4, CO_2, N_2, and O_2) analysis, the methane content in the serum bottle headspace was measured using a gas chromatographer (HP Agilent 6890A; Santa Clara, CA, USA) equipped with a packed column (GS-GASPRO; 30 m × 0.32 mm) and flame ionization detector (FID). The conditions were as follows: an inlet temperature of 100 °C, oven temperature of 50 °C (for 3 min), and detector temperature of 200 °C; He as the carrier gas; and the flow rate of to 1.2 mL min^{-1} with a split ratio of 20:1.

For the methanol analysis, a gas chromatographer (HP Agilent 6890A; head space G1888) equipped with a packed column (INNOWAX 30 m × 0.25 mm × 0.25 μm) and a FID was used. The conditions were an oven temperature of 50 °C (5 min), which was raised to 250 °C (for 3 min), and a detector temperature of 250 °C, using N_2 as the carrier gas and setting the flow rate to 1.2 mL min^{-1} with a split ratio of 10:1.

For the formaldehyde analysis, high-performance liquid chromatography (HPLC; Agilent 1100 series) with a column (Eclipse XDB-C18; 5 μm, 4.6 × 150 mm) was used. The mobile phases were mobile A (D.W 100%) and mobile B (ACN 100%), with a mobile flow of 1.5 mL/min, eluent A:B of 50:50, runtime of 38 min, and detection wavelength of 360 nm.

To measure the extracellular polymeric substances (EPS; bound EPS located close to the cells of microorganisms) in the microbial sludge, the samples were centrifuged (Centrifuge-416; Dongseo Science, Ltd., Dangjin, Korea) at 2700× *g*. Then, the protein was analyzed using the Lowry method, i.e., measured via absorbance at 750 nm with a bovine serum albumin standard for quantification [14]. In addition, polysaccharide was analyzed using the phenol–sulfuric acid method and measured as the absorbance at 490 nm, using a glucose standard for quantification [15,16].

For the microbiological analysis, DNA extraction, polymerase chain reaction (PCR) amplification, and pyrosequencing were performed by ChunLab, Inc. (Seoul, Korea). The 16S rRNA genes for each sample were amplified using barcoded universal primers.

Analyses of water quality and components of treated wastewater (TWW) were based on the American Standard Methods for the Examination of Water and Wastewater and Environmental Protection Agency (EPA) Methods (EPA Method 1613).

2.5. Treatment Process Composition

For the batch test to examine the characteristics of methanol production and changes in nutritive salts in the secondary-treated wastewater mediated by methanotrophs, a 160 mL serum bottle was filled with secondary-treated water inoculated with 50 mL of methanotroph culture, after which 25 mL (22.7%) of the 110 mL headspace was substituted with biogas to adjust the O_2, CH_4, and CO_2 contents to 17.8 mL, 16.9 mL, and 7.4 mL, respectively.

The newly developed combined MOB and aerobic–anoxic reactor (aerobic–MOB–anoxic reactor) for nitrogen and phosphorus removal was designed as shown in Figure 1. The biogas and oxygen consumed in the MOB were injected automatically from a gas tank based on input by a pressure sensor. During the biological process of nitrogen removal in aerobic–anoxic reactors, organic substance removal and nitrification occur in the aerobic reactor, whereas NO_3^- denitrification occurs in the anaerobic reactor. In our system, the carbon source for denitrification in the anaerobic reactor was supplied by the effluent containing methanol produced by the MOB. Table 2 presents the operation conditions for the combined aerobic–MOB–anoxic process.

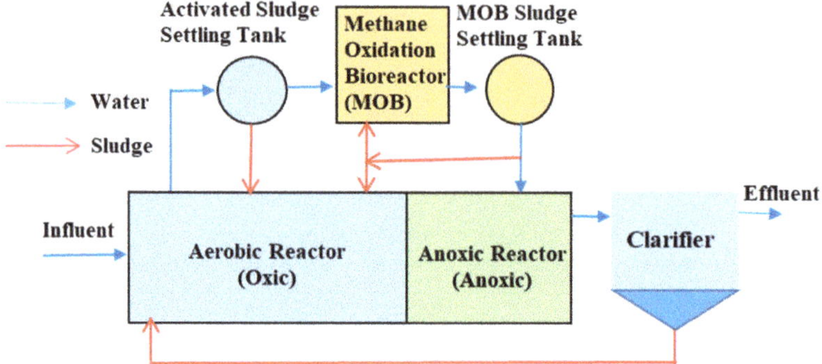

Figure 1. Schematic diagram of the newly developed system combining the aerobic–anoxic process with a methane oxidation bioreactor (MOB).

Table 2. Operation conditions of the aerobic–methane oxidation bioreactor (MOB)–anoxic process.

Reactor	MLSS (mg·L^{-1})	Inflow (L·day^{-1})	HRT (h)
Aerobic	2200–2300	24	4.3
Anoxic	2300–2400	–	1.7
MOB	1000–1200	–	3

MLSS: mixed liquor suspended solids; HRT: hydraulic retention time.

Following primary settling of effluent from the aerobic reactor, the supernatant was transferred to the MOB, and the settled sludge was returned to the aerobic reactor. The MOB effluent underwent another round of settling, after which the supernatant was transferred to the anaerobic reactor. The produced methanol was settled in the anaerobic reactor without a carbon source to enhance the nitrogen removal efficiency, while the settled sludge was returned to maintain the mixed liquor suspended solids (MLSS) content of the MOB at 1000–1200 mg·L^{-1} and residual sludge was sent to the aerobic reactor.

The settling time for the activated sludge settling tank and the sludge settling tank in the MOB were 50 min and 10 min, respectively, as determined based on an analysis of the sludge settling characteristics (see Section 3.3). In the aerobic reactor receiving 24 L·day^{-1} wastewater collected from the influent of the target WWTP, an air stone was used to supply air to enable an adequate supply of dissolved oxygen (DO) in the sludge mixture, while an aerator operating at approximately 150 rpm was installed in the anaerobic reactor to maintain the anaerobic condition. The settled sludge in the final settling tank was returned to the aerobic reactor at 0.5 Q (Q: influent concentration), while the hydraulic retention time (HRT) was set as 6 h and the solids retention time was set as 25 days. The effluent of the aerobic reactor was delivered to the MOB, and after a 3 h reaction (HRT 3 h), the supernatant was returned to the anaerobic reactor at 0.1 Q or 0.2 Q. The MOB including an aerator, a gas supplier, and a pressure gauge was operated at 150 rpm, while biogas was supplied to maintain 25% (*v*/*v*) of the headspace. The mixed liquor suspended solids (MLSS) contents of the reactors were 2300–2400 mg·L^{-1}

for the anaerobic reactor, 2200–2300 mg·L^{-1} for the aerobic reactor, and 1000–1200 mg·L^{-1} for the MOB. All experiments were conducted in the laboratory at 20–25 °C.

3. Results and Discussion

3.1. Methanol Production in the MOB and Characteristics of Nitrogen and Phosphorus Removal

Table 3 presents the methanol and formaldehyde production results using activated sludge or mixed methanotrophs in biogas and secondary-treated wastewater. With an MLSS concentration of 520 mg·L^{-1}, 12.13 mg·L^{-1} (chemical oxygen demand (COD) conversion value 18.20 mg·L^{-1}) and 35.23 mg·L^{-1} (COD conversion value 52.85 mg·L^{-1}) of methanol were produced after 1 h and 3 h of culturing, respectively. Meanwhile, 0.34 mg·L^{-1} (COD conversion value 0.36 mg·L^{-1}) and 1.74 mg·L^{-1} (COD conversion value 1.86 mg·L^{-1}) of formaldehyde were produced after 1 h and 3 h of culturing, respectively. In comparison, substantially less methanol and formaldehyde were produced when activated sludge was used in other identical conditions (i.e., MLSS concentration 520 mg·L^{-1}, culture time).

Table 3. Concentrations of methanol and formaldehyde in the effluent of the methane oxidation bioreactor.

Sludge Inoculant	Time (h)	Methanol (mg·L^{-1})	Formaldehyde (mg·L^{-1})	MLSS (mg·L^{-1})
Activated sludge	3	<0.52 ± 0.02	0.03 ± 0.01	520
Methanotrophs	1	12.13 ± 0.31	0.34 ± 0.02	520
	3	35.23 ± 1.21	1.74 ± 0.08	520

Data represent the means ± standard deviations of three replicates (n = 3).

After 3 h of culturing, the secondary-treated wastewater of the MOB showed a substantial increase in COD, from the initial concentration of 12.42 mg·L^{-1} to 59.84 mg·L^{-1} due to methanol production (Table 4). In contrast, the NH_3-N and NO_3^--N concentrations in the secondary-treated wastewater were drastically reduced by 87.1% and 92.0%, respectively, through the nitrogen-assimilating action of methanotrophs. Furthermore, the PO_4^{-3}-P concentration was reduced by 63.1% compared to the initial concentration (Table 4).

Table 4. Changes in chemical oxygen demand (COD) and nitrogen and phosphorus concentrations via the actions of methanotrophs in the methane oxidation bioreactor using biogas and secondary-treated wastewater.

Parameter	Initial Concentration (mg·L^{-1})	Concentration after 3 h (mg·L^{-1})
COD	12.42 ± 0.22	59.84 ± 0.62
NH_3-N	16.75 ± 0.18	2.16 ± 0.06
NO_3^--N	15.33 ± 0.21	1.23 ± 0.03
PO_4^{-3}-P	1.68 ± 0.02	0.62 ± 0.01

COD: chemical oxygen demand. Data represent the means ± standard deviations of three replicates (n = 3).

Anthony [17] explained such a reduction in nitrogen based on the fact that methanotrophs have a relatively high nitrogen demand during the growth phase, whereby 0.25 mol of nitrogen is consumed to assimilate 1 mol of carbon using methane. This is distinguished from the denitrification mechanism. Herein, methanol is used as the carbon source by methylotrophs and common denitrifying bacteria, as the two mechanisms are assumed to occur concurrently for nitrogen removal via methanotroph-mediated nitrogen assimilation.

Analyzing the nitrogen and phosphorus contents in sludge revealed a higher average total nitrogen (TN) content in the methanotroph sludge (8.53%) than in the activated sludge (5.34%). Similarly, the average total phosphorus (TP) content was much higher in methanotroph sludge (7.8%) than in activated sludge (1.6%). Trotsenko et al. [18] explained that the high phosphorus removal rate in a MOB and higher phosphorus content in methanotroph sludge than activated sludge, as shown in this study, was the result of a high level of intracellular inorganic pyrophosphate (5 mM) in methanotrophs, despite the low level of Adenosine triphosphate (ATP) (0.5 mM).

3.2. Microbial Consortium in the MOB

The microbial consortium of the bacterial sludge in the MOB was comprised of 81.9% methanotrophs, predominated by the genera *Methylomonas*, *Methylococcus*, *Methylomonas_f_uc*, *Methylobacter*, and *Methylosarcina*. Among them, *Methylomonas* was most abundant (48%). Meanwhile, nonmethanotrophic bacterial strains in the genera *Pseudomonas* (5.8%) and *Methylophilus* (6.6%) were relatively abundant (Figure 2). *Pseudomonas* is commonly found in the activated sludge of WWTPs and is mainly isolated from activated sludge for cultures [19–21].

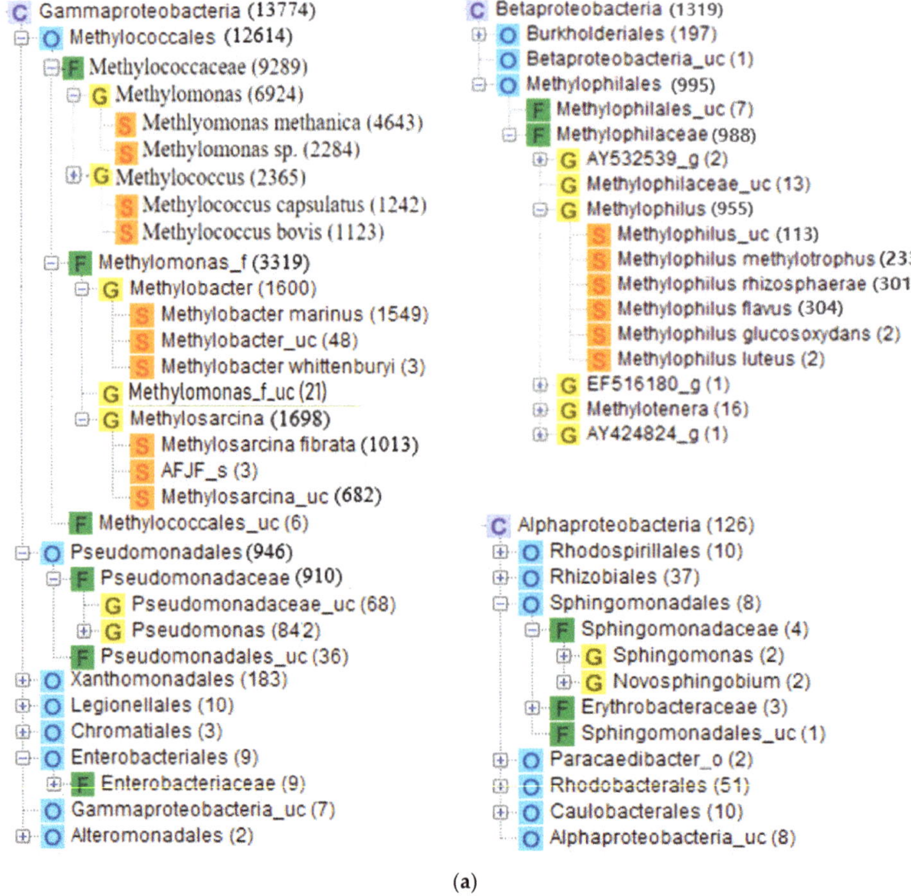

(a)

Figure 2. *Cont.*

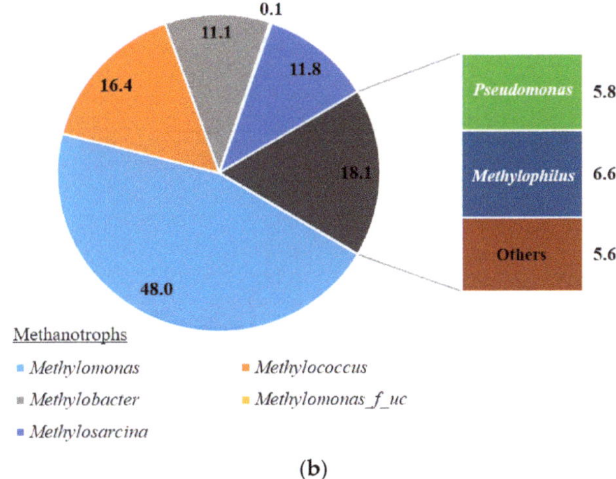

(b)

Figure 2. Microbial consortium in the methane oxidation bioreactor: (**a**) Microbial species found in MOB sludge (output from the CLcommunity program (ChunLab Inc., Seoul, Korea); (**b**) proportions (%) of the most abundant identified taxa at the genus level.

In the MOB sludge, *Methylophilus* was detected in relatively high proportions as a denitrification bacterium when methanol was used as the substrate, and disappeared from the microbial consortium once methanol was depleted. To this end, Osaka et al. [22] discovered an association with *Methylophilus* and *Methylobacillus* of the family Methylophilaceae and the methanol-assimilating denitrification activity of the genus *Aminomonas* using a stable-isotope probing method. Because *Methylophilus falvus* also oxidizes methanol into formaldehyde using pyrroloquinoline quinone-linked methanol dehydrogenase (PQQ-MDH) [23], the denitrification reaction also occurs, and the *Methylophilus* genome only encodes assimilatory denitrification reactions [24]. *Methylophilus flavus*, which was clearly shown to coexist in the present study, is an obligate methanol-using gram-negative bacterium and a strict aerobic bacterium [23]. The bacterium converts methanol into formaldehyde using PQQ-MDH and employs NO_3^- and NH_3 compounds as the carbon source, while assimilating methanol via the ribulose monophosphate pathway. Another dominant strain, *Methylophilus rhizosphaerae*, is also a strictly aerobic gram-negative bacterium that displays an identical mechanism [25].

3.3. Sedimentation Properties of the Methanotrophic Microbial Consortium

Water treatment processes require a sedimentation step to separate treated water and microorganisms. Thus, the sedimentation properties of the MOB sludge were comparatively analyzed with activated sludge in a 1000 mL graduated cylinder. Complete sedimentation of MOB sludge occurred within 5 min (Figure 3), which would require a settling tank retention time of 10 min. By contrast, the activated sludge settling tank retention time would need to be 50 min due to the slow rate of sedimentation of activated sludge.

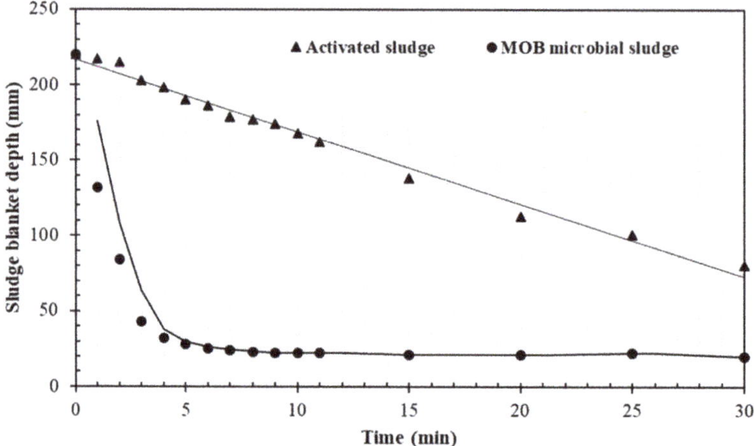

Figure 3. Comparison of the rate of sedimentation between the methane oxidation bioreactor (MOB) sludge and activated sludge. MOB sludge concentration: 1200 mg·L^{-1}; activated sludge concentration: 2300 mg·L^{-1}.

The characteristic of rapid sedimentation brought by MOB sludge was determined by a much higher EPS content than activated sludge (Table 5), resulting in larger particle size in MOB sludge (742 μm) than in activated sludge (107 μm).

Table 5. Comparison of extracellular polymeric substance (EPS) characteristics (mg·g VSS^{-1}) and a particle size between methane oxidation bioreactor (MOB) sludge and activated sludge.

Characteristic	Activated Sludge	MOB Sludge
EPS protein (mg·g VSS^{-1})	76.8 ± 4.9	106.2 ± 9.4
EPS polysaccharide (mg·g VSS^{-1})	31.4 ± 3.2	46.5 ± 2.6
Particle size (μm)	107 ± 12	742 ± 22

Data represent the means ± standard deviations of three replicates (n = 3); VSS: volatile suspended solid.

3.4. Denitrification Efficiency

3.4.1. Aerobic–Anoxic Process

Table 6 presents the characteristics of the aerobic–anoxic process in removing organic substances, nitrogen, and phosphorus using actual WWTP influent compared with the effluent from a general activated sludge process. In the case of the aerobic–anoxic process, the influent COD was 106.3 mg·L^{-1}, which was reduced to 11.5 mg·L^{-1} in the aerobic reactor and 11.4 mg·L^{-1} in the anaerobic reactor, resulting in a total removal efficiency of 89.5% (i.e., 11.1 mg·L^{-1} reduction). In the case of the activated sludge process, the COD removal efficiency was 75.2%. In the aerobic–anoxic process, the influent NH_3-N concentration was 17.2 mg·L^{-1}, of which 96.0% was removed (effluent concentration 0.68 mg·L^{-1}) due to the complete nitrification in the aerobic reactor that had been maintained at a DO concentration of 1.5–1.8 mg·L^{-1}. In contrast, the activated sludge process resulted in only partial nitrification and discharge of 14.98 mg·L^{-1} NH_3-N. The influent NO_3^--N concentration was 1.6 mg·L^{-1} in the aerobic–anoxic process, which increased to 10.6 mg·L^{-1} in the aerobic reactor due to nitrification, and then decreased to a final concentration of 9.9 mg·L^{-1} in the anaerobic reactor, showing negligible removal of NO_3^--N between these two steps. Although 10.7 mg·L^{-1} was transferred from the aerobic reactor to the anaerobic reactor, it is presumed that the lack of a carbon source (the theoretical

COD requirement was approximately 37 mg·L^{-1}; 2.47 g methanol g^{-1} NO_3^--N^{-1} [26]) prevented the denitrification of NO_3^--N. Both processes showed nonsignificant removal rates.

Table 6. Comparison of nutrient removal between the aerobic–anoxic process and the activated sludge process.

Parameter	Influent (mg·L^{-1})	Aerobic–Anoxic Process		Activated Sludge Process	
		Effluent Concentration (mg·L^{-1})	Removal Rate (%)	Effluent Concentration (mg·L^{-1})	Removal Rate (%)
COD	106.3	11.1 ± 1.6	89.56	26.3 ± 2.6	75.25
NH_3-N	17.2	0.68 ± 0.04	96.05	14.98 ± 1.5	10.30
NO_3^--N	1.7	9.9 ± 1.3	41.92 (as TN)	2.7 ± 0.4	2.95 (as TN)
PO_4^{-3}-P	7.34	6.42 ± 0.76	12.53	6.24 ± 0.64	14.98

COD: chemical oxygen demand; TN: total nitrogen. Data represent the means ± standard deviations of 22 experiments (n = 22).

3.4.2. Aerobic–MOB–Anoxic Process

In the combined aerobic–MOB–anoxic process, the settled effluent from the aerobic reactor was transferred to the MOB to provide a denitrification source in the anaerobic reactor, at a rate of 0.1 Q (2.4 L·day^{-1}) or 0.2 Q (4.8 L·day^{-1}), which are 10% and 20%, respectively, of the influent flow rate (24 L·day^{-1}). Table 7 presents the experiment results of the inflow to the anaerobic reactor following the methane oxidation reaction mediated by the injection of biogas.

Table 7. Comparison of nutrient removal in the aerobic–methane oxidation bioreactor (MOB)–anoxic process under supply rates of settled effluent into the MOB.

Parameter	Influent (mg·L^{-1})	0.1 Q [1] (2.4 L·day^{-1})		0.2 Q [1] (4.8 L·day^{-1})	
		Effluent Concentration (mg·L^{-1})	Removal Rate (%)	Effluent Concentration (mg·L^{-1})	Removal Rate (%)
COD	129.3	9.4 ± 1.2	92.7	9.7 ± 0.8	92.5
NH_3-N	27.93	0.12 ± 0.01	99.9	0.11 ± 0.01	99.9
NO_3^--N	1.3	8.4 ± 0.6	71.21 (as TN)	4.1 ± 0.2	85.87 (as TN)
PO_4^{-3}-P	6.49	3.82 ± 0.06	41.14	2.0 ± 0.03	69.47

COD: chemical oxygen demand; TN: total nitrogen. Data represent the means ± standard deviations of 22 experiments (n = 22). [1] Q: influent concentration = 24 L·day^{-1}.

At 0.1 Q (2.4 L·day^{-1}) of influent wastewater from the MOB, the influent COD concentration was 129.3 mg·L^{-1}, which was reduced to 18.1 mg·L^{-1} in the aerobic reactor and then to 11.9 mg·L^{-1} in the anaerobic reactor; therefore, the COD concentration in the final treated water was 9.4 mg·L^{-1} (removal efficiency 92.7%). The influent NH_3-N concentration was 27.93 mg·L^{-1}, which was mostly removed in the aerobic reactor through nitrification. Although the NO_3^--N concentration increased to 23.4 mg·L^{-1} in the aerobic reactor, it was subsequently reduced to 8.4 mg·L^{-1}. The influent PO_4^{-3}-P concentration was 6.49 mg·L^{-1}, but the aerobic and anoxic processes resulted in a final concentration of 3.82 mg·L^{-1}. Although negligible NO_3^--N was removed in the anaerobic reactor without the MOB-treated water as a carbon source (Table 6), when 0.1 Q MOB effluent was provided to the anaerobic reactor, the TN removal efficiency increased substantially from 41.9% to 71.2%. PO_4^{-3}-P showed a similar trend; the reduction efficiency was 12.5% without MOB effluent (Table 6), but was 41.1% when 0.1 Q MOB effluent was supplied (Table 7).

At 0.2 Q (4.8 L·day^{-1}) of MOB effluent, the influent COD concentration of 129.3 mg·L^{-1} was reduced to 17.8 mg·L^{-1} in the aerobic reactor and to 11.6 mg·L^{-1} in the anaerobic reactor; therefore, the final treated water upon discharge had a COD concentration of 9.7 mg·L^{-1} (Table 7). Nitrification in the aerobic reactor removed most of the 27.93 mg·L^{-1} of NH_3-N, and although NO_3^--N initially increased to 21.9 mg·L^{-1} in the aerobic reactor, it was reduced to 5.3 mg·L^{-1} in the anaerobic reactor, resulting in a final concentration of 4.1 mg·L^{-1} in the treated water. Compared to the nitrogen removal trend upon the supply of 0.1 Q, the TN removal efficiency was enhanced by 1.2 times upon increasing the proportion of MOB effluent supply to 0.2 Q. Similarly, the influent PO_4^{-3}-P concentration was 6.49 mg·L^{-1}, but removal via the aerobic and anoxic processes resulted in a final concentration of 2.0 mg·L^{-1}, showing a 1.7-fold increase in removal efficiency compared to the 0.1 Q condition.

3.4.3. Comparison with a Tertiary Advanced WWTP Facility

The aerobic–MOB–anoxic process developed in this study was operated for three months, and the results were compared with the nutrient removal characteristics of an actual tertiary advanced system in a WWTP using the same wastewater influent. For the tertiary advanced treatment to remove nitrogen and phosphorous, the plant runs a water treatment system based on the modified Ludzack–Ettinger process [9,27] and an ultrarapid coagulation (URC) technique for phosphorus removal via condensation filtration. Figure 4 shows the changes in the composition of wastewater influent and effluent during the three-month operation of the MOB and WWTP.

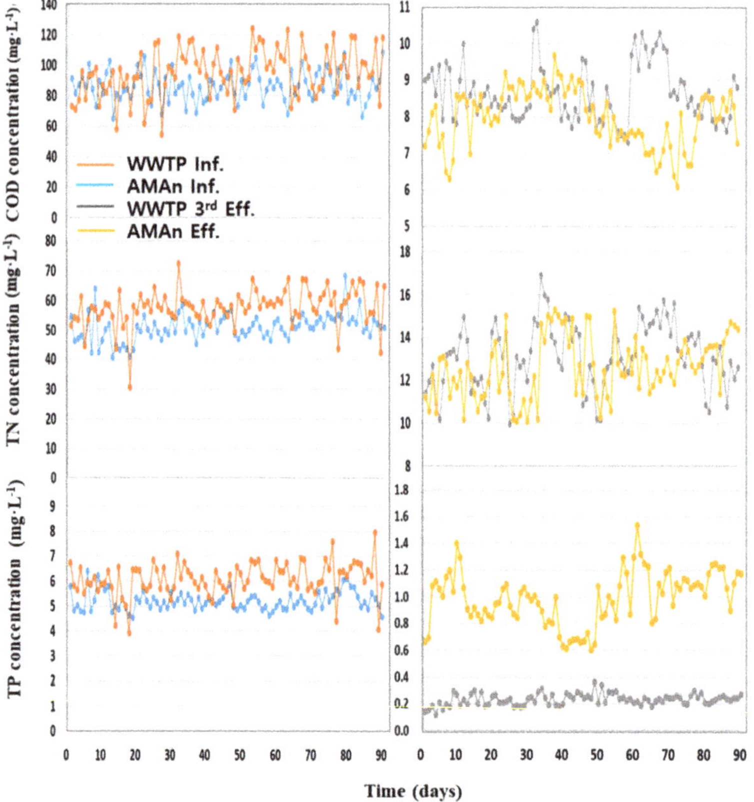

Figure 4. Comparison of the changes in influent water quality based on chemical oxygen demand (COD), total nitrogen (TN), and total phosphorus (TP) removal between the tertiary advanced treatment of a wastewater treatment plant (WWTP) and the long-term operation of the aerobic–methane oxidation bioreactor–anoxic system (AMAn). Orange line stands for WWTP influent (Inf.), blue line AMAn influent, gray line tertiary WWTP effluent, and yellow line AMAn effluent (Eff.).

Compared to the water quality of the WWTP tertiary advanced treatment, the newly developed aerobic–MOB–anoxic process (under the 0.2 Q condition) showed a consistent and similar treatment efficiency with respect to COD and TN removal. In case of TP, the process in this study led to a relatively high removal efficiency of 83.7% (Table 8), even without an additional phosphorus removal system. This is encouraging because the efficiency of the aerobic–MOB–anoxic process was lower than the 95.5% efficiency of the WWTP, which used a separate URC technique for phosphorus removal.

Table 8. Comparison of the total nitrogen and total phosphorus removal rates between the wastewater treatment plant (WWTP) tertiary advanced treatment and the aerobic–MOB–anoxic process.

	COD				Total Nitrogen				Total Phosphorus			
	Influent		Effluent		Influent		Effluent		Influent		Effluent	
	WWTP	AMAn	WWTP	AMAn	WWTP	AMAn	WWTP	AMAn	WWTP	AMAn	WWTP	AMAn
Conc. ($mg{\cdot}L^{-1}$)	86.4	95.9	8.6	8.0	51.0	57.9	13.1	13.7	5.23	6.08	0.23	0.21
SD	9.77	16.63	0.77	0.78	4.56	6.10	1.54	1.23	0.38	0.67	0.04	0.20
Removal rate (%)	90.0	91.7	–	–	74.2	76.3	–	–	95.5	83.7	–	–

COD, chemical oxygen demand; WWTP, tertiary advanced wastewater treatment plant; AMAn, aerobic–MOB–anoxic process; SD, standard deviation. Data represent the means ± SD from WWTP experiments (n = 84) and AMAn experiments (n = 89).

4. Conclusions

We developed a new combined aerobic–MOB–anoxic wastewater treatment system for improved nitrogen and phosphorus removal, and compared the MOB sludge with typical activated sludge and the whole system with a tertiary advanced WWTP using the same wastewater influent. Our system relies on the production of methanol and formaldehyde from biogas and secondary-treated wastewater inoculated with cultured, mixed methanotroph species; the MOB secondary-treated wastewater showed a nearly five-fold increase in COD concentration (to 59.84 $mg{\cdot}L^{-1}$) due to methanol production, whereas NH_3-N and NO_3^--N showed substantial reductions of 87.1% and 92.0%, respectively, due to methanotroph-mediated nitrogen assimilation. PO_4^{-3}-P also showed a 63.1% reduction (to 0.62 $mg{\cdot}L^{-1}$).

Methanotrophs accounted for 81.9% of the microbial consortium of the MOB microbial sludge, including the genera *Methylomonas*, *Methylococcus*, *Methylomonas_f_uc*, *Methylobacter*, and *Methylosarcina*; relatively abundant nonmethanotrophs included the genera *Pseudomonas* and *Methylophilus*. The MOB sludge had a much higher EPS content (protein concentration 76.8 $mg{\cdot}L^{-1}$ vs. 106.2 $mg{\cdot}L^{-1}$) and particle size (742 μm vs. 107 μm) than that of activated sludge, leading to much more rapid sedimentation and shorter retention time in a settling tank.

Compared to the aerobic–anoxic process, the aerobic–MOB–anoxic process (applying 0.2 Q of MOB settled effluent to the anoxic tank) led to substantial improvement in TN removal (41.9% vs. 85.9%) and PO_4^{-3}-P removal (41.1% vs. 69.5%). After three months of operation with actual wastewater, the aerobic–MOB–anoxic process showed a TN removal rate of 76.3%, similar to that of the tertiary advanced WWTP, as well as an 83.7% phosphorus removal rate.

This study confirmed that the new denitrification system combining an aerobic–anoxic process and methane oxidation bioreactor is applicable as an advanced sewage treatment method.

Author Contributions: Conceptualization, I.-T.K. and Y.-S.Y.; Validation, Y.-E.L. and Y.-S.Y.; Formal Analysis, I.-T.K., W.J. and Y.-H.Y.; Investigation, I.-T.K., D.-C.S. and Y.-E.L.; Resources, I.-T.K.; Writing—Original Draft Preparation, I.-T.K.; Writing—Review & Editing, I.-T.K., Y.J., and Y.-E.L.; Supervision, Y.-S.Y.; Project Administration, I.-T.K.

Funding: This work was supported by a Major Project of the Korea Institute of Civil Engineering and Building Technology, grant number 20190152-001.

Conflicts of Interest: The authors declare no conflict of interest.

References

1. Mechsner, K.L.; Hamer, G. *Denitrification by Methanotrophic, Methylotrophic Bacterial Associations in Aquatic Environments*; Plenum Press: New York, NY, USA, 1985; pp. 257–271.
2. Werner, M.; Kayser, R. Denitrification with Biogas as External Carbon Source. *Water Sci. Technol.* **1991**, *23*, 701–708. [CrossRef]
3. Thalasso, F.; Vallecillo, A.; Garcıa-Encina, P.; Fdz-Polanco, F. The use of methane as a sole carbon source for wastewater denitrification. *Water Res.* **1997**, *31*, 55–60. [CrossRef]
4. Houbron, E.P.; Torrijos, M.; Capdeville, B. An Alternative Use of Biogas Applied at the Water Denitrification. *Water Sci. Technol.* **1999**, *40*, 115–122. [CrossRef]
5. Rajapakse, J.P.; Scutt, J.E. Denitrification with natural gas and various new growth media. *Water Res.* **1999**, *33*, 3723–3734. [CrossRef]
6. Bedard, C.; Knowles, R. Physiology, biochemistry, and specific inhibitors of CH4, NH4+, and CO oxidation by methanotrophs and nitrifiers. *Microbiol. Rev.* **1989**, *53*, 68–84. [PubMed]
7. Calsen, H.N.; Joergensen, L.; Degn, H. Inhibition by ammonia of methane utilization in Methylococcus capsulatus (Bath). *Appl. Microbiol. Biotechnol.* **1991**, *35*, 124–127.
8. Applegate, C.S.; Wilder, B.; DeShaw, J.R. Total Nitrogen Removal in a Multichemical Oxidation System. *J. Water Pollut. Control Fed.* **1980**, *52*, 568–577.
9. Ludzack, F.J.; Ettinger, M.B. Controlling Operation to Minize Activated Sludge Effluent Nitrogen. *J. Water Pollut. Control Fed.* **1962**, *34*, 920–931.
10. Higgins, I.J.; Best, D.J.; Hammond, R.C.; Scott, D. Methane-oxidizing microorganisms. capsularus (Bath). Its ability to oxygenate n-alkanes. *Microbiol. Rev.* **1981**, *45*, 556–590.
11. Patel, S.K.; Mardina, P.; Kim, S.Y.; Lee, J.K.; Kim, I.W. Biological Methanol Production by a Type II Methanotroph Methylocystis bryophila. *J. Microbiol. Biotechnol.* **2016**, *26*, 717–724. [CrossRef]
12. Mardina, P.; Li, J.; Patel, S.K.S.; Kim, I.W.; Lee, J.K.; Selvara, C. Potential of Immobilized Whole-Cell Methylocella tundrae as a Biocatalyst for Methanol Production from Methane. *J. Microbiol. Biotechnol.* **2016**, *26*, 1234–1241. [CrossRef] [PubMed]
13. Kim, I.T.; Yoo, Y.S.; Yoon, Y.H.; Lee, Y.E.; Jo, J.H.; Jeong, W.; Kim, K.-S. Bio-Methanol Production Using Treated Domestic Wastewater with Mixed Methanotroph Species and Anaerobic Digester Biogas. *Water* **2018**, *10*, 1414. [CrossRef]
14. Raunkjær, K.; Hvitved-Jacobsen, T.; Nielsen, P.H. Measurement of pools of protein, carbohydrate and lipid in domestic wastewater. *Water Res.* **1994**, *28*, 251–262. [CrossRef]
15. DuBois, M.; Gilles, K.A.; Hamilton, J.K.; Rebers, P.A.; Smith, F. Colorimetric Method for Determination of Sugars and Related Substances. *Anal. Chem.* **1956**, *28*, 350–356. [CrossRef]
16. Wu, Z.; Zhiwei, W.Z.; Zhou, Z.; Yu, G.; Gu, G. Sludge rheological and physiological characteristics in a pilot-scale submerged membrane bioreactor. *Desalination* **2007**, *212*, 152–164. [CrossRef]
17. Anthony, C. *The Biochemistry of Methylotrophs*; Academic Press Ltd.: London, UK, 1982; pp. 235–279.
18. Trotsenko, Y.A.; Shishkina, V.N. Studies on phosphate metabolism in obligate methanotrophs. *FEMS Microbiol. Lett.* **1990**, *87*, 267–271. [CrossRef]
19. Annaduraia, G.; Juangb, R.S.; Lee, D.J. Microbiological degradation of phenol using mixed liquors of Pseudomonas putida and activated sludge. *Waste Manage.* **2002**, *22*, 703–710. [CrossRef]
20. Nishihara, T.; Okamoto, T.; Nishiyama, N. Biodegradation of didecyldimethyl ammonium chloride by Pseudomonas fluorescens TN4 isolated from activated sludge. *J. Appl. Microbiol.* **2000**, *88*, 641–647. [CrossRef]
21. Dhanya, V.; Usharani, M.V.; Jayadev, P. Biodegradation of cyclohexanol and cyclohexanone using mixed culture of Pseudomonas in activated sludge process. *J. Environ. Res. Dev.* **2016**, *11*, 324–331.
22. Osaka, T.; Yoshie, S.; Tsuneda, S.; Hirata, A.; Iwami, N.; Inamori, Y. Identification of acetate- or methanol-assimilating bacteria under nitrate-reducing conditions by stable-isotope probing. *Microb. Ecol.* **2006**, *52*, 253–266. [CrossRef]
23. Gogleva, A.A.; Kaparullina, E.N.; Doronina, N.V.; Trotsenko, Y.A. Methylophilus flavus sp. nov. and Methylophilus luteus sp. nov., aerobic, methylotrophic bacteria associated with plants. *Int. J. Syst. Evol. Microbiol.* **2010**, *60*, 2623–2628. [CrossRef] [PubMed]

24. Beck, D.A.C.; McTaggart, T.L.; Setboonsarng, U.; Vorobev, A.; Kalyuzhnaya, M.G.; Ivanova, N.; Goodwin, L.; Woyke, T.; Lidstrom, M.E.; Chistoserdova, L. The expanded diversity of Methylophilaceae from Lake Washington through cultivation and genomic sequencing of novel ecotypes. *PLoS ONE* **2014**, *9*, e102458. [CrossRef] [PubMed]
25. Madhaiyan, M.; Poonguzhali, S.; Kwon, S.; Sa, T. Methylophilus rhizosphaerae sp. nov., a restricted facultative methylotroph isolated from rice rhizosphere soil. *Int. J. Syst. Evol. Microbiol.* **2009**, *59*, 2904–2908. [CrossRef] [PubMed]
26. McCarty, P.L.; Beck, L.; Amant, P.S. Biological Denitrification of Wastewater by Addition of Organic Materials. In Proceedings of the 24th Industrial Waste Conference, Lafayette, IN, USA, 6–8 May 1969.
27. Metcalf & Eddy Inc. *Wastewater Engineering: Treatment and Reuse*, 4th ed.; Tchobanoglous, G., Burton, F.L., Stensel, H.D., Eds.; McGraw-Hill: New York, NY, USA, 2003.

Article

Cross-Linked Magnetic Chitosan/Activated Biochar for Removal of Emerging Micropollutants from Water: Optimization by the Artificial Neural Network

Amin Mojiri [1,*], Reza Andasht Kazeroon [2] and Ali Gholami [3]

1 Department of Civil and Environmental Engineering, Hiroshima University, 1-4-1 Kagamiyama, Higashihiroshima 739-8527, Japan
2 Faculty of Civil Engineering, University Technology Mara (UiTM), Shah Alam 40450, Selangor, Malaysia; reza.andasht@gmail.com
3 Department of Soil Science, Ahvaz Branch, Islamic Azad University, Ahvaz 6165765675, Iran; ali.gholami54@gmail.com
* Correspondence: amin.mojiri@gmail.com

Received: 18 February 2019; Accepted: 13 March 2019; Published: 17 March 2019

Abstract: One of the most important types of emerging micropollutants is the pharmaceutical micropollutant. Pharmaceutical micropollutants are usually identified in several environmental compartments, so the removal of pharmaceutical micropollutants is a global concern. This study aimed to remove diclofenac (DCF), ibuprofen (IBP), and naproxen (NPX) from the aqueous solution via cross-linked magnetic chitosan/activated biochar (CMCAB). Two independent factors—pH (4–8) and a concentration of emerging micropollutants (0.5–3 mg/L)—were monitored in this study. Adsorbent dosage (g/L) and adsorption time (h) were fixed at 1.6 and 1.5, respectively, based on the results of preliminary experiments. At a pH of 6.0 and an initial micropollutant (MP) concentration of 2.5 mg/L, 2.41 mg/L (96.4%) of DCF, 2.47 mg/L (98.8%) of IBP, and 2.38 mg/L (95.2%) of NPX were removed. Optimization was done by an artificial neural network (ANN), which proved to be reasonable at optimizing emerging micropollutant elimination by CMCAB as indicated by the high R^2 values and reasonable mean square errors (MSE). Adsorption isotherm studies indicated that both Langmuir and Freundlich isotherms were able to explain micropollutant adsorption by CMCAB. Finally, desorption tests proved that cross-linked magnetic chitosan/activated biochar might be employed for at least eight adsorption-desorption cycles.

Keywords: chitosan; diclofenac; ibuprofen; magnetic biochar; naproxen

1. Introduction

Emerging micropollutants or organic micropollutants exist in the environment at trace concentrations, and their impact on the human health and the environment are presently unknown. These pollutants are contained in polycyclic aromatic hydrocarbons (PAH), personal care products, pharmaceuticals, pesticides, industrial chemicals, and metallic trace elements [1]. Pharmaceutical micropollutants are commonly found in various environmental compartments. The growing use of pharmaceuticals is raises questions regarding their potential risk to human health, the environment, and water quality [2]. Diclofenac, ibuprofen, and naproxen are non-steroidal anti-inflammatory drugs (NSAIDs), which are a commonly consumed class of pharmaceuticals [3]. All pharmaceuticals belonging to this group are acidic in nature with pKa values in the range of 3–5 [4].

Among the different non-steroidal anti-inflammatory drugs, diclofenac is widely applied. Diclofenac (Figure 1; 2-((2,6-dichlorophenyl)amino)phenylacetic-acid) has been stated to cause chronic results such as renal and gastrointestinal tissue damage in some vertebrates [5]. Ibuprofen (Figure 1;

2-(4-Isobutylphenyl)propionic acid) is applied for the treatment of pain and inflammation and dropping of a fever [6]. Naproxen (Figure 1) enters aquatic environments chiefly over the effluents of wastewater treatment plants. It is categorized as a high-priority pharmaceutical. Naproxen might affect living organisms and diminish the biodiversity of natural environmental communities because of its biological activities [7]. Mostly, conventional wastewater treatment methods fail to eliminate pharmaceuticals totally from the water [2]. One of the most promising ways to remove emerging micropollutants is by using adsorbents. Biochar and chitosan are low-cost adsorbents which have been previously used in the literature to remove micropollutants from water [8,9].

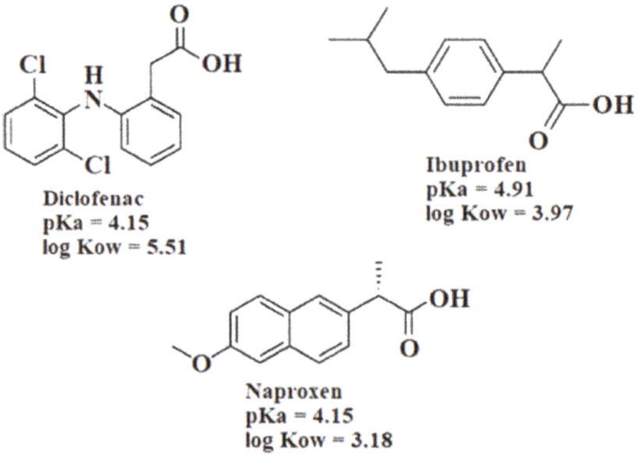

Diclofenac
pKa = 4.15
log Kow = 5.51

Ibuprofen
pKa = 4.91
log Kow = 3.97

Naproxen
pKa = 4.15
log Kow = 3.18

Figure 1. Structures of the studied organic micropollutants [10,11].

Chitosan is one of the biopolymers that is derived from chitin; chitin is a natural amino polysaccharide and is composed primarily of repeating β-(1,4)-2-amino-2-deoxy-d-glucose (or d-glucosamine) units. The benefits of chitosan include its low cost, ease of polymerization and functionalization, and good stability [12]. Amouzgar and Salamatinia [8] stated that chitosan has certain capabilities for removing emerging micropollutants from water. Another low-cost adsorbent is biochar.

Thermochemical decomposition procedures transform biomass materials to syngas, bio-oil, and biochar. Biochar is low cost, environmentally friendly, and can be applied for a variety of purposes [13]. Quesada et al. [14] stated that using biochar as a low-cost material is a promising way to eliminate pharmaceuticals from wastewater. Sizmur et al. [15] stated that the activation process improves the surface area and porosity of biochar, so its adsorption capacity might be increased. Hence, this study aimed to produce a new cross-linked magnetic chitosan/activated biochar to remove emerging micropollutants and to optimize the removal efficiency using an artificial neural network (ANN). This experiment design and its optimization process have not been previously reported in the literature.

2. Materials and Methods

2.1. Materials

In this study, biochar was extracted from agricultural residues. Chitosan (medium molecular weight; code: 07947-52), diclofenac sodium (DCF; $C_{14}H_{10}Cl_2NNaO_2$; 98%; molecular weight = 294.05 g/mol), ibuprofen (IBP; $C_{13}H_{18}O_2$; 98%; molecular weight = 206.3 g/mol), and naproxen (NPX; 98%; $C_{14}H_{14}O_3$; molecular weight = 230.2 g/mol) were obtained from Sigma–Aldrich in the analytical purity and applied in the experiments directly without any further purification. Chloroform, acetone, and methanol (99.5% mass purity) were from Merck.

2.2. *Producing Cross-Linked Magnetic Chitosan/Activated Biochar (CMCAB)*

Based on the method by Liu et al. [16] in the first step, magnetic fluid was prepared by a co-precipitation technique. Fe^{2+} and Fe^{3+} (molar ratio 2:3) solution was placed into a beaker using a stirrer at 55 °C, then NaOH solution was added dropwise with continuous stirring for almost 15 min until the pH got to 9.0. After altering the temperature of the reaction vessels to 65 °C, 0.8 mL Tween 80 was augmented into the mixture using a stirrer for 30–40 min, and the pH value was adjusted to 7.0. After that, the product was washed with distilled water three times and was dispersed in an ultrasonic device for 40 min. Finally, the solution was diluted to gain magnetic fluid (40 g L^{-1}).

In the second step, the activated biochar was produced. Biochar extracted from agricultural residues was done by an activation process with 4 M NaOH for 2 h and then dried for 12 h at 105 °C. Then, the biochar was separated from the NaOH solution via a Buchner filter funnel, heated at 800 °C for 2 h under a 2 L/min nitrogen gas flow, and then let to cool at a rate of 10 °C /min. The activated biochar was washed consecutively with deionized (DI) water and 0.1 M HCl to attain pH 7 and dried again at 105 °C. As a final point, the activated biochar was crushed and sieved through a 200-mesh (74 μm) sieve [17].

Finally, to achieve the cross-linked magnetic chitosan/activated biochar, 5.0 g of chitosan was dissolved in 250 mL 2% acetic solution with stirring. Next, 25 mL of magnetic fluid was added dropwise into the solution with constant stirring for 30 min in a water bath at 50 °C. Then, 5.0 g activated biochar was augmented with continuous stirring for another 60 min. Afterward, 6 mL of glutaraldehyde was injected into the reaction system to produce a gel and the pH of the reaction system was adjusted to 8.0–10.0. As a final point, the mixture was retained in a water bath for 1 h. The precipitate was washed till the pH touched about 7 and was dried at 60 °C and sieved [16]. Table 1 shows the features of the cross-linked magnetic chitosan/activated biochar (CMCAB). The CMCAB features were monitored by the Autosorb (Quantachrome AS1wintm, version 2.02, Quantachrome Instruments, Boynton Beach, FL, USA). In terms of the BET technique, the specific surface area and pore size distribution of CMCAB with the specific surface area and pore size distribution analyzers were determined under the conditions of liquid nitrogen temperature. The zeta potential of the CMCAB was analyzed by the zeta potential meter (Zetasizer nano-ZS90, Malvern Panalytical Ltd, Malvern, UK) at 25 °C in different pH (3–9).

Table 1. Characteristics of the cross-linked magnetic chitosan/activated biochar (CMCAB).

Parameter	Unit	Value
BET surface area	m^2/g	502
Langmuir surface area	m^2/g	796
BJH method cumulative adsorption surface area	m^2/g	12.7
Single point surface area at p/p_o 0.2027	m^2/g	422
Micropore area	m^2/g	217
Single point total pore volume of pores less than 1265.1476 in diameter at p/p_o 0.9845	cc/g	0.4
Micropore volume	cc/g	0.11
Average pore diameter (4 *v/a* by Langmuir)	A	8.9
BJH adsorption average pore diameter (4 *v/a*)	A	31.2

2.3. *Producing the Synthetic Aqueous Solution and Experiment Design*

Stock solutions of organic micropollutants were prepared in acetone, chloroform, or methanol as described by Sühnholz et al. [18]. In this study, the initial concentration of organic micropollutants ranged from 0.5 mg/L [19] to 3 mg/L [20]. The pH was varied from 4 to 8 [21]. Based on preliminary experiments, the adsorption time (h) was fixed at 1.5, which is in line with selected ranges by

Kim et al. [9]. Based on preliminary experiments, the adsorbent dosage was fixed at 1.6 g/L, which is in line with the findings of Wu et al. [22]. Based on preliminary experiments, each run was carried out at room temperature (25 ± 1 °C) using a shaker with 300 rpm shaking speed for all conditions [17,23]. A schematic of the current study is shown in Figure 2.

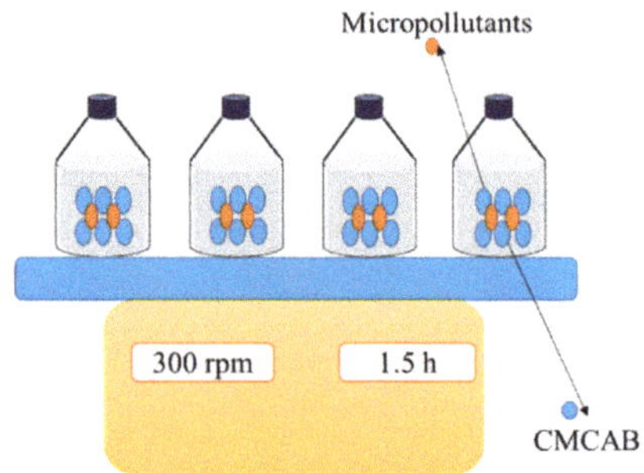

Figure 2. Schematic of experiments.

2.4. Analytical Techniques

All analytical methods were conducted on the basis of the standard methods [24]. The concentrations of emerging micropollutants were tested via ultraviolet spectra and measured by a high-pressure liquid chromatography (HPLC) (LC-20AT, Shimadzu International Trading (Shanghai) Co., Ltd., Tokyo, Japan). The analytical techniques for diclofenac (DCF), ibuprofen (IBP), and naproxen (NPX) were obtained from the literature [25]. The applied mobile phase contained a mixture of acetonitrile and 0.2% formic acid in water (60:40, v/v) at a flow rate of 0.8 mL min^{-1}. The concentrations of DCF, IBP, and NPX were tested using a UV detector at the wavelengths of 200, 200, and 230 nm.

2.5. Optimization Analysis using an Artificial Neural Network (ANN)

The percentage of micropollutants (MP) eliminated from the solution was estimated using Equation (1)

$$\text{Removal}\ \% = \frac{C_i - C_f}{C_i} \times 100 \tag{1}$$

The initial concentration of MP and the final concentration of MP are denoted by C_i and C_f, respectively.

MATLAB R2015a software (R2015a, Mathsworks, Natick, MA, USA) was applied to model the adsorption procedure on the basis of an ANN. Figure 3 displays the topology for the ANN and the variation of parameters in this study. The two neurons in the input layer represent pH (4–8) and micropollutant concentration (0.5–3 mg/L). There were four neurons in the hidden layer and one neuron in the output layer (removal efficiency) for modeling each micropollutant elimination. A total of 50 experimental results applied to model the network were divided randomly into training (60%), validation (20%), and test (20%) sets [26]. The ANN performance was defined based on the values of the mean squared error (MSE) and coefficient of determination (R^2). They were respectively evaluated using Equations (2) and (3). Levenberg–Marquardt (LM) was applied to train the model, and validation was stopped when the maximum validation failures were equal to zero.

$$\text{MSE} = \frac{1}{N}\sum\nolimits_{i=1}^{N}\left(\left|y_{prd,i} - y_{exp,i}\right|\right)^2, \tag{2}$$

$$R^2 = 1 - \frac{\sum_{i=1}^{N}\left(y_{prd,\ i},\ y_{exp,i}\right)}{\sum_{i=1}^{N} y_{prd,i} - y_m}, \tag{3}$$

In Equations (4) and (5), $y_{prd,i}$ refers to the predicted value using the ANN model, $y_{exp,i}$ is the experimental value, N is the number of datapoints, and y_m indicates the average of the experimental values.

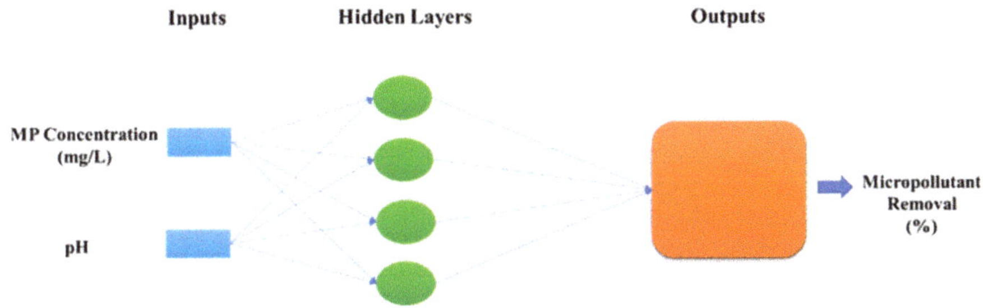

Figure 3. Schematic of an artificial neural network (ANN) design.

2.6. Adsorption Isotherm Study

Batch adsorption studies were done via different dosages (1–7 g/L) of the CMCAB in a fixed MP concentration (2.5 mg/L), pH (6), and adsorption time (30 min). Beakers with working volumes of 100 mL were shaken at 300 rpm for 30 min.

The capacity of adsorption (mg/g) was estimated via the following Equation (4) [27]:

$$q_e = \frac{(C_0 - C_{eq})V}{m_s}, \quad (4)$$

where the initial micropollutant (MP) concentration is denoted by q_e, C_{eq} is the MP concentration (mg L^{-1}) at equilibrium, the volume of solution (L) is represented by V, and m_s is the mass of the adsorbent (g).

2.7. Regeneration and Desorption Study

Regeneration studies were carried out to monitor the economic usability of the CMCAB adsorbent. The adsorbent was regenerated by soaking in 100 mL methanol for 2–3 h in batch experiments and then washed using distilled water in order to consider the desorption and regeneration of the CMCAB. Eight adsorption/desorption cycles were carried out. After every cycle, the residual concentration of MPs was monitored [28].

3. Results and Discussion

The efficiency of the removal of emerging micropollutants via cross-linked magnetic chitosan/activated biochar (CMCAB) is shown in Table 2. Figure 4 shows the FTIR results of CMCAB.

In the FTIR results of chitosan (Figure 4a), peaks 3398 and 2913 can be attributed to O–H and C–H, respectively [29]; peak 1613 may be related to C = O [30]. N–H and CH–OH could explain peaks 1584 and 1401, respectively [29]; and peak 837 is attributed to CH groups [30]. In the FTIR results of activated biochar (Figure 4b), peaks 3207 and 2981 are attributed to O–H and C–H, respectively, while peaks 1608 and 1513 may be related to C = O and C = C, respectively [31]. C–O and O–H could be responsible for peak 1201 [31], and peak 842 is attributed to C–H groups [30]. In the FTIR results of the CMCAB (Figure 4c), peaks 3496 and 2915 are attributed to –OH (or –NH) and C–H, respectively [29]. Peaks C = N and C–O could explain peaks 1638 and 1043, respectively [32,33] and peaks 771 and 573 are attributed to Fe–O [32,33]. The zeta potential of CMCAB is shown in Figure 5. Based on Figure 5, the zeta potential of CMCAB was positive in pH (3) to (5) it is in line with finding of Liu et al. [16] and Zhang et al. [33]. Zeta potentials (mV) were 19, 16 and 1 in pH (3), pH (4) and pH (5), respectively. After that zeta potential became negative which could be supported by findings of Zhang et al. [33].

Zeta potentials (mV) were −5, −6, −8 and −11 in pH (6), pH (7), pH (8), and pH (9), respectively. It should be mentioned that the zero point during the zeta potential testing for CMCAB was reached at 5.2 of pH, which could be supported by findings of Liu et al. [16].

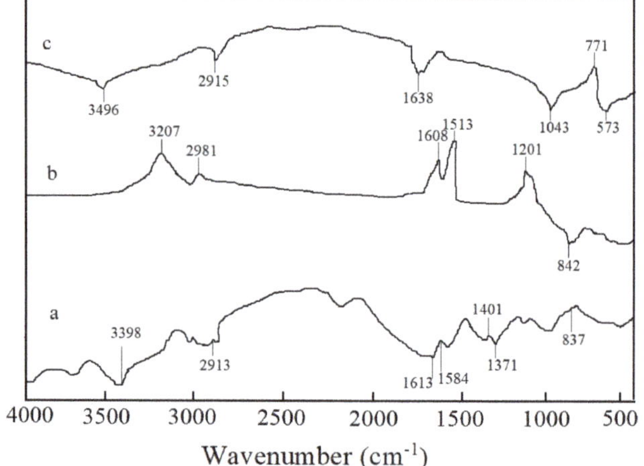

Figure 4. FTIR images of chitosan (**a**), activated Biochar (**b**), and cross-linked magnetic chitosan/activated biochar (CMCAB) (**c**).

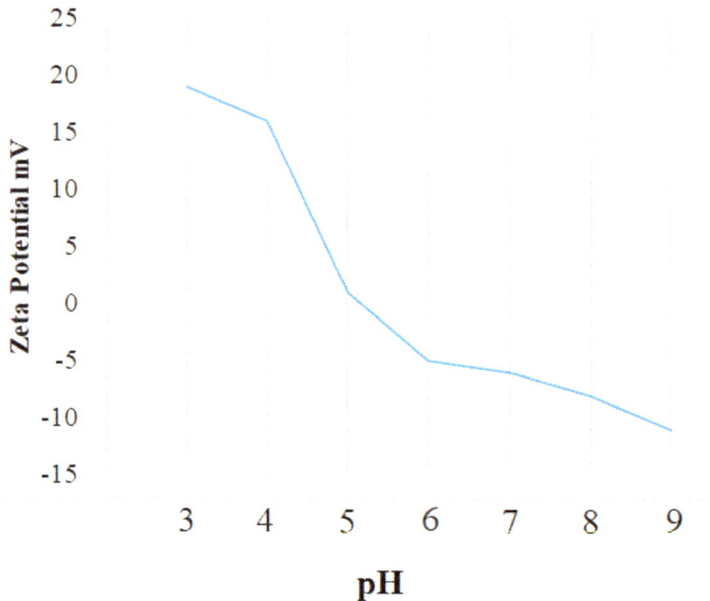

Figure 5. The zeta potential of CMCAB in pH (3–9).

Table 2. Elimination of diclofenac (DCF), ibuprofen (IBP), and naproxen (NPX) by CMCAB.

Run	pH	Initial Concentration (mg/L)	DCF Removal		IBP Removal		NPX Removal	
			(%)	(mg/L)	(%)	(mg/L)	(%)	(mg/L)
1	4.00	0.5	63.7	0.319	65.2	0.326	64.7	0.324
2	4.00	1.0	64.1	0.641	65.8	0.658	64.9	0.649
3	4.00	1.5	64.7	0.971	66.6	0.999	65.8	0.987
4	4.00	2.0	66.2	1.324	66.6	1.332	65.4	1.308
5	4.00	2.5	67.9	1.698	67.5	1.688	66.6	1.665
6	4.00	3.0	66.6	1.998	66.2	1.986	65.7	1.971
7	4.50	0.5	73.2	0.366	74.1	0.371	73.1	0.366
8	4.50	1.0	73.5	0.735	74.8	0.748	73.4	0.734
9	4.50	2.0	73.5	1.470	75.4	1.508	74.9	1.498
10	4.50	2.5	75.9	1.898	75.6	1.890	75.5	1.888
11	4.50	3.0	76.3	2.289	77.1	2.313	75.4	2.262
12	5.00	0.5	81.3	0.407	82.1	0.411	81.6	0.408
13	5.00	1.0	81.6	0.816	82.7	0.827	81.6	0.816
14	5.00	1.5	82.1	1.232	84.9	1.274	81.7	1.226
15	5.00	2.0	82.1	1.642	84.9	1.698	82.2	1.644
16	5.00	2.5	83.0	2.075	84.9	2.123	82.4	2.060
17	5.00	3.0	83.8	2.514	84.7	2.541	81.9	2.457
18	5.50	0.5	87.6	0.438	88.2	0.441	86.9	0.435
19	5.50	1.0	88.2	0.882	88.6	0.886	87.3	0.873
20	5.50	2.0	88.1	1.762	89.1	1.782	87.5	1.750
21	5.50	2.5	89.1	2.228	90.4	2.260	88.2	2.205
22	5.50	3.0	88.6	2.658	90.8	2.724	87.5	2.625
23	6.00	0.5	93.6	0.468	94.8	0.474	91.6	0.458
24	6.00	1.0	93.9	0.939	94.6	0.946	93.2	0.932
25	6.00	1.5	94.7	1.421	97.3	1.460	93.5	1.403
26	6.00	2.0	95.0	1.900	97.8	1.956	94.1	1.882
27	6.00	2.5	96.4	2.410	98.8	2.470	95.2	2.380
28	6.00	3.0	96.1	2.883	98.2	2.946	94.8	2.844
29	6.50	0.5	83.8	0.419	84.0	0.420	82.8	0.414
30	6.50	1.0	84.2	0.842	84.6	0.846	82.6	0.826
31	6.50	2.0	85.1	1.702	86.3	1.726	83.1	1.662
32	6.50	2.5	85.9	2.148	87.4	2.185	83.6	2.090
33	6.50	3.0	85.2	2.556	87.3	2.619	83.2	2.496
34	7.00	0.5	61.0	0.305	62.6	0.313	60.3	0.302
35	7.00	1.0	61.3	0.613	63.5	0.635	60.3	0.603
36	7.00	1.5	61.9	0.929	63.2	0.948	60.8	0.912
37	7.00	2.0	61.9	1.238	64.3	1.286	61.0	1.220
38	7.00	2.5	62.8	1.570	64.8	1.620	61.4	1.535
39	7.00	3.0	62.5	1.875	64.3	1.929	61.4	1.842
40	7.50	0.5	54.7	0.274	52.6	0.263	51.9	0.260

Table 2. *Cont.*

Run	pH	Initial Concentration (mg/L)	DCF Removal		IBP Removal		NPX Removal	
			(%)	(mg/L)	(%)	(mg/L)	(%)	(mg/L)
41	7.50	1.0	54.7	0.547	53.1	0.531	52.2	0.522
42	7.50	2.0	53.6	1.072	52.7	1.054	51.1	1.022
43	7.50	2.5	53.1	1.328	52.7	1.318	51.2	1.280
44	7.50	3.0	52.8	1.584	51.8	1.554	50.7	1.521
45	8.00	0.5	41.7	0.209	42.4	0.212	39.6	0.198
46	8.00	1.0	40.1	0.401	41.6	0.416	39.6	0.396
47	8.00	1.5	39.2	0.588	41.2	0.618	39.1	0.587
48	8.00	2.0	38.7	0.774	40.7	0.814	38.7	0.774
49	8.00	2.5	39.8	0.995	40.9	1.023	39.2	0.980
50	8.00	3.0	40.7	1.221	40.2	1.206	39.6	1.188

3.1. Emerging Micropollutants Removal

Based on Table 2 and Figure 6a, the maximum removal of diclofenac (DCF) was 96.4% (2.41 mg/L) at pH 6 and an initial concentration of 2.5 mg/L, while the minimum removal of DCF was 38.7% (0.77 mg/L) at pH 8 and an initial concentration of 2 mg/L. Liang et al. [34] reported 70% DCF removal via magnetic amine-functionalized chitosan. Lonappan et al. [35] reported 42% to 98% DCF removal in the presence of a high dosage of biochar microparticles (2–20 g/L). Based on Table 2 and Figure 6b, the optimum elimination of ibuprofen (IBP) was 98.8% (2.47 mg/L) at pH 6 and an initial concentration of 2.5 mg/L, and the minimum removal of IBP was 40.2% (1.20 mg/L) at pH 8 and an initial concentration of 3 mg/L. Chakraborty et al. [36] removed 82% to 91% of IBP via bi-directional activated biochar over high contact time (12–18 h). Paradis-Tanguay et al. [37] removed 70% of IBP using chitosan/polyethylene oxide (PEO) electrospun nanofibers. Based on Table 2 and Figure 6c, the maximum removal of naproxen (NPX) was 95.2% (2.38 mg/L) at pH 6 and an initial concentration of 2.5 mg/L, while the minimum removal of NPX was 38.7% (0.77 mg/L) at pH 8 and an initial the concentration of 2 mg/L. Jung et al. [38] reported 97% NPX removal via a combined coagulation/biochar method. Based on Table 2, the removal effectiveness slightly increased with increasing initial concentration of emerging micropollutants from 0.5 mg/L to 2.5 mg/L.

As shown in Table 2 and Figure 6, the elimination efficiencies of the emerging micropollutants increased with increasing the pH from 4 to 6, and maximum micropollutant removal occurred at pH 6, whereas at pH 5–6, the net surface charge is positive and ion repulsion still exists [39]. Then, the removal effectiveness decreased from pH 6 to 8. Gu et al. [40] reported that the pH of a solution has a significant impact on the adsorption procedure because the surface charge of the adsorbent might be changed in the varied pH. Rafati et al. [41] reported that the maximum removal of emerging micropollutants using an adsorption method was reached at pH 6. The diminishing competition of H^+ ions at increasing pH improved the adsorption to reach the maximum removal at pH 6. Besha et al. [42] expressed that elimination of acidic pharmaceuticals such as ibuprofen, naproxen, and diclofenac might be enriched at slightly acidic pH; this is probably because of the hydrophobicity of these compounds.

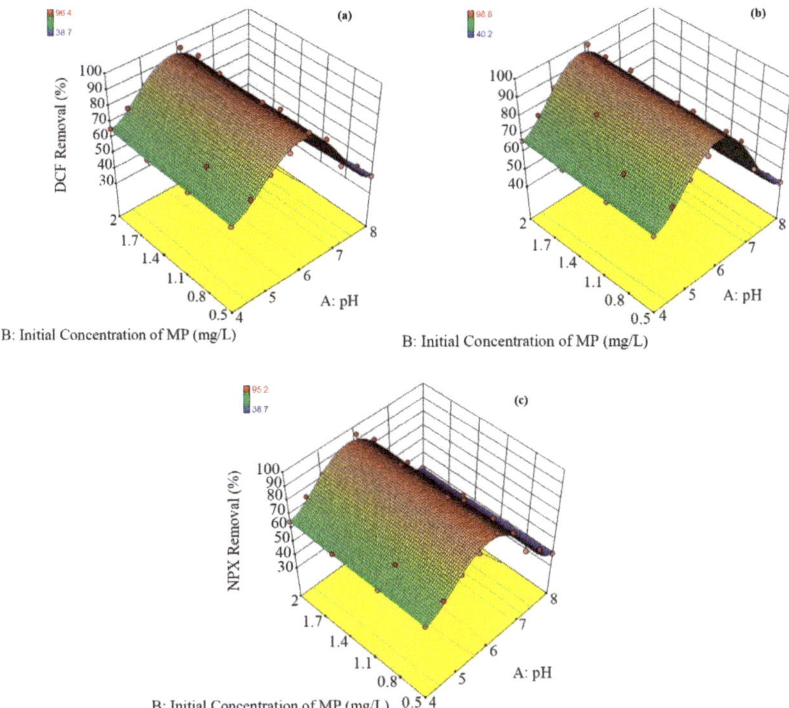

Figure 6. Removal efficiencies for diclofenac (**a**), ibuprofen (**b**), and naproxen (**c**).

3.2. Optimization using an ANN

Artificial neural networks (ANN) are computer techniques on the basis of models of the human brain's biological activities, such as the capability to learn, think, solve issues and remember. Neural network models contain weights and neurons. The neural network contains a combined structure comprising an input layer, intermediate layer (hidden layer), and an output layer. Each layer contains of simple processing features called neurons. The mean square error (MSE) and R^2 values (Table 3) for DCF, IBP, and NPX elimination are shown in Table 3. Figure 7 indicates the best setting of the ANN. Figure 8 displays the change in the MSE values by Levenberg–Marquardt (LM) through selecting various functions such as pure linear, transig, and log sigmoid. This figure also specifies that the training was completed after 68, 25, and 34 epochs for DCF (a), IBP (b), and NPX (c), respectively. These consequences also proved that the ANN model was well-trained at the end of the training phase [43,44].

The high values of R^2 (Figure 9) indicated an excellent agreement between the ANN predicted data and the actual data [43].

Table 3. R^2 and MSE values for the removal of each pollutant in the selection of the best model.

Parameter	R^2			MSE		
	Training	Validation	Test	Training	Validation	Test
For DCF Removal	0.999	0.998	0.998	0.220	0.837	0.212
For IBP Removal	0.998	0.998	0.998	0.697	2.929	0.829
For NPX Removal	0.999	0.999	0.999	0.235	0.276	0.371

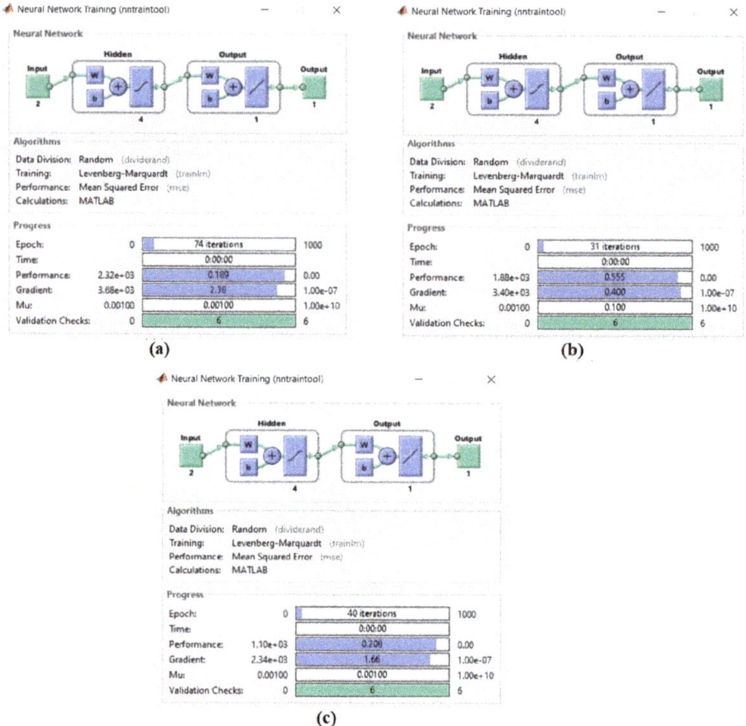

Figure 7. Artificial neural network (ANN) settings for the best model for diclofenac (**a**), ibuprofen (**b**), and naproxen (**c**).

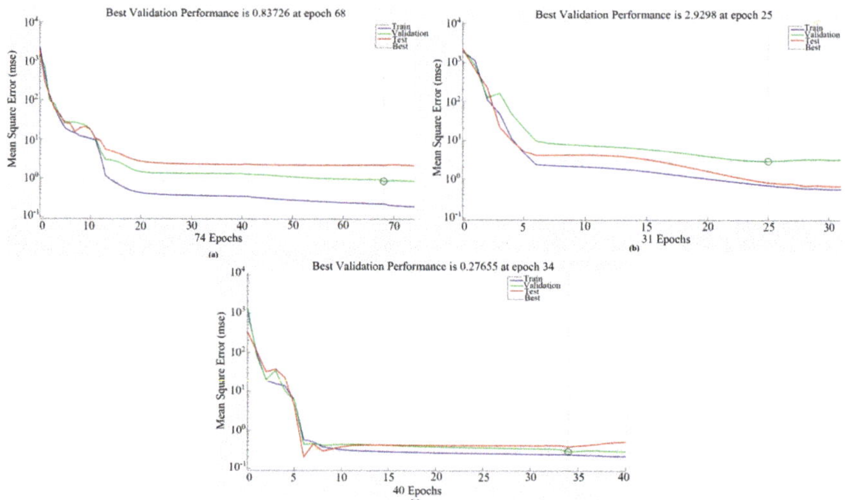

Figure 8. Mean square error (MSE) versus the number of epochs for diclofenac (**a**), ibuprofen (**b**), and naproxen (**c**).

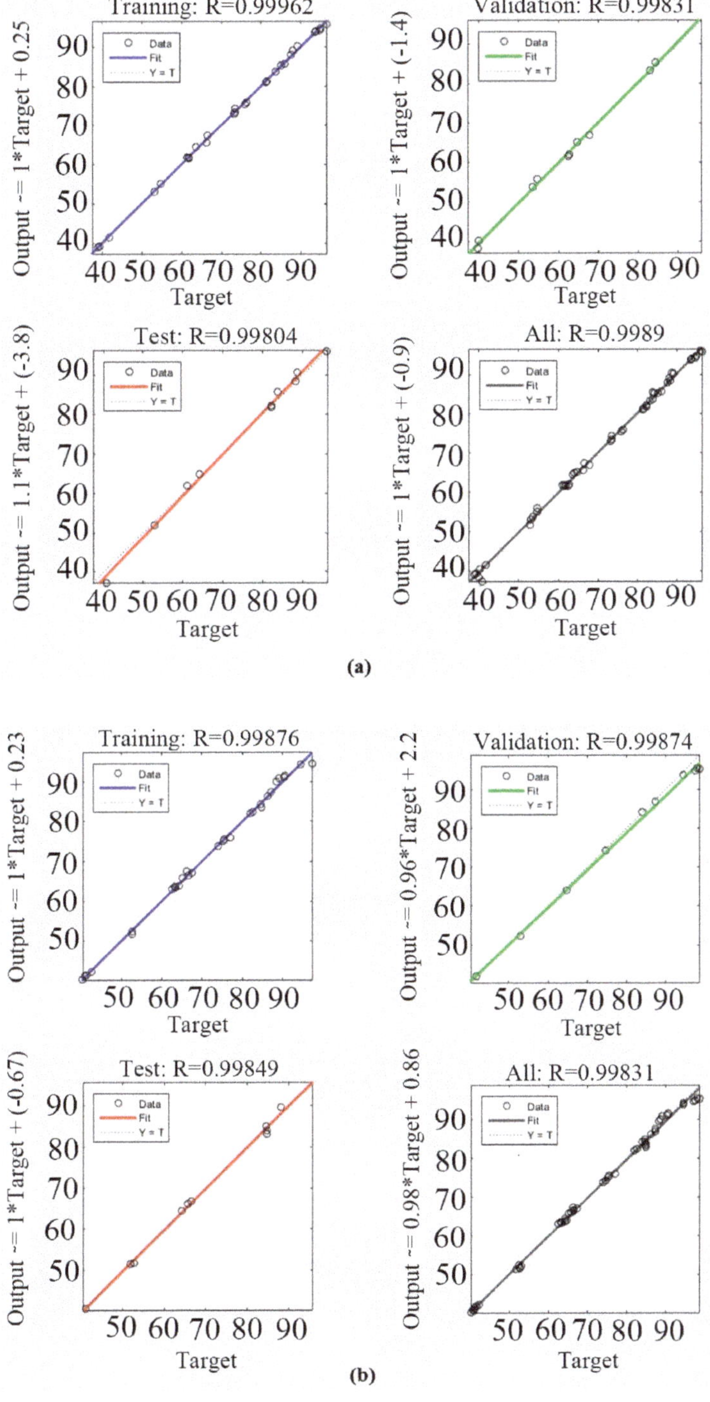

Figure 9. *Cont.*

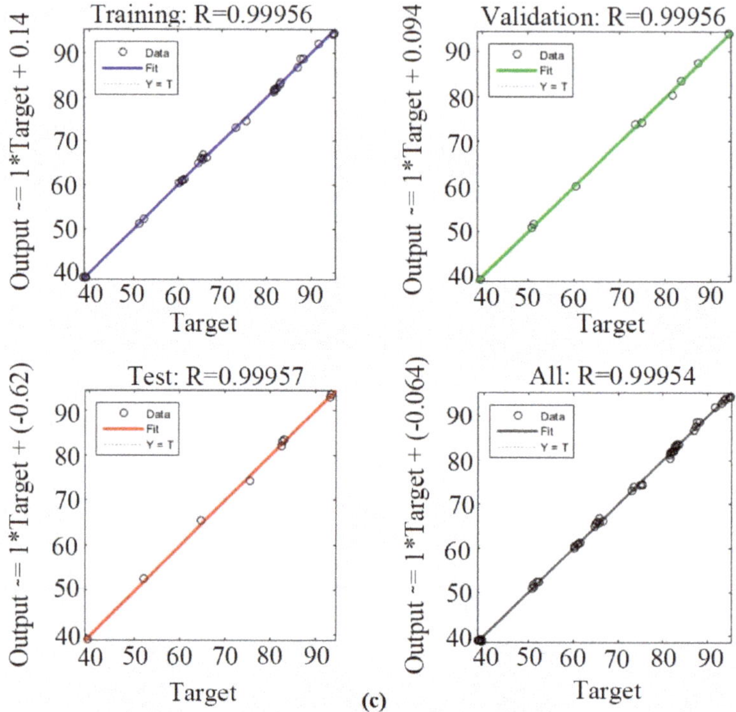

Figure 9. Model prediction versus experimental values for the optimum topology for diclofenac (**a**), ibuprofen (**b**), and naproxen (**c**).

3.3. Adsorption Isotherm

3.3.1. Langmuir Isotherm

The mathematical expression of this isotherm is presented in the following Equation (5):

$$\frac{x}{m} = \frac{abC_e}{(1 + bC_{e})}, \tag{5}$$

where $\frac{x}{m}$ corresponds the mass of adsorbate adsorbed/unit mass of adsorbent (mg adsorbate per g adsorbent), a and b denote empirical constants, and C_e denotes the equilibrium concentration of the adsorbate in the solution following adsorption (mg/L) [27].

Table 4 and Figure 10 display the details of the Langmuir isotherm studies. The R^2 values were 0.932, 0.962, and 0.893 for DCF, IBP, and NPX removal, respectively. Based on Figure 10, with the decrease in values of $(1/C_e)$, the values of $(1/(x/m))$ were increased. The high R^2 values show that elimination of DCF, IBP, and NPX could be explained by the Langmuir isotherm. For the DCF elimination using the Langmuir isotherm model, the values of *b* and Q (mg/g) were 0.77 and 22.1, respectively. Jodeh et al. [45] reported Q = 22.2 during DCF removal using an adsorption method. For IBP removal using the Langmuir isotherm model, the values of *b* and Q (mg/g) were 0.64 and 21.2, respectively. For NPX elimination using the Langmuir isotherm model, the values of b and Q (mg/g) were 0.76 and 33.3, respectively. Values of Q_m = 21.7, b = 0.75, and R^2 = 0.8 were reported by Sun et al. [11] and are in line with the results of the current study. Sun et al. [11] reported Q_m = 33.6, b = 0.75 and R^2 = 0.97 during NPX removal via an adsorption method, which are also in line with the results of the current study.

Table 4. Langmuir and Freundlich isotherms study for DCF, IBP, and NPX removal by CMCAB.

Parameters	Langmuir Isotherm			Freundlich Isotherm		
	Q_m (mg/g)	b (L/mg)	R^2	K_f $(mg/g(L/mg)^{1/n})$	1/n	R^2
DCF	22.1	0.772	0.932	30.27	−4.26	0.943
IBP	21.2	0.643	0.962	54.57	−3.75	0.934
NPX	33.3	0.765	0.893	16.94	−3.03	0.988

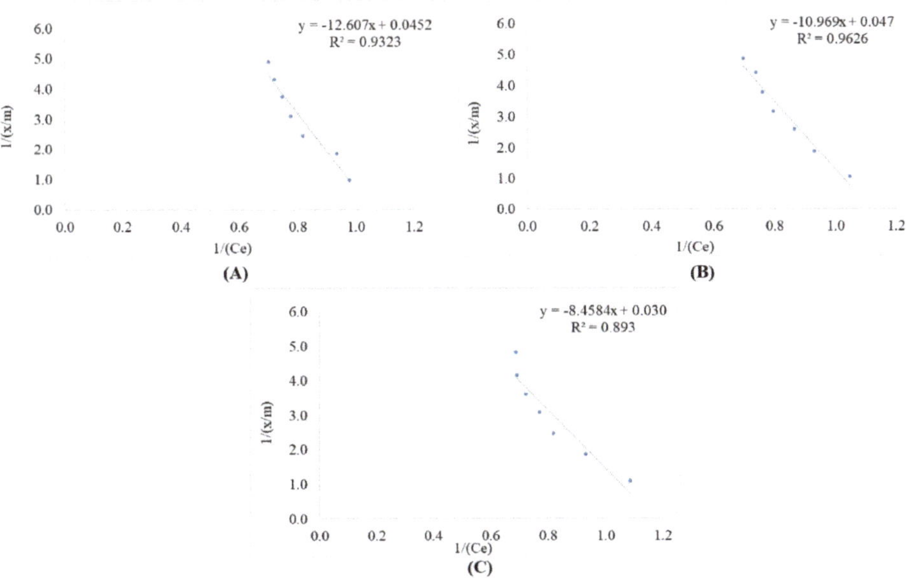

Figure 10. Langmuir isotherm regressions for diclofenac (**A**), ibuprofen (**B**), and naproxen (**C**).

3.3.2. Freundlich Isotherm

The Freundlich isotherm defines the adsorption equilibrium as follows (Equation (6)):

$$q_m = K_f\, C_e^{1/n} \tag{6}$$

where K_f is a fixed variable representing the relative adsorption capability of the adsorbent $(mg^{1-(1/n)}/L^{1/n}/g^{-1})$, and n is a fixed variable signifying adsorption intensity [27].

Table 4 and Figure 11 display the details of the Freundlich isotherm studies. The R^2 values were 0.943, 0.934, and 0.988 for DCF, IBP, and NPX removal, respectively. The high R^2 values show that removal of DCF, IBP, and NPX could fit the Freundlich isotherm. Based on Figure 11, with the increase in values of Log(C_e), the values of Log(x/m) were decreased.

The Freundlich capacity factor (K) and 1/n were 30.27 and −17.27, respectively, for DCF removal. Values of K_f in the range 7.6–63.6 and R^2 in the range 0.92–0.96 were reported by Sathishkumar et al. [46] for DCF removal via an adsorption method. The Freundlich capacity factor (K) and 1/n were 54.57 and −19.41, respectively, for IBP removal. Coimbra et al. [47] reported a K_f value of 55.30 and R^2 of 0.98 for IBP removal by an adsorption method. The Freundlich capacity factor (K) and 1/n were 16.94 and −12.26, respectively, for NPX removal. Mojiri et al. [48] stated that higher 1/n values indicate that the adsorption bond is weak. Increasing the log (C_e) caused decreasing the $\log(\frac{x}{m})$. Thus, $1/n$ (the slope of the line) is negative [48].

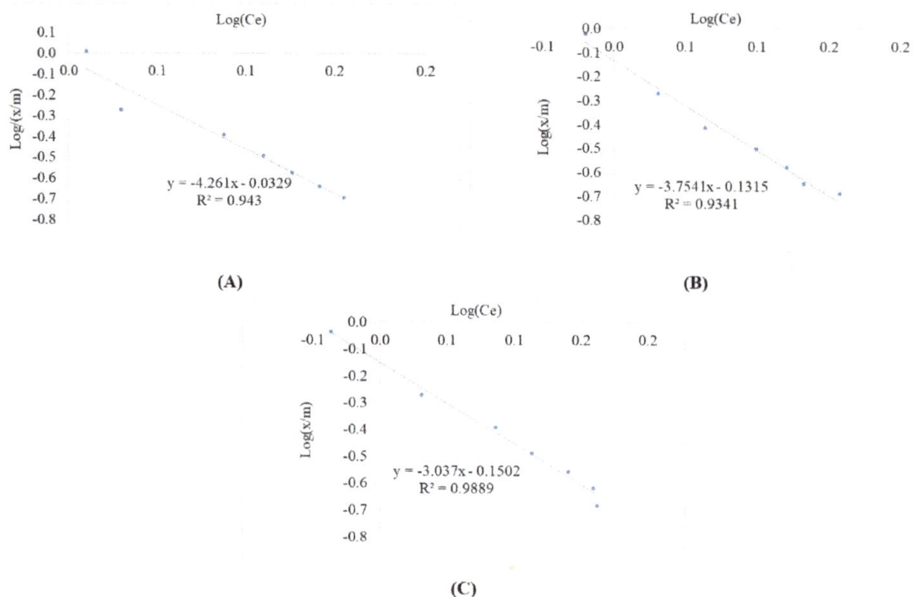

Figure 11. Freundlich isotherm regressions for diclofenac (**A**), ibuprofen (**B**), and naproxen (**C**).

3.4. Regeneration and Desorption Study

Regeneration of adsorbents is a vital procedure in wastewater treatment to decrease the processing cost. Various regeneration methods have been applied for desorption studies, including thermal regeneration and chemical regeneration. Nevertheless, it is vital to select the appropriate pH and desorbents (such as inorganic desorbents NaOH, H_2SO_4, and HCl or organic desorbents ethanol, methanol, and acetic acid) for the chemical desorption procedure [49]. Emerging micropollutants are highly soluble in alcohols due to the presence of hydroxyl groups. Moreover, the low molecular weight alcohols may enrich the effectiveness of emerging micropollutants desorption. Alizadeh Fard and Barkdoll [50] stated that NaOH and HCl could not efficiently desorb micropollutants. In addition, they also stated that methanol's restoration capacity is higher than ethanol's. In this study, after eight (Figure 12) cycles with an initial concentration of 2.5 mg/L, the removal effectiveness of the cross-linked magnetic chitosan/activated biochar remained almost unaffected.

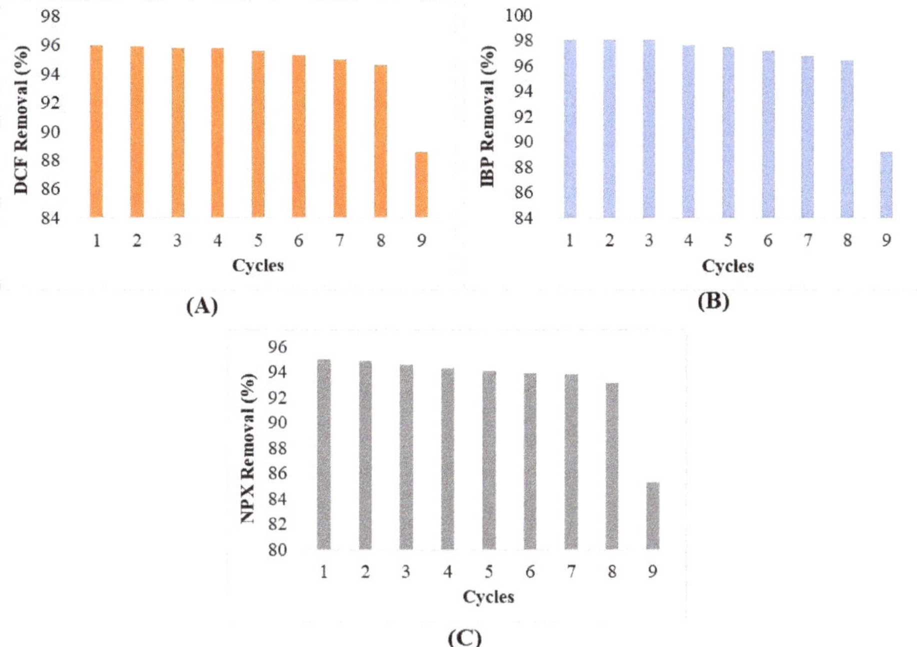

Figure 12. Regeneration results of CMCAB during removal of DCF (**A**), IBP (**B**) and NPX (**C**).

4. Conclusions

Diclofenac (DCF), ibuprofen (IBP), and naproxen (NPX) are anti-inflammatory drugs, which are a frequently consumed class of pharmaceuticals. As these are emerging micropollutants, we evaluated their removal using cross-linked magnetic chitosan/activated biochar (CMCAB). An artificial neural network (ANN) with two independent factors—pH (4–8) and micropollutant concentration (0.5–3 mg/L)—was applied to optimize the elimination efficiency. The main conclusions of this new research are listed below:

1. The optimum elimination values of DCF, IBP, and NPX were 96.4% (2.41 mg/L), 98.8% (2.47 mg/L), and 95.2% (2.38 mg/L) at pH 6.0 and an initial micropollutant concentration of 2.5 mg/L.
2. Based on the ANN, the R^2 values were 0.998, 0.998, and 0.999 for DCF, IBP, and NPX removal, indicating that optimization could be done well by an ANN.
3. Freundlich isotherm could explain the DCF and NPX removal via the CMCAB better than the Langmuir isotherm as indicated by the high R^2 values attained. But IBP removal was better fitted by Langmuir isotherm.
4. Regeneration and desorption studies indicated that CMCAB could be applied for at least eight cycles without changing the performance.

Author Contributions: A.M. was responsible for setting up the experiments, completing most of the experiments, and writing the initial draft of the manuscript; R.A.K. and A.G. modified the manuscript and contributed to the literature search.

Funding: This research received no external funding.

Acknowledgments: The author would like to express their gratitude to the Institute for Infrastructure Engineering and Sustainable Management (IIESM), University Technology Mara (UiTM), Malaysia.

Conflicts of Interest: The authors declare no conflict of interest.

References

1. Grandclement, C.; Seyssiecq, I.; Piram, A.; Wong-Wah-Chung, P.; Vanot, C.; Tiliacos, N.; Roche, N. From the conventional biological wastewater treatment to hybrid processes, the evaluation of organic micropollutant removal: A review. *Water Res.* **2017**, *111*, 297–317. [CrossRef] [PubMed]
2. Mahmoud, W.M.M.; Rastogi, T.; Kummerer, K. Application of titanium dioxide nanoparticles as a photocatalyst for the removal of micropollutants such as pharmaceuticals from water. *Curr. Opin. Green Sustain. Chem.* **2017**, *6*, 1–10. [CrossRef]
3. Landry, K.A.; Boyer, T.H. Diclofenac removal in urine using strong-base anion exchange polymer resins. *Water Res.* **2013**, *47*, 6432–6444. [CrossRef] [PubMed]
4. Wieszczycka, K.; Zembrzuska, J.; Bornikowska, J.; Wojciechowska, A.; Wojciechowska, I. Removal of naproxen from water by ionic liquid-modified polymer sorbents. *Chem. Eng. Res. Des.* **2017**, *117*, 698–705. [CrossRef]
5. Hiew, B.Y.Z.; Lee, L.Y.; Lee, X.J.; Gan, S.; Thangalazhy-Gopakumar, S.; Lim, S.S.; Pan, G.T.; Yang, T.C.K. Adsorptive removal of diclofenac by graphene oxide: Optimization, equilibrium, kinetic and thermodynamic studies. *J. Taiwan Inst. Chem. Eng.* **2018**. [CrossRef]
6. Zhang, L.; Lv, T.; Zhang, Y.; Stein, O.R.; Arias, C.A.; Brix, H.; Carvalho, P.N. Effects of constructed wetland design on ibuprofen removal—A mesocosm scale study. *Sci. Total Environ.* **2017**, 38–45. [CrossRef] [PubMed]
7. Górny, D.; Guzik, U.; Hupert-Kocurek, K.; Wojcieszyńska, D. Naproxen ecotoxicity and biodegradation by Bacillus thuringiensis B1(2015b) strain. *Ecotoxicol. Environ. Saf.* **2019**, *167*, 505–512. [CrossRef] [PubMed]
8. Amouzgar, P.; Salamatinia, B. A Short Review on Presence of Pharmaceuticals in Water Bodies and the Potential of Chitosan and Chitosan Derivatives for Elimination of Pharmaceuticals. *J. Mol. Genet. Med.* **2015**, *S4*, 001. [CrossRef]
9. Kim, E.; Jung, C.; Han, J.; Her, N.; Park, C.M.; Jang, M.; Son, A.; Yoon, Y. Sorptive removal of selected emerging contaminants using biochar in aqueous solution. *J. Ind. Eng. Chem.* **2016**, *36*, 364–371. [CrossRef]
10. Lahti, M.; Oikari, A. Microbial Transformation of Pharmaceuticals Naproxen, Bisoprolol, and Diclofenac in Aerobic and Anaerobic Environments. *Arch. Environ. Contam. Toxicol.* **2011**, *61*, 202–210. [CrossRef]
11. Sun, W.; Li, H.; Li, H.; Li, S.; Cao, X. Adsorption mechanisms of ibuprofen and naproxen to UiO-66 and UiO-66-NH2: Batch experiment and DFT calculation. *Chem. Eng. J.* **2019**, *360*, 645–653. [CrossRef]
12. Yang, Z.; Miao, H.; Rui, Z.; Ji, H. Enhanced Formaldehyde Removal from Air Using Fully Biodegradable Chitosan Grafted β-Cyclodextrin Adsorbent with Weak Chemical Interaction. *Polymers* **2019**, *11*, 276. [CrossRef]
13. Cha, J.S.; Park, S.H.; Jung, S.C.; Ryu, C.; Jeon, J.K.; Shin, M.C.; Park, Y.K. Production and utilization of biochar: A review. *J. Ind. Eng. Chem.* **2016**, *40*, 1–15. [CrossRef]
14. Quesada, H.B.; Alves Baptista, A.T.; Cusioli, L.F.; Seibert, D.; de Oliveira Bezerra, C.; Bergamasco, R. Surface water pollution by pharmaceuticals and an alternative of removal by low-cost adsorbents: A review. *Chemosphere* **2019**, *222*, 766–780. [CrossRef]
15. Sizmur, T.; Fresno, T.; Akgül, G.; Frost, H.; Moreno Jiménez, E. Biochar modification to enhance sorption of inorganics from water. *Biores. Technol.* **2017**, *246*, 34–47. [CrossRef]
16. Liu, S.; Huang, B.; Chai, L.; Liu, Y.; Zeng, G.; Wang, X.; Zeng, W.; Shang, M.; Deng, J.; Zhou, Z. Enhancement of As (V) adsorption from aqueous solution by a magnetic chitosan/biochar composite. *RSC Adv.* **2017**, *7*, 10891–10900. [CrossRef]
17. Kim, S.; Park, C.M.; Jang, A.; Hernández-Maldonadoe, A.J.; Yu, M.; Heo, J.; Yoon, Y. Removal of selected pharmaceuticals in an ultrafiltration-activated biochar hybrid system. *J. Membr. Sci.* **2019**, *570*, 77–84. [CrossRef]
18. Sühnholz, S.; Kopinke, F.D.; Weiner, B. Hydrothermal treatment for regeneration of activated carbon loaded with organic micropollutants. *Sci. Total Environ.* **2018**, 854–861. [CrossRef]
19. Lonappan, L.; Rouissi, T.; Liu, Y.; Brar, S.K.; Surampalli, R.Y. Removal of diclofenac using microbiochar fixed-bed column bioreactor. *J. Environ. Chem. Eng.* **2019**, *7*, 102894. [CrossRef]
20. Zhang, D.Q.; Hua, T.; Gersberg, R.M.; Zhu, J.; Ng, W.J.; Tan, S.K. Carbamazepine and naproxen: Fate in wetland mesocosms planted with Scirpus validus. *Chemosphere* **2013**, *91*, 14–21. [CrossRef]
21. Song, S.; Su, Y.; Adeleye, A.S.; Zhang, Y.; Zhou, X. Optimal design and characterization of sulfide-modified nanoscale zerovalent iron for diclofenac removal. *Appl. Catal. B Environ.* **2017**, *201*, 211–220. [CrossRef]

22. Wu, H.; Feng, Q.; Yang, H.; Alam, E.; Gao, B.; Gu, D. Modified biochar supported Ag/Fe nanoparticles used for removal of cephalexin in solution: Characterization, kinetics and mechanisms. *Coll. Surf. Physicochem. Eng. Asp.* **2017**, *517*, 63–71. [CrossRef]
23. De Luna, M.D.G.; Murniatib-Budianta, W.; Rivera, K.K.P.; Arazo, R.O. Removal of sodium diclofenac from aqueous solution by adsorbents derived from cocoa pod husks. *J. Environ. Chem. Eng.* **2017**, *5*, 1465–1474. [CrossRef]
24. APHA. *Standard Methods for Examination of Water and Wastewater*, 22th ed.; American Public Health Association: Washington, DC, USA, 2012.
25. Madikizelaa, L.M.; Luke Chimuka, L. Synthesis, adsorption and selectivity studies of a polymer imprinted with naproxen, ibuprofen and diclofenac. *J. Environ. Chem. Eng.* **2016**, *4*, 4029–4037. [CrossRef]
26. Tanzifi, M.; Hosseini, S.H.; Kiadehi, A.D.; Olazar, M.; Karimipour, K.; Rezaiemehr, R.; Ali, I. Artificial neural network optimization for methyl orange adsorption onto polyaniline nano-adsorbent: Kinetic, isotherm and thermodynamic studies. *J. Mol. Liq.* **2017**, *244*, 189–200. [CrossRef]
27. Mojiri, A.; Ohashi, A.; Ozaki, N.; Shoiful, A.; Kindaichi, T. Pollutant Removal from Synthetic Aqueous Solutions with a Combined Electrochemical Oxidation and Adsorption Method. *Int. J. Environ. Res. Public Health* **2018**, *15*, 1443. [CrossRef]
28. He, J.; Li, Y.; Cai, X.; Chen, K.; Zheng, H.; Wang, C.; Zhang, K.; Lin, D.; Kong, L.; Liu, J. Study on the removal of organic micropollutants from aqueous and ethanol solutions by HAP membranes with tunable hydrophilicity and hydrophobicity. *Chemosphere* **2017**, *174*, 380–389. [CrossRef]
29. Yasmeen, S.; Kabiraz, M.K.; Saha, B.; Qadir, M.R.; Gafur, M.A.; Masum, S.M. Chromium (VI) Ions Removal from Tannery Effluent using Chitosan-Microcrystalline Cellulose Composite as Adsorbent. *Inter. Res. J. Pure App. Chem.* **2016**, *10*, 1–14. [CrossRef]
30. Queiroz, M.F.; Melo, K.R.T.; Sabry, D.A.; Sassaki, G.L.; Rocha, H.A.O. Does the Use of Chitosan Contribute to Oxalate Kidney Stone Formation? *Mar. Drugs* **2015**, *13*, 141–158. [CrossRef]
31. Jindo, K.; Mizumoto, H.; Sawada, Y.; Sanchez-Monedero, M.A.; Sonoki, T. Physical and chemical characterization of biochars derived from different agricultural residues. *Biogeosciences* **2014**, *11*, 6613–6621. [CrossRef]
32. Yu, Z.; Zhang, X.; Huang, Y. Magnetic Chitosan–Iron (III) Hydrogel as a Fast and Reusable Adsorbent for Chromium (VI) Removal. *Ind. Eng. Chem. Res.* **2013**, *52*, 11956–11966. [CrossRef]
33. Zhang, M.M.; Liu, Y.G.; Li, T.T.; Xu, W.H.; Zheng, B.H.; Tan, X.F.; Wang, H.; Guo, Y.M.; Guo, F.Y.; Wang, S.F. Chitosan modification of magnetic biochar produced from Eichhornia crassipes for enhanced sorption of Cr (VI) from aqueous solution. *RSC Adv.* **2015**, *5*, 46955. [CrossRef]
34. Liang, X.X.; Omer, A.M.; Hu, Z.H.; Wang, Y.G.; Yu, D.; Ouyang, X.K. Efficient adsorption of diclofenac sodium from aqueous solutions using magnetic amine-functionalized chitosan. *Chemosphere* **2019**, *217*, 270–278. [CrossRef] [PubMed]
35. Lonappan, L.; Rouissi, T.; Brar, S.K.; Verma, M.; Surampalli, R.Y. Adsorption of diclofenac onto different biochar microparticles: Dataset—Characterization and dosage of biochar. *Data Brief* **2018**, *16*, 460–465. [CrossRef] [PubMed]
36. Chakraborty, P.; Show, S.; Benerjee, S.; Halder, G. Mechanistic insight into sorptive elimination of ibuprofen employing bi-directional activated biochar from sugarcane bagasse: Performance evaluation and cost estimation. *J. Environ. Chem. Eng.* **2018**, *6*, 5287–5300. [CrossRef]
37. Paradis-Tanguay, L.; Camire, A.; Renaud, M.; Chabot, B. Sorption capacities of chitosan/polyethylene oxide (PEO) electrospun nanofibers used to remove ibuprofen in water. *J. Polym. Eng.* **2018**. [CrossRef]
38. Jung, C.; Oh, J.; Yoon, Y. Removal of acetaminophen and naproxen by combined coagulation and adsorption using biochar: Influence of combined sewer overflow components. *Environ. Sci. Poll. Res.* **2015**, *22*, 10058–10069. [CrossRef]
39. Dewage, N.B.; Fowler, R.E.; Pittman, C.U., Jr.; Mohan, D.; Mlsna, T. Lead (Pb^{2+}) sorptive removal using chitosan-modified biochar: Batch and fixed-bed studies. *RSC Adv.* **2018**, *8*, 25368–25377. [CrossRef]
40. Gu, Y.; Yperman, J.; Carleer, R.; D'Haen, J.; Maggen, J.; Vanderheyden, S.; Vanppelen, K.; Garcia, R.M. Adsorption and photocatalytic removal of Ibuprofen by activated carbon impregnated with TiO_2 by UV–Vis monitoring. *Chemosphere* **2019**, *217*, 724–731. [CrossRef]

41. Rafati, L.; Ehrampoush, M.H.; Rafati, A.A.; Mokhtari, M.; Mahvi, H. Modeling of adsorption kinetic and equilibrium isotherms of naproxen onto functionalized nano-clay composite adsorbent. *J. Mol. Liq.* **2016**, *224*, 832–841. [CrossRef]
42. Besha, A.T.; Gebreyohannes, A.Y.; Tufa, R.A.; Bekele, D.N.; Curcio, E.; Giorno, L. Removal of emerging micropollutants by activated sludge process and membrane bioreactors and the effects of micropollutants on membrane fouling: A review. *J. Environ. Chem. Eng.* **2017**, *5*, 2395–2414. [CrossRef]
43. Ghaedi, A.M.; Karamipour, S.; Vafaei, A.; Baneshi, M.M.; Kiarostami, V. Optimization and modeling of simultaneous ultrasound-assisted adsorption of ternary dyes using copper oxide nanoparticles immobilized on activated carbon using response surface methodology and artificial neural network. *Ultrason. Sonochem.* **2019**, *51*, 264–280. [CrossRef]
44. Uddin, M.K.; Khan Rao, R.A.; Chandra Mouli, K.V.V. The artificial neural network and Box-Behnken design for Cu^{2+} removal by the pottery sludge from water samples: Equilibrium, kinetic and thermodynamic studies. *J. Mol. Liq.* **2018**, *266*, 617–627. [CrossRef]
45. Jodeh, S.; Abdelwahab, F.; Jaradat, N.; Warad, I.; Jodeh, W. Adsorption of diclofenac from aqueous solution using Cyclamen persicum tubers based activated carbon (CTAC). *J. Assoc. Arab Univ. Basic Appl.* **2016**, *20*, 32–38. [CrossRef]
46. Sathishkumar, P.; Arulkumar, M.; Ashokkumar, V.; Yusoff, A.R.M.; Murugesan, K.; Palvannan, T.; Salam, Z.; Nasir Ain, F.; Hadibarata, T. Modified phyto-waste Terminalia catappa fruit shells: A reusable adsorbent for the removal of micropollutant diclofenac. *RSC Adv.* **2015**, *5*, 30950–30962. [CrossRef]
47. Coimbra, R.N.; Escapa, C.; Otero, M. Adsorption Separation of Analgesic Pharmaceuticals from Ultrapure and Waste Water: Batch Studies Using a Polymeric Resin and an Activated Carbon. *Polymers* **2018**, *10*, 958. [CrossRef]
48. Mojiri, A.; Ahmad, Z.; Tajuddin, M.R.; Arshad, M.F.; Barrera, V. Molybdenum (VI) removal from aqueous solutions using bentonite and powdered cockle shell; Optimization by response surface methodology. *Glob. NEST J.* **2017**, *19*, 232–240. [CrossRef]
49. Zhou, Y.; Zhang, L.; Cheng, Z. Removal of organic pollutants from aqueous solution using agricultural wastes: A review. *J. Mol. Liq.* **2015**, *212*, 739–762. [CrossRef]
50. Alizadeh Fard, M.; Barkdoll, B. Using recyclable magnetic carbon nanotube to remove micropollutants from aqueous solutions. *J. Mol. Liq.* **2018**, *249*, 193–202. [CrossRef]

Article

Inactivation and Loss of Infectivity of Enterovirus 70 by Solar Irradiation

Muhammad Raihan Jumat and Pei-Ying Hong *

Biological and Environmental Science and Engineering Division (BESE), Water Desalination and Reuse Center (WDRC), King Abdullah University of Science and Technology (KAUST), Thuwal 23955-6900, Saudi Arabia; raihan.jumat@kaust.edu.sa

* Correspondence: peiying.hong@kaust.edu.sa; Tel.: +966-12-808-2218

Received: 28 November 2018; Accepted: 24 December 2018; Published: 2 January 2019

Abstract: Enterovirus 70 (EV70) is an emerging viral pathogen that remains viable in final treated effluent. Solar irradiation is, therefore, explored as a low-cost natural disinfection strategy to mitigate potential concerns. EV70 was exposed to simulated sunlight for 24 h at a fluence rate of 28.67 $J/cm^2/h$ in three different water matrices, namely, phosphate-buffered saline (PBS), treated wastewater effluent, and chlorinated effluent. In the presence of sunlight, EV70 decreased in infectivity by 1.7 log, 1.0 log, and 1.3 log in PBS, effluent, and chlorinated effluent, respectively. Irradiated EV70 was further introduced to host cell lines and was unable to infect the cell lines. In contrast, EV70 in dark microcosms replicated to titers 13.5, 3.3, and 4.2 times the initial inoculum. The reduction in EV70 infectivity was accompanied by a reduction in viral binding capacity to Vero cells. In addition, genome sequencing analysis revealed five nonsynonymous nucleotide substitutions in irradiated viruses after 10 days of infection in Vero cells, resulting in amino acid substitutions: Lys14Glu in the VP4 protein, Ala201Val in VP2, Gly71Ser in VP3, Glu50Gln in VP1, and Ile47Leu in $3C^{pro}$. Overall, solar irradiation resulted in EV70 inactivation and an inhibition of viral activity in all parameters studied.

Keywords: enteric virus; remediation technology; water quality

---

1. Introduction

Climate change, urbanization, and increasing global population have placed considerable pressure on freshwater supplies [1–3]. Wastewater can be used as an alternative water resource for agriculture irrigation and aquifer recharge but would first require appropriate treatment in wastewater treatment plants (WWTPs). WWTPs act as engineered barriers to treat municipal wastewater to a quality that is sufficiently safe for reuse. In most WWTPs, the final treatment step typically includes the use of chlorine as a disinfectant to reduce the biological activity of remnant pathogens present in the treated effluent [4]. However, each pathogen reacts differently to different disinfectants, and a single disinfection strategy is rarely effective against all pathogens [5]. For example, a WWTP utilizing chlorination as a disinfection strategy was able to inactivate human adenoviruses but not enteroviruses fully from wastewater [6]. Additional disinfection strategies, particularly those that are low-cost and easily accessible, may have to be deployed to further inactivate remnant viral contaminants.

Solar irradiation is a freely accessible, low-cost biocidal strategy that is abundant in many tropical countries and can be used to circumvent this need. The biocidal effect of sunlight works through the effects of ultraviolet A (UV-A) and ultraviolet B (UV-B). UV-A, of wavelengths 320–400 nm, is absorbed by molecular chromophores which, in turn, generate reactive oxidative species (ROS). ROS induce damage to cellular membranes, proteins and nucleic acids, rendering viruses and other pathogens inactive. UV-B, of wavelengths 280–320 nm, functions directly through absorption by nucleic acids and proteins. UV-B can also affect pyrimidines directly, inducing mutagenic and genotoxic effects in the genomes of microbes [7,8].

Several studies have documented the effects of irradiation on viruses. However, the dosage required for viral inactivation varies widely with viral species, particle size, genome type, length, and polarity [9–12]. For instance, numerous studies have investigated the effects of solar irradiation on members of the *Picornaviridae* family, which contain a single positively stranded RNA genome [13]. Heaselgrave et al., reported a 4-log inactivation of polioviruses with solar irradiation ranging from 198 to 1224 J/cm^2 [14,15]. In contrast, Coxsackie viruses required 117–198 J/cm^2 of solar irradiation for a 4-log inactivation, while ECHO viruses required 50–60 J/cm^2 for a 2-log reduction [15,16].

The variation in solar intensity required to inactivate different RNA viruses within the same family shows that susceptibility of viruses to solar irradiation differs at the species level. A species within the *Picornaviridae* family that has not been studied in this aspect is enterovirus 70 (EV70). These viruses are mainly transmitted by the fecal-oral route and cause gastroenteritis. However, it can cause other symptoms, which include hemorrhagic conjunctivitis, diabetes (through infection of islet cells), and central nervous system complications [17–20]. These viruses are acid and heat stable, allowing for their survival in the gastrointestinal tract but inadvertently conferring persistence in WWTPs [21,22]. Infectious EV70 has been detected in the effluents of several WWTPs globally [6,23–25]. This indicates that the existing disinfection procedures employed are not adequate to provide safe water for reuse, and there exists a need to explore the efficacy of solar irradiation as a possible additional disinfection strategy against enterovirus 70 (EV70).

In this study, EV70 was exposed to simulated sunlight irradiation for 24 h at a fluence rate of 28.67 $J/cm^2/h$. Aliquots of the virus were harvested at specific time points followed by determination of its infectious titer and RNA concentration. We employed a focus forming assay to overcome the inability of EV70 to replicate well in cell culture [6,26]. To determine if any damage was incurred on the capsid, viruses were assayed for their binding ability to Vero cells. The viral growth kinetics were also assayed by counting the foci generated over a nine-day infection period. Ten days after infection, the genomes of EV70 were sequenced. The assays revealed that irradiated viruses had inhibited replication and binding and harbored nonsynonymous nucleotide substitutions compared to dark-control viruses. Viruses suspended in a wastewater matrix also experienced a significant reduction in viral activity upon exposure to solar irradiation, albeit not as pronounced as that observed when suspended in a saline buffer. Interestingly, all of the irradiated viruses in this study failed to replicate in cell culture, providing a strong endorsement of sunlight as a low-cost natural disinfection strategy.

2. Materials and Methods

2.1. Cells and Viruses

Enterovirus 70 (EV70) was purchased from American Type Culture Collection (ATCC VR-836, Manassas, VA, USA) and propagated in human embryonic kidney (HEK) 293T cells (ATCC CRL-3216). HEK 293T cells were maintained in 75 cm^2 flasks (Corning Incorporated, Corning, NY, USA) in Dulbecco's modified Eagle's medium (DMEM) supplemented with 10% fetal bovine serum and 1× penicillin and streptomycin (growth medium) (Corning Incorporated, Corning, NY, USA). For the infection study, the HEK 293T cells were seeded in 175 cm^2 flasks till confluency and inoculated with EV70 diluted in 10 mL of DMEM supplemented with 2% FBS and 1× penicillin and streptomycin. Flasks were incubated for 1 h at 37 °C and 5% CO_2. After incubation, another 10 mL of DMEM supplemented with 2% FBS (virus infection media) was added, and the flasks were returned to the incubator. The cells were observed daily for cytopathic effect (CPE). Once CPE was observed, cells were harvested in the supernatant and pelleted by centrifuging at 2000× g for 10 min. The cells then underwent three freeze-thaw cycles to release any intracellular viruses. The lysate was collected in the supernatant harvested earlier. Next, 30% polyethylene glycol (PEG) 8000 in 0.4 M NaOH was added to the total volume of the lysate and supernatant to a final concentration of 15%. This mixture was stirred at 150 rpm overnight at 4 °C. Viruses were pelleted by centrifuging at 10,000× g for 30 min. The pellet was resuspended in 50 mL of sterile 1× phosphate-buffered saline (PBS) or 0.45 µm-filtered effluent or chlorinated effluent wastewater collected from the Wastewater Treatment Plant of King Abdullah University of Science and Technology (KAUST) [6]. Physical parameters of these collected wastewaters are shown in Table S1. Resuspended viruses were immediately used for solar inactivation.

2.2. Simulated Solar Inactivation Trials

Six milliliters of PEG-purified viruses was dispensed into each of the six microcosms. Each microcosm was made from 5 mL glass beakers (solution depth: 2.7 cm) wrapped in black duct tape to prevent unwanted light penetration. Each microcosm contained a magnetic stirrer and was covered at the top with aluminum foil, in the case of dark controls (n = 3), or left open (n = 3). Each of these microcosms was then placed in 50 mL beakers containing 20 mL of water that served to regulate the temperature at 20 °C. The 50 mL beakers of the dark controls were also covered with aluminum foil, while the irradiated microcosms were covered with a glass filter (Newport Corporation, Irvine, CA, USA) that allowed light of wavelengths ≥280 nm to pass through. Each beaker was then placed in an Atlas Suntest® XLS+ photostimulator (Atlas, Chicago, IL, USA) equipped with a xenon arc lamp. The solar irradiation was measured by a spectroradiometer (ILT950, International Light Technologies, Peabody, MA, USA) to determine the UV irradiance as described previously [27]. The irradiance rate at 280–700 nm was ca. 28 J/cm^2/h. The UV irradiance provided by the solar simulator approximated the irradiance measurements of direct noon sunlight measured at two locations on the KAUST campus [27]. Viruses (500 µL) were harvested from the irradiated samples and dark controls after 2 h, 4 h, 6 h, 8 h, 10 h, 12 h, 14 h, 16 h, 20 h, and 24 h (n = 3) of solar inactivation. Virus samples were stored at −80 °C until use.

2.3. Virus Inactivation Kinetics Evaluated by Means of Infectious Assay

Vero cells CCL-81 (American Type Culture Collection, Manassas, VA, USA) were seeded onto 96-well plates at 2 × 10^4 cells/well in growth media overnight at 37 °C with 5% CO_2. Cells were infected with viruses harvested at all time points from the solar inactivation experiment described above. Briefly, viruses were serially diluted (10^0–10^{-4}), and 50 µL of the diluted inoculum was added to a well with confluent Vero cells for 1 h at 37 °C with 5% CO_2. The inoculum was aspirated and replaced with DMEM supplemented with 100 µL of virus infection media. Cells were then returned to the incubator for 16 h. After incubation, cells were washed with 1× sterile PBS and subsequently fixed and permeabilized with 100 µL of ice-cold methanol-acetone (50%-50%) for 10 min. Cells were then washed with 1× PBS three times and incubated with 20 µL of anti-EV antibody (Merck, Cat No. 3321) for 1 h at 37 °C. Cells were then washed with 1× PBS three times and incubated with 20 µL of anti-Mouse (Ms) IgG-FITC (Merck) for 1 h at 37 °C. Cells were washed again with 1× PBS before adding 100 µL of 1× PBS to each of the wells after the last wash. Wells were observed under an epifluorescence microscope for foci (single infected cells). Foci were counted, and the viral titer was estimated using Equation (1):

$$\mathrm{FFU/L} = f \times \frac{10^6\ \mu\mathrm{L}}{50\ \mu\mathrm{L}} \times \frac{1}{\text{Dilution Factor}} \quad (1)$$

where FFU/L is focus-forming units per liter and f is the number of fluorescently labeled cells. The infectious titer of irradiated samples was compared against the dark control. Each viral dilution was inoculated in two wells of a 96-well plate, and titration was carried out in triplicate (n = 3). The results from this focus-forming assay were converted to log and natural log (ln) curves by calculating $\log(N_t/N_0)$ and $\ln(N_t/N_0)$, respectively, where N_t = virus titer at time t and N_0 = virus titer at 0 h (start of experiment). The slopes of the dark-control and irradiated samples (k) were calculated from the ln curves. Prior to the solar inactivation experiments, the absorbance value at 280–700 nm of the PEG-purified virus was determined using a UV-3600 UV–VIS spectrometer

(Shimadzu, Kyoto, Japan). The readings were used to generate correction factors that were applied to the slopes of the decay curves prior to half-life calculations and statistical comparison. The half-lives for each experiment, or the durations needed to reduce the viral titer by half, were calculated using the first-order kinetics Equation (2):

$$\ln(N_t/N_0) = -k^*t \tag{2}$$

where k^* is the corrected slope of the inactivation curve and t is time. Statistics were carried out by simple linear regression analysis of the $\log(N_t/N_0)$ values of dark-control and irradiated samples.

2.4. Virus Inactivation Kinetics Evaluated by Means of RNA Concentration Decay

RNA was extracted from viruses harvested at each time point throughout the solar inactivation trials by an RNeasy Mini kit (Qiagen, Hilden, Germany). Viral RNA was eluted in 35 μL of water, and the concentration of extracted RNA was determined using the Qubit® single-stranded RNA assay kit with the Qubit® fluorometer (Thermo Fisher Scientific, Carlsbad, CA, USA). Decay kinetics and statistical comparisons were carried out similarly as described in the previous section. Specifically, the RNA concentration obtained at the start of the experiment was defined as N_0 and was expressed in $\log(N_t/N_0)$ and $\ln(N_t/N_0)$ equations, where N_t is the RNA concentration of the virus sample harvested at time t and N_0 is the RNA concentration harvested from 0 h (start of experiment).

2.5. Growth Curve Analysis

Vero cells were seeded onto 96-well plates at a cell density of 2×10^4 cells per well in growth media overnight at 37 °C with 5% CO_2. Cells were then inoculated with 50 μL of serial dilutions (10^0–10^{-4}) of T0, D24 or L24 at the multiplicity of infection of 0.1, where T0 is presolar-irradiated EV70 (wild-type), D24 is the dark-control EV70 post solar irradiation, and L24 is EV70 that had undergone 24 h of simulated solar irradiation. Cells were incubated at 37 °C with 5% CO_2 for 1 h for viral absorption. Each dilution was added to two wells of a 96-well plate, and the titration for each virus sample was done in duplicate (n = 2). Virus inoculum was removed, replaced with viral infection medium, and incubated for infection to take place. Cells were fixed with ice cold methanol-acetone (1:1 *v/v*) for 10 min at 1, 3, 5, 7, and 9 days post-infection (dpi). Cells were subsequently labeled with anti-EV and anti-Ms IgG-FITC antibodies, and the corresponding viral titers were calculated as described above.

2.6. Virus Absorption Assay

Vero cells were seeded in 6-well plates at a cell density of 1.2×10^6 cells per well in growth media overnight at 37 °C with 5% CO_2. Cells were then inoculated in 100 μL of T0, D24 or L24 diluted in 500 μL of 1× sterile PBS for 1 h at 4 °C. After binding, cells were washed with 1× sterile PBS and subsequently scraped and collected in 400 μL of 1× sterile PBS. This cell suspension went through three rounds of freeze-thaw to release bound viruses and were then serially diluted (10^0–10^{-4}) and titered in a similar manner as described in Section 2.3.

2.7. EV70 Genome Sequencing

Vero cells were seeded in the same conditions as for the viral absorption assay described in Section 2.6. Cells were then inoculated with similar titers of T0, D24 or L24 in 500 μL of 1× sterile PBS for 1 h at 37 °C with 5% CO_2. The viral inoculum was replaced with DMEM supplemented with 2% FBS and 1× penicillin and streptomycin. Cells were placed in the incubator for 10 days. At 10 days post-infection (dpi), cells were washed, scraped and collected in 400 μL of 1× sterile PBS. RNA was extracted from the cells by the RNeasy Mini kit (Qiagen, Hilden, Germany). RNA was used as the template for fragment Polymerase Chain Reaction (PCR) where the genome was amplified into 18 overlapping fragments of 750 base pairs. Reaction mixes were prepared by adding 25 μL of 2× RT Buffer, 1.5 μL of forward and reverse primers at 10 μM concentration, 1 μL of Life Technologies

SuperScript™ II RT enzyme (Thermo Fisher Scientific, Carlsbad, CA, USA), 18 µL of water and 0.83 µL of template RNA. Primers for each of the fragments are listed in Table S2.

Fragments were amplified by touchdown PCR, which included cDNA synthesis at 55 °C, for 30 min, an initial denaturation step of 94 °C, for 2 min, followed by 15 cycles of 94 °C for 15 s, annealing at 52 °C for 30 s, and extension at 68 °C for 100 s, with the annealing temperature decreasing by 1 °C with each cycle. This was followed by another 30 cycles of 94 °C for 30 s, annealing at 48 °C for 30 s, and extension at 68 °C for 100 s. A final extension at 72 °C for 5 min was performed. Amplicons were run on a 1.2% agarose gel and visualized by SYBR Green (Thermo Fisher Scientific, Carlsbad, CA, USA). Bands corresponding to ~750 bp were extracted using the Wizard® SV Gel and PCR Clean-Up system (Promega, Fitchburg, WI, USA). Purified PCR products were sent to the KAUST Genomics Core lab for Sanger sequencing. PCR sequences were aligned using the SeqMan program of the DNASTAR's Lasergene software package (DNASTAR, Madison, WI, USA). Aligned contigs were saved as consensus sequences for each of the viral samples. Consensus sequences were aligned in BioEdit Sequence Alignment Editor [28] and translated in silico. The amino acid sequences of each generated genome were submitted to the Phyre2 web Portal for 3D structure prediction [29]. The .pdb files generated from Phyre2 were visualized in PyMOL [30].

3. Results

3.1. Viral Inactivation Upon Solar Irradiation

The infectious capacities of irradiated EV70 in phosphate-buffered saline (PBS), as well as in effluent and in chlorinated effluent wastewater matrices were assayed against dark-control EV70 (Figure 1A). After 24 h of irradiation, at a fluence of 688 J/cm^2, the infectivity of dark-control EV70 was reduced by 0.18 ± 0.07 log, 0.11 ± 0.003 log and 0.12 ± 0.03 log in PBS, effluent and chlorinated effluent matrices, respectively (k_{obs} = 0.014, 0.010 and 0.009) (Figure 1A and Table 1A). One-way ANOVA revealed no significant difference between the decay constants of the dark controls in the three different matrices ($p > 0.1$). Linear regression analysis of the slopes of the dark control showed no positive correlation with respect to time, suggesting that dark-control viruses were relatively stable in each of the matrices over a 24 h period ($p > 0.01$).

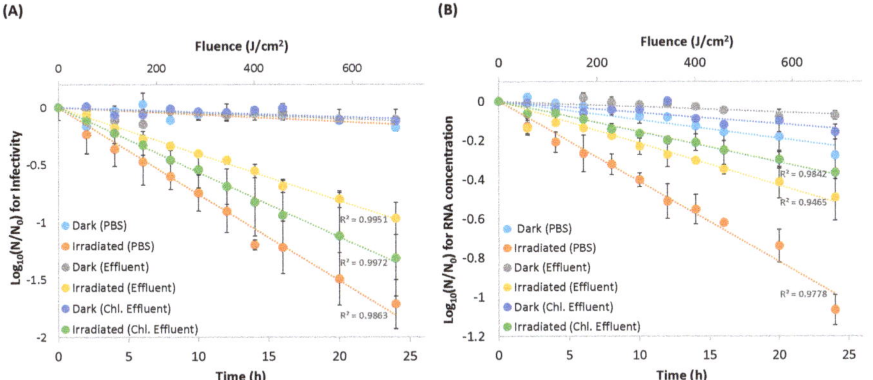

Figure 1. Decay curves of EV70 under simulated solar irradiation (•, •, •) compared to dark-control EV70 (•, •, •) over 24 h for viruses suspended in phosphate-buffered saline (PBS) (n = 3), effluent matrix (n = 2), and chlorinated effluent matrix (n = 2). The irradiance rate at 280–700 nm was 27.86 J/cm^2/h. (**A**) Viruses were harvested at each time point and subsequently tittered. (**B**) RNA was extracted from the viruses harvested at each time point and subsequently quantified.

Table 1. Decay kinetic constants and half-life of (**A**) EV70 infectivity, and (**B**) RNA concentration, under simulated solar irradiation for 24 h in phosphate-buffered saline (PBS) (n = 3), effluent wastewater matrix (n = 2), and chlorinated effluent wastewater matrix (n = 2). k_{obs} = decay constant. $t_{1/2}$ = half-life of decay.

		PBS		Effluent		Chlorinated Effluent	
		k_{obs}	$t_{1/2}$	k_{obs}	$t_{1/2}$	k_{obs}	$t_{1/2}$
(A)	Dark Control	0.014 ± 0.003	52 ± 12 h	0.010 ± 5 × 10^{-5}	70 ± 0.4 h	0.009 ± 0.001	77 ± 3.8 h
	Irradiated EV70	1.4 ± 0.2	30 ± 3 min	0.9 ± 0.1	47 ± 5 min	1.0 ± 0.1	41 ± 3 min
(B)	Dark Control	0.002 ± 0.006	34.8 ± 11.2 h	0.007± 0.001	101.8 ± 8.2 h	0.013 ± 0.006	67.0 ± 29.3 h
	Irradiated EV70	0.77 ± 0.16	57 ± 14 min	0.48 ± 0.05	88 ± 9 min	0.29 ± 0.01	145 ± 7 min

In contrast, irradiated EV70 reduced in infectivity by 1.7 ± 0.2, 1.0 ± 0.1, and 1.3 ± 0.3-logs in PBS, effluent and chlorinated effluent with decay constants of 1.4, 0.9 and 1.0 (Figure 1A and Table 1A). *t*-test analysis between the decay constants of the dark-control and irradiated samples showed that the decay within each matrix was significant ($p < 0.05$). One-way ANOVA of the decay constants of the irradiated samples revealed that the decay in each of the matrices was significantly different from the others, with the decay in PBS being the fastest ($t_{1/2}$ = 30 ± 3 min), followed by the decay in chlorinated effluent ($t_{1/2}$ = 41 ± 3 min), then effluent ($t_{1/2}$ = 47 ± 5 min) ($p < 0.05$) (Figure 1A and Table 1A).

3.2. RNA Decay

After the same dose of simulated solar irradiation, the RNA concentration from dark-control EV70 decreased by 0.3 ± 0.1 log, 0.1 ± 0.02 log, and 0.2 ± 0.04 log in PBS, effluent and chlorinated effluent matrices, respectively (k_{obs} = 0.002, 0.007, and 0.013). One-way ANOVA revealed that the decay constants of the dark control did not differ between the three matrices ($p > 0.05$) (Figure 1B and Table 1B). Linear regression analysis of the dark control showed a positive correlation with respect to time, suggesting that RNA of the dark-control samples was not stable in any matrix ($p < 0.05$).

The RNA concentrations of the irradiated samples decayed by 1.1 ± 0.1 log, 0.5 ± 0.1 log, and 0.4 ± 0.1 logs in PBS, effluent and chlorinated effluent, respectively (k_{obs} = 0.77, 0.48, and 0.29) (Figure 1B and Table 1B). Within each matrix, the decay constant of the irradiated samples differed significantly from the dark control, suggesting that simulated solar irradiation sped up RNA decay in EV70 ($p < 0.05$). The decay constants of the irradiated samples did not differ significantly from each other ($p > 0.05$), but RNA from EV70 in PBS decayed the fastest ($t_{1/2}$ = 57 ± 14 min), followed by EV70 in effluent wastewater matrix ($t_{1/2}$ = 88 ± 9 min) and then EV70 in chlorinated effluent wastewater matrix ($t_{1/2}$ = 145 ± 7 min) (Table 1B).

3.3. Irradiated EV70 Displays Inhibited Viral Replication

To study the replication kinetics of irradiated viruses, Vero cells were infected with the same titer of T0, D24 or L24 for nine days. Focus-forming units were counted throughout this period to produce the growth curves presented in Figure 2. Sixteen hours post-infection (hpi), T0 in PBS replicated to $3.9 \times 10^3 \pm 8.1 \times 10^2$ FFU/mL, and D24 in PBS replicated to $4.7 \times 10^3 \pm 1.1 \times 10^3$ FFU/mL. Both T0 and D24 peaked on the fifth day post-infection (dpi) at titers of $7.7 \times 10^4 \pm 4.1 \times 10^3$ FFU/mL and $6.3 \times 10^4 \pm 1.0 \times 10^4$ FFU/mL, respectively. A titer of $2.9 \times 10^4 \pm 1.0 \times 10^4$ FFU/mL and $3.2 \times 10^4 \pm 1.4 \times 10^4$ FFU/mL was observed for both these viruses at 9 dpi, respectively. This apparent reduction in titer was probably due to the detachment of infected cells from the monolayer nine days after infection. In contrast, L24 in PBS replicated to $6.4 \times 10^3 \pm 1.3 \times 10^3$ FFU/mL 16 hpi and remained relatively similar over the 9-day period. Nor did L24 display a peak at 5 dpi, as seen in T0 and D24 in PBS (Figure 2A).

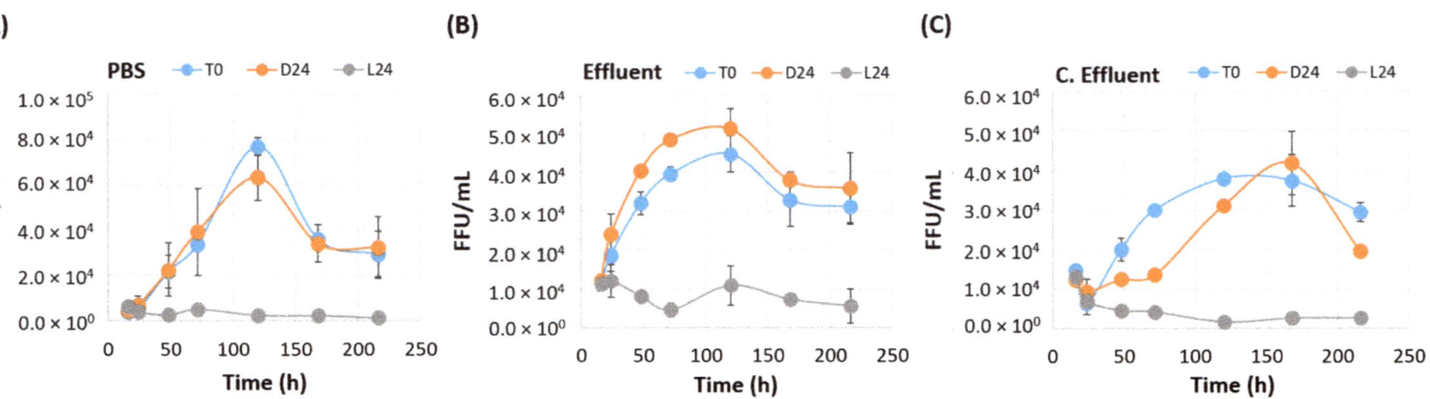

Figure 2. Growth kinetics of nontreated EV70 (T0 •), dark-control EV70 (D24 •), and simulated-solar-irradiated EV70 (L24 •). Confluent monolayers were infected with similar concentrations of each of the viral samples, and the foci formed over a nine-day period were enumerated. (**A**) EV70 resuspended in phosphate-buffered saline (PBS) (n = 3). (**B**) EV70 resuspended in effluent wastewater matrix (n = 2). (**C**) EV70 resuspended in chlorinated effluent wastewater matrix (n = 2).

T0 in effluent matrix replicated to $1.1 \times 10^4 \pm 2.6 \times 10^2$ FFU/mL at 16 hpi and peaked to $4.5 \times 10^4 \pm 4.7 \times 10^3$ FFU/mL at 5 dpi, and this titer reduced to $3.1 \times 10^4 \pm 4.5 \times 10^3$ FFU/mL at 9 dpi. D24 in effluent matrix exhibited a similar growth pattern, with a titer of $1.2 \times 10^4 \pm 1.4 \times 10^3$ FFU/mL at 16 hpi, $5.1 \times 10^4 \pm 5.2 \times 10^3$ FFU/mL at 5 dpi and $3.6 \times 10^4 \pm 9.1 \times 10^3$ FFU/mL at 9 dpi. L24 in effluent matrix replicated to $1.1 \times 10^4 \pm 1.8 \times 10^3$ FFU/mL at 16 hpi. On 5 dpi, when T0 and D24 replicated to peak titers, L24 in effluent matrix only displayed a titer of $1.1 \times 10^4 \pm 5.1 \times 10^3$ FFU/mL. The titer of L24 in effluent matrix did not exceed the titer displayed at 16 hpi throughout the course of the experiment (Figure 2B).

In the chlorinated effluent wastewater matrix, T0 replicated to $1.5 \times 10^4 \pm 1.4 \times 10^2$ FFU/mL at 16 hpi and peaked at approximately 4.0×10^4 FFU/mL approximately 130 hpi. It reached $3.0 \times 10^4 \pm 2.5 \times 10^3$ FFU/mL at 9 dpi. D24 in chlorinated effluent replicated to $1.3 \times 10^4 \pm 1.3 \times 10^3$ FFU/mL at 16 hpi and peaked at $4.2 \times 10^4 \pm 8.1 \times 10^3$ FFU/mL at 7 dpi, before finally reaching a titer of 2.0×10^4 at 9 dpi. In contrast, L24 in chlorinated effluent replicated to $1.3 \times 10^4 \pm 1.6 \times 10^3$ FFU/mL at 16 hpi and decreased to $2.6 \times 10^3 \pm 9 \times 10^1$ FFU/mL by 9 dpi (Figure 2C). T0 and D24 in chlorinated effluent peaked later than in PBS or effluent, possibly due to the presence of residual disinfection byproducts that may have a toxic effect on mammalian cell lines [31].

3.4. Irradiated EV70 Displayed Reduced Binding Capability

Untreated EV70 (T0) and dark-control EV70 harvested at 24 h post-irradiation (D24) displayed similar binding affinities to Vero cells in cell culture in all three matrices (Figure 3) ($p > 0.05$). EV70 in PBS displayed the greatest reduction in binding among the three matrices, 2.6 ± 0.3 log, followed by EV70 in effluent matrix at 1.8 ± 0.3 log and EV70 in chlorinated effluent matrix at 1.3 ± 0.2 log. One-way analysis of variance (ANOVA) revealed that the log reduction values (LRV) of the irradiated samples were significantly different from each other ($p < 0.01$), suggesting that the matrix affected the binding affinity of EV70 to Vero cells in cell culture (Figure 3).

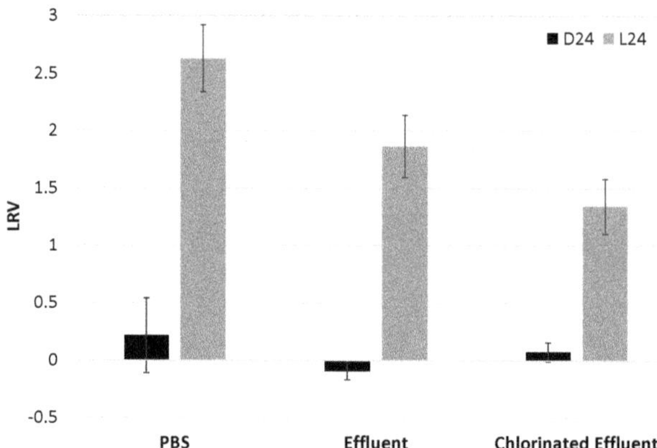

Figure 3. Log reduction values of the binding affinity of dark-control viruses (D24) and solar-irradiated EV70 (L24) with respect to untreated EV70. *Y*-axis represents log reduction value (LRV) with respect to nontreated EV70. D24: dark-control EV70 viruses that were placed in the solar simulator for 24 h but kept in the dark. L24: EV70 viruses that were exposed to simulated solar irradiation for 24 h. The irradiance rate at 280–700 nm was 28 J/cm^2/h. PBS, n = 3; effluent, n = 2; chlorinated effluent, n = 2.

3.5. Irradiated Viruses Select for Five Nonsynonymous Mutations

D24 and L24 from all three water matrices were infected in Vero cells for 10 dpi. Viral RNA was extracted and amplified into 18 overlapping fragments for Sanger-based sequencing. Figure 4 shows the alignment of the nucleotide sequence and the in silico-translated amino acid sequence. Nucleotide numbers here start at the first ATG of the coding sequence of the reference strain EV70 J670/71 (GenBank D00820.1) [32]. D24 in all matrices and in both replicates in PBS yielded the same nucleotide sequence, showing that any sequence difference seen in L24 was an effect of solar irradiation. L24 in PBS displayed five nonsynonymous nucleotide substitutions (Figure 4). A40G and C809T were observed in the VP4 and VP2 genes, respectively. These mutations resulted in conserved-amino-acid substitutions: Lys14Glu and Ala201Val (Figure 4A,B). Both these changes occurred in unstructured motifs of their respective proteins (Figure S1). G1171A was observed in the VP3 gene, which caused a nonconserved-amino-acid substitution of Gly71Ser (Figure 4C). However, this mutation maintained the β-sheet structure of this protein (Figure S1). G1810C was observed in the VP1 gene, resulting in Glu50Gln substitution, which occurred in an unstructured region of the protein (Figure 4D and Figure S1). A4801C in the $3C^{pro}$ gene resulted in the conserved-site substitution of Ile47Leu (Figure 4E). No structural changes were observed due to this mutation (Figure S1). Out of these five nonsynonymous mutations, A40G was also seen in L24 in chlorinated effluent matrix (Figure 4A). L24 in both wastewater matrices harbored a synonymous mutation of G4698A in the $3C^{pro}$ gene, which was not observed in L24 in PBS (Figure 4F).

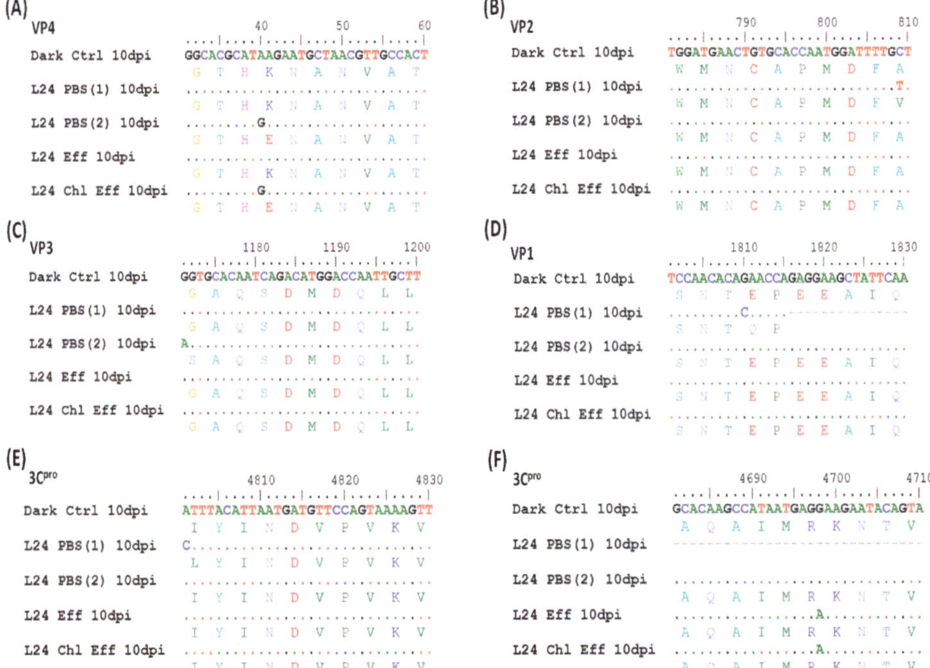

Figure 4. Genome sequence analysis of D24 and L24 (in PBS, effluent and chlorinated effluent) 10 days post-infection (dpi) in Vero cells. The nucleotide sequence of D24 in PBS is taken as the reference strain in this alignment (top row). The second (and other even-numbered) rows represent the predicted amino acid sequence. Identical nucleotide sequences are represented by a dot (.). Six nonsynonymous mutations were observed in the genes coding for (**A**) VP4, (**B**) VP2, (**C**) VP3, (**D**) VP1, and (**E**,**F**) $3C^{pro}$. The number of experiments performed was n = 2 for EV70 in PBS, n = 1 in effluent (Eff) and n = 1 in chlorinated effluent (Chl Eff). Sequence data could only be obtained from one experimental run for solar inactivation in wastewater matrices. Scale (top row) represents the nucleotide sequence of EV70 strain J670/71 (from NCBI reference D00820), with position 1 corresponding to the first nucleotide of the open reading frame.

4. Discussion

Earlier observations of viable and infective viruses in post-treated effluent provided the main impetus for this study [6], as their presence can complicate the reuse of reclaimed waters. To circumvent viral risks, chlorine disinfection is typically performed at the last step of a wastewater treatment process. However, chlorine works with varying effectiveness against different types of viruses [5]. This led to the suggestion of including combinations of various disinfection processes in a single WWTP. However, retrofitting different modular units of disinfection processes may incur additional operating costs. Solar disinfection of treated wastewater was therefore studied to provide a natural, low-cost and abundant disinfection strategy to further inactivate remnant viruses present in the reclaimed waters.

Specifically, EV70 was chosen as a model organism in this study, as infectious enteroviruses were previously found after wastewater treatment in concentrations approximating the infectious dose [6,33]. EV70 has not been studied extensively for its susceptibility to disinfectants due to its lack of plaque-producing capability in cell culture. To overcome this hurdle, a focus-forming assay was employed, which measured viral titer by fluorescently labeling virus-infected cells with virus-specific antibodies. This technique also required shorter duration compared to a traditional plaque assay [26].

We observed that EV70 in PBS experienced a 1.7-log reduction in infectivity after a dose of 688 J/cm^2 (Figure 1). This is consistent with the finding that poliovirus type 2 experienced a 4-log reduction with a simulated solar irradiation of 1224 J/cm^2, which is equivalent to a 2-log reduction at approximately 612 J/cm^2 [14]. Both EV70 and polioviruses are from the *Picornaviridae* family and have similar sizes (approximately 30 nm in diameter), capsid structures and genome lengths (EV70: 7200 nt, poliovirus: 7500 nt) [34,35]. In contrast, other members of *Picornaviridae* require differing doses of solar irradiation to achieve a similar reduction in infectivity. For example, Coxsackie viruses require approximately 58.5–99 J/cm^2, and ECHO viruses require 50–60 J/cm^2 of solar irradiation to achieve a 2-log reduction [15,16]. Both Coxsackie and ECHO viruses have similar sizes (28 nm and 24–30 nm, respectively) and genome lengths (approximately 7400 nt and 7500 nt, respectively) to EV70 [36–38]. The data presented in this study agree with earlier studies that infer the need for varying solar fluence to inactivate different viral species. While the structures of viruses are generally similar within a family, species might differ in protein folding and genome secondary structure, which give rise to differences in susceptibility to solar irradiation [39].

Picornaviridae have a positively stranded RNA genome that is directly translated by host-cell ribosomes [13]. Here, damage to the genome was indicated by the decay in the RNA concentrations in the presence of solar irradiation (Figure 1B). In addition, the reduction in binding capacity of L24 indicated conformational damage to the capsid, stopping it from recognizing the viral receptor on the Vero cells (Figure 3). This reduction in binding was of a larger magnitude than the reduction in infectivity as seen in Figure 1A at 24 h across all three matrices. Since receptor binding is the first step in a virus replication cycle, any irradiation-induced damage to the capsid could result in the inability of the capsid to recognize the receptor on host cells. Hence, this decrease in binding of L24 was most likely the primary cause for the decrease in infectious capacity.

Although EV70 with a damaged capsid may have a reduced binding capability, replication would theoretically still be possible if the interior structure of the viral particle remained undamaged. To test this, we performed a growth curve analysis, which showed that L24 was unable to replicate to similar titers as T0 or D24 even after 9 dpi in all three water matrices (Figure 2). This was observed despite the similar multiplicity of infection between T0, D24, and L24. This information indicates an inability of solar-irradiated EV70 to replicate as effectively as wild-type or dark-control viruses. The capsids of *picornaviruses* undergo a dramatic antigenic alteration before the virus uncoats [40]. Translation is then initiated by the internal ribosomal entry sites in the 5′ untranslated region of the genome, which is composed of five stem-loops (II-VI) [41,42]. Viral translation is also promoted by the binding of host-cell IRES trans-acting factors, such as FBP1-3, hnRNP K and hnRNP A, which recognize the 5′ untranslated region of the viral genome [42–45]. Solar irradiation could induce structural damage to the capsid and genome of EV70, leading to reduced binding and replication capacity. Not only would a structurally damaged capsid fail to bind to the host-cell receptor, but it might fail to undergo the antigenic alteration necessary for uncoating to occur [40]. UV irradiation promotes RNA-protein cross-linking [46]. The formation of covalent bonds between the EV70 genome and the capsid might affect the release of the RNA out of the capsid during the uncoating process. The integrity of the cloverleaf and stem loop structure present in the 5′ untranslated region of the EV70 genome might be negatively affected by solar irradiation. A disintegration of structure in this region of the genome may prevent successful docking of the ribosome and other host-cell translation initiation factors. Lastly, owing to the structural damage to the genome, translation might not proceed as efficiently as in wild-type viruses, producing proteins which might not support viral replication.

To elucidate if mutations did indeed occur in key proteins of EV70, we sequenced the coding region of the genome. Initially, fragment PCR of viruses directly sampled after 24 h of solar irradiation was performed (data not shown). However, this did not yield sufficient concentrations of PCR amplicons for sequencing. To overcome this technical constraint, Vero cells were infected with L24 or D24 for 10 days, and the viral RNA, which had amplified in the course of the infection, was extracted and sequenced. The sequence of the L24 viral genome derived from this experiment is, hence, not a direct product of solar irradiation but was selected for 10 dpi. This genome could be viewed as an 'escape mutant', being the only sequence that had replicated enough to be amplified by PCR. However, this sequence was still unable to replicate as effectively as T0 or D24 (Figure 2).

The irreproducibility of nucleotide substitutions between trials 1 and 2 for L24 in PBS indicate that solar irradiation induces mutations in a random manner. However, four out of the six mutations listed occurred in the capsid genes, which are at the 5′ end of the genome. Positions 40 and 4801 also showed mutations in two of the four irradiated samples (Figure 4A,F). These findings might suggest that the capsid genes, as well as position 4801 in the $3C^{pro}$ gene, are more prone to mutation by solar irradiation compared to the rest of the genome.

The structure of the capsid of bovine enterovirus (BEV), a *picornavirus*, has been determined [47]. The structural proteins of BEV share 48% identity with EV70 [48], and its tertiary structure is collinear with other enteroviruses [49]. Comparisons with the amino acid sequence of BEV's capsid reveal that the amino acid substitutions of EV70 listed in this study did not occur in any of the known functional motifs. However, an earlier study showed that an introduction of a single amino acid substitution at five different positions in the capsid genes resulted in a change in viral tropism [50]. These proteins constitute the capsid and form the depression known as the 'canyon', which recognizes the cellular receptor DAF/CD55 for attachment to the host [51,52]. VP1, which is the most exposed protein of the capsid of *Picornavirus* [53], forms a hydrophobic pocket that allows for myristic acid binding [54] and is believed to be involved in the binding of metal ions [55,56]. VP1 is also believed to have a role in the uncoating of the virus particle [50]. Even though Glu50Gln in VP1 occurred in an unstructured motif (Figure S1), the substitution might alter the charge of VP1.

EV70 with a glutamic acid instead of a lysine at position 14 of VP4 protein replicates poorly in HeLa cells [50]. In this current study, L24 in PBS and chlorinated effluent displayed this substitution (Figure 4), which may have accounted for the poor replication. It is likely that this mutation resulted in a change in the charge of the overall protein, as lysine is typically positive at neutral pH while glutamic acid is negatively charged. This would have resulted in poor binding of EV70 to the host cells. Similarly, even though the amino acid substitution Gly71Ser did not affect the folding of the β-sheet in VP3 (Figure S1), the overall polarity of the protein might have been affected owing to the polar nature of serine as opposed to glycine. Both these substitutions might have synergistically affected viral function.

In addition to assessing changes in the capsid proteins, the $3C^{pro}$ protein of *Picornaviridae* was also assessed since this protein displays a multitude of functions in the infected cell. Initially shown to be a protease that cleaves the functional proteins from the polyprotein precursor, $3C^{pro}$ also cleaves host-cell proteins to shut down host-cell transcription, translation, and nucleo-cytoplasmic trafficking and promote apoptosis (reviewed in [57]). There exist four main functional domains in the $3C^{pro}$

protein: the *N*-terminal domain (aa 12–13), the central domain (aa 82–86), the β-ribbon (aa 123–133) and the C-terminal domain (aa 154–156) [58–60]. The amino acid substitution Ile47Leu occurred in between the *N*-terminal domain and the central domain, an area that lacks any known function (Figure 4). It is important to note that this substitution did not affect the integrity of the β-sheet motif of this protein (Figure S1). However, further investigations into this amino acid substitution should be carried out to determine if the function of the $3C^{pro}$ protein is altered. If this substitution results in a change in the function of the protein, this could explain the reduced ability of L24 to replicate to high titers, as seen in cells infected with T0 and D24 (Figure 2). This could be a result of inadequate cleavage of the viral polyprotein or inadequate suppression of host-cell factors, allowing for the host cell to overcome the viral replication machinery.

In addition to observing a significant impact on the viral infectivity and persistence due to solar irradiation, we also observed that viral inactivation occurred at a slower rate when the viruses were present in wastewater matrix. This concurs with earlier observations [61–63]. Furthermore, out of the six nucleotide mutations found, only 2 were seen in L24 in wastewater matrices, A40G and G4698A, while L24 in PBS had 5 mutations (Figure 4). This indicates that viruses in the wastewater are less susceptible to UV-B [12,16]. Effluent and chlorinated effluent wastewaters used in this study had a total organic carbon (TOC) concentration of 4.2 mg/L and 5.2 mg/L, respectively, while PBS had undetected levels of TOC, as expected (Table S1). These organic compounds can act as radical scavengers [64,65], reduce light intensity [66], or encapsulate viruses with a protective organic coating that makes them more resistant to external environmental stressors when present in wastewaters. The latter has been alluded to by the findings that non-enveloped viruses are stable in wastewaters [67,68]. Alternatively, the high alkalinity in wastewater might favor the reaction between bicarbonates and hydroxyl radicals formed upon solar irradiation. This reaction results in the generation of CO_3●- which reacts slower with organic molecules compared to ●O_2 radicals [27,69,70].

While these reasons could explain the slower inactivation rates of EV70 in wastewater matrices, it is important to note that irradiated viruses, irrespective of matrix, all failed to propagate in cell culture (Figure 4). This indicates that solar irradiation successfully inhibits viral replication in cell culture, preventing the generation of infectious viral progeny in all three water matrices evaluated in this study. This strongly suggests that solar irradiation modifies the replication capacity of EV70 to the point that it might not pose a significant public health threat. Although the data from this study suggest that solar irradiation may serve as a good disinfection technique, its efficacy may be lower in turbid waters due to lower solar penetration and higher light-scattering effect. Operators would also need to create a holding tank that is shallow enough to allow for proper solar penetration and irradiation. This would not be feasible in densely populated places with limited land space. Hence, the use of solar irradiation as an effective, natural, and low-cost disinfection strategy against EV70 would only be feasible for use in low-turbidity waters, presumably in permeates after membrane filtration processes, and in places unconstrained by land availability.

Supplementary Materials: The following are available online at http://www.mdpi.com/2073-4441/11/1/64/s1, Figure S1: Predicted structure of EV70, Table S1: Physical parameters of the matrices used in this study, Table S2: Sequences and names of primers used in this study.

Author Contributions: M.R.J. and P.-Y.H. conceived and planned the experiments. M.R.J. carried out the experiments and performed the analysis. M.R.J. and P.-Y.H. wrote the manuscript together. P.-Y.H. provided reagents and materials. All authors have read and approved the manuscript.

Funding: This research was supported by the KAUST baseline funding BAS/1/1033/-01-01 awarded to P.-Y.H.

Acknowledgments: The authors would like to thank George Princeton Dunsford for access to the KAUST wastewater treatment plant, Moustapha Harb for providing sampling assistance, Nada Al Jassim for training required to operate the solar simulator and Noor Zaouri for assistance in the LC-OCD analysis.

Conflicts of Interest: The authors declare no conflict of interest.

References

1. AghaKouchak, A.; Feldman, D.; Hoerling, M.; Huxman, T.; Lund, J. Water and climate: Recognize anthropogenic drought. *Nature* **2015**, *524*, 409–411. [CrossRef] [PubMed]
2. Wang, J.; Da, L.; Song, K.; Li, B.-L. Temporal variations of surface water quality in urban, suburban and rural areas during rapid urbanization in shanghai, china. *Environ. Pollut.* **2008**, *152*, 387–393. [CrossRef] [PubMed]
3. Liyanage, C.; Yamada, K. Impact of population growth on the water quality of natural water bodies. *Sustainability* **2017**, *9*, 1405. [CrossRef]
4. Scholz, M. Chapter 19—Disinfection. In *Wetlands for Water Pollution Control*, 2nd ed.; Scholz, M., Ed.; Elsevier: Amsterdam, The Netherlands, 2016; pp. 129–136.
5. Zhong, Q.; Carratalà, A.; Ossola, R.; Bachmann, V.; Kohn, T. Cross-resistance of uv- or chlorine dioxide-resistant *echovirus 11* to other disinfectants. *Front. Microbiol.* **2017**, *8*, 1928. [CrossRef]
6. Jumat, M.; Hasan, N.; Subramanian, P.; Heberling, C.; Colwell, R.; Hong, P.-Y. Membrane bioreactor-based wastewater treatment plant in saudi arabia: Reduction of viral diversity, load, and infectious capacity. *Water* **2017**, *9*, 534. [CrossRef]
7. Lo, H.-L.; Nakajima, S.; Ma, L.; Walter, B.; Yasui, A.; Ethell, D.W.; Owen, L.B. Differential biologic effects of cpd and 6-4pp uv-induced DNA damage on the induction of apoptosis and cell-cycle arrest. *BMC Cancer* **2005**, *5*, 135. [CrossRef]
8. Drouin, R.; Therrien, J.P. Uvb-induced cyclobutane pyrimidine dimer frequency correlates with skin cancer mutational hotspots in p53. *Photochem. Photobiol.* **1997**, *66*, 719–726. [CrossRef]
9. Lamont, Y.; Rzeżutka, A.; Anderson, J.; MacGregor, S.; Given, M.; Deppe, C.; Cook, N. Pulsed uv-light inactivation of *poliovirus* and *adenovirus*. *Lett. Appl. Microbiol.* **2007**, *45*, 564–567. [CrossRef]
10. Lizasoain, A.; Tort, L.F.L.; García, M.; Gillman, L.; Alberti, A.; Leite, J.P.G.; Miagostovich, M.P.; Pou, S.A.; Cagiao, A.; Razsap, A.; et al. Human enteric viruses in a wastewater treatment plant: Evaluation of activated sludge combined with uv disinfection process reveals different removal performances for viruses with different features. *Lett. Appl. Microbiol.* **2018**, *66*, 215–221. [CrossRef]
11. Chevrefils, G.; Caron, É.; Wright, H.; Sakamoto, G.; Payment, P.; Barbeau, B.; Cairns, B. Uv dose required to achieve incremental log inactivation of bacteria, protozoa and viruses. *IUVA News* **2006**, *8*, 38–45.
12. Lytle, C.D.; Sagripanti, J.L. Predicted inactivation of viruses of relevance to biodefense by solar radiation. *J. Virol.* **2005**, *79*, 14244–14252. [CrossRef] [PubMed]
13. Racaniello, V.R. *Picornaviridae*: The viruses and their replication. In *Fields virology*, sixth edition; Wolters Klewer/Lippincott Williams Wilkinspp: Philadelphia, PA, USA, 2013; Volume 1, pp. 453–489.
14. Heaselgrave, W.; Patel, N.; Kilvington, S.; Kehoe, S.; McGuigan, K. Solar disinfection of *poliovirus* and *acanthamoeba polyphaga* cysts in water—A laboratory study using simulated sunlight. *Lett. Appl. Microbiol.* **2006**, *43*, 125–130. [CrossRef] [PubMed]
15. Heaselgrave, W.; Kilvington, S. The efficacy of simulated solar disinfection (sodis) against *coxsackievirus*, *poliovirus* and *hepatitis a virus*. *J. Water Health* **2012**, *10*, 531–538. [CrossRef] [PubMed]
16. Gerba, C.P.; Gramos, D.M.; Nwachuku, N. Comparative inactivation of *enteroviruses* and *adenovirus 2* by uv light. *Appl. Environ. Microbiol.* **2002**, *68*, 5167–5169. [CrossRef] [PubMed]
17. Mirkovic, R.R.; Kono, R.; Yin-Murphy, M.; Sohier, R.; Schmidt, N.J.; Melnick, J.L. *Enterovirus type 70:* The etiologic agent of pandemic acute haemorrhagic conjunctivitis. *Bull. World Health Organ.* **1973**, *49*, 341–346.
18. Khetsuriani, N.; Lamonte-Fowlkes, A.; Oberst, S.; Pallansch, M.A. *Enterovirus* surveillance—United States, 1970–2005. *MMWR Surveill Summ* **2006**, *55*, 1–20. [PubMed]
19. Roivainen, M. *Enteroviruses*: New findings on the role of *enteroviruses* in type 1 diabetes. *Int. J. Biochem. Cell Biol.* **2006**, *38*, 721–725. [CrossRef] [PubMed]
20. Ylipaasto, P.; Klingel, K.; Lindberg, A.M.; Otonkoski, T.; Kandolf, R.; Hovi, T.; Roivainen, M. *Enterovirus* infection in human pancreatic islet cells, islet tropism in vivo and receptor involvement in cultured islet beta cells. *Diabetologia* **2004**, *47*, 225–239. [CrossRef] [PubMed]
21. Blomqvist, S.; Savolainen, C.; Raman, L.; Roivainen, M.; Hovi, T. *Human rhinovirus 87* and *enterovirus 68* represent a unique serotype with *rhinovirus* and *enterovirus* features. *J. Clin. Microbiol.* **2002**, *40*, 4218–4223. [CrossRef] [PubMed]
22. Xie, J.; Li, D.; Xie, G.; Hu, Y.; Zhang, Q.; Kong, X.; Guo, N.; Li, Y.; Duan, Z. Inactivation of ev71 by exposure to heat and ultraviolet light. *Bing Du Xue Bao* **2015**, *31*, 500–506. [PubMed]

23. Ottoson, J.; Hansen, A.; Westrell, T.; Johansen, K.; Norder, H.; Stenstrom, T.A. Removal of *noro-* and *enteroviruses, giardia cysts, cryptosporidium oocysts*, and fecal indicators at four secondary wastewater treatment plants in sweden. *Water Environ. Res.* **2006**, *78*, 828–834. [CrossRef] [PubMed]
24. Qiu, Y.; Lee, B.E.; Neumann, N.; Ashbolt, N.; Craik, S.; Maal-Bared, R.; Pang, X.L. Assessment of human virus removal during municipal wastewater treatment in edmonton, canada. *J. Appl. Microbiol.* **2015**, *119*, 1729–1739. [CrossRef] [PubMed]
25. Simmons, F.J.; Xagoraraki, I. Release of infectious human enteric viruses by full-scale wastewater utilities. *Water Res.* **2011**, *45*, 3590–3598. [CrossRef] [PubMed]
26. Zhong, J.; Gastaminza, P.; Cheng, G.; Kapadia, S.; Kato, T.; Burton, D.R.; Wieland, S.F.; Uprichard, S.L.; Wakita, T.; Chisari, F.V. Robust *hepatitis c virus* infection *in vitro*. *Proc. Natl. Acad. Sci. USA* **2005**, *102*, 9294–9299. [CrossRef] [PubMed]
27. Al-Jassim, N.; Mantilla-Calderon, D.; Wang, T.; Hong, P.-Y. Inactivation and gene expression of a virulent wastewater *escherichia coli* strain and the nonvirulent commensal *escherichia coli* dsm1103 strain upon solar irradiation. *Environ. Sci. Technol.* **2017**, *51*, 3649–3659. [CrossRef]
28. Hall, T.A. Bioedit: A user-friendly biological sequence alignment editor and analysis program for windows 95/98/nt. *Nucleic Acids Sympos. Ser.* **1999**, *41*, 95–98.
29. Kelley, L.A.; Mezulis, S.; Yates, C.M.; Wass, M.N.; Sternberg, M.J.E. The phyre2 web portal for protein modeling, prediction and analysis. *Nat. Protoc.* **2015**, *10*, 845. [CrossRef]
30. DeLano, W.L.; Lam, J.W. Pymol: A communications tool for computational models. *Abstr. ACS* **2005**, *230*, U1371–U1372.
31. Wagner, E.D.; Plewa, M.J. Cho cell cytotoxicity and genotoxicity analyses of disinfection by-products: An updated review. *J. Environ. Sci.* **2017**, *58*, 64–76. [CrossRef]
32. Ryan, M.D.; Jenkins, O.; Hughes, P.J.; Brown, A.; Knowles, N.J.; Booth, D.; Minor, P.D.; Almond, J.W. The complete nucleotide sequence of *enterovirus type 70*: Relationships with other members of the *picornaviridae*. *J. Gen. Virol.* **1990**, *71*, 2291–2299. [CrossRef] [PubMed]
33. Cliver, D.O. Experimental infection by waterborne *enteroviruses*. *J. Food Prot.* **1981**, *44*, 861–865. [CrossRef]
34. Hogle, J.M. *Poliovirus* cell entry: Common structural themes in viral cell entry pathways. *Annu. Rev. Microbiol.* **2002**, *56*, 677–702. [CrossRef] [PubMed]
35. Taylor, A.R.; McCormick, M.J. Electron microscopy of *poliomyelitis virus*. *Yale J. Biol. Med.* **1956**, *28*, 589–597. [PubMed]
36. Chen, L.; Yang, H.; Feng, Q.-J.; Yao, X.-J.; Zhang, H.-L.; Zhang, R.-L.; He, Y.-Q. Complete genome sequence of a *coxsackievirus a16* strain, isolated from a fatal case in Shenzhen, Southern China, in 2014. *Gen. Ann.* **2015**, *3*, e00391-15. [CrossRef] [PubMed]
37. Crowell, R.L.; Landau, B.J. A short history and introductory b. In *The Coxsackie B Viruses*; Tracy, S., Chapman, N.M., Mahy, B.W.J., Eds.; Springer: Berlin/Heidelberg, Germany, 1997; pp. 1–11.
38. Dahllund, L.; Nissinen, L.; Pulli, T.; Hyttinen, V.P.; Stanway, G.; Hyypia, T. The genome of *echovirus 11*. *Virus Res.* **1995**, *35*, 215–222. [CrossRef]
39. Silverman, A.I.; Peterson, B.M.; Boehm, A.B.; McNeill, K.; Nelson, K.L. Sunlight inactivation of human viruses and bacteriophages in coastal waters containing natural photosensitizers. *Environ. Sci. Technol.* **2013**, *47*, 1870–1878. [CrossRef] [PubMed]
40. Fenwick, M.L.; Wall, M.J. Factors determining the site of synthesis of *poliovirus* proteins: The early attachment of virus particles to endoplasmic membranes. *J. Cell Sci.* **1973**, *13*, 403–413. [PubMed]
41. Fitzgerald, K.D.; Semler, B.L. Bridging ires elements in mrnas to the eukaryotic translation apparatus. *Biochim. Biophys. Acta* **2009**, *1789*, 518–528. [CrossRef]
42. Lin, J.Y.; Li, M.L.; Huang, P.N.; Chien, K.Y.; Horng, J.T.; Shih, S.R. Heterogeneous nuclear ribonuclear protein k interacts with the *enterovirus 71* 5′ untranslated region and participates in virus replication. *J. Gen. Virol.* **2008**, *89*, 2540–2549. [CrossRef]
43. Huang, H.I.; Chang, Y.Y.; Lin, J.Y.; Kuo, R.L.; Liu, H.P.; Shih, S.R.; Wu, C.C. Interactome analysis of the ev71 5′ untranslated region in differentiated neuronal cells sh-sy5y and regulatory role of fbp3 in viral replication. *Proteomics* **2016**, *16*, 2351–2362. [CrossRef] [PubMed]
44. Hung, C.-T.; Kung, Y.-A.; Li, M.-L.; Brewer, G.; Lee, K.-M.; Liu, S.-T.; Shih, S.-R. Additive promotion of viral internal ribosome entry site-mediated translation by far upstream element-binding protein 1 and an *enterovirus 71*-induced cleavage product. *PLoS Pathog.* **2016**, *12*, e1005959. [CrossRef]

45. Lin, J.-Y.; Shih, S.-R.; Pan, M.; Li, C.; Lue, C.-F.; Stollar, V.; Li, M.-L. Hnrnp a1 interacts with the 5′ untranslated regions of *enterovirus 71* and *sindbis virus* rna and is required for viral replication. *J. Virol.* **2009**, *83*, 6106–6114. [CrossRef]
46. Noah, J.W.; Shapkina, T.; Wollenzien, P. Uv-induced crosslinks in the 16s rrnas of *escherichia coli*, *bacillus subtilis* and *thermus aquaticus* and their implications for ribosome structure and photochemistry. *Nucleic Acids Res.* **2000**, *28*, 3785–3792. [CrossRef] [PubMed]
47. Smyth, M.; Fry, E.; Stuart, D.; Lyons, C.; Hoey, E.; Martin, S.J. Preliminary crystallographic analysis of *bovine enterovirus*. *J. Mol. Biol.* **1993**, *231*, 930–932. [CrossRef] [PubMed]
48. Poyry, T.; Kinnunen, L.; Hyypia, T.; Brown, B.; Horsnell, C.; Hovi, T.; Stanway, G. Genetic and phylogenetic clustering of *enteroviruses*. *J. Gen. Virol.* **1996**, *77*, 1699–1717. [CrossRef] [PubMed]
49. Smyth, M.; Tate, J.; Hoey, E.; Lyons, C.; Martin, S.; Stuart, D. Implications for viral uncoating from the structure of *bovine enterovirus*. *Nat. Struct. Biol.* **1995**, *2*, 224–231. [CrossRef] [PubMed]
50. Kim, M.S.; Racaniello, V.R. *Enterovirus 70* receptor utilization is controlled by capsid residues that also regulate host range and cytopathogenicity. *J. Virol.* **2007**, *81*, 8648–8655. [CrossRef] [PubMed]
51. Olson, N.H.; Kolatkar, P.R.; Oliveira, M.A.; Cheng, R.H.; Greve, J.M.; McClelland, A.; Baker, T.S.; Rossmann, M.G. Structure of a *human rhinovirus* complexed with its receptor molecule. *Proc. Natl. Acad. Sci. USA* **1993**, *90*, 507–511. [CrossRef] [PubMed]
52. Karnauchow, T.M.; Tolson, D.L.; Harrison, B.A.; Altman, E.; Lublin, D.M.; Dimock, K. The hela cell receptor for *enterovirus 70* is decay-accelerating factor (cd55). *J. Virol.* **1996**, *70*, 5143–5152. [PubMed]
53. Smyth, M.S.; Trudgett, A.; Hoey, E.M.; Martin, S.J.; Brown, F. Characterization of neutralizing antibodies to *bovine enterovirus* elicited by synthetic peptides. *Arch. Virol.* **1992**, *126*, 21–33. [CrossRef] [PubMed]
54. Smith, T.J.; Kremer, M.J.; Luo, M.; Vriend, G.; Arnold, E.; Kamer, G.; Rossmann, M.G.; McKinlay, M.A.; Diana, G.D.; Otto, M.J. The site of attachment in *human rhinovirus 14* for antiviral agents that inhibit uncoating. *Science* **1986**, *233*, 1286–1293. [CrossRef] [PubMed]
55. Arnold, E.; Rossmann, M.G. The use of molecular-replacement phases for the refinement of the *human rhinovirus 14* structure. *Acta Cryst.* **1988**, *44*, 270–283. [CrossRef]
56. Kim, S.S.; Smith, T.J.; Chapman, M.S.; Rossmann, M.C.; Pevear, D.C.; Dutko, F.J.; Felock, P.J.; Diana, G.D.; McKinlay, M.A. Crystal structure of *human rhinovirus* serotype 1a (hrv1a). *J. Mol. Biol.* **1989**, *210*, 91–111. [CrossRef]
57. Sun, D.; Chen, S.; Cheng, A.; Wang, M. Roles of the picornaviral 3c proteinase in the viral life cycle and host cells. *Viruses* **2016**, *8*, 82. [CrossRef] [PubMed]
58. Cui, S.; Wang, J.; Fan, T.; Qin, B.; Guo, L.; Lei, X.; Wang, J.; Wang, M.; Jin, Q. Crystal structure of human *enterovirus 71* 3c protease. *J. Mol. Biol.* **2011**, *408*, 449–461. [CrossRef] [PubMed]
59. Franco, D.; Pathak, H.B.; Cameron, C.E.; Rombaut, B.; Wimmer, E.; Paul, A.V. Stimulation of *poliovirus* synthesis in a hela cell-free in vitro translation-rna replication system by viral protein 3cdpro. *J. Virol.* **2005**, *79*, 6358–6367. [CrossRef]
60. Blair, W.S.; Parsley, T.B.; Bogerd, H.P.; Towner, J.S.; Semler, B.L.; Cullen, B.R. Utilization of a mammalian cell-based rna binding assay to characterize the rna binding properties of *picornavirus* 3c proteinases. *RNA* **1998**, *4*, 215–225.
61. Pecson, B.M.; Martin, L.V.; Kohn, T. Quantitative pcr for determining the infectivity of bacteriophage ms2 upon inactivation by heat, uv-b radiation, and singlet oxygen: Advantages and limitations of an enzymatic treatment to reduce false-positive results. *Appl. Environ. Microbiol.* **2009**, *75*, 5544–5554. [CrossRef]
62. Templeton, M.R.; Andrews, R.C.; Hofmann, R. Inactivation of particle-associated viral surrogates by ultraviolet light. *Water Res.* **2005**, *39*, 3487–3500. [CrossRef]
63. Barrett, M.; Fitzhenry, K.; O'Flaherty, V.; Dore, W.; Keaveney, S.; Cormican, M.; Rowan, N.; Clifford, E. Detection, fate and inactivation of pathogenic *norovirus* employing settlement and uv treatment in wastewater treatment facilities. *Sci. Total Environ.* **2016**, *568*, 1026–1036. [CrossRef]
64. Klöcking, R.; Felber, Y.; Guhr, M.; Meyer, G.; Schubert, R.; Schoenherr, J.I. Development of an innovative peat lipstick based on the uv-b protective effect of humic substances. *Mires Peat* **2013**, *11*, 1–9.
65. Westerhoff, P.; Aiken, G.; Amy, G.; Debroux, J. Relationships between the structure of natural organic matter and its reactivity towards molecular ozone and hydroxyl radicals. *Water Res.* **1999**, *33*, 2265–2276. [CrossRef]
66. Sun, C.-X.; Kitajima, M.; Gin, K.Y.-H. Sunlight inactivation of somatic coliphage in the presence of natural organic matter. *Sci. Total Environ.* **2016**, *541*, 1–7. [CrossRef] [PubMed]

67. Symonds, E.M.; Griffin, D.W.; Breitbart, M. Eukaryotic viruses in wastewater samples from the united states. *Appl. Environ. Microbiol.* **2009**, *75*, 1402–1409. [CrossRef]
68. Bofill-Mas, S.; Albinana-Gimenez, N.; Clemente-Casares, P.; Hundesa, A.; Rodriguez-Manzano, J.; Allard, A.; Calvo, M.; Girones, R. Quantification and stability of *human adenoviruses* and *polyomavirus jcpyv* in wastewater matrices. *Appl. Environ. Microbiol.* **2006**, *72*, 7894–7896. [CrossRef] [PubMed]
69. McGuigan, K.G.; Conroy, R.M.; Mosler, H.-J.; Preez, M.d.; Ubomba-Jaswa, E.; Fernandez-Ibañez, P. Solar water disinfection (sodis): A review from bench-top to roof-top. *J. Hazard. Mater.* **2012**, *235–236*, 29–46. [CrossRef]
70. Canonica, S.; Kohn, T.; Mac, M.; Real, F.J.; Wirz, J.; von Gunten, U. Photosensitizer method to determine rate constants for the reaction of carbonate radical with organic compounds. *Environ. Sci. Technol.* **2005**, *39*, 9182–9188. [CrossRef]

Article

Potential Use of *Dimocarpus longan* Seeds as a Flocculant in Landfill Leachate Treatment

Hamidi Abdul Aziz [1,2,*], Nor Aini Rahim [1], Siti Fatihah Ramli [1], Motasem Y. D. Alazaiza [1], Fatehah Mohd Omar [1] and Yung-Tse Hung [3]

1 School of Civil Engineering, Engineering Campus, Universiti Sains Malaysia, Nibong Tebal 14300, Penang, Malaysia; aunie78@gmail.com (N.A.R.); ctfatihah88@gmail.com (S.F.R.); my.azaiza@gmail.com (M.Y.D.A.); cefatehah@usm.my (F.M.O.)

2 Solid Waste Management Cluster, Science and Technology Research Centre, Engineering Campus, Universiti Sains Malaysia, Nibong Tebal 14300, Penang, Malaysia

3 Department of Civil and Environmental Engineering, Cleveland State University, Cleveland, OH 44115, USA; yungtsehung@yahoo.com

* Correspondence: cehamidi@usm.my; Tel.: +60-4599-6215

Received: 19 October 2018; Accepted: 14 November 2018; Published: 16 November 2018

Abstract: Landfill leachate is a highly polluted and generated from water infiltration through solid waste produced domestically and industrially. In this study, a coagulation–flocculation process using a combination of Polyaluminium chloride (PACl) as a coagulant and *Dimocarpus longan* seed powder (LSP) as coagulant aid was used in treating landfill leachate. LSP has been tested as the main coagulant and as coagulant aid with PACl. As the main coagulant, the optimum dosage and pH for PACl were 5 g/L and 6, respectively, with removal efficiencies of 67.44%, 99.47%, and 98% for COD, SS, and color, respectively. For LSP as the main coagulant, results show that LSP is not effective where the removal efficiencies obtained for COD, SS, and color were 39.40%, 22.20%, and 28.30%, respectively, with the optimum dosage of 2 g/L and pH 4. The maximum removal efficiencies of COD, SS, and color were 69.19%, 99.50%, and 98.80%, respectively, when LSP was used as coagulant aid with PACl. Results show that using LSP as coagulant aid was found to be more effective in the removal of COD, SS, and color with less PACl dosage. The PACl dosage was decreased from 5 to 2.75 g/L when LSP was used as a coagulant aid. Cost estimation for using PACl alone and using LSP as the coagulant aid showed a reduction in the cost of approximately 40% of the cost of using PACl alone. Overall, this study confirmed the efficiency of LSP to be used as a natural coagulant aid in leachate treatment.

Keywords: *Dimocarpus longan* seeds; leachate treatment; coagulant–flocculation; polyaluminium chloride

1. Introduction

Landfill is the most widely accepted and prevalent methods for municipal solid waste (MSW) disposal in developing in many countries around the world due to its inherent forte in terms cost saving and simpler operational mechanism [1]. Environmental pollution caused by the landfill leachate has been one of the typical dilemmas of landfilling method [2]. Leachate is the liquid produced when water percolates through solid waste and contains dissolved or suspended materials from various disposed materials and decomposition process. It is often high-strength wastewater with extreme pH, chemical oxygen demand (COD), biochemical oxygen demand (BOD), inorganic salts and toxicity [3,4]. Its composition differs over the time and space within a particular landfill, influenced by a broad spectrum of factors such as waste composition, landfilling practice (solid waste contouring and compacting), local climatic conditions [5], landfill's physicochemical conditions, biogeochemistry and landfill age [6].

The composition and characteristics of the landfill leachate are the main factors determine the choices of the treatment method [7].

To protect groundwater aquifer and adjacent surface water from leachate contamination, the handling and treatment of leachate must be meticulously designed to minimize its potential adverse impacts [8]. Biological treatment is an environmentally friendly method that can be applied for the treatment of young or freshly produced leachate [9]. However, it is ineffective for leachate from older landfills that usually contain high COD and ammonium content, low biodegradability (high COD/BOD ratio) and multiple heavy metal ions [10]. In contrast, physical and chemical methods are more effective for older leachate as compared to young leachate [11,12]. Sometimes, the quantity of short-term leachate could be difficult to predict since it mainly depends on precipitation; however, the quantity of long-term can be predicted more accurately [13].

In general, generated leachate at early stages of waste decomposition is highly rich in BOD_5 and contains a high amount of biodegradable and nonbiodegradable materials such as volatile fatty acids [12]. However, the stabilized leachate from old landfills is often highly polluted with non-biodegradable organic substances, such as fulvic substances and humic-like, which are measured as COD [14]. Furthermore, the stabilized leachate contains a large quantity of inorganic substances, especially ammonium-nitrogen (NH_3-N) [15], which results from the hydrolysis and fermentation of nitrogen-containing fractions of biodegradable refuse substrates. When the bioreactor landfills contain contaminant leachate, collection and in situ recirculation for acceleration of decomposition of readily available organic fractions of wastes, leachate NH_3-N concentrations may accumulate to produce higher levels as compared to the traditional landfills [16].

Coagulation–flocculation is one of the major chemical methods being used for leachate pretreatment [17]. These methods are applied to remove suspended solids and recalcitrant substances such as humic acid, fulvic acid, or undesirable compounds like heavy metals, absorbable organic halides (AOX) and polychlorinated biphenyls (PCBs) from the leachate [18]. This simple method outshines other advanced technologies like membrane and chemical oxidation technologies in terms of leachate pre-treatment application. The treatment mechanism of this method mainly consists of charge neutralization between negatively charged colloids and cationic hydrolysis products followed by an amalgamation of impurities through flocculation [19]. Total suspended solids (TSS), as well as colloidal particles, are the main parameters removed from this process [20]. The major component of colloid particles is organic compounds. Coagulation is generally defined as destabilization of a colloidal suspension or solution by neutralizing the forces that keep them apart. Cationic coagulants give a positive electric charge that reduces the negative charge of the colloids in solution. As a result, particles of colloid to compose large particles (flocs). Usually, fast mixing is required to disperse the coagulant throughout the solution [21].

The chemistry of coagulation–flocculation is mainly based on the electrical characteristics. The majority of the particles present in the leachate are negatively charged (-30 to -40 mV) [22]. Therefore, they tend to replace each other. Most of the coagulant chemicals are usually used to neutralize the negative charge on colloidal particles to prevent the repelling of these particles with each other [4]. The quantity of coagulant that will be added to the leachate is related to the zeta potential which is known as the electrical potential reflecting the voltage difference between the diffuse layer boundary and the dispersant [23]. Therefore, in the case of large zeta potential, more coagulant is needed. The coagulants have positive charges which attracted to the negative particles in solution; hence the combination of negative and positive charges results in a neutral charge and turn the particles no longer repel each other [4].

The commonly used commercial coagulants are aluminum sulfate (alum), polyaluminum chloride (PACl), ferrous sulfate, ferric chlorosulphate, and ferric chloride [24]. Inorganic coagulants are generally effective however there are some drawbacks related to the high amount of metal ions in sludge [25] while on the other hand, natural coagulants were found to produce relatively low sludge volume and are safe to humans when compared to the inorganic coagulant [26]. Natural coagulants have been widely applied in wastewater treatment [27] but it is still not used widely in landfill leachate treatment despite that they are in abundant quantities, relatively less expensive, and environmentally friendly [28].

In Malaysia, there has been a recent upsurge in the food industry where a large number of solid wastes was generated annually and especially *Dimocarpus longan* seeds, which are disposed from food manufacturing factories. *Dimocarpus longan* belongs to Sapindaceae family and goes by many scientific names: Nephelium *Dimocarpus longan* Camp, *Dimocarpus longan* Lam, and Euphoria *Dimocarpus longan* Strand. Currently, *Dimocarpus longan* is consumed as fresh and processed fruits while the seeds, which account for about 17% of the fresh weight of whole fruits, are discarded as waste or burned as fuel [29]. The seeds have been found as a rich source of antioxidant phenolic compounds that promising as functional food ingredients or natural preservatives. Soong and Barlow [30] reported that *Dimocarpus longan* seeds contained high levels of corilagin, gallic acid, and ellagic acid, which have been proven to acquire strong free radical-scavenging activity [31]. The seeds have been shown earlier to contain the hydrolysable tannins (ellagitannins) corilagin and acetonyl-geraniin [32]. Corilagin has been extensively studied for its pharmacological activities in the extract of plants such as Acer nikoense and Phyllanthus amarus and also as a pure isolated compound.

No actual data are available on the production and area of *Dimocarpus longan* in Malaysia. It is mainly cultivated in Penang and Kedah. Obtaining precise data on the production and acreage of this species is relatively difficult due to its small production. Based on the author's knowledge, there are no published studies in the literature regarding the usage of *Dimocarpus longan* seed powder (LSP) in wastewater and landfill leachate treatment. The main goal of this study is to investigate the applicability of composite coagulant made from LSP as a natural coagulant in removing color, COD and Suspended Solids (SS) from stabilized leachate. The main objectives of the study are; (i) to determine the optimum pH and dosage of polyaluminum chloride (PACl) and LSP as the main coagulant in removing COD, SS and color; and (ii) to determine the efficiency of LSP as coagulant aid and PACl as the main coagulant in removing COD, SS, and color.

2. Materials and Methods

2.1. Leachate Sampling and Characterization

Landfill leachate samples were collected from Alor Pongsu Landfill Site (APLS) in Bagan Serai, Perak, Malaysia from January through April 2018. APLS is classified as an anaerobic stabilized landfill. APLS started its operation in the year of 2000. Since its operation started, the landfill received approximately an average of 660,000 metric tons of solid waste per year, which is roughly 200 metric tons per day [33]. The site covers an area of 10 acres of palm oil plantation. Sampling was carried out using the grab sampling method while preservation was done according to Standard Methods for the Examination of Water and Wastewater [34]. The initial characteristics of the six leachate samples obtained were as contained in Table 1. All samples were kept in HDPE (high-density polyethylene) containers with sealed caps. Samples were transported to the laboratory within 1 h and stored in a cold room at 4 °C to minimize biological and chemical reactions prior to any treatability study. Before experiments, leachate samples were conditioned by putting them at room temperature for 2–3 h and homogenized by manual agitation. During the study, the leachate samples were characterized before and after each treatment. The samples were characterized in terms of turbidity, pH, suspended solids (SS), color, COD, manganese (Mn^{2+}), copper (Cu^{2+}), iron (Fe^{3+}), zinc (Zn^{2+}), phosphate (PO_4^{3-}) and ammonia-nitrogen (NH_3-N). All the analytical procedures were performed according to the Standard Method of Water and Wastewater [34]. pH was measured using a portable pH meter (CyberScan pH 510, Eutech, Singapore). Turbidity was measured using a turbidimeter (HACH 2100 N, HACH, Singapore). COD was measured using colorimetric method (5220-D). Heavy metals, NH_3-N, and color were measured using a spectrophotometer (DR/2800, HACH, Singapore).

Table 1. Characteristics of raw leachate.

Parameter	Min	Max	Average	Std. Dev.	Standard
pH	7.89	8.72	8.28	0.22	6.0–9.0
Residual conductivity (μS/cm)	814.00	966.00	880.94	51.19	-
Particle size, d (μm)	0.45	94.66	61.15	23.36	-
COD (mg/L)	3016.67	3055.00	3036.82	14.34	400
BOD_5 (mg/L)	107.33–176		130.92	48.55	20
Turbidity (NTU)	228.75	337.50	306.25	34.59	-
Suspended solids (mg/L)	591.67	866.67	745.00	99.67	50
Manganese (mg/L)	4.17	7.50	5.83	1.49	0.2
Iron (mg/L)	4.00	4.92	4.47	0.35	5.0
Copper (mg/L)	3.00	5.08	4.00	0.61	0.2
Zinc (mg/L)	0.83	1.75	1.15	0.26	2.0
Ammonia-nitrogen (mg/L)	737.50	875.00	794.00	47.82	5.0
Phosphate (mg/L)	43.75	62.50	53.25	6.28	-
Color (PtCO)	4525.00	7150.00	5517.50	794.75	100 ADMI

2.2. Preparation of Dimocarpus Longan Seed Powder (LSP)

The extraction method was adapted from Katayon et al. [35] with some modification by heating distilled water for 30 min with 100 °C using a hot plate and stirrer. Firstly, 500 g of fresh *Dimocarpus longan* seeds were peeled and separated from its aril. Then, they were washed using tap water to remove dirt before being air-dried for 48 h. After that, the layer, which is called the seeds coat covering the seeds, was removed. The seeds were again air-dried for another 48 h to ensure that it was completely dry before turning into powder form. The dried seeds were ground using a ring mill for 15 s until it became a fine powder. Finally, the seed powder was kept in a dry place to be used for experiments to be carried out later.

2.3. Coagulation–Flocculation

The current study investigated the coagulation–flocculation process using a combination of PACl as coagulant and LSP as a coagulant aid. A hydrolyed solution of PACl with the formula of [Al (OH)x Cly] (where x is in the range 1.35–1.65, and y = 3 − x) and pH 2.3–2.9 due to the presence of hydrochloric acid was supplied by Hasrat Bestari Sdn Bhd, Penang, Malaysia. An 18% solution of PACl was used as a stock solution throughout the experiments. Coagulation–flocculation experiments were carried out using jar test apparatus (SW6 Stuart Bibby Scientific Limited, Staffordshire, UK). Leachate samples were allowed to reach room temperature (approximately 3 h) before testing, and they were also thoroughly agitated to resuspend any settled solids. The leachate sample volume per beaker was 500 mL. The time and speed for rapid and slow mixing were set with an automatic controller. The jar test consisted of three subsequent stages: (1) rapid mixing stage with speed of 120 rpm for 3 min, (2) slow mixing stage with speed of 20 rpm for 15 min, and (3) final settling time for 45 min. During rapid mixing, the coagulant was added into the beakers while the impellers were maintained at fast speed. After a certain rapid mixing period, the stirrers were set to a slower speed for another period of time. After that, the stirrers were stopped, and the samples were left for final settling. Then, the samples were withdrawn using plastic syringe from 10 cm below the surface for the analytical determinations. Analyses were undertaken in triplicates. A 500 mL of leachate samples were filled into six beakers and agitated simultaneously while varying the rotational speed and allowing simulation of different mixing intensities and resulting flocculation process [36].

A preliminary coagulant performance study using jar test was conducted to determine optimum pH and dosage for PACl and LSP. It was noteworthy that the preliminary optimum pH studies were performed first with controlled coagulant dosages and the results were carried over to the preliminary optimum dosage studies as controlled pH since the pH would have a major impact on coagulant dosage. Different dosages and pH were investigated in this study for PACl and LSP as coagulant aid for removing COD, color, and TSS. The examination of pH effect was performed by adjusting the pH value of leachate samples between 5 and 9 using solutions of 0.1 N sulphuric acid (H_2SO_4) and 0.1 N sodium hydroxide (NaOH). The removal efficiency was investigated by using LSP as coagulant aid and PACl. Zeta potential test was conducted to enhance the results of the jar test and justify the removal mechanisms of the coagulation process. Zeta potential can present a measure of the net surface charge on the particle and potential distribution at the interface. Consequently, zeta potential serves as an important parameter in the description of the electrostatic interaction between particles in dispersed systems and the properties of the dispersion as affected by this electrical phenomenon [37]. In this study, the surface charge was evaluated by using Malvern Zetasizer Nano ZS. Measurements were taken at 25 °C with distilled water as the dispersal medium.

3. Results and Discussion

3.1. Characteristics of Leachate

The physicochemical parameters of landfill leachate are listed in Table 1. The leachate is categorized as stabilized leachate since its BOD_5/COD ratio < 0.1. The BOD_5/COD ratio indicates the degree of biodegradation and landfill age. For example, young leachate has BOD_5/COD ratio up to 0.83 during the acidogenic phase and decrease to 0.05 for old landfills during methanogenic phase [38]. The low BOD_5 and BOD_5/COD values for stabilized leachate agreed with the literature [11,15]. The high concentration of SS (745 mg/L) indicated the presence of organic and inorganic solids. A considerable concentration of ammonia nitrogen was found which is attributed to the decomposition of nitrogenous substances in refuse and the release of soluble nitrogen from solid wastes [15]. The dissolved organics mainly contributed a greater concentration of color (5517 Pt-Co). These organic compounds may be present in the form of recalcitrant material mainly composed of humic-like substances. A low value of BOD_5 means low biodegradability while the presence of high concentration of NH_3-N indicates high leachate toxicity [39].

3.2. Characteristics of LSP

Figure 1 illustrates the particle size distribution of LSP using Mastersizer analysis (Malvern Panalytical Ltd., Westborough, MA, USA). Results show that d_{10}, d_{50}, and d_{90} were recorded at 5.317 μm, 13.087 μm, and 32.460 μm, respectively. Fourier transformed infrared spectroscopy (FTIR) was used to investigate the structure of LSP and the analysis of their functional groups as shown in Figure 2. The FTIR spectrums of LSP show a weak intensity at 3435 cm^{-1}, due to the O-H stretching and also overlap with a primary amine and aliphatic primary amine due to N-H stretching. At 2989 cm^{-1} the functional group is under carboxylic acid which is bonded by strong O-H stretching. This band also overlaps with medium C-H stretching under a functional group of an alkane. Under wavenumber 2591 cm^{-1}, Aldehyde with a medium bond of C-H, ariel together with Thiol is weak in the intensity of S-H stretching. A weak aromatic compound with C-H bonding was found in 1867 cm^{-1}. Alkene shows at 1639 cm^{-1}, with strong and medium bond due to C=C stretching. At 1526 cm^{-1} the functional group is under nitro compound which is bonded with strong N-O stretching while at 1276 cm^{-1} there is a strong intensity due to the stretching of C-F bond under fluor compound function. At 1136 cm^{-1} wavenumber, there is a strong intensity in stretching of sulfone with a strong bond of S=O. At wavenumber of 1020 cm^{-1}, strong intensity of C=O under a functional group of alkyl aryl-ether, medium intensity of C-N stretching with a functional group of amine and also a strong intensity of C-O with vibration group on stretching and vinyl ether functional group. Surface morphology for LSP was investigated using a scanning electron microscope (SEM) with different magnifications as shown in Figure 3. They grouped the oval granular of LSP together and formed into a clod of an elliptical. It also had a cloudy or velvety like coating surface. SEM shows that the surface texture of the LSP was rough and there was an accumulation of fine particles with irregular geometric shapes spotted on the surface.

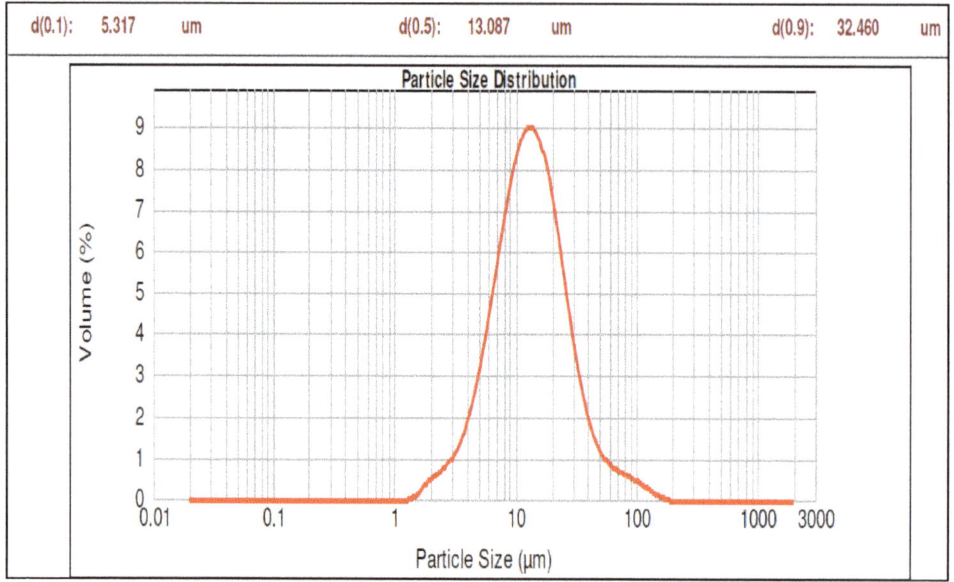

Figure 1. Particle size distribution of LSP (Longan seeds powder).

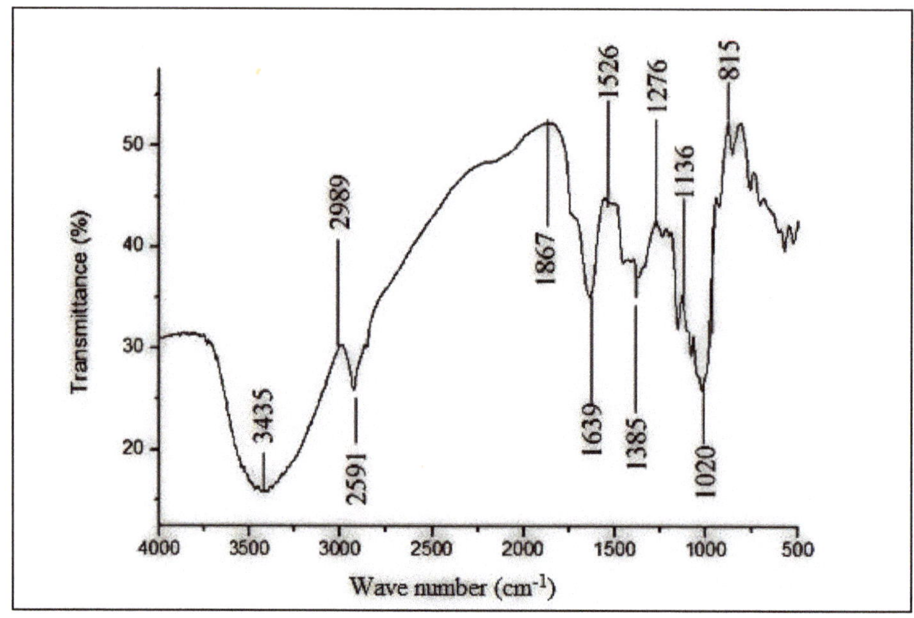

Figure 2. Fourier-transform infrared (FTIR) analysis of LSP.

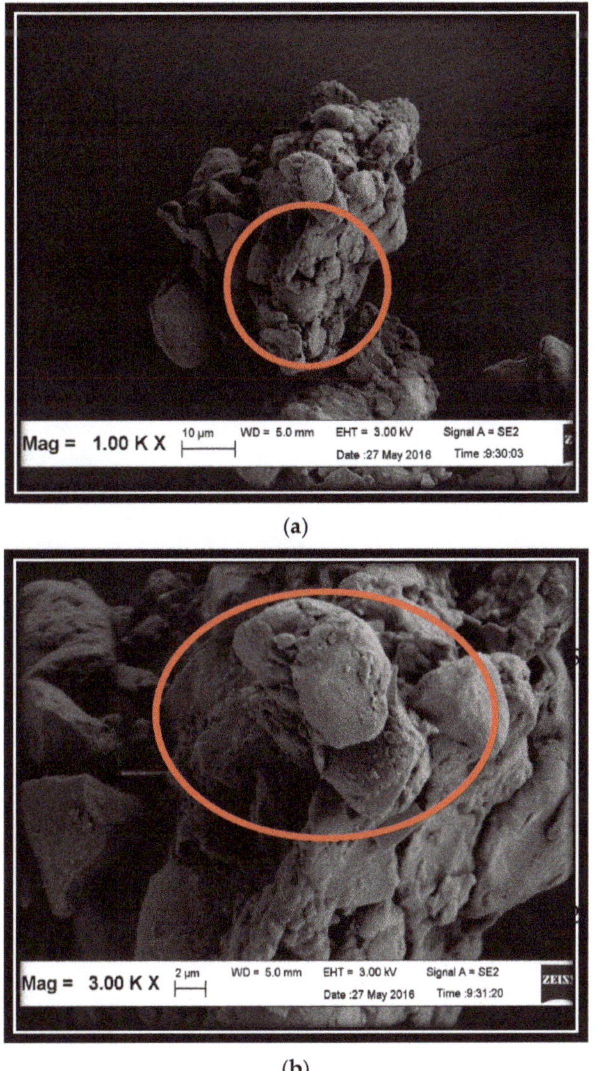

Figure 3. SEM images of LSP at (**a**) 1000 magnification (**b**) 3000 magnification.

3.3. Determination of Zeta Potential and Particle Size od LSP as a Function of pH

Figure 4 illustrates the zeta potential of LSP in conjunction with pH. It had a negative charge over the same pH range and reached the point of zero charges (PZC) at pH 7.

LSP has a negative level at surface starting from pH 2 to pH 6. However, at pH 7, the surface charges for LSP had become absolutely neutral and gradually turn negative when it was at pH 12. Under this condition, it could be said that the LSP was anionic coagulants, and the main mechanism governing the aggregation of the constituent was bridging [40]. Figure 5 illustrates the zeta potential and z-average particle size variation in conjunction with pH. It had a negative charge over the same pH range and reached PZC at pH 7.

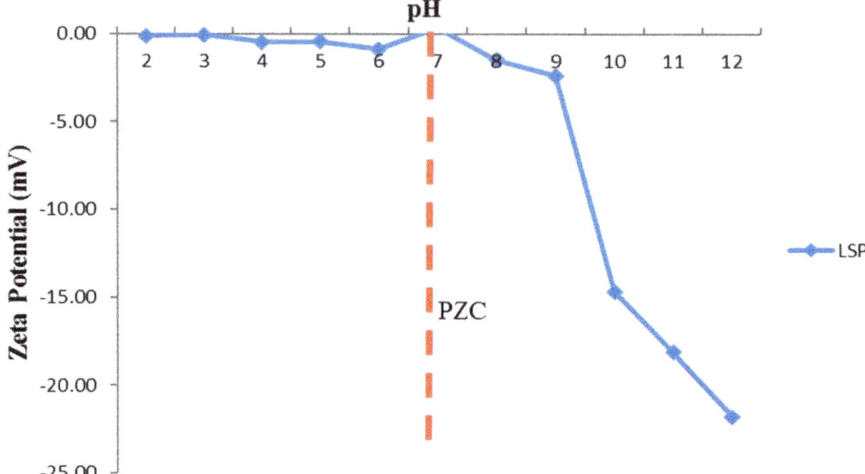

Figure 4. Effect of pH on the zeta potential values of LSP.

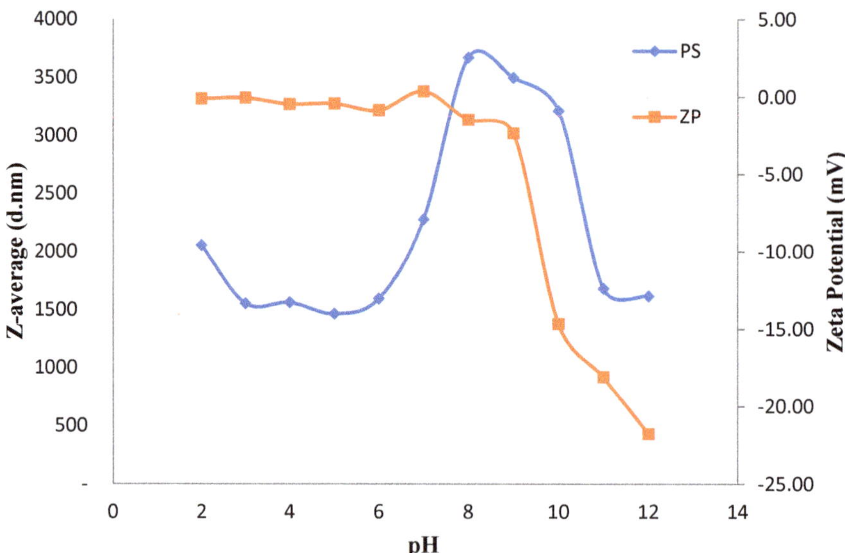

Figure 5. Zeta potential and z-average particle size variation with different pH values (PS is the particle size, ZP is the zeta potential).

3.4. PACl as a Main Coagulant (Optimum pH and Dosage)

The preliminary coagulant performance study for PACl showed that the PACl is very effective as a main coagulant at pH 6 (data not shown). Therefore, the process of the determination of the optimum dosage for PACl was conducted by adjusting the pH value to pH 6. Figure 6 shows the major range of PACl dose to remove pollutants at a constant value of pH 6.

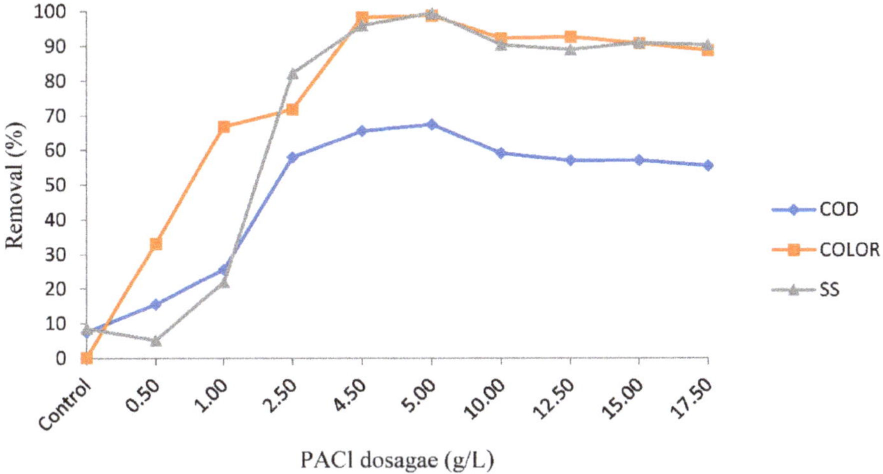

Figure 6. The effect of PACl in major range of dosage on the removal of COD, color and SS at pH 6.

The observations in Figure 6 show that the highest pollutant removal by PACl was in the range of 4.25–7.5 g/L PACl dosage. By taking this range as a reference, further coagulation tests on removal was conducted within this major range to get the optimum coagulant dose for PACl. The dosages varied from 4.25 to 7.5 g/L at a constant pH 6. Figure 7 shows the effect of PACl major range of dosage on the removal of COD, color, and SS at a constant pH 6. Based on the results, the optimum dosage of PACl was 5.0 g/L which achieved highest removal efficiencies of 67.44%, 98.73%, and 99.47% for COD, color, and SS, respectively.

The usage of PACl provided a better removal of color and SS. At the beginning of the experiments, leachate samples have an initial black color due to the presence of a humic substance [41]. The most effective removal of color occurs at the dosage of 5.0 g/L with 98.73%. The removal efficiency decreased gradually even though PACl dosage still added. This observation is most likely because when a large amount of coagulant dosage has been added over the optimum dosage, the surface of the particle's charge reversed due to of continuous absorption of mono and polynuclear hydrolysis species of PACl. The colloidal particles cannot be removed by perikinetic flocculation as they became positively charged particles [42]. The charge neutralization theory can explain this behavior. When a coagulant is added to the landfill leachate at optimum pH, colloid destabilization occurred when positively charged metal ions encounter with negatively charged colloids neutralizes the charge. The removal of particles will only take place more effectively when the more metal-based coagulant is added, as explained by the Schulze–Hardy rule [43]. As a result, when an extra dosage is added, colloids start to absorb the excessive positive charges which remain in the solution and become positively charged. Therefore, the electrical repulsions between positively charged colloids and metal ions occur. The colloids become stable again as the result, weaken the ability of coagulant to remove contaminants. According to Baghvand et al. [44], overdosing of coagulant will disturb the development process. Thus, the right amount of dose should be added to any wastewater treatment.

COD removal for PACl recorded a high removal of 67.44% at 5.0 g/L of PACl. This is because at a higher concentration of coagulant dosage, the flocs produced have a good consistency and in a better structure than at a lower dosage [45]. Below the optimum dosage, the removal of COD by PACl is not effective due to the fact that at lower concentrations of dose, a smaller floc is produced, and it influenced the velocity of the sludge [46]. SS also has the same pattern of removal, where the highest removal of SS was 99.47% at 5.0 g/L of PACl.

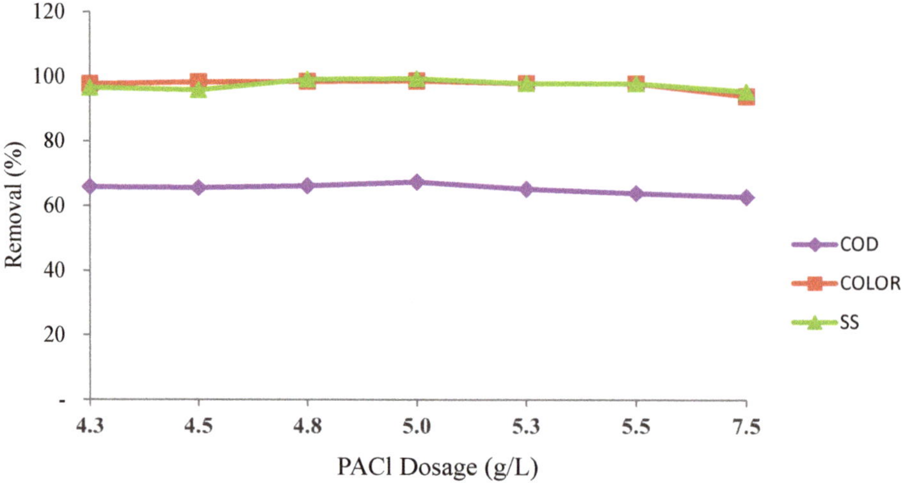

Figure 7. The effect of PACl major range of dosage on the removal of COD, color and SS at a constant pH 6.

3.5. Removal Efficiency of SS, Color, and COD Using LSP

A preliminary coagulant test has been conducted using LSP to determine the optimum pH. From the observation, LSP showed its effectiveness at pH 4 (data not shown). Therefore, the subsequent tests were carried out at pH 4. Different dosages of LSP were used to investigate its effect on the removal of color, SS, and COD at constant pH 4 as shown in Figure 8. Results show that the maximum removal efficiencies of color, SS, and COD were 28.3%, 11.2% and 15.1%, respectively at 2 g/L of LSP dosage.

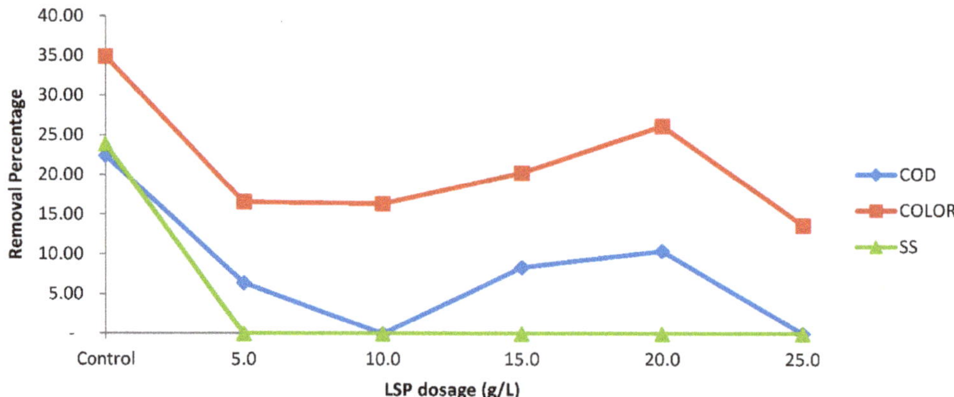

Figure 8. The effect of LSP in major dosages on the removal of color, SS, and COD at a constant pH 4.

Therefore, the optimum dosage of LSP was 2 g/L. It was found that at a higher concentration of LSP, the removal efficiencies of color, SS and COD was decreasing gradually until 5 g/L dosage. Thereafter, the removal efficiencies fluctuated for COD and color and remained at zero for SS.

3.6. PACl as the Main Coagulant with LSP as a Coagulant Aid

Optimum dosage of PACl (5 g/L) was added to different dosages of LSP by fixing the same conditions of jar test for the slow and rapid mixing followed by settling for 2 h at the optimum pH 6 for PACl. LSP dosage as coagulant aid, (0 g/L as a control sample) with 0.5, 1.0, 2.0, 4.0, 6.0, 8.0 and 10.0 g/L were used for pollutant removal and for reducing the dosage of primary coagulant.

Figure 9 shows the effect of LSP as a coagulant aid to remove COD, color, and SS at a different concentration of PACl. In the PACl predetermined test, 5.0 g/L was used as optimum dosage and resulted in removal efficiencies of 67.44%, 98.73%, and 99.47% for COD, color, and SS. However, when LSP was used as a coagulant aid, the results were better than the use of only PACl in terms of reducing the dosage of PACl and increase the removal efficiencies. The performance of COD removal has been improved from 67.44% to 69.19% at 2.75 g/L of PACl with the combination of 2 g/L of LSP as a coagulant aid when compared with 5 g/L PACl alone. In the coagulation test for the optimum dosage, it was found that PACl alone was able to remove 98% color at 5.0 g/L concentration. From the graph, it shows that as LSP dose increased from 1.0 to 10.0 g/L; the removal rates of color also almost similar for all LSP dosages. The maximum removal efficiency of color (98.80%) was obtained at LSP dosage of 2 g/L and 2.75 g/L PACl. Furthermore, the highest removal of SS was detected at the same combination of 2 g/L LSP and 2.75 g/L PACl with 99.50% removal efficiency.

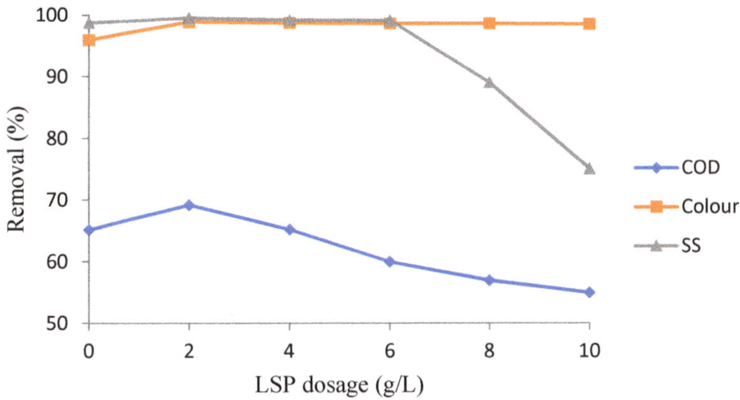

Figure 9. The effect of LSP as coagulant aid on the removal of color, SS, and COD at a constant pH 6 and using 2.75 g/L of PACl dosage.

Figure 10 shows a comparison between the efficiency of using PACl in landfill leachate treatment alone with two different dosages (5 g/L and 2.75 g/L) and using a combination of PACl as a coagulant in conjunction with LSP as a coagulant aid. From the graph, it is clear that the removal efficiency of COD, color, and SS was better when the PACl was used in conjunction with LSP as a coagulant aid. The dose of metal coagulant can be reduced without affecting the removal performance when polyelectrolyte is used as a coagulant aid because polyelectrolyte has higher charge density and molecular weight which act as an important role in coagulation. This is due to the addition of coagulant aid could help to form bigger flocs and produced more particles sediment, thus increasing the sedimentation rate [47]. Formation of flocs became quicker when LSP is used as a coagulant aid.

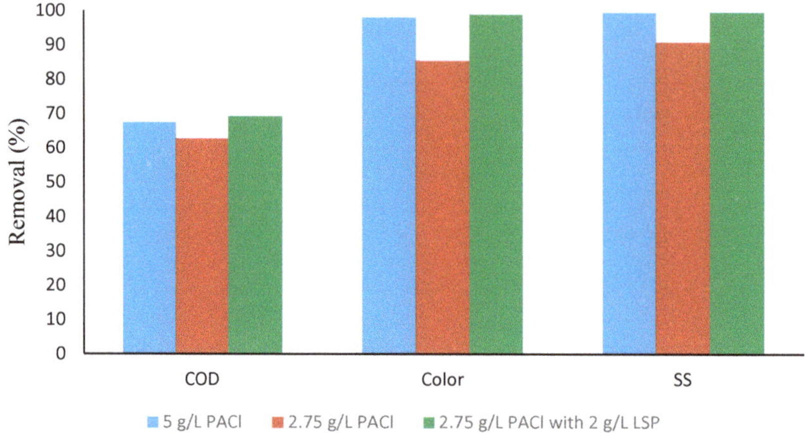

Figure 10. Comparison of using PACl alone and PACl in conjunction with LSP as a coagulant aid.

3.7. Coagulant Cost Estimation

Generally, the leachate treatment cost depends on factors, such as landfill design, the quantity of leachate, level or degree of treatment needed, and final removal method for residues and effluent. Obtaining data on the cost of leachate treatment is difficult because it requires the cooperation of the company in charge. Therefore, on the basis of the chemicals used, the costs of both coagulants are estimated. Table 2 showed the cost comparisons when the LSP was used as the coagulant aid. From the comparisons, a cost of RM 7800 was found when 5 g/L of PACl was used alone, whereas the use of 2 g/L of LSP as coagulant aid only cost RM 4646, a reduction of approximately 40% of the cost.

Table 2. Cost estimation for PACl and LSP.

Coagulant	Price of Chemical (RM)	Optimum Concentration Used	Amount of Chemical to Treat 1 m^3 of Leachate/Day	The Cost to Treat 1 m^3 of Leachate (RM) *	Total (RM) *
5000 mg/L PACl	300/L	13 mL/500 mL	26 L	7800	7800
2750 mg/L PACl	300/L	7.6 mL/500 mL	15.2 L	4560	4646
2000 mg/L LSP	4.30/kg	10 g/500 mL	20 kg	86	

* 1 USD = 4 RM.

4. Conclusions

The application of the coagulation–flocculation process to landfill leachate was examined in this study using longan seed powder (LSP) as a natural coagulant aid. LSP was not effective when used as a main coagulant. Compared to when PACl was used alone, a slight improvement of COD, color, and SS removal efficiencies were obtained when PACl was used as a main coagulant and LSP as a coagulant aid. The maximum removal efficiencies of COD, color, and SS were 69.19%, 98.80%, and 99.50%, respectively. In addition, using of LSP as coagulant aid was able to reduce the PACl dosage from 5 g/L when used alone to 2.75 g/L when used with conjunction of LSP. A cost estimation for using LSP as a coagulant aid showed a reduction in the cost of using PACl alone of approximately 40% from the cost of PACl. Overall, this study confirmed that LSP is an effective material to be used as a natural coagulant aid for landfill leachate treatment.

Author Contributions: H.A.A. conceived and designed the experiments, N.A.R. performed the experiments, M.Y.D.A. wrote and revised the paper, S.F.R. and F.M.O. conducted the data analysis, and Y.-T.H revised and proofread the paper.

Funding: This work was funded by Universiti Sains Malaysia under iconic grant scheme [Grant No. 1001/CKT/870023] for research associated with the Solid Waste Management Cluster, Engineering Campus, Universiti Sains Malaysia.

Conflicts of Interest: The authors declare no conflicts of interest.

References

1. Zamri, M.F.M.A.; Kamaruddin, M.A.; Yusoff, M.S.; Aziz, H.A.; Foo, K.Y. Semi-aerobic stabilized landfill leachate treatment by ion exchange resin: Isotherm and kinetic study. *Appl. Water Sci.* **2017**, *7*, 581–590. [CrossRef]
2. Aziz, H.A.; Alias, S.; Adlan, M.N.; Asaari, A.; Zahari, M.S. Colour removal from landfill leachate by coagulation and flocculation processes. *Bioresour. Technol.* **2007**, *98*, 218–220. [CrossRef] [PubMed]
3. Fang, C.; Chu, Y.; Jiang, L.; Wang, H.; Long, Y.; Shen, D. Removal of phthalic acid diesters through a municipal solid waste landfill leachate treatment process. *J. Mater. Cycles Waste Manag.* **2018**, *20*, 585–591. [CrossRef]
4. Aziz, H.A.; Yusoff, M.S.; Adlan, M.N.; Adnan, N.H.; Alias, S. Physico-chemical removal of iron from semi-aerobic landfill leachate by limestone filter. *Waste Manag.* **2004**, *24*, 353–358. [CrossRef] [PubMed]
5. Wang, Z.-P.; Zhang, Z.; Lin, Y.-J.; Deng, N.-S.; Tao, T.; Zhuo, K. Landfill leachate treatment by a coagulation–photooxidation process. *J. Hazard. Mater.* **2002**, *95*, 153–159. [CrossRef]
6. Wang, Y.; Gong, B.; Lin, Z.; Wang, J.; Zhang, J.; Zhou, J. Robustness and microbial consortia succession of simultaneous partial nitrification, ANAMMOX and denitrification (SNAD) process for mature landfill leachate treatment under low temperature. *Biochem. Eng. J.* **2018**, *132*, 112–121. [CrossRef]
7. Kulikowska, D.; Klimiuk, E. The effect of landfill age on municipal leachate composition. *Bioresour. Technol.* **2008**, *99*, 5981–5985. [CrossRef] [PubMed]
8. Umar, M.; Aziz, H.A.; Yusoff, M.S. Trends in the use of Fenton, electro-Fenton and photo-Fenton for the treatment of landfill leachate. *Waste Manag.* **2010**, *30*, 2113–2121. [CrossRef] [PubMed]
9. Agamuthu, P.; Fauziah, S. Solid waste landfilling: Environmental factors and health. In Proceedings of the EU-Asia Solid Waste Management Conference, Perak, Malaysia, 28–29 October 2008.
10. Dia, O.; Drogui, P.; Buelna, G.; Dubé, R. Hybrid process, electrocoagulation-biofiltration for landfill leachate treatment. *Waste Manag.* **2018**, *75*, 391–399. [CrossRef] [PubMed]
11. Uygur, A.; Kargı, F. Biological nutrient removal from pre-treated landfill leachate in a sequencing batch reactor. *J. Environ. Manag.* **2004**, *71*, 9–14. [CrossRef] [PubMed]
12. Ahn, D.-H.; Chung, Y.-C.; Chang, W.-S. Use of coagulant and zeolite to enhance the biological treatment efficiency of high ammonia leachate. *J. Environ. Sci. Health Part A* **2002**, *37*, 163–173. [CrossRef]
13. Aziz, H.A.; Daud, Z.; Adlan, M.N.; Hung, Y.-T. The use of polyaluminium chloride for removing colour, COD and ammonia from semi-aerobic leachate. *Int. J. Environ. Eng.* **2009**, *1*, 20–35. [CrossRef]
14. Bashir, M.J.K.; Aziz, H.A.; Yusoff, M.S.; Huqe, A.; Mohajeri, S. Effects of ion exchange resins in different mobile ion forms on semi-aerobic landfill leachate treatment. *Water Sci. Technol.* **2010**, *61*, 641–649. [CrossRef] [PubMed]
15. Al-Hamadani, Y.A.; Yusoff, M.S.; Umar, M.; Bashir, M.J.; Adlan, M.N. Application of psyllium husk as coagulant and coagulant aid in semi-aerobic landfill leachate treatment. *J. Hazard. Mater.* **2011**, *190*, 582–587. [CrossRef] [PubMed]
16. Postacchini, L.; Ciarapica, F.E.; Bevilacqua, M. Environmental assessment of a landfill leachate treatment plant: Impacts and research for more sustainable chemical alternatives. *J. Clean. Prod.* **2018**, *183*, 1021–1033. [CrossRef]
17. Bashir, M.J.K.; Aziz, H.A.; Yusoff, M.S. New sequential treatment for mature landfill leachate by cationic/anionic and anionic/cationic processes: Optimization and comparative study. *J. Hazard. Mater.* **2011**, *186*, 92–102. [CrossRef] [PubMed]
18. Onay, T.T.; Pohland, F.G. In situ nitrogen management in controlled bioreactor landfills. *Water Res.* **1998**, *32*, 1383–1392. [CrossRef]
19. Chiang, L.-C.; Chang, J.-E.; Chung, C.-T. Electrochemical oxidation combined with physical–chemical pretreatment processes for the treatment of refractory landfill leachate. *Environ. Eng. Sci.* **2001**, *18*, 369–379. [CrossRef]

20. Ishak, A.R.; Hamid, F.S.; Mohamad, S.; Tay, K.S. Stabilized landfill leachate treatment by coagulation-flocculation coupled with UV-based sulfate radical oxidation process. *Waste Manag.* **2018**, *76*, 575–581. [CrossRef] [PubMed]
21. Duan, J.; Gregory, J. Coagulation by hydrolysing metal salts. *Adv. Colloid Interface Sci.* **2003**, *100*, 475–502. [CrossRef]
22. Kurniawan, T.A.; Lo, W.-H.; Chan, G.Y. Physico-chemical treatments for removal of recalcitrant contaminants from landfill leachate. *J. Hazard. Mater.* **2006**, *129*, 80–100. [CrossRef] [PubMed]
23. Amokrane, A.; Comel, C.; Veron, J. Landfill leachates pretreatment by coagulation-flocculation. *Water Res.* **1997**, *31*, 2775–2782. [CrossRef]
24. Luo, K.; Pang, Y.; Li, X.; Chen, F.; Liao, X.; Lei, M.; Song, Y. Landfill leachate treatment by coagulation/flocculation combined with microelectrolysis-Fenton processes. *Environ. Technol.* **2018**. [CrossRef] [PubMed]
25. Kaszuba, M.; Corbett, J.; Watson, F.M.; Jones, A. High-concentration zeta potential measurements using light-scattering techniques. *Philos. Trans. R. Soc. Lond. A Math. Phys. Eng. Sci.* **2010**, *368*, 4439–4451. [CrossRef] [PubMed]
26. Lee, M.R. Treatment of Landfill Leachate in Coagulation-Flocculation Method by Using Micro Zeolite and Micro Sand. Ph.D. Thesis, Faculty of Civil Engineering, Universiti Tun Hussein Onn Malaysia, Parit Raja, Malaysia, 2013.
27. Fosso-Kankeu, E.; Waanders, F.; Mulaba-Bafubiandi, A.; Mishra, A. Flocculation performances of polymers and nanomaterials for the treatment of industrial wastewaters. *Smart Mater. Waste Water Appl.* **2016**, *7*, 213–235.
28. Jadhav, M.V.; Mahajani, Y. A comparative study of natural coagulants in flocculation of local clay suspensions of varied turbidities. *Int. J. Civ. Environ. Eng.* **2013**, *35*, 26–39.
29. Santos, T.R.; Silva, M.F.; Nishi, L.; Vieira, A.M.; Klein, M.R.; Andrade, M.B.; Vieira, M.F.; Bergamasco, R. Development of a magnetic coagulant based on *Moringa oleifera* seed extract for water treatment. *Environ. Sci. Pollut. Res.* **2016**, *23*, 7692–7700. [CrossRef] [PubMed]
30. Meraz, K.A.S.; Vargas, S.M.P.; Maldonado, J.T.L.; Bravo, J.M.C.; Guzman, M.T.O.; Maldonado, E.A.L. Eco-friendly innovation for nejayote coagulation–flocculation process using chitosan: Evaluation through zeta potential measurements. *Chem. Eng. J.* **2016**, *284*, 536–542. [CrossRef]
31. Zheng, G.; Xu, L.; Wu, P.; Xie, H.; Jiang, Y.; Chen, F.; Wei, X. Polyphenols from longan seeds and their radical-scavenging activity. *Food Chem.* **2009**, *116*, 433–436. [CrossRef]
32. Soong, Y.Y.; Barlow, P.J. Isolation and structure elucidation of phenolic compounds from longan (*Dimocarpus longan* Lour.) seed by high-performance liquid chromatography–electrospray ionization mass spectrometry. *J. Chromatogr. A* **2005**, *1085*, 270–277. [CrossRef] [PubMed]
33. Rangkadilok, N.; Sitthimonchai, S.; Worasuttayangkurn, L.; Mahidol, C.; Ruchirawat, M.; Satayavivad, J. Evaluation of free radical scavenging and antityrosinase activities of standardized longan fruit extract. *Food Chem. Toxicol.* **2007**, *45*, 328–336. [CrossRef] [PubMed]
34. Zaidi, S.; Jain, P.; Saleem, S.; Rao, K. Extraction and Purification of Total Phenolic Content from seeds of *Dimocarpus longan* fruit. *Res. Rev. J. Pharm.* **2018**, *4*, 19–21.
35. Katayon, S.; Noor, M.M.M.; Asma, M.; Ghani, L.A.; Thamer, A.; Azni, I.; Ahmad, J.; Khor, B.; Suleyman, A. Effects of storage conditions of *Moringa oleifera* seeds on its performance in coagulation. *Bioresour. Technol.* **2006**, *97*, 1455–1460. [CrossRef] [PubMed]
36. Zawawi, M.H.; Abustan, I. Detection of groundwater aquifer using resistivity imaging profiling at Beriah Landfill Site, Perak, Malaysia. *Adv. Mater. Res.* **2011**, *250–253*, 1852–1855. [CrossRef]
37. APAH. *Standard Methods for the Examination of Water and Waste Water*, 21st ed.; American Public Health Association: Washington, DC, USA, 2005.
38. Makhtar, S.M.Z.; Ab Wahab, M.; Selimin, M.T.; Mohamed, N.C. Landfill Leachate Treatment by a Coagulation–Photocatalytic Process. In Proceedings of the 2011 International Conference on Environment and Industrial Innovation (ICEII 2011), Kuala Lumpur, Malaysia, 4–5 June 2011.
39. Li, L.C.; Tiau, Y. *Zeta Potential*; Mercel Dekker: New York, NY, USA, 1997.
40. Salem, Z.; Hamouri, K.; Djemaa, R.; Allia, K. Evaluation of landfill leachate pollution and treatment. *Desalination* **2008**, *220*, 108–114. [CrossRef]

41. Bashir, M.J.K.; Aziz, H.A.; Yusoff, M.S.; Adlan, M.N. Application of response surface methodology (RSM) for optimization of ammoniacal nitrogen removal from semi-aerobic landfill leachate using ion exchange resin. *Desalination* **2010**, *254*, 154–161. [CrossRef]
42. Awang, N.A.; Aziz, H.A. Hibiscus rosa-sinensis leaf extract as coagulant aid in leachate treatment. *Appl. Water Sci.* **2012**, *2*, 293–298. [CrossRef]
43. Jamali, H.A.; Mahvi, A.H.; Nabizadeh, R.; Vaezi, F.; Omrani, G.A. Combination of coagulation–flocculation and ozonation processes for treatment of partially stabilized landfill leachate of Tehran. *World Appl. Sci. J.* **2009**, *5*, 9–15.
44. Baghvand, A.; Zand, A.D.; Mehrdadi, N.; Karbassi, A. Optimizing coagulation process for low to high turbidity waters using aluminum and iron salts. *Am. J. Environ. Sci.* **2010**, *6*, 442–448. [CrossRef]
45. Rui, L.M.; Daud, Z.; Latif, A.A.A. Treatment of Leachate by coagulation-flocculation using different coagulants and polymer: A review. *Int. J. Adv. Sci. Eng. Inf. Technol.* **2012**, *2*, 114–117. [CrossRef]
46. Sugimoto, T.; Cao, T.; Szilagyi, I.; Borkovec, M.; Trefalt, G. Aggregation and Charging of Sulfate and Amidine Latex Particles in the Presence of Oxyanions. *J. Colloid Interface Sci.* **2018**, *524*, 456–464. [CrossRef] [PubMed]
47. Abood, A.R.; Bao, J.; Abudi, Z.N.; Zheng, D.; Gao, C. Pretreatment of nonbiodegradable landfill leachate by air stripping coupled with agitation as ammonia stripping and coagulation–flocculation processes. *Clean Technol. Environ. Policy* **2013**, *15*, 1069–1076. [CrossRef]

Review

The Application of Modified Natural Polymers in Toxicant Dye Compounds Wastewater: A Review

Siti Aisyah Ishak [1], Mohamad Fared Murshed [1,*], Hazizan Md Akil [2], Norli Ismail [3], Siti Zalifah Md Rasib [2] and Adel Ali Saeed Al-Gheethi [4]

1 School of Civil Engineering, Engineering Campus, Universiti Sains Malaysia, Nibong Tebal 14300, Seberang Perai Selatan, Pulau Pinang, Malaysia; syahisha@gmail.com
2 School of Material Engineering, Engineering Campus, Universiti Sains Malaysia, Nibong Tebal 14300, Seberang Perai Selatan, Pulau Pinang, Malaysia; hazizan@usm.my (H.M.A.); zalifahmadrasib@gmail.com (S.Z.M.R.)
3 School of Industrial Technology, Universiti Sains Malaysia, Gelugor 11800, Pulau Pinang, Malaysia; norlii@usm.my
4 Micro-pollutant Research Centre (MPRC), Department of Water and Environmental Engineering, Faculty of Civil & Environmental Engineering, Universiti Tun Hussein Onn Malaysia, Parit Raja 86400, Batu Pahat, Johor, Malaysia; adel@uthm.edu.my
* Correspondence: cefaredmurshed@usm.my; Tel.: +60-45996285

Received: 12 June 2020; Accepted: 11 July 2020; Published: 17 July 2020

Abstract: The utilization of various types of natural and modified polymers for removing toxicant dyes in wastewater generated by the dye industry is reviewed in this article. Dye wastewater contains large amounts of metals, surfactants, and organic matter, which have adverse effects on human health, potentially causing skin diseases and respiratory problems. The removal of dyes from wastewaters through chemical and physical processes has been addressed by many researchers. Currently, the use of natural and modified polymers for the removal of dyes from wastewater is becoming more common. Although modified polymers are preferred for the removal of dyes, due to their biodegradability and non-toxic nature, large amounts of polymers are required, resulting in higher costs. Surface-modified polymers are more effective for the removal of dyes from the wastewater. A survey of 80 recently published papers demonstrates that modified polymers have outstanding dye removal capabilities, and thus have a high applicability in industrial wastewater treatment.

Keywords: natural and modified polymer; biodegradability; toxicant dyes; industrial wastewater treatment

---

1. Introduction

The wastewater generated from different manufacturing processes poses serious problems for organisms and aquacultures, due to the high toxicity of these wastes, which contain different types of pollutants, such as plastic, leather, ink, fabric, palm oil, soap, pulp, and paper. These wastes are disposed of directly (with partial treatment) into the environment and natural water systems. Short-term exposure to these pollutants causes tremors and nervous system disorders, while long-term exposure causes thyroid dysfunction, weight loss, and generalized hypoxia [1]. Therefore, the treatment of the dye-containing wastewater before the final disposal is an urgent matter, not only to meet international standards, but also to protect the biodiversity in nature and to ensure the availability of pure water for future generations. The dye-containing wastewaters have been of great interest to researchers during the past several years, primarily due to the high tectorial values of the dyes, where the discharge of less than 1 ppm of a dye into the water might cause significant changes in the water's physical and chemical characteristics. The traditional treatment methods used for the treatment of wastewaters depend mainly on chemical, physical, and biological processes (Table 1), which contribute effectively to improving the quality of the effluent parameters, such as the chemical oxygen demand (COD), biochemical oxygen demand (BOD), total suspension solids (TSS), and turbidity. Unfortunately, these methods are insufficient to remove the dyes from the wastewater. Coagulation/flocculation is a potential alternative, and a highly efficient method of removing dyes from dye-containing wastewaters.

Table 1. Advantages and disadvantages of the physical, chemical, and biological processes in dye wastewater treatment.

Process	Advantages	Disadvantages	References
Physical Process			
Ion exchange	Good surface area, effective sorbent	Not effective for all dyes, derived from petroleum-based materials, sensitive to particle charge	[2]
Adsorption	Effective for dye removal	Eco-friendly disposal of spent adsorbents, requires pretreatment, and costly maintenance	[3]
Membrane filtration	Able to remove all types of dyes	Requires proper pretreatment for SS removal, membrane fouling, expensive	[4]
Irradiation	Effective in lab settings	Requires large amounts of O_2	[5]
Electrokinetic coagulation	Reasonable cost	High sludge problem	[6]
Chemical process			
Cucurbituril	Good sorption capacity for different types of dyes	Expensive process	[7]
Oxidation	Simple application	H_2O_2 needs to be activated	[8]
Chemical coagulation-flocculation	Can remove dye molecules from dyebath effluent	High sludge production	[9]
Ozonation	Efficient for colour removal	Aldehyde by-product	[10]
Fenton's reagent	able to treat soluble and insoluble colour pigments	Results in large amounts of precipitate	[11]
Biological process			
Aerobic	Economically attractive and eco-friendly	Slow process	[12]
Anaerobic	Decolourized azo and water-soluble dyes	Produces methane and hydrogen sulphide	[13]
Decolourization by mixed cultures	Able to remove colour in 30 h	Lower efficacy, due to aerobic conditions	[14]
Decolourization by white-rot fungus	Degrades dyes using enzymes	Unpredictable enzyme production	[15]

The conventional coagulation/flocculation process, using inorganic polymers (synthetic or semi-synthetic), such as alum and ferrous sulphate ($FeSO_4$), could increase the environmental pollution levels by introducing non-biodegradable compounds [16]. Therefore, many researchers have shifted to using natural coagulants for wastewater treatment due to the advantages of these coagulants over chemical agents, particularly their low toxicity, low residual sludge production, and biodegradability [17]. Natural coagulants are of great interest to scientists, since they are natural, low-cost products, characterized by their environmentally friendly behaviour, and are presumed to be safe for human health [18]. However, in many of these studies, the utilization of natural coagulants was associated with the addition of natural polymers, in order to enhance the floc size by attracting smaller particles to generate much larger flocs, and, in some of the studies, natural polymers were used, without adding any coagulant, due to the high efficacies of natural polymers in the flocculation process (direct flocculation) [19,20].

The use of natural polymers (plant or animal sources) is a promising method for treating wastewater and removing dyes, due to the chemical structure and the composition of the polymers, such as the presence of many functional groups, which contribute effectively towards the removal of dyes from the wastewater. In addition, natural polymers are non-toxic, low-cost, renewable, biodegradable, and biocompatible [21]. Natural polymers are synthesized from plant products, such as starch, guar gum, gum acacia, locust bean gum, pectin, nirmali seeds (Strychnos potatorum), and drumstick trees (Moringa oleifera), as well as from non-plant sources, such as alginates, carrageenans, chitin, chitosan, bacteria, algae, and fungi [22–28]. Nonetheless, in many cases, natural polymers are not sufficient to remove the dyes from highly complex dye-containing wastewaters, containing different types of pollutants, such as heavy metals, which have a negative effect on the attraction of dyes to natural polymers. Therefore, these polymers should be subjected to a modification process, involving chemical or physical treatment, in order to increase their efficiency in removing dyes from complex wastewaters. It is vital to modify the polymers according to the target application with tailor-made specifications, designed using blending, grafting, curing or derivatization methods. The natural polymer can be chemically modified by mineral acids, bases, salts of weak acids, enzymes, acetylation, saponification, concentrated ammonium systems, and primary aliphatic amines [29]. The physical modification process includes the blending of two or more types of the polymer at an ambient temperature or elevated temperature. Polymer grafting involves the monomer being covalently bonded onto the polymeric chain, which requires a longer time compared to curing. Curing forms, a coat of oligomers mixture onto the substrate using physical forces. In derivatization, the substitution of a simple molecule with a reactive group on the polymeric chain occurs to provide additional functional groups.

The current review article discusses the application of modified natural polymers in the removal of dyes from dye-containing wastewaters. The characteristics of natural polymers from plant and non-plant sources are reviewed. The feasibility of using modified natural polymers as an alternative technology for the removal of dyes from wastewater is investigated. The main aim of this article is to summarize the characteristics of dye-containing wastewaters, as well as the recent research concerning the application of modified natural polymers for the removal of dyes from different wastewaters. A comparison of several publications on the application of natural polymers has been compiled for this purpose. The authors recommend that the reported removal capacities of natural polymers be taken as a response to specific conditions, instead of maximum removal capacities.

2. Characteristics of Dyes

Over 100,000 types of commercial dye, for a total of more than 7×10^5 tons of dyestuff, are produced by the textile industries around the world annually [30]. Dyes found in the wastewater are primarily used in industrial activities, such as the textile industry and food processing. These wastes contain different types of chemicals, such as dyestuff, bleaching agents, finishing chemicals, starch, thickening agents, surface active chemicals, wetting and dispensing agents, as well as metal salts, which are used during each stage of textile production [31]. Several types of heavy metals have negative impacts on human health, as listed in Table 2 [32].

Table 2. Classes of different dyes [32].

Class	Substrates	Method of Application	Chemical Types
Acid	Nylon, wool, silk, paper, inks, and leather	Usually from neutral to acidic dyebaths	Azo (including premetallized), anthraquinone, triphenylmethane, azine, xanthene, nitro, and nitroso
Azoic components and compositions	Cotton, rayon, cellulose acetate, and polyester	Fibre impregnated with coupling component and treated with a solution of stabilized diazonium salt	Azo
Base	Paper, polyacrylonitrile, modified nylon, polyester, and inks	Applied from acidic dyebaths	Cyanine, hemicyanine, diazahemicyanine, diphenylmethane, triarylmethane, azo, azine, xanthene, acridine, oxazine, and anthraquinone
Direct	Cotton, rayon, paper, leather, and nylon	Applied from neutral or slightly alkaline baths, containing additional electrolytes	Azo, phthalocyanine, stilbene, and oxazine
Disperse	Polyester, polyamide, acetate, acrylic, and plastics	Fine aqueous dispersions often applied by using high-temperature/pressure, or lower-temperature carrier methods; dye may be padded on cloth and baked on, or thermofixed	Azo, anthraquinone, styryl, nitro, and benzodifuranone
Mordant	Wool, leather, and anodized aluminium	Applied in conjunction with Cr salts	Azo and anthraquinone
Oxidation bases	Hair, fur, and cotton aromatic	Amines and phenols oxidized on the substrate	Aniline black and indeterminate structures
Reactive	Cotton, wool, silk, and nylon	Reactive site on dye reacts with functional group on fibre, to bind dye covalently under influence of heat and pH (alkaline)	Azo, anthraquinone, phthalocyanine, formazan, oxazine, and basic
Solvent	Plastics, gasoline, varnishes, lacquers, stains, inks, fats, oils, and waxes	Dissolution in the substrate	Azo, triphenylmethane, anthraquinone, and phthalocyanine
Sulphur	Cotton and rayon	Aromatic substrate vatted with sodium sulphide and reoxidized to insoluble sulphur-containing products on fibre	Indeterminate structures
Vat	Cotton, rayon, and wool	Water-insoluble dyes, solubilised by reducing them with sodium hydrogen sulphite, then exhausted on fibre and reoxidized	Anthraquinone (including polycyclic quinones) and indigoids

Dyes are classifieds into anionic dyes, cationic dyes, and non-ionic dyes. Cationic dyes are a category of basic dyes, while anionic dyes are known as disperse dyes, and are comprized of acid dyes, as well as direct and reactive dyes. Cationic dyes are water-soluble, with a positive charge and high color visibility. Anionic dyes carry a negative charge and differ from cationic dyes in terms of water-solubility, structure, and ionic substituents [30]. Several different types of dye have been universally utilized in textile industries, such as azo, triphenylmethane, perylene, anthraquinone, and indigoid dyes [33]. The different types of dyes are listed in Table 3 [34]. It can be noted that many of the dyes are used in different industrial applications.

Table 3. List of heavy metal toxicities [34].

Heavy Metals	Toxicities
Arsenic (As)	Skin manifestations, visceral cancers, vascular disease
Cadmium (Cd)	Kidney damage, renal disorder, human carcinogen
Chromium (Cr)	Headache, diarrhea, nausea, vomiting, carcinogenic
Copper (Cu)	Liver damage, Wilson's disease, insomnia
Nickel (Ni)	Dermatitis, nausea, chronic asthma, coughing, human carcinogen
Zinc (Zn)	Depression, lethargy, neurological signs, and increased thirst
Lead (Pb)	Damage to the fetal brain, kidney diseases, as well as circulatory system and nervous system diseases
Mercury (Hg)	Rheumatoid arthritis, kidney diseases, as well as circulatory system, and nervous system diseases

In addition, the dyes are subjected to chemical and physical processes during their application, which might result in the production of different, unknown secondary chemical substrates in the generated wastewater. This point is of critical concern among scientists, since new unknown chemicals are being released into the environment. Thousands of types of synthetic dyes are commercialized to obtain multicolor fabrics [32]. It is estimated that the concentration of dye effluent can be in the range of 10 to 250 mg/L [33]. However, it depends on the specific dye industry. The highest concentration of dye effluent recorded from the reactive dye industry reached 7000 mg/L [35]. Another concern associated with the traditional treatment of the dye-containing wastewater lies in the dye's unknown degradation pathway, which has resulted in the release of secondary toxic by-products into the environment and the natural wastewater, rather than being removed from the partially treated dye wastewater. The formation of chlorinated compounds and phthalic acid esters (PAEs) in the treated dye wastewater has been reported in the literature, for example, [36]. Dyes released into the environment can have acute effects on organisms, based on their level of toxicity [37]. The presence of dyed water can even be at concentrations as low as 1 mg/L [38]. This may affect the amount of light penetrating the water, reducing photosynthesis rates. Thus, it is compulsory for raw wastewater to be treated before being released into the environment. The standards and the guidelines for the disposal of textile wastewater are illustrated in Table 4. Most of the countries listed showed similar numbers for permissible pollutant concentrations to those listed by the United Sates Environmental Protection Agency (US EPA), with the exception of Nigeria. These pollutants were SS, Hg, As, Cn, Cu, Mn, Sn, Zn, B, Fe, Ag, Al, Se, Ba, F, formaldehyde, phenol, sulphide, oil and grease, ammoniacal nitrogen, and color. This shows that the environmental regulation in Nigeria is not strict, at least in terms of water and wastewater sustainability. Amendments to the regulations should be considered, in order to provide clean water for future generations. Jordan and Bangladesh did not list the permissible concentrations of certain pollutants, such as Cr^3, Sn, Ag, Al, Ba, formaldehyde, and color, according to their regulations. One possible reason is that these countries follow US EPA guidelines only for several ion's discharge limits. The list shows the different concentrations of ion discharge limits except for As, Pb, Ag, Al, Se, and Ba.

Table 4. Permissible standard limits for industrial wastewater in different countries.

Parameter	Unit	Malaysia [39]	Jordan [40]	Bangladesh [41]	Nigeria [42]	Singapore [43]	US EPA [44]
Temperature	°C	40	Summer: 40, Winter: 45	Summer: 40 Winter: 45	<40	45	40
pH Value	-	5.5–9.0	6.0–9.0	6.0–9.0	6.0–9.0	6.0–9.0	6.0–8.5
BOD	mg/L	50	50	50	50	50	40
SS	mg/L	100	150	150	NL	50	TSS: 50
Hg	mg/L	0.05	0.01	0.01	NL	0.05	0.005
Cd	mg/L	0.02	0.05	0.05	<1	0.1	0.01
Cr	mg/L	0.05	0.5	0.5	<1	1	0.1
Cr^3	mg/L	0.1	NL	NL	<1	1	0.5
As	mg/L	0.1	0.2	0.2	NL	0.1	0.2
Cyanide (Cn)	mg/L	0.1	0.1	0.1	NL	0.1	0.2
Lead (Pb)	mg/L	0.5	0.1	0.1	<1	0.1	0.1
Copper (Cu)	mg/L	1.0	0.5	0.5	NL	0.1	3.0
Manganese (Mn)	mg/L	1.0	5	5	NL	5	NL
Nickel (Ni)	mg/L	1.0	1	1	<1	1	3.0
Sn	mg/L	1.0	NL	NL	NL	10	NL
Zinc (Zn)	mg/L	2.0	5	5	NL	1	2.0
Boron (B)	mg/L	4.0	2	2	NL	5	NL
Iron (Fe)	mg/L	5.0	2	2	NL	10	3.0
(xix) Silver (Ag)	mg/L	1.0	NL	NL	NL	0.1	NL
(xx) Aluminium (Al)	mg/L	15	NL	NL	NL	NL	NL
(xxi) Selenium (Se)	mg/L	0.5	0.05	0.05	NL	0.5	0.05
(xxii) Barium (Ba)	mg/L	2.0	NL	NL	NL	2	NL
(xxiii) Fluoride (F)	mg/L	5.0	7	7	NL	NL	2.0
(xxiv) Formaldehyde	mg/L	2.0	NL	NL	NL	NL	NL
(xxv) Phenol	mg/L	1.0	1	1	NL	0.2	1.0
(xxvi) Free Chloride	mg/L	2.0	600	600	600	NL	1500
(xxvii) Sulphide	mg/L	0.50	1	1	NL	0.2	2.0
(xxviii) Oil and Grease	mg/L	10	10	10	NL	10	10
(xxix) Ammoniacal Nitrogen	mg/L	20	50	50	NL	NL	50
(xxx) Colour	ADMI *	200	NL	NL	NL	7 Lovibond Units	456 nm: 7 m^{-1},525 nm: 5 m^{-1}, 620 nm: 3 m^{-1}

Table 4. *Cont.*

Parameter	Unit	Malaysia [39]	Jordan [40]	Bangladesh [41]	Nigeria [42]	Singapore [43]	US EPA [44]
(xxxi) COD:							
(a) Pulp and paper industry							
(i) Pulp mill	mg/L	200	200	200	150	100	250
(ii) Paper mill (recycled)	mg/L	350	NL	NL	NL	NL	NL
(iii) Pulp and paper mill	mg/L	250	NL	NL	NL	NL	NL
		300	NL	NL	NL	NL	NL
(b) Textile industry	mg/L	250	NL	NL	NL	NL	NL
(c) Fermentation and distillery industry	mg/L	400	NL	NL	NL	NL	NL
(d) Other industries	mg/L	200	NL	NL	NL	NL	NL

* NL: not listed.

3. Application of Polymers in Wastewater Treatment

A polymer, whether grafted as a polysaccharide base, is highly efficient at binding and linking particles to itself and vice versa during collisions, resulting in the formation of larger, more settled flocs [45]. Polymers used in flocculation and coagulation might be inorganic or organic, and might be generated from natural resources, such as tannin, pectin, sodium alginate, chitosan, cellulose, gums and mucilages, which are derived from polysaccharides and proteins [46], or synthesized, such as acrylamide based poly-(2-methacryloyloxyethyl)-trimethylammonium chloride [47–49]. However, most of the previous studies focus extensively on the utilization of natural polymers, due to their high biodegradability. Many of the natural polymers discussed in the literature have been extracted from Moringa oleifera, Strychnos potatorum, Pseudomonas plecoglossicida, Spirogyra sp., and Aspergillus niger [50–54]. The extraction of polymers from agro-waste, such as guar gum, pectin, tannin, and locust bean gum, is explored due to these being environmentally safe, natural compounds from renewable resources, and not producing unintended hazardous wastes [55]. The high efficiency of the natural polymers in the removal of dyes from wastewater lies in the presence of different functional groups, such as the carboxyl, hydroxyl, phosphate, amine functional groups, which can bind to the cationic charges on the dye molecules by using electrostatic force [56]. Moreover, the renewable resources are abundant, and the aspect of biodegradability attracts many researchers [57].

The studies concerning the application of natural polymers in water and wastewater treatment are listed in Table 5. Most of them were applied as adsorbents or flocculants during the coagulation and flocculation process. Natural polymers usually work best in an acidic environment. However, some studies used a pH of 8 during the adsorption, to remove the methylene blue dye with the aid of acrylic acid [58]. About 20 mg of effective material was able to remove the dye color up to 45%. The highest removal rate, at 99.2%, was accomplished at pH 2 using pectin in 34.32 mg/L of Crystal Ponceau 6R dye [59]. A natural polymer extracted from animal waste, known as chitosan, was also able to remove 99% of the Duasyn Direct dye at pH 3.4, in combination with other materials, such as polyacrylamide and bentonite, as coagulants [60]. The cellulose/polyaniline (Ce/Pn) nanocomposite removed more than 90% of Remazol Brilliant Blue R (RBBR), Reactive Orange 16 (RO), Remazol Brilliant Violet 5R (RBVR), and Reactive Black 5 (RB) from the synthetic Remazol dye effluent. In contrast, 70.23% and 80.78% of the RBBR dyestuff was removed by using a chitosan-poly (acrylic acid) conjugate in an acidic environment, at pH 4 and pH 5, respectively [61]. RBBR dyes were successfully removed (100%) by using a combination of chitosan and cross-linked chitosan.

Natural adsorbents can also minimize the reaction time. Studies of alginates showed a 50% removal of methylene blue and methyl orange, which only required 10 and 17 min to achieve, respectively [62]. Previously, researchers used an alum and Acanthocereus tetragonus (a cactus species) to treat synthetic water, containing 100 to 500 ppm of Congo Red and Direct Blue dye [62]. The results indicated that using A. tetragonus as a coagulant resulted in 90% (up to 96%) color removal, while only 80% of color was removed by the alum. In addition, two types of plant organisms, namely Moringa oleifera seeds and Grewia venusta peel, were investigated for the treatment of synthetic dyes, namely, indigo carmine (reactive dye) and methyl orange dye [63]. The M. oleifera seeds were more effective than the G. venusta peel, with 99% and 85% dye removal, respectively. Both plants also showed optimum performances in acidic environments. The G. venusta peel, however, required harsh acidic conditions, at pH 2, to achieve the highest removal rate. In actual textile wastewater, plant extraction was not able to decolorize the pollution when using a single coagulant. Additional assistance was required to bind the dye particles together. Inorganic iron was added to the okra mucilage at the optimum pH of 6, in order to remove 93.57% of the colorant [64]. Textile wastewater is known to be a complicated waste to process, due to the presence of both cationic and anionic charges.

Table 5. Treatment of several types of wastewater using natural polymers.

Natural Polymer	Types Of Wastewater	Treatment Process	Condition	Type Of Dye	Colour Removal	References
Pectin	Textile wastewater	Coagulation and flocculation	pH 5, 427.4 mg/l $MgCl_2$ and 21.9 mg/l pectin	-NA-	54.20%	[57]
	Synthetic dye	Adsorption	pectin dose 20 mg; pH 8 20 mg/l of Methylene blue	Methylene blue dye	45.00%	[58]
	Synthetic dye	Adsorption	pH 2, 247.4 mg/l dose; 34.32 mg/l dye concentration; 540-min time	Crystal Ponceau 6 R dye	99.20%	[59]
	Synthetic dye	Adsorption	no pH adjustment; 20 mg of beads was added to 50 mL of the dye solution	Methylene blue dye	1550.3 mg/g for Pectin bead and 2307.9 mg/g for pectin/cellulose microfiber bead	[60]
Chitosan	Synthetic dye	Coagulation & flocculation	pH 4.0, coagulant dose of 25 mg/l, flocculation time of 60 min and temperature of 340 K	Congo Red (CR) dye removal-	94.50%	[61]
	Synthetic dye	Adsorption	chitosan beads, 1-butyl-3-methylimidazolium acetate and 1-butyl-3-methylimidazolium (pH 4.0, dose of 0.008 g, and agitation time of 20 min)	Malachite Green (MG) dye	8.07 mg g^{-1} and 0.24 mg g^{-1}	[65]
	Palm oil mill effluent	Coagulation and flocculation	3 g/L alum + 0.4 g/L chitosan pH 4.51, 250-rpm rapid mixing speed for 3 min, 30-rpm slow mixing speed for 30 min, and 60 min settling time.	-	95.24%	[60]
Cellulose/Polyaniline (Ce/Pn) Nanocomposite	Synthetic dye	Adsorption process		Remazol dye effluent	95.90%, 91.90%, 92.70%, and 95.70% of RBBR, RO, RV, and RBK, respectively.	[61]
Alginate	Synthetic dyes	Activated carbon		Methylene blue and Methyl orange dyes	50.00% Methylene blue in 10 min and Methyl orange in 17 min	[62]

Table 5. *Cont.*

Natural Polymer	Types Of Wastewater	Treatment Process	Condition	Type Of Dye	Colour Removal	References
Acanthocerous Tetragonus (Cactus)	Synthetic dyes	Coagulation	pH 6 at dose of 5 mg/L pH 4 at dose of 6 mg/L	Congo red dye Direct blue dye	96.00% 90.00%	[62]
Moringa Oleifera Seeds (Mos)	Synthetic dyes	Adsorption	pH 5 at MOS dose of 0.5 g and 150 mg/L concentration of dye	Indigo carmine (reactive dye)	31.25 mg g^{-1}	[63]
GrewiaVenusta Peel (Gvp)	Synthetic dye	Adsorption	pH 2 at 0.5 g of GVP and 150 mg/L dye concentration	Methyl orange dye	85.00% 188.68 mg g^{-1}	[63]
Okra Mucilage (Abelmoschus Esculentus)	Textile wastewater (during washing and finishing processes)	Coagulation and flocculation	pH 6 at 3.20-mg/L dose of Okra, and 88.0 mg/L of Fe3+	-	93.57% 5.78 mg g^{-1}	[64]
Tannin	Synthetic dye	Coagulation and flocculation	pH 2.9 (Bentonite+ anionic flocculant) pH 2.2 (Bentonite+ anionic PAM) pH 3.4 (Bentonite+ cationic PAM) pH 2.8 (Bentonite + cationic flocculant)	Methylene blue Crystal violet Duasyn direct dye Acid black 2	>90.00% 89.00% 99.00% 83.00%	[66]
	Manufacturing wastewater treatment	Coagulation and flocculation	-	-	99.00%	[66]
Gums C. Javahikai Seed Gum (Cj)	Synthetic dyes	Coagulation and flocculation		Direct dyes	>70.00%	[67]

Various monomers are tabulated and listed, along with some details on the dye removal rates from previous studies, in Table 6. Cationic monomers, such as poly-(2-methacryloyloxyethyl)-trimethylammonium chloride (PDMC), diallyldimethyl ammonium chloride, diethanolamine, and polyethylenimine, were very promising in thermal conditions and certain solvents. Grafting by using the opposite charges on the polymer's main backbone has a greater potential to increase the dye removal rate. A previous study showed an amphoteric grafting branch on chitosan using two monomers, known as carboxymethyl and poly-(2-methacryloyloxyethyl)-trimethylammonium chloride, for the removal of different charges of a dye molecule (cationic and anionic dyes) [68]. The existence of a quaternary ammonium group increased the cationic charges in the acidic environment, but a weak anionic character was displayed when the pH was above its isoelectric point. As the result, the bridging between these polymers and dyes became stronger. Some monomers made the polymer sensitive to pH in both acidic and basic environments. Diallyldimethyl ammonium chloride was grafted onto carboxymethyl cellulose (CMC), which was already cross-linked with mono-chloroacetic acid (MCA) and epichlorohydrin (ECH), and demonstrated a good performance in a methylene blue dye reduction, at over 98.54% removal in acidic conditions, and 83.07% removal in basic conditions, within a mere 20 min [69]. Using a PDMC monomer grafted onto carboxymethyl chitosan resulted in a 90% removal of Acid Green 25 at pH 4 and 98% of Basic Bright Yellow at pH 11. The quaternary ammonium salt was attracted to the anionic Acid Green 25 dye, producing a strong electrostatic attraction in an acidic environment, which is known as the neutralization effect. The presence of anionic charges on carboxymethyl chitosan assisted with the binding between the molecule of dye, known as Basic Bright Yellow dye, and anionic charges in the alkaline phase. A study on the alteration of bentonite with polyethylenimine to form an electrostatic charge and a hydrogen bond with the Amino Black dye, was successfully executed in an acidic environment (pH 3), at an adsorption rate of 264.5 mg/g [70]. The results showed that the mixture, containing 30% of polyethylenimine and 70% bentonite, was successfully grafted onto a molecule of (3-Glycidyloxypropyl) trimethoxysilane, with the aid of an epoxy bond, over a period of 24 h at 60 °C, in the presence of nitrogen gas.

Table 6. Examples of monomers introduced in graft copolymerization.

Monomers	Classes	Molecular Structure	Removal Percentage	References
poly(2-methacryloyloxyethyl) trimethylammonium chloride (PDMC)	Cationic		90% Acid Green 25 at pH 4 98% Basic Bright Yellow at pH 11	[69]
Diallyldimethyl ammonium Chloride	Cationic		>98.54% of methylene blue and 83.07% of anionic dye orange II pH = 4 and 11	[68]
Diethanolamine	Cationic		99.6% of Congo red, 99.8% of methyl blue, 97.5% of sunset yellow, and 81.2% of neutral red	[71]
Polyethylenimine	Cationic		Adsorption rate of 264.5 mg/g for amino black dye at pH 3	[70]
Acrylamide	Non-ionic		>99% of methylene blue at pH 10	[72]
Acrylic acid	Anionic		99.5% of Congo red and 98.7% of methylthionine chloride	[69]
Poly(glycidyl methacrylate)	Anionic		98.5% using 200 mg/L of methyl orange and methylene blue dyes	[73]
Triochloroaceticac	Anionic		Nearly 99% Rhodamine B, malachite green, and anionic dye orange were at concentration 222.6, 190.6, and 40 mg g^{-1}, respectively, at pH < 3	[74]

The next type of monomer, known as non-ionic monomers (such as acrylamide), have been widely grafted onto many types of polymers. One of the important steps for improving the efficiency during water treatment was to select the polymer based on the best grafting ratio. An ultrasound-assisted method was applied in order to increase the grafting efficiency since it could reduce the polymerization time. A previous study showed that sodium alginate (SAG) grafted with polyacrylamide (PAM) was able to achieve a high color removal with a grafting efficiency of 75% [72]. The highest efficiency was attained at pH 10, with a 99% adsorption of methylene blue. Recently, an in-situ ultrasonic wave-assisted polymerization was explored as a substitute for proper physical emulsion mixing. In one of the studies, the ultrasonic system was applied to anionic monomers, such as acrylic, triochloroacetic acid, and poly (glycidyl methacrylate), and showed an affinity with cationic dyes. In addition, a monomer can also help increase the pure water flux by using an acrylic monomer to reduce the graft density, which can enhance the hydrophilicity. The fabrication of polypropylene composite hollow fibre membranes with acrylic monomers demonstrated good dye retentions, with a 99.5% and 98.7% removal of Congo Red and methylthionine chloride, respectively [75]. A recent study explored the modification of a magnetic adsorbent, using poly (glycidyl methacrylate (PGMA)) microspheres, cross-linked with ethylene glycol dimethacrylate (EGDMA), in the decolorization of dyes [73]. The polymerization techniques were modified, to some extent. A modified multi-step swelling polymerization method was employed with iminodiacetic acid (IDA) used to produce carboxyl groups, and the magnetic traits were successfully embedded inside the microsphere's pore using in-situ chemical co-precipitation. The coating microspheres were then exposed to ultraviolet (UV) radiation. The results indicated a good adsorption rate, and the decolorization rate reached 98.5%. The decolorization efficiency was more than 80%, despite the adsorption–desorption cycle being run ten times. Other monomers, such as trichloroacetic acid, remove 99% of cationic dyes, such as malachite green and rhodamine B [74]. High adsorption capacities were recorded for rhodamine B, at 222.6 mg/g, and 190.6 mg/g for malachite green. A low adsorption rate was detected for anionic orange dye, at 40 mg/g, due to a smaller number of cationic charges on the adsorbent.

Current research focuses on the extraction of microbial polymers, since it is easy to carry out, and cost effective for industries. In order to obtain highly effective polymers for use in the removal of dyes, researchers have focused on the polymers generated from indigenous microbes, such as Pseudomonas pseudoalcaligenes, Pseudomonas plecoglossicida, and Staphylococcus aureus, in fawn dyes, mediblue, whale dyes, and mixed dyes [76], because these dyes have a high resistance to decolorization. This is possibly due to the acidic nature of these dyes, which makes it difficult for them to be absorbed by microbial polymers [77]. The indigenous microbes might have adopted and developed a resistance mechanism in order to survive in these dyes, therefore, these organisms exhibit a high dye removal efficiency. In a recent study, one bacteria species (Brevibacillus laterosporus) and one yeast species (Galactomyces geotrichum) were immobilized in a stainless-steel sponge and in polyurethane foam. The microbial consortia successfully decolorized 50 mg/L of Remazol Red dye in the stainless-steel sponge and in the polyurethane foam in 11 h and 15 h, respectively [78]. Immobilization by using calcium alginate and polyvinyl alcohol produced more consistent results but required more time to complete the decolorization process at 20 h in the stainless-steel sponge and 24 h in the polyurethane foam. In another study, nine different bacterial strains from textile wastewater and sludge were isolated, which resulted in one Planococcus sp. with decolorization abilities being found in textile wastewater [79]. The Planococcus sp. decolorization ability was increased to 78% by combining it with a 55% peptone and a 60% dextrose solution (in a nitrogen and a carbon source, respectively).

4. Graft Polymer (Coagulant/Flocculant)

New developments in the polymer research have drawn attention towards graft polymers, also known as grafted copolymers. The advantages of graft polymers include their non-toxicity, high biodegradability in nature, and low cost (Figure 1). Moreover, their high molecular weights, as well as the existence of new branching on the molecular chains, makes them more suitable for removing dyes from wastewater. Natural graft polymers are defined as additional polymers inserted into the backbone of a natural polymer, in order to alter the molecular chain. The alteration extends the natural polymer's length, thus improving the adsorption of molecules with opposite charges in the solution [80]. Moreover, the existing branches of natural polymers have been modified in many studies, by inserting an acetyl group into the chitosan, resulting in more functional groups being added to the polymeric chain, thus improving the absorbance capacity of the polymers for azo dyes from wastewater [81]. The formation of the carboxylic group on the chitosan's polymeric chain was able to remove cationic and anionic dyes, which in this case involved the synthetic methylene blue and methyl orange dyes. As reported by the previous study, the grafted surface of graphene oxide showed a good potential for dyes due to the presence of carboxylic and hydroxyl groups that produce a colloidal dispersion in the aqueous medium due to its hydrophilic nature [82]. To conclude, the existence of the carboxylic group was indeed helpful in removing the dye particles.

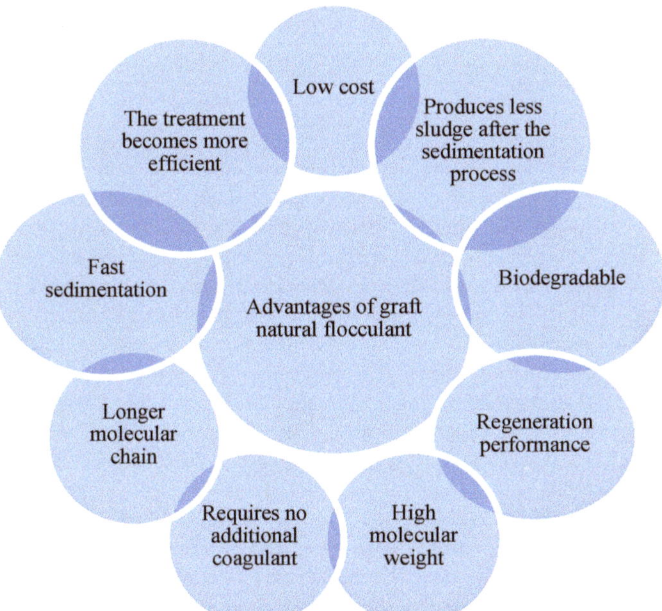

Figure 1. Benefits of grafting a natural polymer.

In order to better understand graft polymers, further studies on their mechanisms are required. The two mechanisms involved in graft polymers are charged neutralization and bridge aggregation. Several insoluble complexes form at higher speeds during rapid mixing, indicating that neutralization has occurred. Subsequently, bridging occurred, resulting in the aggregation of the insoluble complexes and contributing to the formation of larger flocs, due to the increase in the molecular weight [83]. The links in the large flocs settled down rapidly, followed by those in the smaller flocs. A study showed that the bridging effect was more beneficial to the flocculation of grafted natural polymers than for linear polymers [84]. Figure 2 demonstrates the differences in natural polymer branching before and after modification [83]. Different particles are adsorbed onto the grafted natural polymer's chain to form bridges, which link to the opposite charges and occupy freer binding sites. Natural graft polymers carry more binding sites as the length of the modified natural polymer is longer than the original length [85].

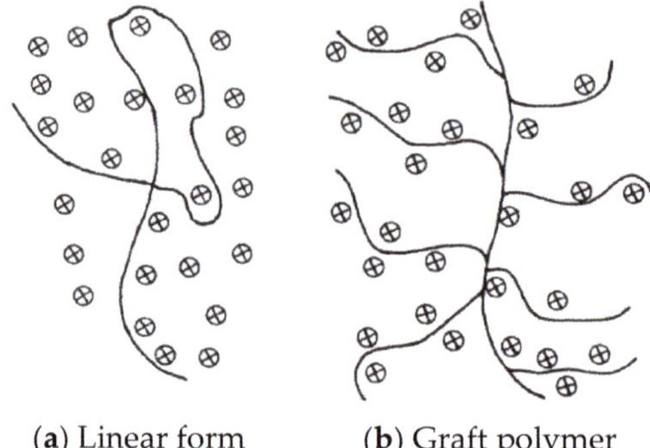

Figure 2. (**a**) Linear polymer represents the coil form bound to the oppositely charged particles, while (**b**) is the graft polymer presenting the comb form, able to bind to a large number of oppositely charged particles in the solution [83].

4.1. Factors Affecting the Efficiency of Polymer Coagulants

4.1.1. Type of Coagulant

The current trends in wastewater coagulation research focus mainly on the utilization of biodegradable polymers, such as chitosan, Moringa oleifera, nirmali seeds, and cactuses [86]. Natural polymers are used as coagulants because they are derived from renewable resources, biodegradable, cheap, non-toxic, and able to minimize the sludge production at the end of the treatment, as well as prevent the health risks associated with the utilization of alum, such as Alzheimer's disease [87]. Nonetheless, the study authors stated that synthetic organic polymers, such as diallyldimethyl ammonium chloride, as well as copolymers of quaternized dimethylaminoethyl acrylate or methacrylate, exhibited a better performance in the coagulation processes, due to charge densities and molecular masses [88]. The advantages of the synthetic organic polymers include the slow degradation activity, compared to that of the natural polymers with a longer life span, as well as the high charge density and molecular mass, which make them effective coagulants [89]. The disadvantages of synthetic polymers were not only their hydrophilic nature, but also their high cost [90]. This resulted in extra care being required during storage and transportation. Other drawbacks included limited molecular weight and dosage scale, which narrowed down the application range, and the presence of poisonous monomers, which are non-biodegradable and can be hazardous to the environment [91]. Although the wastewater treatment becomes more effective with an increase in dosage, problems arise when synthetic polymers are used due to unreacted chemicals making up the monomer unit (e.g., formaldehyde), as well as unreacted monomers (e.g., diallyldimethylammonium chloride and acrylamide) and reaction by-products.

4.1.2. Coagulant Concentrations and Mixing Conditions

A sufficient amount of coagulant should be dispersed completely in the wastewater, with the optimum mixing speed and time, in order to get the maximum contact between the coagulant and the suspended particles. One of the main factors affecting a coagulant's efficiency is its concentration. It has been demonstrated that coagulants with a cationic charge neutralize the suspensions and destabilize the colloids. Increasing a polymer's concentration improves its performance, however, a high dosage has a negative effect on the coagulation processes and restabilizes the colloids, reversing the charge and reducing the removal rate [92]. A previously published study proved that increasing the organic coagulant concentration increased the COD removal rate, which, in this case, involved a combination of Moringa oleifera and potassium chloride. Using aluminium sulphate had the opposite effect [93]. The removal of organic matter declined as the concentration of hydrolysed metal salt increased.

The hydrolysed metal salt can contribute to coagulation by adsorption [94]. However, the reversal of charge on the colloidal particle is the reason why the removal rate decreased. The aim of adding the salt to the natural polymer was to aggregate the particles, by means of a double layer compression [10]. The excessive addition of salt may also reduce the removal rate, due to the reduction in protein solubility (the salting-out effect) and the effect of hydration [95].

On the other hand, the mixing processes can also increase the removal rate efficiencies. The two phases that involve the mixing parameters in the coagulation are rapid mixing and slow mixing. The speed limit and the duration of time also play a role. The purpose of the rapid mix phase is to disperse the coagulant well, in order to stimulate particle collisions by using a power paddle. Rapid mixing was applied in a study, at speeds ranging from 80 rpm to 400 rpm and for time periods ranging from 0.1 min to 8 min. By varying rapid mixing, the formation, the breakage, and the regeneration of floc can be determined. A study showed that each coagulant has its own rapid mix condition [96]. An aluminium-based coagulant, for example, required the minimum time period to form a larger floc during the rapid mix phase, but the maximum time was needed when using a cationic polyelectrolyte [97]. Somehow, the regrowth of flocs after longer time periods was possible when using the cationic polyelectrolyte. The alum-based coagulant, on the other hand, had an irreversible effect on the floc recovery after the breakage. In agreement with other studies, mixing at 120 rpm for half a minute is required, in order to form larger flocs of highly turbid water [98]. The study also found that rapid mixing was connected to slow mixing, due to the flocs' resistance throughout the slow mixing phase. The time requirement during mixing did not appear to be the primary factor in removal effectiveness. The study found that the efficiency of coagulation, in terms of color and turbidity, had an indirect effect on the time [99]. However, the application of these factors during the process can provide some additional data. The following phase of slow mixing can take place at speeds ranging from 10 rpm to 60 rpm, for time periods ranging from approximately 5 min to 30 min. The investigation indicated that a slow mixing intensity had a beneficial effect on the charge neutralization coagulation when compared to sweep flocculation. Based on the optimum conditions, longer time periods are required during slow mixing, in order to produce larger flocs. During charge neutralization, an extended slow mixing phase can help to boost the coagulation performance, in case of inadequate rapid mixing [100]. This is in contrast with the sweep flocculation mechanism, where a shorter period of time is required for the slow mixing phase if the rapid mixing is excessively long. The mixing process should be followed by a sufficiently long settlement process [101]. Having larger flocs can shorten the time required for the settlement. Thus, the formation of larger flocs is important in the coagulation process.

4.1.3. Functional Groups

Polymeric coagulants contain several types of functional groups with negative charges, such as hydroxyl (OH^-), amine ($NH3^-$), phosphate (PO^{4-}), and carboxyl (COO^-). These groups have bridging effects on the particles with opposite charges in the water and the wastewater [102]. In nirmali seeds, the presence of OH^- groups along the galactomannan and the galactan molecular chains provide abundant attachment sites for interparticle bridging. The presence of the hydroxyl group, indicated by a broad wave between 3100 and 3500 cm^{-1}, can be observed for natural polymers. Previous studies on different polymer analyses of functional groups are summarized in Table 7.

Table 7. Summary of the functional groups on selected polymers.

Polymers	Functional Groups
Chitosan	Chitosan was identified by the presence of OH groups, with NH band overlapping around 3300 cm^{-1}; with additional NH stretching vibrations at 1649 cm^{-1} and 1578 cm^{-1}, and C-O bonds appearing at 1419 cm^{-1} and 1378 cm^{-1}, due to the presence of remaining acetylated moieties of chitin [103].
Tannin	Tannin can be recognized by the vibration of phenolic hydroxyl group, identified by peaks at 3200–3700 and 1325 cm^{-1}. The C-O aromatic ring and the C-C stretching vibration were observed at 1204 and 1536 cm^{-1}, respectively. The absorption peak between 1000 and 1150 cm^{-1} was associated with ethers (C–O–C) [103].
Moringa olefera	Moringa oleifera spectrum showed the absorption of OH stretching at 3400 cm^{-1}; symmetric stretching of C-H at 2925 cm^{-1}; carboxylic group (COO^{-}) stretching bands at 1610 cm^{-1} and 1451 cm^{-1}, and carbonyl group (C-O) stretching at d 1070 cm^{-1}) [104].
Zinc oxide nanoparticles	The absorption of a large number of hydroxyl groups is indicated by the broad peak between 3500 and3100 cm^{-1}; and Zn-O bond stretching is indicated by the lower absorption peak < 600 cm^{-1} [105].
Polyacrylamide	Polyacrylamide spectrum demonstrates broad absorption bands at 3421 and 3192 cm^{-1}, indicating N-H stretching vibrations. The absorbance peaks at 1655 cm^{-1} correspond to the asymmetric C = O stretching vibrations [106].
Graphene oxide	Common absorption bands, observed at 3387 cm^{-1}, are ascribed to –OH stretching vibrations. The C = O groups (–COOH), the C = C stretching, the C-O-C stretching, and the C-O (C-OH) stretching peak can be seen at 1730, 1621, 1224, and 1049 cm^{-1}, respectively [106].
Sodium alginate	Hydroxyl absorption peak was recorded at 3421 cm^{-1}; asymmetric and symmetric stretching vibrations of carboxylic groups were observed at the peaks near 1621 cm^{-1} and 1395 cm^{-1} [106].
Carboxymethyl cellulose	Carboxymethyl cellulose (CMC) can be identified by the wide spectra associated with ether bonds at absorption peaks between 1000 and 1100 cm^{-1}. The absorption band at 1588 cm^{-1} was attributed to the scissoring mode of the carboxylic groups (COO-) [73].
Pectin	Pectin absorption band were indicated by five bands at 1019, 1052, 1076, 1104, and 1149 cm^{-1}. Meanwhile, other bands at 1350–1750 were associated with carboxylic groups and C = O stretching in the protonated carboxyl group. In addition, two bands, representing symmetric and asymmetric stretching modes in carboxylic group at 1600–1650 cm^{-1} and 1400–1450 cm^{-1}, respectively. In addition, the ether and the C-C bond included in the molecule of pectin were indicated by the absorption bands between 1100 and 1200 cm^{-1} [107].
Locust bean gum	The previous spectrum showed a broad peak at 3311 cm^{-1}, which is assigned to the O-H group. The presence of sharp peaks at 1004 cm^{-1} was attributed to C-O-H vibrations [108].

Biopolymeric flocculants, such as pectin, have similarly negative charged particles (OH and COO) [109] with longer chains, resulting in electrostatic repulsion, due to the chains stretching out [110,111]. Studies on Polydiallyldimethylammonium chloride (PDADMAC) grafted onto a locust bean gum showed the adsorption of amino groups reacting with anionic dye particles, indicated at 1474 and 3022 cm^{-1} on the spectrum, increased with time, until the adsorption equilibrium was reached at approximately 600 min [112]. Other studies on the existence of phenolic groups in tannin structures showed these groups undergoing deprotonation to produce phenoxide effortlessly, thus expanding the oxygen atom's electron density [103]. The efficiency of the coagulation process is enhanced by the presence of additional phenolic compounds in the polymer structure [81]. The studies have shown that phenolic and amine groups are accessible in commercial tannin, which is a cationic polymer, with single tertiary amine groups present in each monomer [113]. Natural polymers can be grafted onto the surface of graphene oxide (GO) for the removal of dyes as GO has show a good potential for the removal of dye from aqueous solutions, as investigated by experimental and computational methods [82]. The computational evidence was illustrated using visual molecular dynamic programmes as can be seen in Figure 3. The surface area of graphene oxide posed a more negative charge in low pH with the absence of salt, which resulted in a protonated carboxylic group since no sodium ion (Na^+) can attach to the surface. Otherwise, an anionic group (CH_3COO^-, SCN^-, SO_4^{2-}, NO_3^-) was able to attach to the hydrophobic graphene oxide. A previous study claimed that the higher surface area assisted the graphene oxide's ability to interact with direct blue Indosol dyes (a subcategory of anionic dyes) at pH < 4, due to several factors, such as the presence of hydroxyl, carboxylic, and oxygen groups, as well as the active sites on the dye particles, which may have resulted in stronger chemical bonds between the graphene oxide and the dyes [114]. Instead, Direct Red 81 was added to solutions with higher pH values (>7.5), resulting in nearly 100% removal. The new hydroxyl group will interact with amine groups from dye particles in basic conditions, increasing the adsorption observed in studies. In addition, multiple layers of neutralized sulphonate groups in the dye are formed, due to the electrostatic interaction between the oxygen-containing functional groups on the graphene oxide. In addition, the pi interaction and the bonding with the hydrogen molecules also help to increase the adsorption rate.

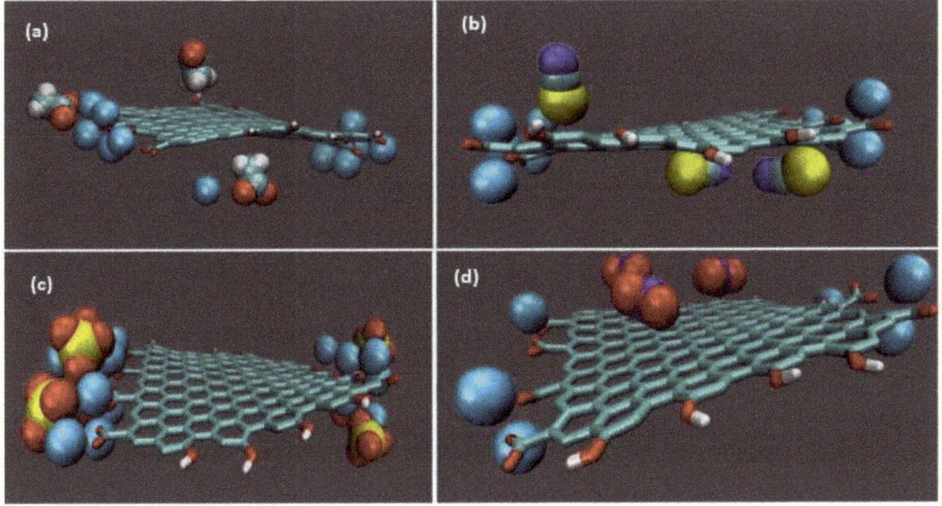

Figure 3. Visual molecular dynamics simulation of the attachment between graphene oxide (GO): (**a**) CH_3COO^-; (**b**) SCN^-; (**c**) SO_4^{2-} and (**d**) NO_3^-. Color coding: C-cyan, O-red, N-dark blue, H-white, S-yellow, Na-light blue [82].

4.1.4. Molecular Weight (MW) of the Flocculant/Coagulant Aid

Different polymers have different molecular weights, based on their molecular structures. The molecular weight of the flocculant added after the coagulant has a significant impact on the process. Polymers with higher molecular weights contribute to efficient toxin removal, due to the mechanisms involved, such as charge neutralization, bridging, and electrostatic patch. Negatively charged particles are destabilized by the Van der Waals forces, due to the presence of elements with high molecular weights and numerous positive charges [110]. As soon as the Van der Waals forces are balanced out by repulsive electrostatic forces, flocs begin to develop. In addition, larger loops and ends form as the molecular weight increases; consequently, more sites are available to attract suspended particles [115]. Polymers with a minimum molecular weight of 800,000 daltons are more suitable for bridging [116]. A study of anionic polyacrylamides with different molecular weights indicates that a larger equivalent size (resulting from a higher molecular weight being added at an extremely rapid mixing rate), results in a faster settling time [117]. However, a higher dosage is required to increase the density and to shorten the settling time, by using a lower molecular weight. Polymers with higher molecular weights and more branched chains demonstrate a better color removal performance and faster settling rates than linear chains with small numbers of active sites for the pollutant to bind [118]. A high resistance and large flocs simulated the separation of the pollutant particles out of the solution and their subsequent sinking to the bottom. The adsorption of the target pollutant requires a strong adhesion between solid–liquid interfaces, which is influenced by the polymer's molecular weight [119]. In another study, lignin- [2-(methacryloyloxy) ethyl] trimethyl ammonium chloride (METAC), with a high molecular weight, was more successful at removing Reactive Orange 16 than Reactive Black 5, though both dyes are anionic azo dyes [120]. The polymerization of polyacrylamides, with gelatine used as a tested stabilizer, shows a lower dye removal potential when using a low molecular weight grafted polymer [121]. The Congo Red dye removal was compared when using a copolymer (PAB) before and after grafting with dextran (DAB), and the experiment showed a 68.1% removal rate when using DAB, compared to 40.9% without the DAB, due to the higher molecular weight of the DAB grafted polymer [122]. Using a commercial polymer, such as polyaluminium chloride (PAC), achieves a 48% removal rate. This indicates that the molecular weight does influence the effectiveness of the coagulation, as well as the flocculation performance.

4.1.5. Type of Charge Density

A polymer's charge density is categorized into low, medium, and high, based on the percentage mole of the ionic group (10%, 25%, and 50–100%, respectively) [123]. A low charge density enhances the polymer's bridging effect. The effectiveness of the charge density ranges from 5% to 15% and can improve shear resistance with higher molecular weights. Introducing graft copolymers results in an improved stability and extends the biodegradability, to some extent [124]. In addition, re-flocculation is incomplete when the charge density is below 12. The optimum flocculant concentration has been found to be dependent on the ionic strength [125]. Adding cationic charge polymers significantly enhances the coagulating capabilities. This effect increases in the order of monovalent < bivalent < trivalent. For example, adding the trivalent ion results in a stronger floc structure, and increases the floc size, the density and the shear resistance to a greater extent compared to monovalent ions [126]. One study, comparing four different lignin-based polymers extracted from pulping sludges, showed an excellent removal rate of disperse dye wastewater when higher-charge-density polymers were used. Particles form larger agglomerates when high-charge-density polymers are utilized, while looser molecules, despite the abundance of active sites for the dye molecule to attach, undergo slower reactions, as a consequence of the steric bulk [118].

5. Removal of Dyes by Using Modified Natural Polymers

Modified natural polymers have gained more attention in the recent years. Different types of compounds have been used to modify natural polymers for the purpose of removing dyes from wastewaters—for example, ammonia, formaldehyde, zinc oxide nanoparticles, lignosulfonate, carboxymethylstarch, polyacrylamide and various monomers [127]. The studies concerning modified natural polymers are presented in Table 8.

Table 8. Compilation of toxicant removal from several types of wastewater using modified polymers.

Graft Polymer (Modified)	Additional Compound	Types of Wastewater	Treatment Process	Response	Percentage Removal (%)	References
Amine Modified Tannin Gel Mw: Na	Aqueous ammonia	Synthetic wastewater, using Brilliant green dye	Adsorption	Colour removal	94.05	[47]
Tannin Extracts (*Acacia Mearnsii* De Wild And *Schinopsis Balansae*) Mw: Na	Clarotan with diethanolamine and formaldehyde	Guadiana River, in Badajoz (south-western Spain)	Coagulation and flocculation	Turbidity removal	>90.00	[128]
Chitosan—TiO_2 Nanoparticles	Titanium dioxide nano particles	Simulated textile wastewater	Coagulation and flocculation	Removal of Cd, Cu, and Pb	<97.00	[129]
Chitosan Modified With CHPATC	Cationic moiety N-3-chloro-2-hydroxypropyl trimethyl ammonium chloride (CHPTAC) in presence of sodium hydroxide	Simulated laundry wastewater	Coagulation and flocculation	Colour Turbidity	76.20 90.94	[130]
Cationic Amylopectin Mw: 6.89×10^4 g/Mol	-	Synthetic reactive black dye	Adsorption	Colour removal	78.00	[67]
Cationic Amylose Mw: 2.64×10^6 g/Mol					50.00	
Cationic Chitosan Mw: 1.2×10^5 g/Mol					>80.00	
Cationic Glycogen 6.81×10^6 g/Mol					96.20	
Cationic Guar Gum Mw: 6.6×10^5 g/Mol					70.00	
Cationic Starch Mw: 4.32×10^5 g/Mol					65.00	
Cationic Tamarind Kernel Polysaccharide Mw: 6.12×10^6 g/Mol					82.00	

Table 8. *Cont.*

Graft Polymer (Modified)	Additional Compound	Types of Wastewater	Treatment Process	Response	Percentage Removal (%)	References
Cassia Javahikai Seed Gum-Grafted Polyacrylamide (Cjg), Mw: Na	Polyacrylamide PAM	Textile wastewater effluent	Coagulation and flocculation	Colour removal Total suspended solid	35.00 80.00	[67]
Tannin-Based Hydrogel Mw: Na	Grafted Copolymer of Allyl Glycidyl Ether with Acrylamide	Synthetic heavy metal	Adsorption	Lead (Pb) removal	99.00 at (0.5 mmol/L)	[131]
Carboxymethyl Chitosan Grafted Polyacrylamide (Cmc-G-Pam) Mw: Na	Grafted polyacrylamide	Synthetic dye Methyl orange (anionic dye) Basic right yellow (cationic dye)	Coagulation and flocculation	Colour removal	93.00	[132]
Lignosulfonate–Acrylamide–Chitosan	Grafted with lignosulfonate and acrylamide	Acid blue 115 Reactive black 5 Methyl orange	Flocculation	Colour removal	>95.0 >95.0 >50.0	[133]
Hydroxypropyl Methyl Cellulose, Grafted With Polyacrylamide (Hpmc-G-Pam) Mw: Na	Grafted polyacrylamide	Raw mine wastewater	Coagulation and flocculation	Turbidity	95.00	[83]
Hydrolysed Polyacrylamide-Grafted Carboxymethyl starch (Hyd. Cms-G-Pam) Mw: Na	Grafted Carboxymethyl starch	Textile wastewater	Coagulation and flocculation	Colour removal	88.18	[28]
Polyacrylamide-Grafted Sodium Alginate Mw:Na	Grafted sodium alginate	Synthetic dye	Adsorption	Colour removal	99.00	[134]
Chitosan-Acrylamide-Fulvic Acid (Camfa)	Grafted with acrylamide and fulvic acid	Methylene Blue Acid blue 113, reactive black 5 and methyl orange	Adsorption	Colour removal	99.00 97.00 91.60	[135]

Table 8. *Cont.*

Graft Polymer (Modified)	Additional Compound	Types of Wastewater	Treatment Process	Response	Percentage Removal (%)	References
Polyethylenimine-Grafted Cellulose Mw: Na	Grafted with Cellulose	Silk printing and dyeing wastewater	Flocculation	Total suspended solid (TSS) COD reductionTurbidity Total suspended solid (TSS) COD reduction Turbidity	38.20 73.40 95.70 87.40 96.20 79.90	[136]
Bamboo Pulp Cellulose Grafting Polyacrylamide (Bpc-G-Pam)	Grafted with polyacrylamide	Cationic and disperse dye	Flocculation	Colour removal	87.20 97.00	[137]
Carboxymethylcellulose-G-Poly[(2-Methacryloyloxyethyl) Trimethyl Ammonium Chloride] (Cmc-G-Pdmc)	graft poly[(2-methacryloyloxyethyl) trimethyl ammonium chloride]	Machining wastewater Acid green 25	Coagulation and flocculation	Colour removal	93.50	[69]
Pafc-Starch-G-P(Am-Dmdaac)	graft copolymer with acrylamide and dimethyl diallyl ammonium chloride	Brilliant blue KN-R Yellow M-3RE Dark blue M-2GE Green KE-4B	Coagulation and flocculation	Colour removal	97.30 89.70 83.00 85.00	[138]

Previously, researchers would come up with effective grafting procedures involving natural polymers. For example, an amine-modified tannin gel effectively removed brilliant green (at 94.05%) in neutral pH conditions, using the external surface adsorption mechanism [47]. In another study, a tannin-based polymer was grafted onto tannin extracted from Schinopsis balansae and Acacia mearnsii de Wild, by running a Mannich base reaction [139]. A tannin extract, modified with Clarotan and diethanolamine, was first acknowledged for its potential use as a dye or a surfactant in wastewater and river water [71]. Based on the wastewater simulation test, about 92% of the dye could be removed from a 100 mg/L stock solution, and the surfactant concentration could be reduced from 50 to 7.5 mg/L using a 150 mg/L stock solution. Modified tannin can perform well in water and wastewater treatment processes, as shown by previous studies [140,141]. Another study, focusing on lead adsorption, showed the ability of tannin-based hydrogels to absorb metal elements on their surfaces well [131]. When grafted onto a well-known chitosan, (3-chloro 2-hydroxypropyl) trimethylammonium chloride was used to reduce 1000 mg/L of the melanoidin dye at pH 3 using a 3 g/L dose chitosan-g-CHPTAC, which was able to remove up to 76.2% of the color and 90.14% of the turbidity [130]. A study by Sanghi et al. [67] focuses on grafting polyacrylamides onto different types of polysaccharides, such as amylose, amylopectin, starch, tamarind kernel, guar gum, glycogen, and chitosan. Grafted glycogen, having the highest molecular weight (6.81 × 106 g/mol) and radius of gyration, was the most effective, adsorbing 96.2% of the dyes, due to more branching taking place on the glycogen's backbone. This showed that the molecular weight is related to the effectiveness of the treatment. In a previous study, carboxyl methyl chitosan-graft-polyacrylamide (CMC-g-PAM) exhibited a high efficiency (above 90%) in removing anionic and cationic dyes. An investigation of ternary graft polymers revealed a high decolorization performance when using grafted chitosan for removing anionic and neutral dyes but demonstrated a low efficiency for the removal of cationic dyes [132].

Recent studies have demonstrated that acrylamide grafted onto sodium alginate successfully removes the methylene blue dye, with a removal rate of 99%. The sodium alginate alone did not remove any of the dye at all. However, there are some limitations of using modified natural polymers to treat real textile wastewater, due to the presence of different organic and inorganic complex chemicals. For example, grafted carboxymethyl starch was only able to remove 88.18% of the detected color at 520 nm [28]. The percentage removal of a synthetic dye can reach nearly 100%, but not in real textile wastewater. Grafted cellulose was used for the treatment of silk printing and dyeing wastewater, achieving a 95.7% removal of the COD [136]. The attachment of cellulose onto hyperbranched polyethylenimine resulted in a highly effective removal of ammonia nitrogen, total iron, and total phosphorus, at the original pH. The effluent pH did not require any adjustment, being approximately neutral. This characteristic is an additional benefit of polymer grafting, as the pH values during coagulation and flocculation treatments are independent of one another. Another study reported a wide pH range (from pH 5 to pH 9) available for the treatment of Acid Blue 113 and Reactive Black 5, which happened to remove more than 90% of the dye colors [135]. In addition, increasing the grafting ratio also increases the floc size and compactness, resulting in a lower dosage being required, in order to achieve a higher color removal efficiency. A cellulose-based flocculant, combined with poly-(2-methacryloyloxyethyl)-trimethylammonium chloride, showed the best removal rate for an anionic dye (97.3%). Increasing the grafting ratio also improved the color removal rate. Moreover, other advantages of grafted polymers depend on the dosage applied during the process, which can be reduced up to 50% in synthetic wastewater using PAFC-Starch-g-p (AM-DMDAAC) [138]. With increasing environmental awareness, a recent study demonstrated the outstanding biodegradability of cellulose, extracted from bamboo pulp, grafted with polyacrylamide, at 66.5% and 67.6% after 45 d and 90 d, respectively, in a soil-extracting solution [137]. It was also successful at removing organic dyes, such as cationic and disperse dye solutions, with an average removal rate of 97%. Overall, grafted natural polymers showed a superior dye removal performance, compared to natural polymers on their own.

6. Conclusions

Over the past several decades, high levels of toxicants have been produced as the result of dye wastewater treatment involving harmful chemicals. Even though factors like turbidity, color, COD, BOD, and the levels of heavy metals have been reduced to meet the permissible standards, the sludge produced as the result of the treatment still comes into contact with toxic materials. Thus, an effective solution is required in order to improve water quality. Grafted natural polymers could replace commercial polymers, with an additional incentive of reduced costs. The characteristics of effective coagulants can be enhanced by using specific types of polymers, the concentration and mixing conditions, functional groups, higher molecular weight, and charge density according to the target dyes pollutant. The chemical modification of polymers provided the opportunity to explore beyond conventional applications. A thorough understanding of the polymer and its chemical modification has a vast potential to be the future trend for the use of cosmetics, pharmaceutical, food, leather, paper, and textile industries.

Author Contributions: Conceptualization, S.A.I., H.M.A. and A.A.S.A.-G.; methodology, S.A.I. and S.Z.M.R.; resources, S.A.I.; writing—original draft preparation, S.A.I.; writing—review and editing, S.A.I.; supervision, M.F.M., A.A.S.A.-G., and N.I.; funding acquisition, M.F.M. All authors have read and agreed to the published version of the manuscript.

Funding: This research was funded by RUI grant no. 1001/PAWAM/814259.

Acknowledgments: The authors would like to thank the Malaysian government for providing Mybrain15 scholarship and thankful for the kind support of staff of the Universiti Sains Malaysia.

Conflicts of Interest: The authors declare no conflict of interest.

References

1. Uppal, H.; Tripathy, S.S.; Chawla, S.; Sharma, B.; Dalai, M.K.; Singh, S.P.; Singh, S.; Singh, N. Study of cyanide removal from contaminated water using zinc peroxide nanomaterial. *J. Environ. Sci.* **2017**, *55*, 76–85. [CrossRef] [PubMed]
2. Crini, G.; Lichtfouse, E.; Wilson, L.D.; Morin-Crini, N. Adsorption-Oriented Processes Using Conventional and Non-Conventional Adsorbents for Wastewater Treatment. In *Green Adsorbents for Pollutant Removal*; Springer: Cham, Switzerland, 2018; pp. 23–71.
3. Othmani, A.; Kesraoui, A.; Boada, R.; Seffen, M.; Valiente, M. Textile Wastewater Purification Using an Elaborated Biosorbent Hybrid Material (Luffa–Cylindrica–Zinc Oxide) Assisted by Alternating Current. *Water* **2019**, *11*, 1326. [CrossRef]
4. Zhan, Y.; Wan, X.; He, S.; Yang, Q.; He, Y. Design of durable and efficient poly (arylene ether nitrile)/bioinspired polydopamine coated graphene oxide nanofibrous composite membrane for anionic dyes separation. *Chem. Eng. J.* **2018**, *333*, 132–145. [CrossRef]
5. Arcanjo, G.S.; Mounteer, A.H.; Bellato, C.R.; da Silva, L.M.N.; Dias, S.H.B.; da Silva, P.R. Heterogeneous photocatalysis using TiO_2 modified with hydrotalcite and iron oxide under UV–visible irradiation for color and toxicity reduction in secondary textile mill effluent. *J. Environ. Manag.* **2018**, *211*, 154–163. [CrossRef]
6. Xu, Y.; Li, Z.; Su, K.; Fan, T.; Cao, L. Mussel-inspired modification of PPS membrane to separate and remove the dyes from the wastewater. *Chem. Eng. J.* **2018**, *341*, 371–382. [CrossRef]
7. Rai, A.; Chauhan, P.S.; Bhattacharya, S. Remediation of Industrial Effluents. In *Water Remediation*; Springer: Singapore, 2018; pp. 171–187.
8. Maria, G.; Yolanda, M.D.; Roberto, T.P.; Manuel, L.J.; Gumersindo, F.; Teresa, M.M. Textile Wastewater Treatment by Advanced Oxidation Processes: A Comparative Study. In *Life Cycle Assessment of Wastewater Treatment*; CRC Press: Boca Raton, FL, USA, 2018.
9. Saritha, V.; Srinivas, N.; Vuppala, N.S. Analysis and optimization of coagulation and flocculation process. *Appl. Water Sci.* **2018**, *7*, 451–460. [CrossRef]
10. Ghuge, S.P.; Saroha, A.K. Catalytic ozonation of dye industry effluent using mesoporous bimetallic Ru-Cu/SBA-15 catalyst. *Process Saf. Environ.* **2018**, *118*, 125–132. [CrossRef]

11. Nasuha, N.; Ismail, S.; Hameed, B.H. Activated electric arc furnace slag as an effective and reusable Fenton-like catalyst for the photodegradation of methylene blue and acid blue 29. *J. Environ. Manag.* **2017**, *196*, 323–329. [CrossRef]
12. Manavi, N.; Kazemi, A.S.; Bonakdarpour, B. The development of aerobic granules from conventional activated sludge under anaerobic-aerobic cycles and their adaptation for treatment of dyeing wastewater. *Chem. Eng. J.* **2017**, *312*, 375–384. [CrossRef]
13. Rodríguez, R.; Espada, J.J.; Molina, R.; Puyol, D. Life Cycle Analysis of Anaerobic Digestion of Wastewater Treatment Plants. In *Life Cycle Assessment of Wastewater Treatment*; Taylor & Francis: Oxfordshire, UK, 2018; p. 13.
14. Hameed, B.B.; Ismail, Z.Z. Decolorization, biodegradation and detoxification of reactive red azo dye using non-adapted immobilized mixed cells. *Biochem. Eng. J.* **2018**, *137*, 71–77. [CrossRef]
15. Yesilada, O.; Birhanli, E.; Geckil, H. Bioremediation and Decolorization of Textile Dyes by White Rot Fungi and Laccase Enzymes. In *Mycoremediation and Environmental Sustainability*; Springer: Cham, Switzerland, 2018; pp. 121–153.
16. Huang, X.; Sun, S.; Gao, B.; Yue, Q.; Wang, Y.; Li, Q. Coagulation behavior and floc properties of compound bioflocculant–polyaluminum chloride dual-coagulants and polymeric aluminum in low temperature surface water treatment. *J. Environ. Sci.* **2015**, *30*, 215–222. [CrossRef] [PubMed]
17. Thakur, S.S.; Choubey, S. Use of Tannin based natural coagulants for water treatment: An alternative to inorganic chemicals. *Int. J. ChemTech Res.* **2014**, *6*, 3628–3634.
18. Mishra, A.; Bajpai, M. Flocculation behaviour of model textile wastewater treated with a food grade polysaccharide. *J. Hazard. Mater.* **2015**, *118*, 213–217. [CrossRef]
19. Lee, C.S.; Robinson, J.; Chong, M.F. A review on application of flocculants in wastewater treatment. *Process Saf. Environ. Prot.* **2014**, *92*, 489–508. [CrossRef]
20. Mishra, A.; Srinivasan, R.; Bajpai, M.; Dubey, R. Use of polyacrylamide-grafted Plantago psyllium mucilage as a flocculant for treatment of textile wastewater. *Colloid Polym. Sci.* **2004**, *282*, 722–727. [CrossRef]
21. Anasatakis, K.; Kalderis, D.; Diamadopoulos, E. Flocculation behavior of mallow and okra mucilage in treating wastewater. *Desalination* **2009**, *249*, 786–791. [CrossRef]
22. Rakhshaee, R.; Panahandeh, M. Stabilization of a magnetic nano-adsorbent by extracted pectin to remove methylene blue from aqueous solution: A comparative studying between two kinds of cross-likened pectin. *J. Hazard. Mater.* **2011**, *189*, 158–166. [CrossRef] [PubMed]
23. Junior, O.M.C.; Barros, M.A.S.D.; Pereira, N.C. Study on coagulation and flocculation for treating effluents of textile industry. *Acta Sci. Technol.* **2012**, *35*, 83–88. [CrossRef]
24. Elieh-Ali-Komi, D.; Hamblin, M.R. Chitin and chitosan: Production and application of versatile biomedical nanomaterials. *Int. J. Adv. Res.* **2016**, *4*, 411.
25. Das, M.; Adholeya, A. Potential Uses of immobilized bacteria, fungi, algae, and their aggregates for treatment of organic and inorganic pollutants in wastewater. *Water Chall. Solut. Glob. Scale* **2015**, *1206*, 319–337. [CrossRef]
26. Jia, H.; Yuan, Q. Removal of nitrogen from wastewater using microalgae and microalgae–bacteria consortia. *Cogent Environ. Sci.* **2016**, *2*, 1275089. [CrossRef]
27. Robinson, T.; McMullan, G.; Marchant, R.; Nigam, P. Remediation of dyes in textile effluent: A critical review on current treatment technologies with a proposed alternative. *Bioresour. Technol.* **2001**, *77*, 247–255. [CrossRef]
28. Sen, G.; Ghosh, S.; Jha, U.; Pal, S. Hydrolyzed polyacrylamide grafted carboxymethylstarch (Hyd. CMS-g-PAM): An efficient flocculant for the treatment of textile industry wastewater. *Chem. Eng. J.* **2011**, *171*, 495–501. [CrossRef]
29. Chong, M.F. Direct flocculation process for wastewater treatment. In *Advance in Water Treatment and Pollution Prevention*; Springer: Dordrecht, The Netherlands, 2012.
30. Kudaibergenov, S.; Koetz, J.; Nuraje, N. Nanostructured hydrophobic polyampholytes: Self-assembly, stimuli-sensitivity, and application. *Adv. Compos. Hybrid Mater.* **2018**, *1*, 649–684. [CrossRef]
31. Markies, P.R.; Moonen, J.P.; Colin, P.O.; Evers, R.W.; Van Den Beucken, F.J.; Peng, K.U.S. Aqueous Inkjet Primering Composition Providing Both Pinning and Ink Spreading Functionality. Patent Application No. 15/938,893, 28 March 2018.
32. Hunger, K. *Industrial Dyes: Chemistry, Properties, Applications*; John Wiley & Sons: Hoboken, NJ, USA, 2003.

33. Ghaly, A.E.; Ananthashankar, R.; Alhattab, M.V.V.R.; Ramakrishnan, V.V. Production, characterization and treatment of textile effluents: A critical review. *J. Chem. Eng. Process Technol.* **2014**, *5*, 1–19.
34. Yaseen, D.A.; Scholz, M. Textile dye wastewater characteristics and constituents of synthetic effluents: A critical review. *Int. J. Environ. Sci. Technol.* **2019**, *16*, 1193–1226. [CrossRef]
35. Mishra, P.; Soni, R. Analysis of Dyeing and Printing Wastewater of Balotara Textile Industries. *Int. J. Chem. Sci.* **2016**, *14*, 1929–1938.
36. Liang, J.; Ning, X.A.; Kong, M.; Liu, D.; Wang, G.; Cai, H.; Sun, J.; Zhang, Y.; Lu, Z.; Yuan, Y. Elimination and ecotoxicity evaluation of phthalic acid esters from textile-dyeing wastewater. *Environ. Pollut.* **2017**, *231 Pt 1*, 115–122. [CrossRef]
37. Rawat, D.; Mishra, V.; Sharma, R.S. Detoxification of azo dyes in the context of environmental processes. *Chemosphere* **2016**, *155*, 591–605. [CrossRef]
38. Amini, M.; Arami, M.; Mahmoodi, N.M.; Akbari, A. Dye removal from colored textile wastewater using acrylic grafted nanomembrane. *Desalination* **2011**, *267*, 107–113. [CrossRef]
39. *Environmental Quality Act 1974 (Act 127), Regulations, Rules & Orders*; MDS Publishers Sdn. Bhd.: Kuala Lumpur, Malaysia, 2012.
40. Al-Daghistani, H. Bio-Remediation of Cu, Ni and Cr from rotogravure wastewater using immobilized, dead, and live biomass of indigenous thermophilic Bacillus species. *Internet J. Microbiol.* **2012**, *10*, 1–10.
41. Dey, S.; Islam, A. A review on textile wastewater characterization in Bangladesh. *Resour. Environ.* **2015**, *5*, 15–44.
42. FEPA. *Guidelines and Standards of Environmental Pollution Control in Nigeria*; Federal Environmental Protection Agency: Lagos, Nigeria, 1991.
43. Tortajada, C.; Joshi, Y.K. Water quality management in Singapore: The role of institutions, laws and regulations. *Hydrol. Sci. J.* **2014**, *59*, 1763–1774. [CrossRef]
44. United State Environmental Protection Agency. *Guidelines for Water Reuse*; US EPA: Washington, DC, USA, 2004.
45. Babel, S.; Kurniawan, T.A. Low-cost adsorbents for heavy metals uptake from contaminated water: A review. *J. Hazard. Mater.* **2003**, *97*, 219–243. [CrossRef]
46. Shen, Y.; Yang, R.; Liao, Y.; Ma, J.; Mao, H.; Zhao, S. Tannin modified aminated silica as effective absorbents for removal of light rare earth ions in aqueous solution. *Desalin. Water Treat.* **2016**, *57*, 18529–18536. [CrossRef]
47. Akter, N.; Hossain, M.A.; Hassan, M.J.; Amin, M.K.; Elias, M.; Rahman, M.M.; Asiri, A.M.; Siddiquey, I.A.; Hasnat, M.A. Amine modified tannin gel for adsorptive removal of Brilliant Green dye. *J. Environ. Chem. Eng.* **2016**, *4*, 1231–1241. [CrossRef]
48. Ho, Y.C.; Norli, I.; Abbas, F.M.A.; Morad, N. Enhanced turbidity removal in water treatment by using emerging vegetal biopolymer composite: A characterization and optimization study. *Desalin. Water Treat.* **2016**, *57*, 1779–1789. [CrossRef]
49. Bongiovani, M.C.; Camacho, F.P.; Coldebella, P.F.; Valverde, K.C.; Nishi, L.; Bergamasco, R. Removal of natural organic matter and trihalomethane minimization by coagulation/flocculation/filtration using a natural tannin. *Desalin. Water Treat.* **2016**, *57*, 5406–5415. [CrossRef]
50. Jung, Y.; Kwon, M.; Kye, H.; Abrha, Y.W.; Kang, J.W. Evaluation of Moringa oleifera seed extract by extraction time: Effect on coagulation efficiency and extract characteristic. *J. Water Health* **2018**, *16*, 904–913. [CrossRef]
51. Salehizadeh, H.; Yan, N.; Farnood, R. Recent advances in polysaccharide bio-based flocculants. *Biotechnol. Adv.* **2018**, *36*, 92–119. [CrossRef]
52. Arunkumar, P.; Sadish Kumar, V.; Saran, S.; Bindun, H.; Devipriya, S.P. Isolation of active coagulant protein from the seeds of Strychnos potatorum–a potential water treatment agent. *Environ. Technol.* **2018**, *40*, 1624–1632. [CrossRef]
53. Roy, U.; Sengupta, S.; Banerjee, P.; Das, P.; Bhowal, A.; Datta, S. Assessment on the decolourization of textile dye (Reactive Yellow) using Pseudomonas sp. immobilized on fly ash: Response surface methodology optimization and toxicity evaluation. *J. Environ. Manag.* **2018**, *223*, 185–195. [CrossRef]
54. Mathur, M.; Gola, D.; Panja, R.; Malik, A.; Ahammad, S.Z. Performance evaluation of two Aspergillus spp. for the decolourization of reactive dyes by bioaccumulation and biosorption. *Environ. Sci. Pollut. Res.* **2018**, *25*, 345–352. [CrossRef] [PubMed]
55. Jeyaseelan, K.; Murugesan, A.; Palanisamy, S.B. Biopolymer Technologies for Environmental Applications. In *Nanoscience and Biotechnology for Environmental Applications*; Springer: Cham, Switzerland, 2019; pp. 55–83.
56. Gunatilake, S.K. Methods of removing heavy metals from industrial wastewater. *Methods* **2015**, *1*, 14.

57. Du, Q.; Wei, H.; Li, A.M. Multi-functional natural polymer based water treatment agents. *Environ. Chem.* **2018**, *37*, 1293–1310.
58. Aisyah, S.I.; Norfariha, M.S.; Azlan, M.A.; Norli, I. Comparison of Synthetic and Natural Organic Polymers as Flocculant for Textile Wastewater Treatment. *Iran. J. Energy Environ.* **2014**, *5*, 436–445.
59. Onukwuli, O.D.; Obiora-Okafo, I.A.; Omotioma, M. Characterization and Colour Removal from An Aqueous Solution Using Bio-Coagulants: Response Surface Methodological Approach. *J. Chem. Technol. Metall.* **2019**, *54*, 77–89.
60. Jagaba, A.H.; Abubakar, S.; Lawal, I.M.; Latiff, A.A.A.; Umaru, I. Wastewater Treatment Using Alum, the Combinations of Alum-Ferric Chloride, Alum-Chitosan, Alum-Zeolite and Alum-Moringa Oleifera as Adsorbent and Coagulant. *Int. J. Eng. Manag.* **2018**, *2*, 67–75. [CrossRef]
61. Janaki, V.; Vijayaraghavan, K.; Oh, B.T.; Ramasamy, A.K.; Kamala-Kannan, S. Synthesis, characterization and application of cellulose/polyaniline nanocomposite for the treatment of simulated textile effluent. *Cellulose* **2013**, *20*, 1153–1166. [CrossRef]
62. Vijayaraghavan, G.; Kumar, P.V.; Chandrakanthan, K.; Selvakumar, S. Acanthocereus tetragonus an effective natural coagulant for decolorization of synthetic dye wastewater. *J. Mater. Environ. Sci.* **2017**, *8*, 3028–3033.
63. Agbahoungbata, M.Y.; Fatombi, J.K.; Ayedoun, M.A.; Idohou, E.; Sagbo, E.V.; Osseni, S.A.; Aminou, T. Removal of reactive dyes from their aqueous solutions using Moringa oleifera seeds and Grewia venusta peel. *Desalin. Water Treat.* **2016**, *57*, 22609–22617. [CrossRef]
64. Freitas, T.K.F.S.; Oliveira, V.M.; De Souza, M.T.F.; Geraldino, H.C.L.; Almeida, V.C.; Fávaro, S.L.; Garcia, J.C. Optimization of coagulation-flocculation process for treatment of industrial textile wastewater using okra (A. esculentus) mucilage as natural coagulant. *Ind. Crops Prod.* **2015**, *76*, 538–544. [CrossRef]
65. Naseeruteen, F.; Hamid, N.S.A.; Suah, F.B.M.; Ngah, W.S.W.; Mehamod, F.S. Adsorption of malachite green from aqueous solution by using novel chitosan ionic liquid beads. *Int. J. Biol. Macromol.* **2018**, *107*, 1270–1277. [CrossRef] [PubMed]
66. Aboulhassan, M.A.; Souabi, S.; Yaacoubi, A.; Baudu, M. Coagulation efficacy of a tannin coagulant agent compared to metal salts for paint manufacturing wastewater treatment. *Desalin. Water Treat.* **2016**, *57*, 19199–19205. [CrossRef]
67. Sanghi, R.; Bhattacharya, B.; Singh, V. Use of Cassia javahikai seed gum and gum-g-polyacrylamide as coagulant aid for the decolorization of textile dye solutions. *Bioresour. Technol.* **2006**, *97*, 1259–1264. [CrossRef]
68. Wu, H.; Yang, R.; Li, R.; Long, C.; Yang, H.; Li, A. Modeling and optimization of the flocculation processes for removal of cationic and anionic dyes from water by an amphoteric grafting chitosan-based flocculant using response surface methodology. *Environ. Sci. Pollut. Res.* **2015**, *22*, 13038–13048. [CrossRef]
69. Lin, Q.; Gao, M.; Chang, J.; Ma, H. Adsorption properties of crosslinking carboxymethyl cellulose grafting dimethyldiallylammonium chloride for cationic and anionic dyes. *Carbohydr. Polym.* **2016**, *151*, 283–294. [CrossRef]
70. Du, S.; Wang, L.; Xue, N.; Pei, M.; Sui, W.; Guo, W. Polyethyleneimine modified bentonite for the adsorption of amino black 10B. *J. Solid State Chem.* **2017**, *252*, 152–157. [CrossRef]
71. Liu, M.; Chen, Q.; Lu, K.; Huang, W.; Lü, Z.; Zhou, C.; Gao, C. High efficient removal of dyes from aqueous solution through nanofiltration using diethanolamine-modified polyamide thin-film composite membrane. *Sep. Purif. Technol.* **2017**, *173*, 135–143. [CrossRef]
72. Rocher, V.; Siaugue, J.M.; Cabuil, V.; Bee, A. Removal of organic dyes by magnetic alginate beads. *Water Res.* **2008**, *42*, 1290–1298. [CrossRef]
73. Yu, B.; Yang, B.; Li, G.; Cong, H. Preparation of monodisperse cross-linked poly (glycidyl methacrylate)-Fe_3O_4-diazoresin magnetic microspheres with dye removal property. *J. Mater. Sci.* **2017**, *53*, 6471–6481. [CrossRef]
74. Zeng, L.; Xiao, L.; Long, Y.; Shi, X. Trichloroacetic acid-modulated synthesis of polyoxometalate@ UiO-66 for selective adsorption of cationic dyes. *J. Colloid Interface Sci.* **2018**, *516*, 274–283. [CrossRef]
75. Shao, H.; Qi, Y.; Liang, S.; Qin, S.; Yu, J. Polypropylene composite hollow fiber ultrafiltration membranes with an acrylic hydrogel surface by in situ ultrasonic wave-assisted polymerization for dye removal. *J. Appl. Polym. Sci.* **2019**, *136*, 47099. [CrossRef]
76. Buthelezi, S.P.; Olaniran, A.O.; Pillay, B. Textile dye removal from wastewater effluents using bioflocculants produced by indigenous bacterial isolates. *Molecules* **2012**, *17*, 14260–14274. [CrossRef]
77. Pearce, C.I.; Lloyd, J.R.; Guthrie, J.T. The removal of colour from textile wastewater using whole bacterial cells: A review. *Dye. Pigment.* **2003**, *58*, 179–196. [CrossRef]

78. Kurade, M.B.; Waghmode, T.R.; Xiong, J.Q.; Govindwar, S.P.; Jeon, B.H. Decolorization of textile industry effluent using immobilized consortium cells in upflow fixed bed reactor. *J. Clean. Prod.* **2019**, *213*, 884–891. [CrossRef]
79. Chanwala, J.; Kaushik, G.; Dar, M.A.; Upadhyay, S.; Agrawal, A. Process optimization and enhanced decolorization of textile effluent by Planococcus sp. isolated from textile sludge. *Environ. Technol. Innov.* **2019**, *13*, 122–129. [CrossRef]
80. Aziz, H.A.; Rahim, N.A.; Ramli, S.F.; Alazaiza, M.Y.; Omar, F.M.; Hung, Y.T. Potential use of Dimocarpus longan seeds as a flocculant in landfill leachate treatment. *Water* **2018**, *10*, 1672. [CrossRef]
81. León, O.; Muñoz-Bonilla, A.; Soto, D.; Pérez, D.; Rangel, M.; Colina, M.; Fernández-García, M. Removal of anionic and cationic dyes with bioadsorbent oxidized chitosans. *Carbohydr. Polym.* **2018**, *194*, 375–383. [CrossRef]
82. Borthakur, P.; Boruah, P.K.; Hussain, N.; Sharma, B.; Das, M.R.; Matić, S.; Minofar, B. Experimental and molecular dynamics simulation study of specific ion effect on the graphene oxide surface and investigation of the influence on reactive extraction of model dye molecule at water–organic interface. *J. Phys. Chem. C* **2016**, *120*, 14088–14100. [CrossRef]
83. Das, R.; Ghorai, S.; Pal, S. Flocculation characteristics of polyacrylamide grafted hydroxypropyl methyl cellulose: An efficient biodegradable flocculant. *Chem. Eng. J.* **2013**, *229*, 144–152. [CrossRef]
84. Rahangdale, D.; Kumar, A. Derivatized Chitosan: Fundamentals to Applications. In *Biopolymer Grafting*; Elsevier: Amsterdam, The Netherlands, 2018; pp. 251–284.
85. Abiola, O.N. Polymers for Coagulation and Flocculation in Water Treatment. In *Polymeric Materials for Clean Water*; Springer: Cham, Switzerland, 2019; pp. 77–92.
86. Patel, H.; Vashi, R.T. Removal of Congo Red dye from its aqueous solution using natural coagulants. *J. Saudi Chem. Soc.* **2012**, *16*, 131–136. [CrossRef]
87. Liaquat, L.; Sadir, S.; Batool, Z.; Tabassum, S.; Shahzad, S.; Afzal, A.; Haider, S. Acute aluminum chloride toxicity revisited: Study on DNA damage and histopathological, biochemical and neurochemical alterations in rat brain. *Life Sci.* **2019**, *217*, 202–211. [CrossRef] [PubMed]
88. Lourenço, A.D.S. Development of Novel Flocculants to Treat Oily Waters Using Health-Friendly Processes. Ph.D. Thesis, Universidade de Coimbra, Coimbra, Portugal, 2018.
89. Li, H.; Hu, C.; Yu, H.; Chen, C. Chitosan composite scaffolds for articular cartilage defect repair: A review. *RSC Adv.* **2018**, *8*, 3736–3749. [CrossRef]
90. Crini, G.; Lichtfouse, E.; Wilson, L.D.; Morin-Crini, N. Conventional and non-conventional adsorbents for wastewater treatment. *Environ. Chem. Lett.* **2018**, *17*, 195–213. [CrossRef]
91. Wilts, E.M.; Herzberger, J.; Long, T.E. Addressing water scarcity: Cationic polyelectrolytes in water treatment and purification. *Polym. Int.* **2018**, *67*, 799–814. [CrossRef]
92. Grenda, K.; Arnold, J.; Hunkeler, D.; Gamelas, J.A.F.; Rasteiro, M.G. Tannin-based coagulants from laboratory to pilot plant scales for coloured wastewater treatment. *BioResources* **2018**, *13*, 2727–2747. [CrossRef]
93. Dotto, J.; Fagundes-Klen, M.R.; Veit, M.T.; Palácio, S.M.; Bergamasco, R. Performance of different coagulants in the coagulation/flocculation process of textile wastewater. *J. Clean. Prod.* **2019**, *208*, 656–665. [CrossRef]
94. Freitas, T.F.K.S.; Almeida, C.A.; Manholer, D.D.; Geraldino, H.C.L.; de Souza, M.T.F.; Garcia, J.C. Review of utilization plant-based coagulants as alternatives to textile wastewater treatment. In *Detox Fashion*; Springer: Singapore, 2018; pp. 27–79.
95. Benalia, A.; Derbal, K.; Panico, A.; Pirozzi, F. Use of Acorn Leaves as a Natural Coagulant in a Drinking Water Treatment Plant. *Water* **2019**, *11*, 57. [CrossRef]
96. Mhaisalkar, V.A.; Paramasivam, R.; Bhole, A.G. Optimizing physical parameters of rapid mix design for coagulation-flocculation of turbid waters. *Water Res.* **1991**, *25*, 43–52. [CrossRef]
97. Yukselen, M.A.; Gregory, J. The effect of rapid mixing on the break-up and re-formation of flocs. *Environ. Clean Technol.* **2004**, *79*, 782–788. [CrossRef]
98. Ahmed, S.; Ayoub, G.; Al-Hindi, M.; Azizi, F. The effect of fast mixing conditions on the coagulation–flocculation process of highly turbid suspensions using liquid bittern coagulant. *Desalin. Water Treat.* **2015**, *53*, 3388–3396.
99. Ramphal, S.; Sibiya, S.M. Optimization of time requirement for rapid mixing during coagulation using a photometric dispersion analyzer. *Procedia Eng.* **2014**, *70*, 1401–1410. [CrossRef]

100. Zhang, Z.; Liu, D.; Hu, D.; Li, D.; Ren, X.; Cheng, Y.; Luan, Z. Effects of Slow-Mixing on the Coagulation Performance of Polyaluminum Chloride (PACI). *Chin. J. Chem. Eng.* **2013**, *21*, 318–323. [CrossRef]
101. Bernard, K.N.M.; Sylvere, N.K.; Patrice, K.G.; Joseph, K.G. Coagulation and Sedimentation of Concentrated Laterite Suspensions: Comparison of Hydrolyzing Salts in Presence of *Grewia* spp. Biopolymer. *J. Chem.* **2019**, *2019*, 1–9.
102. Hamza, M.F.; Wei, Y.; Mira, H.I.; Adel, A.H.; Guibal, E. Synthesis and adsorption characteristics of grafted hydrazinyl amine magnetite-chitosan for Ni (II) and Pb (II) recovery. *Chem. Eng. J.* **2019**, *362*, 310–324. [CrossRef]
103. Yang, J.; Li, M.; Wang, Y.; Wu, H.; Zhen, T.; Xiong, L.; Sun, Q. Double cross-linked chitosan composite films developed with oxidized tannic acid and ferric ions exhibit high strength and excellent water resistance. *Biomacromolecules* **2019**, *20*, 801–812. [CrossRef]
104. Ranote, S.; Kumar, D.; Kumari, S.; Kumar, R.; Chauhan, G.S.; Joshi, V. Green synthesis of Moringa oleifera gum-based bifunctional polyurethane foam braced with ash for rapid and efficient dye removal. *Chem. Eng. J.* **2019**, *361*, 1586–1596. [CrossRef]
105. Tian, H.; Fan, H.; Ma, J.; Liu, Z.; Ma, L.; Lei, S.; Long, C. Pt-decorated zinc oxide nanorod arrays with graphitic carbon nitride nanosheets for highly efficient dual-functional gas sensing. *J. Hazard. Mater.* **2018**, *341*, 102–111. [CrossRef] [PubMed]
106. Shan, C.; Wang, L.; Li, Z.; Zhong, X.; Hou, Y.; Zhang, L.; Shi, F. Graphene oxide enhanced polyacrylamide-alginate aerogels catalysts. *Carbohydr. Polym.* **2019**, *203*, 19–25. [CrossRef]
107. Borazan, A.A.; Acikgoz, C. Effect of Quince Variety on The Quality Of Pectin: Chemical Composition And Characterization. *Int. J. Pharm. Chem. Biol. Sci.* **2017**, *7*, 393–400.
108. Sirajo, A.Z.; Vishalakshi, B. Microwave assisted synthesis of poly (diallyldimethylammonium chloride) grafted locust bean gum: Swelling and dye adsorption studies. *Indian J. Adv. Chem. Sci. S* **2016**, *1*, 88–91.
109. Zahrim, A.Y.; Tizaoui, C.; Hilal, N. Evaluation of several commercial synthetic polymers as flocculant aids for removal of highly concentrated CI Acid Black 210 dye. *J. Hazard. Mater.* **2010**, *182*, 624–630. [CrossRef] [PubMed]
110. Wolf, G.; Schneider, R.M.; Bongiovani, M.C.; Morgan Uliana, E.; Garcia Do Amaral, A. Application of coagulation/flocculation process of dairy wastewater from conventional treatment using natural coagulant for reuse. *Chem. Eng. Trans.* **2015**, *43*, 2041–2046. [CrossRef]
111. Ho, Y.C.; Norli, I.; Alkarkhi, A.F.; Morad, N. New vegetal biopolymeric flocculant: A degradation and flocculation study. *Iran. J. Energy Environ.* **2014**, *5*, 2–3. [CrossRef]
112. Razali, M.A.A.; Ahmad, Z.; Ariffin, A. Treatment of pulp and paper mill wastewater with various molecular weight of PolyDADMAC induced flocculation with polyacrylamide in the hybrid system. *Adv. Chem. Eng. Sci.* **2012**, *2*, 490. [CrossRef]
113. Graham, N.; Gang, F.; Fowler, G.; Watts, M. Characterisation and coagulation performance of a tannin-based cationic polymer: A preliminary assessment. *Colloids Surf. A Physicochem. Eng. Asp.* **2008**, *327*, 9–16. [CrossRef]
114. De Assis, L.K.; Damasceno, B.S.; Carvalho, M.N.; Oliveira, E.H.; Ghislandi, M.G. Adsorption capacity comparison between graphene oxide and graphene nanoplatelets for the removal of colored textile dyes from wastewater. *Environ. Technol.* **2019**, *41*, 2360–2371. [CrossRef]
115. Zhao, H.Z.; Sun, Y.; Xu, L.N.; Ni, J.R. Removal of Acid Orange 7 in simulated wastewater using a three-dimensional electrode reactor: Removal mechanisms and dye degradation pathway. *Chemosphere* **2010**, *78*, 46–51. [CrossRef]
116. Torres, M.A.; Vieira, R.S.; Beppu, M.M.; Arruda, E.J.; Santana, C.C. Production of chemically modified chitosan microspheres by a spraying and coagulation method. *Mater. Res.* **2007**, *10*, 347–352. [CrossRef]
117. Costine, A.; Cox, J.; Travaglini, S.; Lubansky, A.; Fawell, P.; Misslitz, H. Variations in the molecular weight response of anionic polyacrylamides under different flocculation conditions. *Chem. Eng. Sci.* **2018**, *176*, 127–138. [CrossRef]
118. Guo, K.; Gao, B.; Wang, W.; Yue, Q.; Xu, X. Evaluation of molecular weight, chain architectures and charge densities of various lignin-based flocculants for dye wastewater treatment. *Chemosphere* **2019**, *215*, 214–226. [CrossRef]

119. Liang, G.; Nguyen, A.V.; Chen, W.; Nguyen, T.A.; Biggs, S. Interaction forces between goethite and polymeric flocculants and their effect on the flocculation of fine goethite particles. *Chem. Eng. J.* **2018**, *334*, 1034–1045. [CrossRef]
120. Wang, S.; Kong, F.; Fatehi, P.; Hou, Q. Cationic High Molecular Weight Lignin Polymer: A Flocculant for the Removal of Anionic Azo-Dyes from Simulated Wastewater. *Molecules* **2018**, *23*, 2005. [CrossRef] [PubMed]
121. Qiu, J.; Wang, H.; Du, Z.; Cheng, X.; Liu, Y.; Wang, H. Preparation of polyacrylamide via dispersion polymerization with gelatin as a stabilizer and its synergistic effect on organic dye flocculation. *J. Appl. Polym. Sci.* **2018**, *135*, 46298. [CrossRef]
122. Zhao, C.; Zheng, H.; Sun, Y.; Zhang, S.; Liang, J.; Liu, Y.; An, Y. Evaluation of a novel dextran-based flocculant on treatment of dye wastewater: Effect of kaolin particles. *Sci. Total Environ.* **2018**, *640*, 243–254. [CrossRef] [PubMed]
123. Bolto, B.; Gregory, J. Organic polyelectrolytes in water treatment. *Water Res.* **2007**, *41*, 2301–2324. [CrossRef]
124. Ghosh, S.; Sen, G.; Jha, U.; Pal, S. Novel Biodegradable Polymeric Flocculant Based on Polyacrylamide-grafted Tamarind Kernel Polysaccharide. *Bioresour. Technol.* **2011**, *101*, 9638–9644. [CrossRef]
125. Wang, W.; Chen, Z.; Wu, K.; Liu, Z.; Yang, S.; Yang, Q.; Dzakpasu, M. Coagulation performance of cucurbit [8] uril for the removal of azo dyes: Effect of solution chemistry and coagulant dose. *Water Sci. Technol.* **2018**, *78*, 415–423. [CrossRef] [PubMed]
126. Wen, Y.; Zheng, W.; Yang, Y.; Cao, A.; Zhou, Q. Influence of Al3+ addition on the flocculation and sedimentation of activated sludge: Comparison of single and multiple dosing patterns. *Water Res.* **2015**, *75*, 201–209. [CrossRef] [PubMed]
127. Wu, H.; Liu, Z.; Yang, H.; Li, A. Evaluation of chain architectures and charge properties of various starch-based flocculants for flocculation of humic acid from water. *Water Res.* **2016**, *96*, 126–135. [CrossRef]
128. Lugo, L.; Martín, A.; Diaz, J.; Pérez-Flórez, A.; Celis, C. Implementation of Modified *Acacia* Tannin by Mannich Reaction for Removal of Heavy Metals (Cu, Cr and Hg). *Water* **2020**, *12*, 352. [CrossRef]
129. Razzaz, A.; Ghorban, S.; Hosayni, L.; Irani, M.; Aliabadi, M. Chitosan nanofibers functionalized by TiO_2 nanoparticles for the removal of heavy metal ions. *J. Taiwan Inst. Chem. Eng.* **2016**, *58*, 333–343. [CrossRef]
130. Momeni, M.M.; Kahforoushan, D.; Abbasi, F.; Ghanbarian, S. Using Chitosan/CHPATC as coagulant to remove color and turbidity of industrial wastewater: Optimization through RSM design. *J. Environ. Manag.* **2018**, *211*, 347–355. [CrossRef]
131. Sán Beltrán-Heredia, J.; Sánchez-Martín, J.; Martín-García, L. Multiparameter quantitative optimization in the synthesis of a novel coagulant derived from tannin extracts for water treatment. *Water Air Soil Pollut.* **2012**, *223*, 2277–2286. [CrossRef]
132. Yang, Z.; Yang, H.; Jiang, Z.; Cai, T.; Li, H.; Cheng, R. Flocculation of both anionic and cationic dyes in aqueous solutions by the amphoteric grafting flocculant carboxymethyl chitosan-graft-polyacrylamide. *J. Hazard. Mater.* **2013**, *254*, 36–45. [CrossRef]
133. He, K.; Lou, T.; Wang, X.; Zhao, W. Preparation of lignosulfonate–acrylamide–chitosan ternary graft copolymer and its flocculation performance. *Int. J. Biol. Macromol.* **2015**, *81*, 1053–1058. [CrossRef]
134. Feira, J.M.C.D.; Klein, J.M.; Forte, M.M.D.C. Ultrasound-assisted synthesis of polyacrylamide-grafted sodium alginate and its application in dye removal. *Polímeros* **2018**, *28*, 139–146. [CrossRef]
135. Lou, T.; Wang, X.; Song, G.; Cui, G. Synthesis and flocculation performance of a chitosan-acrylamide-fulvic acid ternary copolymer. *Carbohydr.Polym.* **2017**, *170*, 182–189. [CrossRef]
136. Zhang, L.W.; Hua, J.R.; Zhu, W.J.; Liu, L.; Du, X.L.; Meng, R.J.; Yao, J.M. Flocculation performance of hyperbranched polyethylenimine-grafted cellulose in wastewater treatment. *ACS Sustain. Chem. Eng.* **2018**, *6*, 1592–1601. [CrossRef]
137. Cai, T.; Li, H.; Yang, R.; Wang, Y.; Li, R.; Yang, H.; Cheng, R. Efficient flocculation of an anionic dye from aqueous solutions using a cellulose-based flocculant. *Cellulose* **2015**, *22*, 1439–1449. [CrossRef]
138. Zhou, L.; Zhou, H.; Yang, X. Preparation and performance of a novel starch-based inorganic/organic composite coagulant for textile wastewater treatment. *Sep. Purif. Technol.* **2019**, *210*, 93–99. [CrossRef]
139. Beltrán-Heredia, J.; Sánchez-Martín, J.; Gómez-Muñoz, M.C. New coagulant agents from tannin extracts: Preliminary optimisation studies. *Chem. Eng. J.* **2010**, *162*, 1019–1025. [CrossRef]

140. Sánchez-Martín, J.; Beltrán-Heredia, J.; Solera-Hernández, C. Surface water and wastewater treatment using a new tannin-based coagulant. Pilot plant trials. *J. Environ. Manag.* **2010**, *91*, 2051–2058. [CrossRef]
141. Peng, Z.; Zhong, H. Adsorption of heavy metal ions from aqueous solution by tannin-based hydrogel. *Asian J. Chem.* **2014**, *26*, 3605–3608. [CrossRef]

Review

Metallic Iron for Environmental Remediation: Starting an Overdue Progress in Knowledge

Rui Hu [1,*], Huichen Yang [2], Ran Tao [2], Xuesong Cui [1], Minhui Xiao [1], Bernard Konadu Amoah [1], Viet Cao [3], Mesia Lufingo [4], Naomi Paloma Soppa-Sangue [1,5], Arnaud Igor Ndé-Tchoupé [6], Nadège Gatcha-Bandjun [7], Viviane Raïssa Sipowo-Tala [5], Willis Gwenzi [8,*] and Chicgoua Noubactep [2,*]

1 School of Earth Science and Engineering, Hohai University, Fo Cheng Xi Road 8, Nanjing 211100, China; cuixuesong@hhu.edu.cn (X.C.); xiaominhui@hhu.edu.cn (M.X.); bernardowuraku@yahoo.com (B.K.A.); soppanaomipaloma@yahoo.fr (N.P.S.-S.)

2 Angewandte Geologie, Universität Göttingen, Goldschmidtstraße 3, D-37077 Göttingen, Germany; huichen.yang@geo.uni-goettingen.de (H.Y.); ran.tao@geo.uni-goettingen.de (R.T.)

3 Hung Vuong University, Nguyen Tat Thanh Str., Viet Tri, Phu Tho 29000, Vietnam; caoviet@hvu.edu.vn

4 Department of Water and Environmental Science and Engineering, Nelson Mandela African Institution of Science and Technology, Arusha P.O. Box 447, Tanzania; lufingom@nm-aist.ac.tz

5 Faculty of Health Sciences, Campus of Banekane, Université des Montagnes, Bangangté P.O. Box 208, Cameroon; sipoworais@yahoo.fr

6 Department of Chemistry, Faculty of Sciences, University of Douala, Douala B.P. 24157, Cameroon; ndetchoupe@gmail.com

7 Faculty of Science, Department of Chemistry, University of Maroua, Maroua BP 46, Cameroon; nadegegatcha@yahoo.fr

8 Biosystems and Environmental Engineering Research Group, Department of Soil Science and Agricultural Engineering, University of Zimbabwe, Mt. Pleasant, Harare P.O. Box MP167, Zimbabwe

* Correspondence: rhu@hhu.edu.cn (R.H.); wgwenzi@yahoo.co.uk or wgwenzi@agric.uz.ac.zw (W.G.); cnoubac@gwdg.de (C.N.)

Received: 22 January 2020; Accepted: 21 February 2020; Published: 27 February 2020

Abstract: A critical survey of the abundant literature on environmental remediation and water treatment using metallic iron (Fe^0) as reactive agent raises two major concerns: (i) the peculiar properties of the used materials are not properly considered and characterized, and, (ii) the literature review in individual publications is very selective, thereby excluding some fundamental principles. Fe^0 specimens for water treatment are typically small in size. Before the advent of this technology and its application for environmental remediation, such small Fe^0 particles have never been allowed to freely corrode for the long-term spanning several years. As concerning the selective literature review, the root cause is that Fe^0 was considered as a (strong) reducing agent under environmental conditions. Subsequent interpretation of research results was mainly directed at supporting this mistaken view. The net result is that, within three decades, the Fe^0 research community has developed itself to a sort of modern knowledge system. This communication is a further attempt to bring Fe^0 research back to the highway of mainstream corrosion science, where the fundamentals of Fe^0 technology are rooted. The inherent errors of selected approaches, currently considered as countermeasures to address the inherent limitations of the Fe^0 technology are demonstrated. The misuse of the terms "reactivity", and "efficiency", and adsorption kinetics and isotherm models for Fe^0 systems is also elucidated. The immense importance of Fe^0/H_2O systems in solving the long-lasting issue of universal safe drinking water provision and wastewater treatment calls for a science-based system design.

Keywords: adsorption capacity; decentralized water supply; electrochemical reaction; inconsistent view; sand filtration; wastewater treatment; zero-valent iron

1. Introduction

Metal corrosion is one of the most important problems in industry, transport and agriculture [1–4]. Understanding metal corrosion comprises its detection (e.g., analytical, visual), its monitoring (e.g., mass loss, H_2 evolution) and its long-term characterization under various field conditions [4,5]. The corrosion research aims at determining the durability of metallic structures under operational conditions (e.g., oil and gas pipelines, tanks) and revealing the mechanisms of corrosion process [5,6]. This mechanism can be chemical, electrochemical or mixed [1]. Various tools have been used to characterize the corrosion resistance of different metals under various application conditions [4]. The overall result is the availability of integrated approaches to assess and predict the corrosion processes and thus the longevity of metallic structures (e.g., buried pipes) [3,4]. However, the frequency and sudden nature of metallic pipe failures worldwide indicate the inadequacy of current knowledge related to the longevity of buried metallic pipes.

Metallic iron (Fe^0) used as a reactive material in subsurface permeable reactive barriers is comparable to iron pipes with three major differences: (i) corrosion is welcome because it is a rather useful process [7–9], (ii) a reactive wall is ideally permanently water saturated, and (iii) the length of used particles (<5 cm) is tiny compared to pipes which are up to 12 m in length. On the one hand, Fe^0 specimens used in water treatment comprise steel wool with thickness varying between 25 and 90 μm [10,11]. On the other hand, the length of these particles is comparable to the wall thickness of iron pipes (2–4 mm). There has been no real system analysis for remediation Fe^0 materials with the aim to outline the differences making their peculiar characteristics. In addition, traceably deriving the longevity of remediation Fe^0 specimens from Fe^0 pipes is impossible because of the differences highlighted.

The remediation Fe^0 was termed as zero-valent iron (ZVI). This acronym is perhaps the first problem of this still innovative technology. A literature research with "zero-valent iron" as keyword would never reveal the ancient literature on Fe^0 for water treatment [12–16]. Indeed, no research group until 2017 has referenced a single article from the ancient use of Fe^0 in water treatment [17,18]. In chemistry, metallic elements are characterized by their oxidation state, the one of Fe^0 is zero (0). Upon oxidation, Fe^0 is transformed to Fe^{II}, Fe^{III} or Fe^{IV} species. Under environmental conditions, only Fe^{II} and Fe^{III} species are stable. The oxidation of Fe^0 to Fe^{II} is a redox process characterized by an electrode potential whose value is ™0.44 V [1]. According to the first principle of chemical thermodynamics, Fe^0 can be oxidized by oxidizing agents from each redox couple having a higher electrode potential (E^0 > ™0.44 V). Water (H_2O or H^+) is a relevant oxidizing agent for Fe^0 under environmental conditions. The electrode potential for the redox couple H^+/H_2 is 0.00 V. Fe^0 immersed in (contaminated or polluted) water is corroded to form H/H_2 and Fe^{II} (and mixed Fe^{II}/Fe^{III}) species which are stand-alone reducing agents [19–26]. Clearly, it is not surprising that selected species undergo reductive transformations in an Fe^0/H_2O system [19,26–29]. The application of metallic iron in water treatment dates back to the 1890s (Table 1). Suspended matter then left to settle

Table 1. Historical application of metallic iron in water treatment (modified from Mwakabona et al. [17]).

Age/Time	Material	Usage	Target Impurities	Application	Reference
Pre–1850	Old iron nails	Shaken with impure water	Suspended matter then left to settle	Used in West England	[30,31]
1850–1900	Iron plates/wires (1857)	Suspended in flowing impure	Suspended matter water	Tested but not widely applied	[13,31,32]
	Magnetic carbide/ Iron oxide	Filter media	Suspended matter/microbes	Widely applied household scale filters	[12,13,31]

Table 1. *Cont.*

Age/Time	Material	Usage	Target Impurities	Application	Reference
	Polarite	Filter media	Suspended matter/colour/microbes	Household filters	[32,33]
	Spongy iron	Filter media	Suspended matter/colour/microbes/chemical contaminants	Household and large scale filters	[30,31,34,35]
	Iron fillings	Agitated in the evolving purifier	Suspended matter/colour/microbes	Large scale in treatment plants	[31,34,35]
1900–1950	Iron fillings	Agitated in the revolving purifier	Suspended matter/colour/microbes/chemical contaminants	Large scale in treatment plants	[16,31,35,36]
1950–1990	Steel wool	Filter media	Radionuclides	Widely tested household scale filters	[37]
	Iron fillings	Filter media	Selenium from agricultural drainage water	Field scale	[38]
	Iron fillings	Static or dynamic	Halogenated hydrocarbons	Concept	[39]
Post–1990s	Iron fillings, steel wool	Filter media	Phosphate from agricultural drainage water	Field scale	[40,41]
	Iron fillings	Filter media	Domestic wastewater	Field scale	[42]
	Al^0/Fe^0 composite	Filter media or batch systems	Pathogen and chemical removal	Laboratory demonstration	[43]
	Iron fillings	Filter media	Pathogen removal	Laboratory demonstration	[44]
	Iron nails	Filter media	Kanchan Arsenic filter	Field scale	[45,46]
	Composite iron matrix	Filter media	SONO Arsenic filter	Field scale	[47]
	Metallic iron	Reactive media in PRBs	Chemical contaminants	Field scale	[22,23,48,49]
	Metallic iron	Filter media or batch systems	All classes of contaminants	Concept	[50]
	Metallic iron	Filter media	*E. coli* and *Listeria monocytogenes* removal from surface water for irrigation	Field scale	[51,52]

Table 1 presents a summary of the historical applications of metallic iron in water treatment. The history of Fe^0 application in environmental remediation has been the subject of several papers by our research group. Comprehensive research on Fe^0 for water treatment revealed that the generation of iron hydroxides and oxides (iron corrosion products or FeCPs) is the root cause of contaminant removal in Fe^0/H_2O systems [22,23,25,26,53,54]. For example, arsenic [55], carbon tetrachloride [29], chromium [56], fluoride [57,58], hexachloroethane [59] , methylene blue [60], methyl orange [61], Orange II [62], phosphate [63], selenium [64] and zinc [65] are all removed in Fe^0/H_2O systems despite their differences in charge, redox-reactivity and size.

A critical research review article from 2008 [54] established that adsorption, co-precipitation and size-exclusion were the fundamental mechanisms of contaminant removal in Fe^0/H_2O systems. Yet to date, 13 years later, researchers are still trying to establish the mechanisms by which aqueous contaminants are removed in the presence of Fe^0 [59,63,64,66,67]. Unfortunately, none of them has proven the alternative concept [53,54,68] wrong, and most of them are considering Fe^0 as the electron donor for observed transformations (electrochemical reaction) [69–71]. Moreover, "contaminant reduction" and "contaminant removal" are mostly randomly interchanged. Therefore, a better argumentation is still needed to convincingly explain the in-depth knowledge on the mechanisms causing contaminant removal in Fe^0/H_2O systems. Another important application of Fe^0 materials entails using Fe^0 to induce a pH shift and/or enhance microbial processes involved in methane production in anaerobic digesters [72–75]. This aspect is not considered herein as it is not focused on contaminant removal as discussed in Section 2.

This paper presents a profound analysis of the Fe^0/H_2O system and derives the leading causes and factors influencing its efficiency for water treatment. Relevant factors include: (i) the Fe^0 specimen including its form and size (intrinsic reactivity), (ii) the water chemistry including the nature of contaminants, the presence of dissolved O_2 and the pH value, (iii) the contact time (flow velocity or mixing intensity), and (iv) the Fe^0 amount and its proportion in the reactive mixture (thickness of the reactive layer or number of columns). Contaminants are explicitly considered within water chemistry, one of the four main groups of factors influencing iron corrosion. The approaches conventionally adopted to investigate these factors, the major findings, the limitations and the knowledge gaps are presented. It is argued that no progress in knowledge is possible before the research community agrees on the key issue that Fe^0 is not a reducing agent (under field conditions).

2. The Fe^0/H_2O System

2.1. *Overview of Fundamental Aspects*

There is a transfer of electrons from the Fe^0 body (solid state) to the Fe^0/H_2O interface whenever a piece of a reactive Fe^0 specimen is immersed in an aqueous solution (Fe^0/H_2O system) [1,5,6,76]. This occurs because Fe^0 is not stable under environmental conditions or because the redox couple H^+/H_2 (E^0 = 0.00) is higher than that of Fe^{II}/Fe^0 (E^0 = ™0.44) in the electrochemical series [1,22–26]. Equation (1) reveals that the oxidative dissolution of Fe^0 by protons (H^+) and Equation (1a) considers that protons are from water ($H_2O \Leftrightarrow H^+ + HO^-$). $Fe(OH)_2$ from Equation (1b) tends to polymerize and precipitate but can also be oxidized to even lower soluble $Fe(OH)_3$ by dissolved O_2 for example (Equation (2a)). $Fe(OH)_2$ and $Fe(OH)_3$ are polymerized and further transformed to various hydroxides and oxides (FeCPs) (Equation (3)) [1,21,77]. The different iron corrosion products (FeCPs) depict different adsorptive affinities for dissolved species [77–79]. Equation (4) summarizes the process of aqueous iron corrosion.

$$Fe^0 + 2\,H^+ \Rightarrow Fe^{2+} + H^2 \tag{1a}$$

$$Fe^0 + 2\,H_2O \Rightarrow Fe(OH)_2 + H_2 \tag{1b}$$

$$4\,Fe^{2+} + O_2 + 4\,H^+ \Rightarrow 4\,Fe^{3+} + 2\,H_2O \tag{2a}$$

$$4\,Fe(OH)_2 + O_2 + 2\,H_2O \Rightarrow 4\,Fe(OH)_3 \tag{2b}$$

$$Fe(OH)_2, Fe(OH)_3 \Rightarrow FeO, Fe_3O_4, Fe_2O_3, FeOOH \tag{3}$$

$$Fe^0 + H_2O + (O_2) \Rightarrow H_2 + \text{iron hydroxides and oxides} \tag{4}$$

In summary, Equation (4) recalls that immersing a reactive Fe^0 in water can be universally used to generate H_2, Fe^{2+} and various Fe^{II}, Fe^{III} and Fe^{II}/Fe^{III} hydroxides and oxides. Equation (1) demonstrates that Fe^0 is a scavenger of humidity (H_2O), while Equation (2) demonstrates the O_2 scavenging nature of Fe^0. These two scavenging characteristics have been exploited in several industrial applications [80–82]. For example, Fe^0 is used as desiccant in food packaging [81]. In the Fe^0 remediation literature, the best illustration for the nature of the Fe^0/H_2O system as generator of FeCPs is perhaps the excellent research article by Furukawa et al. [83]. These authors used several analytical tools to demonstrate the presence of ferrihydrite, green rust, magnetite and lepidocrocite in an Fe^0/H_2O system. More importantly, they conclude that an Fe^0/H_2O system is a temporally and spatially heterogeneous geochemical environment. Concerning the spatial heterogeneity, Furukawa et al. [83] specified that magnetite (Fe_3O_4) is generated in the vicinity of Fe^0, whereas ferrihydrite ($Fe(OH)_3$) precipitates away from the Fe^0 surface. This conclusion corroborates ancient findings [84] and recalls that even under oxic external conditions, there is a progressive O_2 depletion culminating into anoxic conditions in the vicinity of Fe^0. In this context, Stratmann and Müller [85] clearly demonstrated that oxygen is reduced by Fe^{II} species within the oxide scale (chemical reaction), while Fe^0 is oxidized by water (electrochemical reaction).

This section has recalled that, before Fe^0 complete depletion, an Fe^0/H_2O system is a dynamic and heterogeneous system containing Fe^0 and all its corrosion products (Equation (4)). Accordingly, even the most accurate measurements and the most precise observations are just a static snap-shots of dynamic processes within the system. The situation is exacerbated by the evidence that, the processes occur over an enormous range of timescales ranging from some few minutes or hours in laboratory investigations to months and years in field applications [86–88]. This communication insists on the fact that the frequency of discrepant reports in the scientific literature is rooted on an insufficient system analysis. In this regard, an accurate system analysis constitutes the theory of the system. The theory of the Fe^0/H_2O system, in turn, is like a guide to constrain the choice of the model. The theory that some contaminants are reduced by electrons from Fe^0 (electrochemical mechanism) is flawed as water, even present as humidity or moisture do corrode iron [1]. In fact, even deionized water corrodes iron [89]. Figure 1 depicts the interactions of contaminants with solid phases in a pure adsorbent system and the Fe^0-based system.

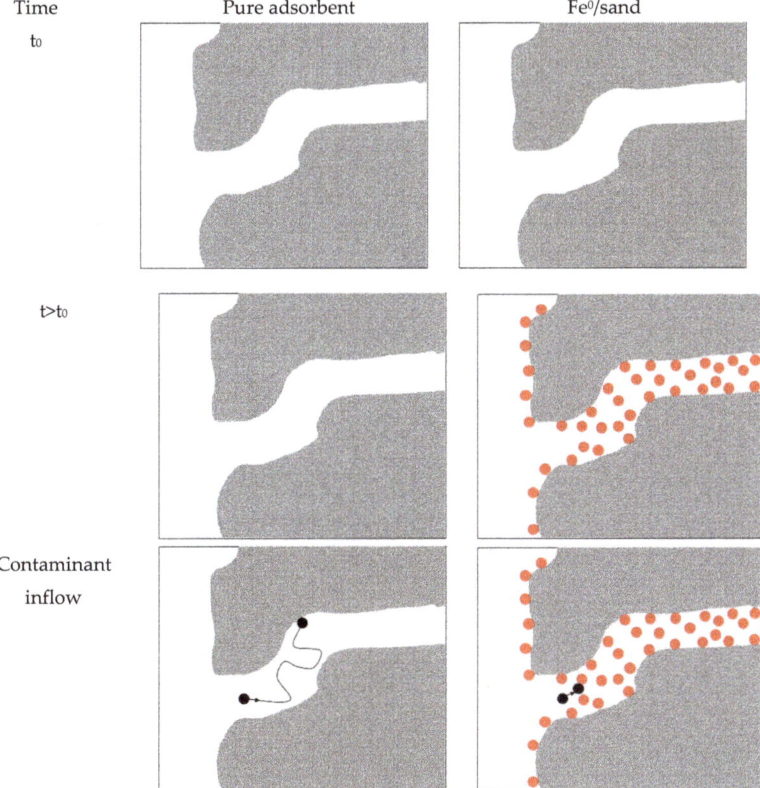

Figure 1. Schematic diagram comparing the interactions of contaminants (black points) with solid phases in a pure adsorbent system (left) and the Fe^0-based system (right). The red points represent iron corrosion products (FeCPs) which are either coated on solids or suspended in the pore solution.

2.2. Oxide Scale on Fe^0 and the Decontamination Process

The universal oxide scale on iron metal (at pH > 4.5) is still regarded by the majority of active researchers on remediation Fe^0 as a disturbing factor compromising the electron transfer from the metal body (reactivity loss) [90–92]. This view contradicts the evidence that a lag time between the start of experiments and reductive transformation is reported in the literature [93,94]. This time corresponds to the quantitative generation of FeCPs, which act as contaminant scavengers. This means that relevant reducing agents are generated in situ. Another problem of the Fe^0 literature is that "contaminant removal" and "contaminant reduction" are randomly interchanged, while no real mass balance of the contaminants has been presented [53,54,95]. On the contrary, contaminants that are not recovered are assumed to be chemically reduced [68–70].

A look at the mechanism of oxide scale formation reveals that it cannot be electronically conductive. In fact, the initial scale is very porous and cannot transfer electrons because air and water are not electronically conductive (an aqueous solution can be ionic conductive—electrolyte). In subsequent stages, available pores are filled with nascent FeCPs, but they are never uniform and the oxide scale is a mixture of iron hydroxides and oxides [5]. An oxide scale made up of Fe_3O_4 alone would have been electronically conductive. However, such an Fe_3O_4 scale cannot exist under natural conditions (immersed Fe^0). All other FeCPs are at best semi-conductors and cannot relay electrons from Fe^0 under natural conditions. Clearly, reports justifying the reductive efficiency of Fe^0/H_2O systems using the semi-conductive nature of FeCPs are mistaken [96]. This assertion encompasses the Fe^0/pyrite/H_2O system, whose efficiency is mainly justified by the semi-conductive nature of FeS species [97–99].

The oxide scale on Fe^0 is definitively a diffusion barrier for all dissolved species, including the pollutants. It is also the contaminant scavenger such that electrochemical corrosion of immersed Fe^0 induces the generation of contaminant scavengers and other reducing agents. Thus the generation of solid FeCPs is a necessary process which has the (perceived negative) side effect of being expansive. Thus, designing an efficient and sustainable system requires answering the question: how long can FeCPs be generated to satisfactorily treat water while keeping a reasonable hydraulic conductivity (permeability)?

2.3. Chemical Aspects

As discussed earlier (Section 2.1), Fe^{2+} from Equation (1) is transformed to ferrous hydroxide ($Fe(OH)_2$) and ferric hydroxide ($Fe(OH)_3$) which have a strong tendency to form colloids of particles that normally carry a positive charge [62,100,101]. These minerals are further transformed to other Fe^{II}/Fe^{III} minerals (e.g., Fe_2O_3, FeOOH, green rust) exhibiting different affinities to dissolved species. The nature of oxides in each individual system depends on the intrinsic reactivity of the used Fe^0 material and the environmental conditions [5,79]. For example, two different Fe^0 specimens corroding under the same environment will not necessarily produce the same iron oxides, because the composition of the oxide scale depends on the relative kinetics of Fe^0 dissolution and Fe hydroxide precipitation, which in turn depends on the solution chemistry, including the pH value, dissolved ions and the salinity [5,6,102]. On the other hand, in-situ generated free Fe^{2+} are adsorbed to the surface of available minerals to form the so-called structural Fe^{II} with a reducing power far larger than that of the free Fe^{2+} ($E^0 < 0.77$ V) and sometimes stronger than Fe^0 ($E^0 < -0.44$ V) [103]. The availability of several reducing agents in the Fe^0/H_2O system, and especially from structural Fe^{II}, partly stronger than Fe^0 implies that the electrochemical series of metal alone cannot predict the chemistry of the system.

2.4. Physical Aspects

The volumetric expansive nature of iron corrosion is the most important physical phenomenon occurring in Fe^0/H_2O systems [104]. There is expansion because the parent metal (Fe^0) produces in-situ both: (i) H_2, occupying a volume about 3100 times larger [26,105], and (ii) each solid oxide and hydroxide is at least twice larger in volume than Fe^0 ($V_{oxide} > V_{iron}$) [106,107]. For example, the specific density of magnetite (Fe_3O_4) is about one half that of iron (Fe^0). It implies that after corrosion a space twice larger than the initial space is occupied [108–110]. While it can be assumed that H_2 escapes from each open system, no free expansion of oxides in porous systems (e.g., water filters, reactive walls) can be assumed [26]. External or internal free expansion occurs in metallic pipes [111,112] and on the walls of steel canister for radioactive waste repositories [110,113]. On the contrary, free expansion cannot be expected in steel-reinforced concrete structures [108,109]. Table 2 presents a comparison of iron corrosion parameters in Fe^0 remediation to that of water pipes and reinforced concrete. Accordingly, considering expansive iron corrosion, which culminates into permeability loss is an essential design parameter for porous Fe^0/H_2O systems (Fe^0 filters). For each Fe^0 filter, the temporal production of both H_2 and oxide is decisive for the long-term efficiency and the permeability of the system [105,107,114,115].

Table 2. Comparison of iron corrosion parameters in Fe^0 remediation to that of water pipes and reinforced concrete.

Parameter	Water Piping	Reinforced Concrete	Fe^0 Remediation
Iron corrosion	Destructive	Destructive	Constructive
Environment	Unknown soil	Known concrete	Known mixture
Fe size	several meters	some cm	<5 mm
	some cm thick	some mm	<2 mm
Rust expansion	Minor or no issue	major issue	major issue
Corrosion effect	Pipe damage	cracking	permeability loss
Linear corrosion	Conservative	non applicable	Absurd
Knowledge status	Quite good	Quite good	Rather poor
Service life	>50 years	>50 years	Up to 30 years
Reference	Enning et al. [2]	Caré et al. [108]	Hu et al. [26]

Another key feature in investigating the remediation Fe^0/H_2O system is that, at pH > 4.5, the Fe^0 surface is permanently covered by an oxide scale. The oxide scale acts both as: (i) conduction barrier for electrons from the metal body, and (ii) physical barrier for dissolved species, including O_2 and pollutants of concern [22–26,53,54]. The net result is that Fe^0 is oxidized by water (electrochemical reaction—Equation (1)), and O_2 and dissolved contaminants are reduced by reducing species present in the oxide scale (Fe^{II} and Fe^{II}/Fe^{III} species, H_2) (chemical reaction). Again, any reasoning based on the electrochemical series of elements is a misuse of chemical thermodynamic, as the physics of the system, specifically the expansive nature of iron corrosion and it electronically non-conductive nature are simply ignored.

2.5. Kinetic Aspects

Aqueous corrosion of Fe^0 materials under environmental conditions (pH > 4.5) is an electrochemical process involving iron dissolution at the anode and H_2 evolution at the cathode (Equation (1)). This electrochemical reaction is accompanied by the formation of an oxide scale on Fe^0 which is not protective as a rule [6,116–118]. In general, oxide scale growth and its protectiveness depend primarily on the precipitation rate of iron hydroxides [5,102]. As the Fe^0 surface corrodes under the initial scale, corrosion continuously undermines the scale. Voids are created and are progressively filled up by the ongoing hydroxide precipitation. The relative rate of (i) Fe^0 oxidative dissolution and (ii) hydroxide precipitation in the Fe^0 vicinity determine the protectiveness of the oxide scale. According to Nesic [5], when the rate of hydroxide precipitation exceeds the rate of Fe^0 dissolution, a dense protective oxide scale is formed. Conversely, when Fe^0 dissolution undermines the new oxide scale faster than hydroxide precipitation can fill in the voids, a porous and non-protective scale forms.

There are several factors influencing the iron corrosion rate, including the intrinsic reactivity of a material and the solution chemistry. The solution chemistry includes the presence of dissolved O_2, contaminants and other (mostly) ubiquitous species. In particular, two different Fe^0 specimens may exhibit different degree of protectiveness under the same operational conditions. The most important feature to consider is that the corrosion rate is never linear. Thus, a linear extrapolation of the initial corrosion rates of Fe^0 specimens in engineered systems can give an inaccurate estimation of its service life. However, for remediation systems, the non-linear nature of the corrosion kinetics implies a decrease in efficiency. Again, results from pipe corrosion or wall corrosion cannot be transferred to remediation Fe^0/H_2O systems. However, it is certain that the semi-permeable nature of the oxide scale allows significant and continuous corrosion over long time periods ("rust never rest"). In this context, Roh et al. [119] reported of buried iron pieces from World War I still corroding in the subsurface. Thus, the objective of the Fe^0 remediation is to couple this long-term corrosion with efficient contaminant removal [7,88,118,120,121].

2.6. Investigating the Fe^0/H_2O System

The extent and kinetics of Fe^0 oxidative dissolution are the result of the following: (i) the nature of metal (intrinsic reactivity), and (ii) the interactions between Fe^0 and the environment in which it is placed [76,122–125]. Therefore, (i) a change of material, and/or (ii) a change in environment results in changes in the rate and the extent of corrosion. In this study, the influence of the environment is considered at micro-scale, specifically what is happening on the Fe^0 surface, in its vicinity or over short distances (within the oxide scale). The oxide/water interface is also considered, but the volume of the solution is not. The aqueous phase is regarded as a reservoir of pollutants and co-solutes, while being the principal Fe^0 oxidizing agent [126,127].

Laboratory and pilot-scale investigations are usually conducted in order to obtain reliable information on the interactions of metallic devices with particular operational environments [128,129]. Laboratory tests are designed to simulate some relevant field situations. Laboratory studies are mostly aimed at obtaining data in a more convenient way and in a shorter time [123,124,130]. They also provide mechanistic information for field applications. However, short-term laboratory experiments are always a simplification and this should be borne in mind when interpreting achieved results [76,129,131,132]. Given the diversity of operational parameters that have been proven to influence iron corrosion from individual studies, one can be overwhelmed by their number and the fact that each material is unique in its corrosion behaviour [11,124,125,133]. Therefore, a first attempt toward a systematic investigation of relevant influencing factors goes through the consideration of the electrochemical nature of aqueous iron corrosion.

Fe^0 is a good conductor of electricity, and its electrochemical aqueous corrosion depends on the conductive nature of the solution in which it is immersed [1,76]. Understanding the corrosion processes helps in selecting Fe^0 materials necessary for designing sustainable systems. Therefore, the best way to design a sustainable system is to consider the fundamentals of iron corrosion at the design stage. This section highlights the knowledge of corrosion principles and the importance of the environment and materials for field design. The four compartments necessary for continuous Fe^0 corrosion are: (i) an anodic region on Fe^0 where the metal is oxidized and releases Fe^{2+} (leaving two electrons behind), (ii) an electrolyte to transport released Fe^{2+} away from the anode, (iii) a cathode where the simultaneous reductive transformation coupled to iron oxidation occurs, and (iv) the Fe^0 body transporting electron from the anode to the cathode. In an Fe^0/H_2O system, the anode and the cathode are different sites on the same Fe^0 specimen. In the conventional remediation technology, the size of Fe^0 particles is generally small (<5 mm). The four compartments must be electronically connected for the electrochemical process to proceed. This means that if Fe^0 is covered by an oxide scale, the scale must be conductive to warrant the transfer of electrons from the metal body to the oxidizing agent within the oxide scale or at the oxide/H_2O interface.

A Fe^0/H_2O system is made up of two interfaces: (i) Fe^0/oxide and (ii) oxide/H_2O. Because the oxide scale is never electronically conductive and is a diffusion barrier to many species, only water can quantitatively reach the Fe^0 surface. The net result is that Fe^0 oxidative dissolution is an electrochemical reaction (water is reduced) but all other observed/reported chemical reduction occur within the oxide film or at the oxide/H_2O interface [53,54,127]. For the Fe^0/H_2O remediation system, it means that contaminants are not reduced by electrons from Fe^0. This has important implications on the operating principles of Fe^0/H_2O systems: first, using the electrochemical series to predict the reductive transformation of any species has been a mistake, and second, using the stoichiometry of any electrochemical reaction involving Fe^0 is also a mistake. Accordingly, contaminants are 'just' dissolved species, capable of modifying: (i) the conductive properties of the electrolyte (H_2O), (ii) the ion conductivity of the oxide scale, and (iii) formation and the transformation of the oxide scale.

3. An Overview of the Mistakes of Past Efforts

3.1. Contaminant Removal Mechanisms

The major mistake of the Fe^0 remediation literature has been to consider that relevant contaminants are reduced by an electrochemical reaction (electrons from the metal) [134–136]. This section summarizes the extent of the confusion using a paper by Kamolpornwijit and Liang [137]. This paper is selected because it considered past efforts in understanding the service life of Fe^0 filters and performed long-term experiments (>400 days). Kamolpornwijit and Liang [137] considered porosity loss during nitrate (NO_3^-) removal in an Fe^0 barrier. The following reactions were considered:

$$Fe^0 + 2\,H_2O \Rightarrow Fe^{2+} + H_2 + OH^- \tag{5}$$

$$4\,Fe^0 + NO_3^- + 7\,H_2O \Rightarrow 4\,Fe^{2+} + NH_4^+ + 10\,OH^- \tag{6}$$

$$5\,Fe^0 + 2\,NO_3^- + 6\,H_2O \Rightarrow 5\,Fe^{2+} + N_2 + 12\,OH^- \tag{7}$$

$$10\,Fe^{2+} + 2\,NO_3^- + 6\,H_2O \Rightarrow 10\,Fe^{3+} + N_2 + 12\,OH^- \tag{8}$$

Equation (5) corresponds to Equation (1) and represents aqueous iron corrosion, an electrochemical reaction. Direct electron transfer from Fe^0 to NO_3^- yielding NH_4^+ (Equation (6)) or N_2 (Equation (7)) are only possible if the oxide scale on Fe^0 is electronically conductive. Because an electronically conductive oxide scale does not exist under immersed conditions, Equations (6) and (7) are wrong. In discussing their results, Kamolpornwijit and Liang [137] considered that the H_2 volume they measured corresponds to the stoichiometry of Equation (5). Comparing this H_2 volume to the N_2 volume after Equation (7), they concluded that the contribution of water to the corrosion of iron is minimal. In chemical terms, this is erroneous as Equation (5) shows that 5 moles of Fe^0 release only 1 mole of N_2. Whether N_2 escapes from the system or not, the 5 moles of Fe^{2+} are further oxidized and or precipitated within the Fe^0 filter and filling the initial porosity. In reality, it is 10 moles of Fe^0 that are needed to release one mole of N_2 (Equation (8)).

One major problem of the Fe^0 literature has been the discussion of the reaction mechanism without mass balance considerations [31,95,138]. Complete mass balance analysis comprises the one of iron, which is admittedly a difficult task as the system is highly dynamic. However, the analysis made herein for nitrate reduction clearly demonstrates the wrongness of the view that Fe^0 is corroded by NO_3^-. On the other hand, the importance of FeCPs in inducing porosity loss is demonstrated. If the discussion of the porosity loss starts with the assumption that one mole of N_2 corresponds to 10 moles of corroded and expanded Fe^0, a better evaluation of changes in the porosity will be achieved. Kamolpornwijit and Liang [137] is also an excellent illustration on how in-situ generated FeCPs are not properly considered while exotic species like $CaCO_3$ or the flow regime (convection versus laminar) are given key role in pore filling and permeability loss. In essence, if a contaminated water is rich in carbonates a pre-treatment unit for their removal can be used such that decontamination units are

HCO_3^- free. Thus, given the expansive nature of iron corrosion, loss of permeability and porosity even occurs in aqueous systems without $CaCO_3$.

Another important feature from Kamolpornwijit and Liang [137] is that, in case nitrate is reduced to NH_4^+, regardless of the real mechanisms, NH_4^+ must be removed from the aqueous phase. This removal occurs by adsorption, co-precipitation and size-exclusion [53,54]. These three mechanisms represent the fundamental mechanisms of contaminant removal in Fe^0/H_2O systems [26,67]. It is essential to recall that chemical reduction and even chemical precipitation are not relevant contaminant removal mechanisms in the concentration ranges relevant to natural waters [111,139]. In particular, for safe drinking water provision, physical methods (e.g., adsorption, filtration, ion exchange) are always mandatory to cope with the stringent regulations. As an example, water defluoridation by chemical precipitation yields an equilibrium fluoride concentration of about 8 mg L^{-1} according to the solubility of CaF_2 [57,58]. This value is far larger than the maximum permissible contamination level of 1.5 mg L^{-1} for drinking water. Coming back to the Fe^0/H_2O system, available results show limited removal extent for fluoride while various removal extents for several contaminants depicting no redox reactivity in the systems has been documented. For this reason, a more rational approach is to consider that all species can be removed regardless of their redox potential, and identify exceptions on a case by case basis. In this regard, recent results by Hildebrant et al. [140] have confirmed that fluoride removal is low, but less reactive materials (EDTA test) comparatively remove more fluoride. This last observation corroborates the view that there is no single Fe^0 material for all situations and calls for more systematic investigations to identify appropriate Fe^0 materials for specific remediation applications.

The lack of systematic investigations has also affected the selection of Fe^0 materials used for pathogen removal [141–146]. Using various Fe^0 materials and very different experimental conditions, discrepant results have been reported [141,146]. Lu et al. [142] can be regarded as a perfect reflection of the state-of-the-art knowledge. The same authors investigated the mechanism of nitrate reduction and iron cycling by an iron-reducing bacteria strain (strain CC76) and metallic iron. They reported that the strain CC76 was able to utilize Fe^{2+} (from iron corrosion) as electron donor for the nitrate removal. More importantly, they observed that Fe^0 inhibited the growth of strain CC76 in the early stage of the operation. This observation corresponds to CC76 removal by Fe^0. However, the authors also reported that after the initial stage, strain CC76 was able "to tolerate" the presence of Fe^0, meaning that no or less removal was achieved. This phase corresponds to a decrease in the kinetics of iron corrosion as discussed above and has been described in microbially-influenced corrosion [2]. If we recall the work of Kamolpornwijit and Liang [137] primarily investigating nitrate removal in an abiotic Fe^0/H_2O system, it then becomes apparently clear that the major source of discrepancy in the literature is the insufficient system analysis. Like with chemical contaminants, all pathogens will be removed in a well-designed system. For a filtration system, the key questions to address are: (i) which Fe^0 material, in what amount, and for which contaminant and water? and (ii) with which filtration depth, for which flow velocity and for which operational duration? Another key aspect to consider is the nature of the aggregates (e.g., gravel, MnO_2, pumice, sand) to be mixed with Fe^0 and the Fe^0 proportion in the mixture [107,114]. These aspects are yet to be addressed in the Fe^0 remediation literature, but are critical for the design and field application of the technology. Table 3 presents a synthesis of current knowledge including the mistakes, and proposed refinements for future studies on Fe^0 remediation.

Table 3. Summary of synthesis of current knowledge and proposed refinements for future studies on Fe^0 remediation.

Previous Study	Synthesis	Refinement for Future Studies
	Electrochemical reduction	Chemical reduction
	Protons as concurrent Fe^0 oxidizer	Protons as sole oxidizing agents for Fe^0
Decontamination mechanism	Adsorption as possible mechanism	Adsorption as fundamental mechanism
	Co-precipitation as possible mechanism	Co-precipitation as fundamental mechanism
	Size-exclusion rarely considered	Size-exclusion as fundamental mechanism
Driving force	E_0 value = −0.44 V	Affinity to iron oxides and hydroxides
Selectivity	E_0 value of the contaminant	Surface charge and size
	H_2 evolution	Fe^0 dissolution in 1,10-Phenanthroline
Material selection	Removal efficiency for selected species	Fe^0 dissolution in EDTA
	Physical characterization tools	Not really useful
	Fe^0 dissolution in complexing agents	Phen test as candidate for standard method
Batch experiments	Shaken or stirred	Quiescent or shaken at < 100 rpm
	Lasting for some hours to days	Lasting for some weeks to months
	Mostly lasting for some few months	Lasting for at least six months
	Many contain only Fe^0	Never contain more than 50% (vol/vol) Fe^0
Column/field experiments	Some accelerated column experiments	Never artificially accelerated iron corrosion
System modelling	Based on the stoichiometry of an electrochemical reduction reaction	Consider protons as the sole oxidizers of Fe^0 Consider porosity loss related to each Fe^0 atom

3.2. *Reactivity and Efficiency*

In the Fe^0 remediation literature the terms "reactivity" and "efficiency" are mostly randomly interchanged. This has introduced confusion in the evaluation of independent results. In an effort to resolve this confusion the notion of "electron efficiency" has been introduced [147,148]. The electron efficiency (in %) is defined as the proportion of electrons from Fe^0 oxidation that is used for the target reduction reaction, for example for nitrate reduction by Kamolpornwijit and Liang [137] (Section 3.1). In this approach, electrons used to reduce water (Eq. 1) or dissolved O_2 are considered as "excess electrons" or avoidable electron wastage [148]. In other words, the main reaction is devalued to a side effect. Contrary to this still prevailing approach [92], the present work and related ones [149–153] recall that aqueous iron corrosion (Equation (1)) is not an unwanted reaction leading to an extra consumption of Fe^0, but Equation (1) produces Fe^{II} species and H_2, which are responsible for the documented chemical reduction and FeCPs which are responsible for contaminant removal and permeability loss [105–107].

Each Fe^0 material is characterized by its intrinsic reactivity while each system is characterized by its efficiency for water treatment [154–156]. According to Miyajima and Noubactep [156], the efficiency is the expression of reactivity in a given system. This means that changing the Fe^0 material modifies the efficiency. Conversely, an Fe^0 material proven efficient in a system could be useless in another system. In the quest of more efficient systems for water treatment, several Fe^0 materials and groups of materials (e.g., bimetallics, iron nails, nano-Fe^0, scrap iron, sponge iron, steel wool) have been tested and used for water treatment (Table 1). These efforts have been rendered difficult by a mistaken system analysis [157,158] and the evidence that most comparative works are based on testing materials for the removal of individual contaminants [159–162]. Lufingo et al. [11] recently discussed approaches that can be considered as universal as they are based on the stoichiometry of Equation (1): (i) H_2 evolution [163,164] and (ii) Fe^{2+} production [11,159]. They concluded that iron dissolution in a dilute solution of 1,10-Phenanthroline (2 mM) (Phen test) is the best available tool to characterize the intrinsic reactivity of Fe^0 materials (Section 4).

3.3. Misuse of Adsorption Isotherms and Kinetic Models

Several studies investigating contaminant removal by Fe^0 materials have applied isotherm and kinetic models initially developed for materials where contaminant removal occurs via adsorption [131,132,165,166]. A detailed discussion of the various adsorption isotherm and kinetic models, including assumptions and equations are presented in earlier reviews [132]. A comprehensive review of the limitations associated with the applications of such models to Fe^0 systems is the subject of another paper, hence is beyond the scope of the current study. Briefly, the extension of such models to Fe^0 materials is problematic for the following reasons: first, unlike adsorbents such as activated carbon whose surface area is readily available for adsorption at the beginning of the experiment (time t_0 = 0; Figure 1), Fe^0 systems are highly dynamic and equilibrium is rarely achieved in such systems (i.e., rust never rests), while the total reactive sites contributing to contaminant removal is not exactly known (Figure 1). Thus, most assumptions for such models are not valid for Fe^0 materials. Second, as discussed earlier, contaminant removal in Fe^0 systems does not occur solely via adsorption. In view of this, the application of adsorption models to Fe^0 systems constitute another fundamental error in Fe^0 literature. Yet despite a number of reviews highlight the mistakes, misuse and inconsistencies in the use of adsorption kinetic and isotherm models [53,54,67–71,131,132], the problem persists. Thus there is need to develop models for contaminant removal in Fe^0 systems that takes into account the iron corrosion phenomena and the fundamental mechanisms responsible for contaminant removal as highlighted in the current review.

Reference [157] severely questioned the validity of specific rate constants (k_{SA}) in Fe^0/H_2O systems. These rate constants are based on the stoichiometry of contaminant reduction by electrons from the metal body (Fe^0) and are not considering the interactions of contaminants within the oxide scale [167]. As discussed in Section 2.2, the Fe^0 surface is not accessible to all contaminants. Even k_{obs} values (which are normalized to the surface area to obtain k_{SA}) currently used in the literature are erroneous as they are rooted on the wrong reaction. However, the most severe mistake has been to use the adsorption capacity for Fe^0 which is considered a reducing agent. Moreover, an adsorption capacity is determined in a system where the reactive agent is not completely depleted.

4. Selection and Characterization of Fe^0 Materials

As discussed above, the scientific reason for using all Fe^0 materials (e.g., granular iron and bimetallics, iron filings, iron nails, iron wire, nano-Fe^0, scrap iron, steel wool) in water treatment is the electrode potential of the redox electrode Fe^{II}/Fe^0: $E^0 = -0.44$ V. Clearly, all reactive Fe^0 materials have the same redox potential. The intrinsic reactivity of individual Fe^0 specimens depends on a myriad of factors from which some are not readily accessible to the researcher. Relevant influencing factors include: alloying elements, Fe^0 form, manufacturing processes, metallography, Fe^0 grain size, surface area, and surface oxidation state. Accordingly, each Fe^0 material has its own intrinsic reactivity which should be characterized in order to better understand how it is influenced by operational conditions to induce the intended remediation goal.

The long history of using Fe^0 for technical chemical applications, including water treatment reveals that there has always been efforts to select appropriate materials for individual applications.

For example, the porous "spongy iron" (sponge iron or direct reduced iron) was proven more suitable in filtration systems than dense materials [12,13,168]. Similarly, multi-metallic systems and nanoscale materials were recently developed to address (recalcitrant) contaminants that were less sensitive to treatment with granular materials [169]. For completeness, it should be stated that there is no Fe^0 material for all situations such that one should rationally select appropriate materials for each specific application. For example, while treating water in fluidized systems (Anderson Process), dense materials were better than sponge iron [13,168]. Similarly, Hildebrant et al. [140] reported that Fe^0 materials of low reactivity according to the EDTA test exhibited a better efficiency for fluoride removal. The question arises how to select the right Fe^0 specimen for a given application? This calls for the development of standardized protocols for selection and characterization of Fe^0 materials for various applications.

It is unfortunate that material selection has not received the due attention given its central role for the technology. Lufingo et al. [11] recently gave an overview of the available tools for the characterization of the intrinsic reactivity of Fe^0 specimens. These authors based their work on an excellent review article by Li et al. [162] and insisted on the quantification of Fe^{2+} from Equation (1). According to Lufingo et al. [11], quantifying Fe^{2+} or H_2 evolution from Equation (1) are the best tools to characterize the intrinsic reactivity. However, quantifying both Fe^{2+} and H_2 in natural systems suffers from the high reactivity of those primary corrosion products within the system, including their adsorption onto FeCPs and their action as own reducing agents. Clearly, measured amounts of Fe^{2+} and H_2 represent an excess quantity and cannot be strictly used to quantify iron corrosion (at pH values >4.5). These considerations clearly show that the H_2 evolution method [163,164] is an approximation, while the EDTA method [170] is disturbed by dissolved O_2. All other methods are contaminant-specific, and thus of low value [133,160,161]. Lufingo et al. [11] then proposed iron dissolution in a dilute (2 mM) 1,10-Phenanthroline solution (Phen test) as a facile method free from the inherent shortcomings of all available methods.

In terms of affordability and applicability, the Phen test is currently the best available method to characterize the intrinsic reactivity of Fe^0 specimens. The test lasts for less than 36 hours and characterizes the initial kinetics of Fe^0 dissolution. It is suggested as a candidate for a standard method and is immediately useful for material selection (screening) and quality control [11]. Each Fe^0 specimen is characterized by its k_{Phen} value which reflects its initial dissolution at the pH value of natural waters but without the interaction of the oxide scale. The k_{Phen} values for nine steel wool specimens (Fe^0 SW) presented by Lufingo et al. [11] were such that $0.07 \leq k_{Phen}$ ($\mu g\ h^{-1}$) ≤ 1.30. This shows a ratio of reactivity of 18.5 for Fe^0 SW specimens which are often tested as a uniform class of materials compared to granular materials for example. Hildebrandt et al. [10,140] presented the k_{EDTA} values for 13 reactive Fe^0 SW and one granular Fe^0 as $3.7 \leq k_{EDTA}$ ($\mu g\ h^{-1}$) ≤ 130.8. The lowest k_{EDTA} value corresponds to granular Fe^0, suggesting a reactivity ratio of up to 36 between Fe^0 SW and granular Fe^0. No data comparing k_{Phen} values for Fe^0 SW and granular Fe^0 are yet available. However, it is certain that the Phen test provides a confidence Fe^0 screening tool and is non-contaminant-specific. Therefore, if a significant body of k_{Phen} data and data for characterized Fe^0 specimens for the removal of selected representative contaminants (or contaminant groups) are made available, then site-specific treatability studies would then be required only to fine-tune design criteria for the optimal performance of remediation Fe^0/H_2O systems.

Our research group has long recognized the need for systematic characterization of Fe^0 materials in terms of their intrinsic reactivity and efficiency. The idea followed by this research group as summarized herein is to characterize the Fe^0 reactivity and the efficiency of Fe^0/H_2O systems in a pollutant-independent-manner. The alternative is to agree on probing pollutants while using standard protocols. A reference Fe^0 material would also be necessary. The use of EDTA and Phen tests to characterize the intrinsic reactivity has already been discussed. The efficiency of the Fe^0/H_2O system is characterized using the methylene blue (MB) discoloration method (MB method) [60]. Herein, the low affinity of MB for FeCPs is used to trace the abundance of in-situ generated iron oxides in comparatively long-term experiments [62,101,159,171]. The Phen test provides a reliable guidance in selecting Fe^0 from the large catalog of available materials and to control the quality of newly manufactured ones. On the other hand, the MB method provides a reliable guidance for the characterization of the efficiency of Fe^0/H_2O systems. Selected probing agents enable the discussion of the results, and orange II, methyl orange and reactive red 120 were positively tested in this regard [61,62]. Recent results by Hildebrandt et al. [140] suggest that fluoride could be the next candidate for affordable probing reagent. In fact, it was found that fluoride removal is more efficient by low reactive Fe^0 specimens.

5. Future Perspectives and Potential Applications

5.1. Investigating the Fe^0/H_2O System

The findings on which future research should be rooted must be based on the evidence that the oxide scale on iron is a diffusion barrier and shall never been disturbed in ways that are not reproduced under field situations [22,23,53,54,56,124,126,127,171–175]. This has been the motivation for adopting quiescent bath experiments as a more suitable approach to investigate processes occurring in Fe^0 permeable reactive barriers some two decades ago [20]. Then and now, most research groups consider the shaking intensity as a relevant operational parameter to be investigated almost in all instances without a quiescent system as a reference. However, there is no given convincing reason for the chosen mixing intensities tested in these experiments [124,125].

In 2005, Devlin and Allin [173] designed a glass-encased magnet batch reactor (GEM reactor) to investigate the impacts of selected anions on the efficiency of granular Fe^0 in removing aqueous contaminants using 4-chloronitrobenzene as a probe molecule. This design aimed to achieve a better comparability between results of different experiments by fixing the stirring method and the stirring rates [169,173]. Using this approach, all experiments should be conducted in the GEM reactor to ensure that the granular iron remains stationary while the solution is stirred. Noubactep et al. [124,127] later demonstrated that, while using a rotary shaker, the shaking intensity should never be larger than 100 rpm. Given that a stirring device (including the GEM reactor) can be difficult to acquire by low-income laboratories, quiescent experiments can be adopted as a rule [171].

Quiescent batch experiments have also been adopted for the determination of the initial corrosion rate of Fe^0 in EDTA (k_{EDTA}) and Phen (k_{Phen}). Herein, the experiments are stopped before solution saturation, meaning that Fe^0 specimens are characterized under conditions where no oxide scale is available. Contrary to the prevalent approach [152–164], Fe^0 specimens are not characterized for any contaminant removal efficiency [176,177]. Solution saturation corresponds to [Fe] values equivalent to the stoichiometry of iron complexation (e.g.,[Fe] = 112 mg L^{-1} for 2 mM EDTA). Lufingo et al. [11,120,121] recently compared the EDTA and the Phen test and established the superiority of the Phen test, which is additionally more affordable.

Another key feature of the remediation Fe^0/H_2O system is related to the oxide scale. The omnipresent oxide scale on Fe^0 is positively charged under natural conditions (pH > 5.0) [62,100,101,178]. Thus, the Fe^0/H_2O system is an ion-selective one and preferentially removes negatively charged species like bacteria [142,146]. One original idea has been to characterize the discoloration of methylene blue (MB) by Fe^0/sand/H_2O systems (MB method). The MB method is grounded on an historical work by Mitchell et al. [178], who observed that sand adsorbs less MB when it is coated with iron oxide. Thus, mixing the same mass of different Fe^0 specimens with the given mass of a sand specimen, and allowing them to equilibrate for the long-time in a MB solution enable the differentiation of the Fe^0 reactivity of various materials [62,101,176,177]. As a rule, the most reactive material produces the largest amount of iron hydroxides, which, in turn, in-situ coat sand and thus discolors MB the least. MB is thus not a model contaminant, but an operational tracer [171,177]. To ease the interpretation of results, methylene orange (MO) [61,176] or Orange II [62,101] can be used as anionic dyes since they have molecular sizes similar to that of MB. Phukan et al. [62,101,176] additionally used reactive red 120 which is also an anionic dye, but is much larger in size that MO and Orange II.

One key advantage of the MB method is that its enables a visual observation of preferential flow within a reasonable time scale (some weeks) [62]. In designing an Fe^0-based filtration systems, a decrease in hydraulic conductivity (permeability loss) is expected, and an early contaminant breakthrough will be observed due to preferential flow [179–183]. Mineral precipitation is particularly intense in the entrance zone of the filter [181], but the created preferential pathways are extended throughout the whole water column [179,181]. The investigation of the process of creation and extension of preferential pathways has been analytically very challenging [181–183]. On the basis of observations using the MB method [176,177], changes generated in the entrance zone can be better followed and considered in modelling efforts. This last aspect is crucial for the development of the technology as it has been convincingly demonstrated that models currently predicting the service life of Fe^0 filters are rooted on a wrong premise [114,115].

5.2. Designing the Next Generation Filter

The determination of the amount of a given Fe^0 which long-term corrosion kinetics would allow a designed Fe^0/aggregate system to satisfactorily treat a given water for a certain time frame (e.g., 12 months) can be regarded as a routine work. This routine engineering application is however complicated by the complexity of the iron corrosion process and mistakes in past research as outlined herein. One major thinking mistake has been to consider that admixing Fe^0 with sand (the most used aggregate) would alter the decontamination kinetics [90,183,184]. Thus, admixing with sand has been mainly considered as an economic tool to safe Fe^0 costs and the negative effects on the resulting system discussed [185]. It has been recently demonstrated that only hybrid Fe^0 systems are sustainable [105–107].

Several hybrid systems have been successfully tested for water treatment including Fe^0/activated carbon [184], Fe^0/Fe_3O_4 [185], Fe^0/MnO_2 [186], Fe^0/pyrite [99] and Fe^0/sand [187]. While inert sand alone has clearly improved the efficiency of Fe^0 systems [188], the efficiency of other tested aggregates was attributed to the specific materials. The Fe^0/sand is regarded as in-situ sand coating [141]. It is certain that other materials will be similarly covered by FeCPs. Thus, it is questionable whether the systems really operate as described.

Rephrasing Notter [189], the success of an Fe^0 filter depends on four main factors: (i) the quantity and quality of the water to be produced (e.g., daily), (ii) the intrinsic reactivity of used Fe^0, (iii) the nature of the contaminant(s), and (iv) the availability of Fe^0 material (to renew exhausted systems). This key principle remains unchanged, one century later. All is needed are systematic investigations take each Fe^0 and each water source as a stand-alone design parameter.

The laboratory procedures used to characterize and evaluate Fe^0-based systems in batch and column experiments vary considerably among studies. For example, some studies use agitated batch experiments to investigate contaminant removal by Fe^0 [126,127], while others used batch experiments operated in quiescent mode, which is closer to field conditions [61]. This makes direct comparison of results among studies problematic, and could lead to misleading conclusions about the performance of Fe^0 materials. Therefore, there is need to develop standardized protocols for the evaluation of Fe^0 material in both batch and column experiments. In the case of batch experiments, such protocols should include specifying the particle/grain size, tests for determination of material reactivity (e.g., EDTA, Phen tests), liquid/solid ratio, sampling frequency and duration of experiment. For column experiments, such protocols should specify filter depth, duration of experiment, grain size of filter material and chemical properties of test solution to be used, while accounting for potential interference among solutes. Besides developing dedicated pristine Fe^0 materials for the water treatment industry, scope also exists to use Fe^0 material generated as wastes from other industries. Similarly, iron oxide-rich spent sludge from Fe^0-based water treatment systems can be used as raw materials in other industries such as the production of pigments (e.g., iron oxide red) [190], and even as filter material in the construction industry. This cyclic flow of iron materials between the water treatment industry and other industrial processes could form part of a circular economy. Filter wastes may also be regenerated and recycled to new Fe^0 material, and then used in other industry when they contain toxic contaminants such as As or U. The recycling of filter wastes in other industries may require detailed environmental risk assessments, including the evaluation of contaminant leaching and potential ecotoxicological effects.

5.3. Field Applications of Fe^0-Based Systems

Fe^0-based systems present unprecedented opportunities for wastewater treatment and safe drinking water provision especially in low-income countries, including those in Africa (Table 4) [144,157,191–201]. In fact, the use of Fe^0-based systems (e.g., the Bishof Process) for clean water provision has a long history dating back to the 19th century [13,14,202]. The history of Fe^0-based drinking water treatment systems is discussed in detail in earlier review papers [17,18,122]. Recently, our research group has proposed the integration of Fe^0-based systems in rainwater harvesting systems as a low-cost technology for decentralized drinking water provision, in what is known as the Kilimanjaro Concept [192–194].

Table 4. Potential field applications of Fe^0-based systems in drinking water and wastewater treatment. The given reference refers to the oldest known application.

Field of application	Remarks	References
	A: Safe drinking water provision:	
1.Centralized safe drinking water systems	Both filter beds and fluidized beds are used	[168]
2. Decentralized water treatment for small communities	Steel wool is used against radionuclides	[37,154]
3. Household filters against arsenic	Traditional filters are amended with iron nails	[45,46]
4. Household filters against pathogens	Biosand filters are amended with Fe^0 materials	[141]
5. Community-scale Fe^0-based systems against arsenic	Natural water equilibrates with iron nails and flocs are filtered on gravel	[191]
6. Decentralized rainwater harvesting systems for drinking provision	Fe^0 filters are efficient to remediate expected contaminants	[192–194]
	B: Wastewater treatment systems:	
1. Decentralized domestic wastewater treatment	A Fe^0 unit in implemented mostly to remove PO_4^{3-}	[42,195]
2. Wastewater for agriculture and aquaculture	Iron filings are used in filter beds to remove Se	[196,197]
3. Industrial wastewaters/effluents	Sponge iron is used to precipitate Cu and Pb from industrial wastes	[198]
4. *Constructed Wetlands for Wastewater Treatment*	Granular iron is added to reactive materials	[79,199]
5. Urban stormwater treatment	Fe^0 is amended to other media to optimize the treatment of runoff water	[200,201]

Fe^0-based systems also have potential applications in domestic and industrial wastewater treatment systems (Table 4). For example, the Harza Process [27,203] has been used to remove Se from agricultural drainage water. Rahman et al. [200] and Fronczyk et al. [201] proposed the amendment of treatment media for runoff infiltration trenches/pits with granular Fe^0. In addition, Fe^0-based systems have been used to treat both domestic [204] and industrial wastewaters [205,206], including acid mine drainage [24,207]. Available studies suggest that Fe^0 can be added as filter media in constructed wetlands designed to treat urban stormwater and industrial wastewaters [199]. However, a lot remains to be done to further develop and disseminate Fe^0-based technologies for wastewater treatment and decentralized safe water provision in developing countries, which such low-cost technologies are most needed. Considering that the bulk of studies are limited to laboratory scale applications, there is need to optimize the Fe^0-based systems and evaluate them under field conditions.

6. Summary and Conclusions

The corrosion of iron in remediation Fe^0/H_2O systems is an electrochemical process, coupling Fe^0 oxidative dissolution to the reduction of water (protons) and to no other available oxidizing agent, including dissolved O_2. This is because the universal oxide scale on Fe^0 acts as diffusion barrier to dissolved species and a conduction barrier to electrons from the metal body. In other words, water is the sole chemical which can remove electrons from the Fe^0 surface. Fe^0 oxidation and water reduction must not necessarily occur at the same locality. The spatial separation of oxidative (anodic) and reductive (cathodic) reactions is possible as the metal body allows the free flow of electrons from anodic to cathodic sites. The tendency of Fe^0 to give off electrons (Equation (1)) is the same for all Fe^0-based materials ($E^0 = -0.44$ V). This makes material selection and characterization critical in designing sustainable Fe^0/H_2O systems.

The need to characterize Fe^0 materials in terms of intrinsic reactivity and efficiency is critical for the design and operation Fe^0/H_2O systems, a subject that has been addressed by our research group. Specifically, EDTA and Phen tests were used to characterize the intrinsic reactivity of Fe^0 materials. In this regard, the Phen test is considered an affordable and appropriate method that provides a reliable guidance in selecting FeO from the large catalog of available Fe^0 materials and to control the quality of newly manufactured ones. The efficiency of the Fe^0/H_2O system is characterized using the methylene blue (MB) discoloration method, while other probing agents investigated include orange II, methyl orange and reactive red 120.

The most characteristic issue of remediation Fe^0 is the small size (<5 mm) of used materials. Assuming uniform corrosion, the corrosion rates for progressive Fe^0 oxidation should be normalized to the individual particles. In other words, expression like mmol year $^{-1}$ should be expressed as mmol year^{-1} particle^{-1} or mmol year^{-1} grain^{-1}. The next important issue will be to consider the non-linear kinetics of the corrosion rate such that the service life of a designed system can be deduced knowing the size of used particles and the long-term corrosion rate. Once this is known, considering the expansive nature of iron corrosion would help to design sustainable systems. The choice of the admixing aggregates (e.g., gravel, MnO_2, pumice, sand) and the mixing ratios are to be investigated on a case-by-case basis.

A better understanding of the long-term corrosion of relevant Fe^0 materials under site-specific conditions is envisioned to ultimately aid in the design of affordable, applicable and efficient remediation Fe^0/H_2O systems. Applications of Fe^0-based systems include; (i) a large variety of water treatment systems, (ii) household and small-scale water treatment plants, including rainwater harvesting systems for drinking water supply, (iii) decentralized domestic wastewater treatment, (iv) urban stormwater, agricultural and industrial wastewater treatment, and (v) as filter media in constructed wetlands. Addressing the key knowledge gaps highlighted here, and extending Fe^0-based systems to other application domains such as wastewater treatment for agriculture are focal research areas in our group, which brings together collaborators from various countries.

Author Contributions: R.H., H.Y., R.T., X.C., M.X., B.K.A., V.C., M.L., N.P.S.-S., A.I.N.-T., N.G.-B., V.R.S.-T., W.G. and C.N. contributed equally to manuscript compilation and revisions. All authors have read and agreed to the published version of the manuscript.

Funding: The work is supported by the Ministry of Education of the People's Republic of China through the Program "Research on Mechanism of Groundwater Exploitation and Seawater Intrusion in Coastal Areas" (Project Code 20165037412).

Acknowledgments: We acknowledge support by the German Research Foundation and the Open Access Publication Funds of the Göttingen University.

Conflicts of Interest: The authors declare no conflict of interest.

References

1. Landolt, D. *Corrosion and Surface Chemistry of Metals*, 1st ed.; EPFL Press: Lausanne, Switzerland, 2007; p. 615.
2. Enning, D.; Garrelfs, J. Corrosion of iron by sulfate-reducing bacteria: New views of an old problem. *Appl. Environ. Microbiol.* **2014**, *80*, 1226–1236. [CrossRef] [PubMed]
3. Wasim, M.; Shoaib, S.; Mubarak, M.N.; Inamuddin Asiri, A.M. Factors influencing corrosion of metal pipes in soils. *Environ. Chem. Lett.* **2018**, *16*, 861–879. [CrossRef]
4. Enikeev, M.R.; Potemkin, D.I.; Enikeeva, L.V.; Enikeev, A.R.; Gubaydullin, I.M. Quantification of non-uniform distribution and growth of corrosion products at steel-concrete interface Analysis of corrosion processes kinetics on the surface of metals. *Chem. Eng.* **2020**, *383*, 123–131.
5. Nesic, S. Key issues related to modelling of internal corrosion of oil and gas pipelines—A review. *Corros. Sci.* **2007**, *49*, 4308–4338. [CrossRef]
6. Lazzari, L. *General Aspects of Corrosion, Chapter 9.1*; Encyclopedia of Hydrocarbons, Istituto Enciclopedia Italian: Rome, Italy, 2008; Volume V.
7. Wilson, E.K. Zero-Valent metals provide possible solution to groundwater problems. *Chem. Eng. News* **1995**, *73*, 19–22. [CrossRef]
8. Fairweather, V. When toxics meet metal. *Civ. Eng.* **1996**, *66*, 44–48.
9. Tratneyk, P.G. Putting corrosion to use: Remediating contaminated groundwater with zero-Valent metals. *Chem. Ind.* **1996**, *13*, 499–503.
10. Hildebrant, B. Characterizing the reactivity of commercial steel wool for water treatment. *Freib. Online Geosci.* **2018**, *53*, 1–80. [CrossRef]
11. Lufingo, M.; Ndé-Tchoupé, A.I.; Hu, R.; Njau, K.N.; Noubactep, C. A novel and facile method to characterize the suitability of metallic iron for water treatment. *Water* **2019**, *11*, 2465. [CrossRef]
12. Bischof, G. *The Purification of Water: Embracing the Action of Spongy Iron on Impure Water*; Bell and Bain: Glasgow, Scotland, 1873; p. 19.
13. Devonshire, E. The purification of water by means of metallic iron. *J. Frankl. Inst.* **1890**, *129*, 449–461. [CrossRef]
14. Tucker, W.G. The purification of water by chemical treatment. *Science* **1892**, *20*, 34–38. [CrossRef] [PubMed]
15. Baker, M.N. Sketch of the history of water treatment. *J. Am. Water Works Assoc.* **1934**, *26*, 902–938. [CrossRef]
16. van Craenenbroeck, W. Easton & Anderson and the water supply of Antwerp (Belgium). *Ind. Archaeol. Rev.* **1998**, *20*, 105–116.
17. Mwakabona, H.T.; Ndé-Tchoupé, A.I.; Njau, K.N.; Noubactep, C.; Wydra, K.D. Metallic iron for safe drinking water provision: Considering a lost knowledge. *Water Res.* **2017**, *117*, 127–142. [CrossRef]
18. Noubactep, C. Metallic iron for water treatment: Lost science in the West. *Bioenergetics* **2017**, *6*, 149. [CrossRef]
19. Gould, J.P. The kinetics of hexavalent chromium reduction by metallic iron. *Water Res.* **1982**, *16*, 871–877. [CrossRef]
20. Noubactep, C. *Untersuchungen zur passiven in-situ-Immobilisierung von U(VI) aus Wasser*; Akad. Buchh.; TU Bergakademie Freiberg: Freiberg, Germany, 2003; (ISSN: 1433-1284).
21. Chaves, L.H.G. The role of green rust in the environment: A review. *Rev. Bras. Eng. Agríc. Ambient.* **2005**, *9*, 284–288. [CrossRef]
22. Gheju, M. Hexavalent chromium reduction with zero-valent iron (ZVI) in aquatic systems. *Water Air Soil Pollut.* **2011**, *222*, 103–148. [CrossRef]
23. Ghauch, A. Iron-based metallic systems: An excellent choice for sustainable water treatment. *Freib. Online Geosci.* **2015**, *38*, 1–80.
24. Hu, R.; Cui, X.; Gwenzi, W.; Wu, S.; Noubactep, C. Fe0/H2O systems for environmental remediation: The scientific history and future research directions. *Water* **2018**, *10*, 1739. [CrossRef]
25. Hu, R.; Noubactep, C. Redirecting research on Fe0 for environmental remediation: The search for synergy. *Int. J. Environ. Res. Public Health* **2019**, *16*, 4465. [CrossRef] [PubMed]

26. Hu, R.; Gwenzi, G.; Sipowo, R.; Noubactep, C. Water treatment using metallic iron: A tutorial review. *Processes* **2019**, *7*, 622. [CrossRef]
27. Anderson, M.A. *Fundamental Aspects of Selenium Removal by Harza Process*; Rep San Joaquin Valley Drainage Program; US Department of the Interior: Sacramento, CA, USA, 1989.
28. Myneni, S.C.B.; Tokunaga, T.K.; Brown, G.E., Jr. Abiotic selenium redox transformations in the presence of Fe(II,III) oxides. *Science* **1997**, *278*, 1106–1109. [CrossRef]
29. Jiao, Y.; Qiu, C.; Huang, L.; Wu, K.; Ma, H.; Chen, S.; Ma, L.; Wu, L. Reductive dechlorination of carbon tetrachloride by zero-valent iron and related iron corrosion. *Appl. Catal. B Environ.* **2009**, *91*, 434–440. [CrossRef]
30. Swete, H. The purification of river water by agitation and metallic iron, as conducted with the water of the river severn at Worcester. *J. R. Soc. Promot. Health* **1892**, *13*, 309–315. [CrossRef]
31. Guinochet, E. *Les Eaux D'alimentation: Épuration, Filtration, Sterilization*; J.B. Baillière et Fils: Paris, France, 1894; p. 380.
32. Bischof, G. On putrescent organic matter in potable water. *Proc. R. Soc. Lond.* **1877**, *26*, 258–261.
33. Bischof, G. On putrescent organic matter in potable water II. *Proc. R. Soc. Lond.* **1878**, *27*, 152–156.
34. Burton, C.E. *The Water Supply of Towns and the Construction of Waterworks: A Practical Treatise for the Use of Engineers and Students of Engineering*; Crosby Lockwood: London, UK, 1898; p. 486. [CrossRef]
35. Anderson, W. The Antwerp Waterworks. *Minutes Proc. Inst. Civ. Eng.* **1883**, *72*, 24–44.
36. Anderson, W. The revolving purifier: An apparatus which effects purification of river water and sewage effluent. In *the 1885 Anderson Patent*; Revolving Purifier Co: London, UK, 1889.
37. Lauderdale, R.A.; Emmons, A.H. A method for decontaminating small volumes of radioactive water. *J. Am. Water Works Assoc.* **1951**, *43*, 327–331. [CrossRef]
38. Harza Engineering Co. *Selenium Removal Study*; Report to Panoche Drainage District; Harza Engineering Co.: Firebaugh, CA, USA, 1986.
39. Khudenko, B.M. Feasibility evaluation of a novel method for destruction of organics. *Water Sci. Technol.* **1991**, *23*, 1873–1881. [CrossRef]
40. James, B.R.; Rabenhorst, M.C.; Frigon, G.A. Phosphorus sorption by peat and sand amended with iron oxides or steel wool. *Water Environ. Res.* **1992**, *64*, 699–705. [CrossRef]
41. Erickson, A.J.; Gulliver, J.S.; Weiss, P.T. Enhanced sand filtration for storm water phosphorus removal. *J. Environ. Eng. ASCE* **2007**, *133*, 485–497. [CrossRef]
42. Wakatsuki, T.; Esumi, H.; Omura, S. High performance and N, P removable on-site domestic wastewater treatment system by multi-soil-layering method. *Water Sci. Technol.* **1993**, *27*, 31–40. [CrossRef]
43. Bojic, A.; Purenovic, M.; Kocic, B.; Perovic, J.; Ursic-Jankovic, J.; Bojic, D. *The Inactivation of Escherichia Coli by Microalloyed Aluminium Based Composite*; Facta Universitatis; University of Niš: Niš, Serbia, 2001; Volume 2, pp. 115–124.
44. You, Y.; Han, J.; Chiu, P.C.; Jin, Y. Removal and inactivation of waterborne viruses using zerovalent iron. *Environ. Sci. Technol.* **2005**, *39*, 9263–9269. [CrossRef]
45. Ngai, T.K.K.; Murcott, S.; Shrestha, R.R.; Dangol, B.; Maharjan, M. Development and dissemination of Kanchan™ Arsenic Filter in rural Nepal. *Water Sci. Technol. Water Supply* **2006**, *6*, 137–146. [CrossRef]
46. Ngai, T.K.K.; Shrestha, R.R.; Dangol, B.; Maharjan, M.; Murcott, S.E. Design for sustainable development—Household drinking water filter for arsenic and pathogen treatment in Nepal. *J. Environ. Sci. Health A* **2007**, *42*, 1879–1888. [CrossRef]
47. Hussam, A.; Munir, A.K.M. A Simple and Effective Arsenic Filter Based on Composite Iron Matrix: Development and Deployment Studies for Groundwater of Bangladesh. *J. Environ. Sci. Health* **2007**, *42*, 1869–1878. [CrossRef]
48. Fu, F.; Dionysiou, D.D.; Liu, H. The use of zero-valent iron for groundwater remediation and wastewater treatment: A review. *J. Hazard. Mater.* **2014**, *267*, 194–205. [CrossRef]
49. Colombo, A.; Dragonetti, C.; Magni, M.; Roberto, D. Degradation of toxic halogenated organic compounds by iron-containing mono-, bi-and tri-metallic particles in water. *Inorg. Chim. Acta* **2015**, *431*, 48–60. [CrossRef]
50. Noubactep, C.; Schöner, A.; Woafo, P. Metallic iron filters for universal access to safe drinking water. *Clean Soil Air Water* **2009**, *37*, 930–937. [CrossRef]

51. Ingram, D.T.; Callahan, M.T.; Ferguson, S.; Hoover, D.G.; Shelton, D.R.; Millner, P.D.; Camp, M.J.; Patel, J.R.; Kniel, K.E.; Sharma, M. Use of zero-valent iron biosand filters to reduce E. coli O157:H12 in irrigation water applied to spinach plants in a field setting. *J. Appl. Microbiol.* **2012**, *112*, 551–560. [CrossRef] [PubMed]
52. Marik, C.M.; Anderson-Coughlin, B.; Gartley, S.; Craighead, S.; Kniel, K.E. The efficacy of zero valent iron-sand filtration on the reduction of Escherichia coli and Listeria monocytogenes in surface water for use in irrigation. *Environ. Res.* **2019**, *173*, 33–39. [CrossRef] [PubMed]
53. Noubactep, C. Processes of contaminant removal in "Fe0–H2O" systems revisited. The importance of co-precipitation. *Open Environ. Sci.* **2007**, *1*, 9–13. [CrossRef]
54. Noubactep, C. A critical review on the mechanism of contaminant removal in Fe0–H2O systems. *Environ. Technol.* **2008**, *29*, 909–920. [CrossRef]
55. Neumann, A.; Kaegi, R.; Voegelin, A.; Hussam, A.; Munir, A.K.M.; Hug, S.J. Arsenic removal with composite iron matrix filters in Bangladesh: A field and laboratory study. *Environ. Sci. Technol.* **2013**, *47*, 4544–4554. [CrossRef]
56. Gheju, M.; Balcu, I. Sustaining the efficiency of the Fe(0)/H_2O system for Cr(VI) removal by MnO_2 amendment. *Chemosphere* **2019**, *214*, 389–398. [CrossRef]
57. Heimann, S. Testing granular iron for fluoride removal. *Freib. Online Geosci.* **2018**, *52*, 1–80.
58. Heimann, S.; Ndé-Tchoupé, A.I.; Hu, R.; Licha, T.; Noubactep, C. Investigating the suitability of Fe0 packed-beds for water defluoridation. *Chemosphere* **2018**, *209*, 578–587. [CrossRef]
59. Rodrigues, R.; Betelu, S.; Colombano, S.; Masselot, G.; Tzedakis, T.; Ignatiadis, I. Elucidating the dechlorination mechanism of hexachloroethane by Pd-doped zerovalent iron microparticles in dissolved lactic acid polymers using chromatography and indirect monitoring of iron corrosion. *Environ. Sci. Pollut. Res.* **2019**, *26*, 7177–7194. [CrossRef]
60. Miyajima, K. Optimizing the design of metallic iron filters for water treatment. *Freib. Online Geosci.* **2012**, *32*, 1–60.
61. Gatcha-Bandjun, N.; Noubactep, C.; Loura, B.B. Mitigation of contamination in effluents by metallic iron: The role of iron corrosion products. *Environ. Technol. Innov.* **2017**, *8*, 71–83. [CrossRef]
62. Phukan, M. Characterizing the Fe0/sand system by the extent of dye discoloration. *Freib. Online Geosci.* **2015**, *40*, 1–70.
63. Sleiman, N.; Deluchat, V.; Wazne, M.; Mallet, M.; Courtin-Nomade, A.; Kazpard, V.; Baudu, M. Phosphate removal from aqueous solution using ZVI/sand bedreactor: Behavior and mechanism. *Water Res.* **2016**, *99*, 56–65. [CrossRef] [PubMed]
64. Liang, L.; Yang, W.; Guan, X.; Li, J.; Xu, Z.; Wu, J.; Huang, Y.; Zhang, X. Kinetics and mechanisms of pH-dependent selenite removal by zero valent iron. *Water Res.* **2013**, *47*, 5846–5855. [CrossRef] [PubMed]
65. Kishimoto, N.; Iwano, S.; Narazaki, Y. Mechanistic consideration of zinc ion removal by zero-valent iron. *Water Air Soil Pollut.* **2011**, *221*, 183–189. [CrossRef]
66. Yu, J.; Hou, X.; Hu, X.; Yuan, H.; Wang, J.; Chen, C. Efficient degradation of chloramphenicol by zero-valent iron microspheres and new insights in mechanisms. *Appl. Catal. B Environ.* **2019**, *256*, 117876. [CrossRef]
67. Noubactep, C. The fundamental mechanism of aqueous contaminant removal by metallic iron. *Water SA* **2010**, *36*, 663–670. [CrossRef]
68. Noubactep, C. Flaws in the design of Fe(0)-based filtration systems? *Chemosphere* **2014**, *117*, 104–107. [CrossRef]
69. Noubactep, C. Research on metallic iron for environmental remediation: Stopping growing sloppy science. *Chemosphere* **2016**, *153*, 528–530. [CrossRef]
70. Noubactep, C.; Makota, S.; Bandyopadhyay, A. Rescuing Fe0 remediation research from its systemic flaws. *Res. Rev. Insights* **2017**. [CrossRef]
71. Noubactep, C. The operating mode of Fe0/H2O systems: Hidden truth or repeated nonsense? *Fresenius Environ. Bull.* **2019**, *28*, 8328–8330.
72. Oleszkiewicz, J.A.; Sharma, V.K. Stimulation and inhibition of anaerobic processes by heavy metals—A review. *Biol. Wastes* **1990**, *31*, 45–67. [CrossRef]
73. Shen, C.F.; Kosaric, N.; Blaszczyk, R. The effect of selected heavy metals (Ni, Co and Fe) on anaerobic granules and their extracellular polymeric substance (EPS). *Water Res.* **1993**, *27*, 25–33. [CrossRef]
74. Zhang, Y.; An, X.; Quan, X. Enhancement of sludge granulation in a zero valence iron packed anaerobic reactor with a hydraulic circulation. *Process Biochem.* **2011**, *46*, 471–476. [CrossRef]

75. Liu, Y.; Wang, Q.; Zhang, Y. Zero valent iron significantly enhances methane production from waste activated sludge by improving biochemical methane potential rather than hydrolysis rate. *Sci. Rep.* **2015**, *5*, 8263. [CrossRef]
76. Hammonds, P. An Introduction to Corrosion and its Prevention (Chapter 4). *Compr. Chem. Kinet.* **1989**, *28*, 233–279.
77. Sikora, E.; Macdonald, D.D. The passivity of iron in the presence of ethylenediaminetetraacetic acid I. General electrochemical behavior. *Electrochem. Soc.* **2000**, *147*, 4087–4092. [CrossRef]
78. Antia, D.D.J. Desalination of water using ZVI, Fe0. *Water* **2015**, *7*, 3671–3831. [CrossRef]
79. van Genuchten, C.M.; Behrends, T.; Stipp, S.L.S.; Dideriksen, K. Achieving arsenic concentrations of <1 μg/L by Fe(0) electrolysis: The exceptional performance of magnetite. *Water Res.* **2020**, *168*, 115–170.
80. Tewari, G.; Jayas, D.S.; Jeremiah, L.E.; Holley, R.A. Absorption kinetics of oxygen scavengers. *Int. Food Sci. Technol.* **2002**, *37*, 209–217. [CrossRef]
81. Polyakov, V.A.; Miltz, J. Modeling of the humidity effects on the oxygen absorption by iron-based scavengers. *Food Sci.* **2010**, *75*, 91–99. [CrossRef] [PubMed]
82. Xu, B.; Jia, M.; Men, J. Preparation, characterization and deoxygenation performance of modified sponge iron. *Arab. J. Sci. Eng.* **2014**, *39*, 31–36. [CrossRef]
83. Furukawa, Y.; Kim, J.-W.; Watkins, J.; Wilkin, R.T. Formation of ferrihydrite and associated iron corrosion products in permeable reactive barriers of zero-valent iron. *Environ. Sci. Technol.* **2002**, *36*, 5469–5475. [CrossRef] [PubMed]
84. Wranglen, G. Review article on the influence of sulfide inclusions on the corradability of iron and steel. *Corros. Sci.* **1969**, *9*, 585–602. [CrossRef]
85. Stratmann, M.; Müller, J. The mechanism of the oxygen reduction on rust-covered metal substrates. *Corros. Sci.* **1994**, *36*, 327–359. [CrossRef]
86. Phillips, D.H.; Van Nooten, T.; Bastiaens, L.; Russell, M.I.; Dickson, K.; Plant, S.; Ahad, J.M.E.; Newton, T.; Elliot, T.; Kalin, R.M. Ten year performance evaluation of a field-scale zero-valent iron permeable reactive barrier installed to remediate trichloroethene contaminated groundwater. *Environ. Sci. Technol.* **2010**, *44*, 3861–3869. [CrossRef]
87. Wilkin, R.T.; Acree, S.D.; Ross, R.R.; Puls, R.W.; Lee, T.R.; Woods, L.L. Fifteen-year assessment of a permeable reactive barrier for treatment of chromate and trichloroethylene in groundwater. *Sci. Total Environ.* **2014**, *468–469*, 186–194. [CrossRef]
88. Wilkin, R.T.; Lee, T.R.; Sexton, M.R.; Acree, S.D.; Puls, R.W.; Blowes, D.W.; Kalinowski, C.; Tilton, J.M.; Woods, L.L. Geochemical and isotope study of trichloroethene degradation in a zero-valent iron permeable reactive barrier: A twenty-two-year performance evaluation. *Environ. Sci. Technol.* **2019**, *53*, 296–306. [CrossRef]
89. Luo, P.; Bailey, E.H.; Mooney, S.J. Quantification of changes in zero valent iron morphology using X-ray computed tomography. *J. Environ. Sci.* **2013**, *25*, 2344–2351. [CrossRef]
90. Henderson, A.D.; Demond, A.H. Impact of solids formation and gas production on the permeability of ZVI PRBs. *J. Environ. Eng.* **2011**, *137*, 689–696. [CrossRef]
91. Henderson, A.D.; Demond, A.H. Long-term performance of zero-valent iron permeable reactive barriers: A critical review. *Environ. Eng. Sci.* **2007**, *24*, 401–423. [CrossRef]
92. Guan, X.; Sun, Y.; Qin, H.; Li, J.; Lo, I.M.C.; He, D.; Dong, H. The limitations of applying zero-valent iron technology in contaminants sequestration and the corresponding countermeasures: The development in zero-valent iron technology in the last two decades (1994–2014). *Water Res.* **2015**, *75*, 224–248. [CrossRef] [PubMed]
93. Schreier, C.G.; Reinhard, M. Transformation of chlorinated organic compounds by iron and manganese powders in buffered water and in landfill leachate. *Chemosphere* **1994**, *29*, 1743–1753. [CrossRef]
94. Hao, Z.-W.; Xu, X.-H.; Wand, D.-H. Reductive denitrification of nitrate by scrap iron filings. *J. Zhejiang Univ. Sci.* **2005**, *6*, 182–186. [CrossRef] [PubMed]
95. Lee, G.; Rho, S.; Jahng, D. Design considerations for groundwater remediation using reduced metals. *Chem. Eng.* **2004**, *21*, 621–628. [CrossRef]
96. Yoshino, H.; Kawase, Y. Kinetic modeling and simulation of zero-valent iron wastewater treatment process: Simultaneous reduction of nitrate, hydrogen peroxide, and phosphate in semiconductor acidic wastewater. *Ind. Eng. Chem. Res.* **2013**, *52*, 17829–17840. [CrossRef]

97. Du, M.; Zhang, Y.; Hussain, I.; Du, X.; Huang, S.; Wen, W. Effect of pyrite on enhancement of zero-valent iron corrosion for arsenic removal in water: A mechanistic study. *Chemosphere* **2019**, *233*, 744–753. [CrossRef]
98. Lü, Y.; Li, J.; Li, Y.; Liang, L.; Dong, H.; Chen, K.; Yao, C.; Li, Z.; Li, J.; Guan, X. The roles of pyrite for enhancing reductive removal of nitrobenzene by zero-valent iron. *Appl. Catal.* **2019**, *242*, 9–18. [CrossRef]
99. Chen, K.; Han, L.; Li, J.; Lü, Y.; Yao, C.; Dong, H.; Wang, L.; Li, Y. Pyrite enhanced the reactivity of zero valent iron for reductive removal of dyes. *Chem. Technol. Biotechnol.* **2020**. [CrossRef]
100. Sato, N. Surface oxides affecting metallic corrosion. *Corros. Rev.* **2001**, *19*, 253–272. [CrossRef]
101. Phukan, M.; Noubactep, C.; Licha, T. Characterizing the ion-selective nature of Fe0-based filters using three azo dyes in batch systems. *J. Environ. Chem. Eng.* **2016**, *4*, 65–72. [CrossRef]
102. Whitman, W.G. Corrosion of iron. *Chem. Rev.* **1926**, *2*, 419–435. [CrossRef]
103. White, A.F.; Peterson, M.L. Reduction of aqueous transition metal species on the surfaces of Fe(II)-containing oxides. *Geochim. Cosmochim. Acta.* **1996**, *60*, 3799–3814. [CrossRef]
104. Pilling, N.B.; Bedworth, R.E. The oxidation of metals at high temperatures. *J. Inst. Metals* **1923**, *29*, 529–591.
105. Noubactep, C. Metallic iron for environmental remediation: Prospects and limitations. In *A Handbook of Environmental Toxicology: Human Disorders and Ecotoxicology*; D'Mello, J.P.F., Ed.; CAB International: Wallingford, UK, 2019; Chapter 36; pp. 531–544.
106. Caré, S.; Crane, R.; Calabrò, P.S.; Ghauch, A.; Temgoua, E.; Noubactep, C. Modeling the permeability loss of metallic iron water filtration systems. *CLEAN—Soil Air Water* **2013**, *41*, 275–282. [CrossRef]
107. Domga, R.; Togue-Kamga, F.; Noubactep, C.; Tchatchueng, J.B. Discussing porosity loss of Fe0 packed water filters at ground level. *Chem. Eng. J.* **2015**, *263*, 127–134. [CrossRef]
108. Caré, S.; Nguyen, Q.T.; L'Hostis, V.; Berthaud, Y. Mechanical properties of the rust layer induced by impressed current method in reinforced mortar. *Cem. Concr. Res.* **2008**, *38*, 1079–1091. [CrossRef]
109. Zhao, Y.; Ren, H.; Dai, H.; Jin, W. Composition and expansion coefficient of rust based on X-ray diffraction and thermal analysis. *Corros. Sci.* **2011**, *53*, 1646–1658. [CrossRef]
110. Pusch, R.; Kasbohm, J.; Knutsson, S.; Yang, T.; Nguyen-Thanh, L. The role of smectite clay barriers for isolating high-level radioactive waste (HLW) in shallow and deep repositories. *Procedia Earth Planet. Sci.* **2015**, *15*, 680–687. [CrossRef]
111. Snoeyink, V.L.; Jenkins, D. *Water Chemistry*; John Wiley & Sons: Hoboken, NJ, USA, 1980; p. 480.
112. Sarin, P.; Snoeyink, V.L.; Bebee, J.; Jim, K.K.; Beckett, M.A.; Kriven, W.M.; Clement, J.A. Iron release from corroded iron pipes in drinking water distribution systems: Effect of dissolved oxygen. *Water Res.* **2004**, *38*, 1259–1269. [CrossRef]
113. Grauer, R. *Ueber Moegliche Wechselwirkungen Zwischen Verfuellmaterial und Stahlbehälter in Endlager C. Interner Bericht 86-01, NAGRA*; Baden, Switzerland, 1986; Available online: https://www.nagra.ch/de (accessed on 27 February 2020).
114. Moraci, N.; Lelo, D.; Bilardi, S.; Calabrò, P.S. Modelling long-term hydraulic conductivity behaviour of zero valent iron column tests for permeable reactive barrier design. *Can. Geotech. J.* **2016**, *53*, 946–961. [CrossRef]
115. Noubactep, C. Predicting the hydraulic conductivity of metallic iron filters: Modeling gone astray. *Water* **2016**, *8*, 162. [CrossRef]
116. Dickerson, R.E.; Gray, H.B.; Haight, G.P., Jr. *Chemical Principles*, 3rd ed.; Benjamin/Cummings Inc.: London, UK, 1979; p. 944.
117. Shi, C.; Wei, J.; Jin, Y.; Kniel, K.E.; Chiu, P.C. Removal of viruses and bacteriophages from drinking water using zero-valent iron. *Sep. Purif. Technol.* **2012**, *84*, 72–78. [CrossRef]
118. Noubactep, C. The suitability of metallic iron for environmental remediation. *Environ. Prog. Sustain. Energy* **2010**, *29*, 286–291. [CrossRef]
119. Roh, Y.; Lee, S.Y.; Elless, M.P. Characterization of corrosion products in the permeable reactive barriers. *Environ. Geol.* **2000**, *40*, 184–194. [CrossRef]
120. Hu, R.; Cui, X.; Xiao, M.; Qiu, P.; Lufingo, M.; Gwenzi, W.; Noubactep, C. Characterizing the suitability of granular Fe0 for the water treatment industry. *Processes* **2019**, *7*, 652. [CrossRef]
121. Hu, R.; Ndé-Tchoupé, A.I.; Lufingo, M.; Xiao, M.; Nassi, A.; Noubactep, C.; Njau, K.N. The impact of selected pre-treatment procedures on iron dissolution from metallic iron specimens used in water treatment. *Sustainability* **2019**, *11*, 671. [CrossRef]
122. Nanseu-Njiki, C.P.; Gwenzi, W.; Pengou, M.; Rahman, M.A.; Noubactep, C. Fe0/H2O filtration systems for decentralized safe drinking water: Where to from here? *Water* **2019**, *11*, 429. [CrossRef]

123. Noubactep, C.; Meinrath, G.; Merkel, J.B. Investigating the mechanism of uranium removal by zerovalent iron materials. *Environ. Chem.* **2005**, *2*, 235–242. [CrossRef]
124. Noubactep, C.; Schöner, A.; Meinrath, G. Mechanism of uranium (VI) fixation by elemental iron. *J. Hazard. Mater.* **2006**, *132*, 202–212. [CrossRef]
125. Noubactep, C.; Kurth, A.M.; Sauter, M. Evaluation of the effects of shaking intensity on the process of methylene blue discoloration by metallic iron. *J. Hazard. Mater.* **2009**, *169*, 1005–1011. [CrossRef] [PubMed]
126. Noubactep, C. Investigating the processes of contaminant removal in Fe0/H2O systems. *Korean J. Chem. Eng.* **2012**, *29*, 1050–1056. [CrossRef]
127. Noubactep, C.; Schöner, A.; Sauter, M. Significance of oxide-film in discussing the mechanism of contaminant removal by elemental iron materials. In *Photo-Electrochemistry & Photo-Biology for the Sustainability*; Union Press: Osaka, Japan, 2012; pp. 97–122.
128. Gillham, R.W.; O'Hannesin, S.F. Enhanced degradation of halogenated aliphatics by zero-valent iron. *Groundwater* **1994**, *32*, 958–967. [CrossRef]
129. Richardson, J.P.; Nicklow, J.W. In situ permeable reactive barriers for groundwater. *Soil Sedim. Contam.* **2002**, *11*, 241–268. [CrossRef]
130. Wang, T.-H.; Li, M.-H.; Teng, S.-P. Bridging the gap between batch and column experiments: A case study of Cs adsorption on granite. *Hazard. Mater.* **2009**, *161*, 409–415. [CrossRef] [PubMed]
131. Tien, C. Remarks on adsorption manuscripts revised and declined: An editorial. *Purif. Technol.* **2007**, *54*, 277–278. [CrossRef]
132. Tran, H.N.; You, S.-J.; Hosseini-Bandegharaei, A.; Chao, H.-P. Mistakes and inconsistencies regarding adsorption of contaminants from aqueous solutions: A critical review. *Water Res.* **2017**, *120*, 88–116. [CrossRef]
133. Miehr, R.; Tratnyek, P.G.; Bandstra, Z.J.; Scherer, M.M.; Alowitz, J.M.; Bylaska, J.E. Diversity of contaminant reduction reactions by zerovalent iron: Role of the reductate. *Environ. Sci. Technol.* **2004**, *38*, 139–147. [CrossRef]
134. Matheson, L.J.; Tratnyek, P.G. Reductive dehalogenation of chlorinated methanes by iron metal. Environ. *Sci. Technol.* **1994**, *28*, 2045–2053. [CrossRef]
135. Weber, E.J. Iron-mediated reductive transformations: Investigation of reaction mechanism. *Environ. Sci. Technol.* **1996**, *30*, 716–719. [CrossRef]
136. O´Hannesin, S.F.; Gillham, R.W. Long-term performance of an in situ "iron wall" for remediation of VOCs. *Groundwater* **1998**, *36*, 164–170. [CrossRef]
137. Kamolpornwijit, W.; Liang, L. Investigation of gas production and entrapment in granular iron medium. *Contam. Hydrol.* **2006**, *82*, 338–356. [CrossRef] [PubMed]
138. Lavine, B.K.; Auslander, G.; Ritter, J. Polarographic studies of zero valent iron as a reductant for remediation of nitroaromatics in the environment. *Microchem. J.* **2001**, *70*, 69–83. [CrossRef]
139. Howe, K.J.; Hand, D.W.; Crittenden, J.C.; Trussell, R.R.; Tchobanoglous, G. *Principles of Water Treatment*; John Wiley & Sons, Inc.: Hoboken, NJ, USA, 2012; p. 647.
140. Hildebrant, B.; Ndé-Tchoupé, A.I.; Lufingo, M.; Licha, T.; Noubactep, C. Steel wool for water treatment: Intrinsic reactivity and defluoridation efficiency. *Processes* **2020**, *8*, 265. [CrossRef]
141. Bradley, I.; Straub, A.; Maraccini, P.; Markazi, S.; Nguyen, T.H. Iron oxide amended biosand filters for virus removal. *Water Res.* **2011**, *45*, 4501–4510. [CrossRef] [PubMed]
142. Lu, J.; Lian, T.; Su, J. Effect of zero-valent iron on biological denitrification in the autotrophic denitrification system. *Res. Chem. Intermed.* **2018**, *44*, 6011–6022. [CrossRef]
143. Bradshaw, R.N. *Assessment of Zero-Valent Iron Capabilities to Reduce Food-Borne Pathogens Via Filtration and Residual Activities in Irrigation Water;* Master of Public Health, University of Maryland: College Park, MD, USA, 2017.
144. Chopyk, J.; Kulkarni, P.; Nasko, D.J.; Bradshaw, R.; Kniel, K.E.; Chiu, P.; Sharma, M.; Sapkota, A.R. Zero-valent iron sand filtration reduces concentrations of virus-like particles and modifies virome community composition in reclaimed water used for agricultural irrigation. *BMC Res. Notes* **2019**, *12*, 223. [CrossRef]
145. Kulkarni, P.; Raspanti, G.A.; Bui, A.Q.; Bradshaw, R.N.; Sapkota, A.R. Zerovalent iron-sand filtration can reduce the concentration of multiple antimicrobials in conventionally treated reclaimed water. *Environ. Res.* **2019**, *172*, 301–309. [CrossRef]

146. Tepong-Tsindé, R.; Ndé-Tchoupé, A.I.; Noubactep, C.; Nassi, A.; Ruppert, H. Characterizing a newly designed steel-wool-based household filter for safe drinking water provision: Hydraulic conductivity and efficiency for pathogen removal. *Processes* **2019**, *7*, 966. [CrossRef]
147. Liu, H.; Wang, Q.; Wang, C.; Li, X. Electron efficiency of zero-valent iron for groundwater remediation and wastewater treatment. *Chem. Eng. J.* **2013**, *215–216*, 90–95. [CrossRef]
148. Wu, C.-Q.; Fu, D.-T.; Chen, X. Nitrobenzene degradation by micro-sized iron and electron efficiency evaluation. *Chem. Pap.* **2014**, *68*, 1350–1357. [CrossRef]
149. Noubactep, C. Relevant reducing agents in remediation Fe0/H2O systems. *Clean—Soil Air Water* **2013**, *41*, 493–502. [CrossRef]
150. Noubactep, C.; Temgoua, E.; Rahman, M.A. Designing iron-amended biosand filters for decentralized safe drinking water provision. *CLEAN—Soil Air Water* **2012**, *40*, 798–807. [CrossRef]
151. Makota, S.; Nde-Tchoupe, A.I.; Mwakabona, H.T.; Tepong-Tsindé, R.; Noubactep, C.; Nassi, A.; Njau, K.N. Metallic iron for water treatment: Leaving the valley of confusion. *Appl. Water Sci.* **2017**, *7*, 4177–4196. [CrossRef]
152. Naseri, E.; Ndé-Tchoupé, A.I.; Mwakabona, H.T.; Nanseu-Njiki, C.P.; Noubactep, C.; Njau, K.N.; Wydra, K.D. Making Fe0-based filters a universal solution for safe drinking water provision. *Sustainability* **2017**, *9*, 1224. [CrossRef]
153. Ebelle, T.C.; Makota, S.; Tepong-Tsindé, R.; Nassi, A.; Noubactep, C. Metallic iron and the dialogue of the deaf. *Fresenius Environ. Bull.* **2019**, *28*, 8331–8340.
154. Lacy, W.J. Removal of radioactive material from water byslurrying with powdered metal. *J. Am. Water Works Assoc.* **1952**, *44*, 824–828. [CrossRef]
155. Banerji, T.; Chaudhari, S. A cost-effective technology for arsenic removal: Case study of zerovalent iron-based iit bombay arsenic filter in west bengal. In *Water and Sanitation in the New Millennium*; Nath, K., Sharma, V., Eds.; Springer: New Delhi, India, 2017.
156. Miyajima, K.; Noubactep, C. Characterizing the impact of sand addition on the efficiency of granular iron for water treatment. *Chem. Eng.* **2015**, *262*, 891–896. [CrossRef]
157. Noubactep, C. On the validity of specific rate constants (kSA) in Fe0/H2O systems. *J. Hazard. Mater.* **2009**, *164*, 835–837. [CrossRef]
158. Noubactep, C. An analysis of the evolution of reactive species in Fe0/H2O systems. *J. Hazard. Mater.* **2009**, *168*, 1626–1631. [CrossRef]
159. Btatkeu-K, B.D.; Miyajima, K.; Noubactep, C.; Caré, S. Testing the suitability of metallic iron for environmental remediation: Discoloration of methylene blue in column studies. *Chem. Eng. J.* **2013**, *215–216*, 959–968.
160. Kim, H.; Yang, H.; Kim, J. Standardization of the reducing power of zero-valent iron using iodine. *J. Environ. Sci. Health A* **2014**, *49*, 514–523. [CrossRef] [PubMed]
161. Li, S.; Ding, Y.; Wang, W.; Lei, H. A facile method for determining the Fe(0) content and reactivity of zero valent iron. *Anal. Methods* **2016**, *8*, 1239–1248. [CrossRef]
162. Li, J.; Dou, X.; Qin, H.; Sun, Y.; Yin, D.; Guan, X. Characterization methods of zerovalent iron for water treatment and remediation. *Water Res.* **2019**, *148*, 70–85. [CrossRef] [PubMed]
163. Reardon, J.E. Anaerobic corrosion of granular iron: Measurement and interpretation of hydrogen evolution rates. *Environ. Sci. Technol.* **1995**, *29*, 2936–2945. [CrossRef] [PubMed]
164. Velimirovic, M.; Carniatoc, L.; Simons, Q.; Schoups, G.; Seuntjens, P.; Bastiaens, L. Corrosion rate estimations of microscale zerovalent iron particles via direct hydrogen production measurements. *J. Hazard. Mater.* **2014**, *270*, 18–26. [CrossRef]
165. Allen-King, R.M.; Halket, R.M.; Burris, D.R. Reductive transformation and sorption of cis-and trans-1, 2-dichloroethene in a metallic iron–water system. *Environ. Toxicol. Chem.* **1997**, *16*, 424–429. [CrossRef]
166. Mamindy-Pajany, Y.; Hurel, C.; Marmier, N.; Roméo, M. Arsenic (V) adsorption from aqueous solution onto goethite, hematite, magnetite and zero-valent iron: Effects of pH, concentration and reversibility. *Desalination* **2011**, *281*, 93–99. [CrossRef]
167. Johnson, T.L.; Scherer, M.M.; Tratnyek, P.G. Kinetics of halogenated organic compound degradation by iron metal. *Environ. Sci. Technol.* **1996**, *30*, 2634–2640. [CrossRef]
168. Anderson, W. On the purification of water by agitation with iron and by sand filtration. *J. Soc. Arts* **1886**, *35*, 29–38. [CrossRef]

169. Ghauch, A.; Abou Assi, H.; Baydoun, H.; Tuqan, A.M.; Bejjani, A. Fe0-based trimetallic systems for the removal of aqueous diclofenac: Mechanism and kinetics. *Chem. Eng. J.* **2011**, *172*, 1033–1044. [CrossRef]
170. Noubactep, C.; Meinrath, G.; Dietrich, P.; Sauter, M.; Merkel, B. Testing the suitability of zerovalent iron materials for reactive Walls. *Environ. Chem.* **2005**, *2*, 71–76. [CrossRef]
171. Btatkeu-K, B.D.; Tchatchueng, J.B.; Noubactep, C.; Caré, S. Designing metallic iron based water filters: Light from methylene blue discoloration. *J. Environ. Manag.* **2016**, *166*, 567–573. [CrossRef] [PubMed]
172. Gheju, M.; Balcu, I. Removal of chromium from Cr(VI) polluted wastewaters by reduction with scrap iron and subsequent precipitation of resulted cations. *J. Hazard. Mater.* **2011**, *196*, 131–138. [CrossRef] [PubMed]
173. Devlin, J.F.; Allin, K.O. Major anion effects on the kinetics and reactivity of granular iron in glass-encased magnet batch reactor experiments. *Environ. Sci. Technol.* **2005**, *39*, 1868–1874. [CrossRef]
174. Pierce, E.M.; Wellman, D.M.; Lodge, A.M.; Rodriguez, E.A. Experimental determination of the dissolution kinetics of zero-valent iron in the presence of organic complexants. *Environ. Chem.* **2007**, *4*, 260–270. [CrossRef]
175. Noubactep, C.; Licha, T.; Scott, T.B.; Fall, M.; Sauter, M. Exploring the influence of operational parameters on the reactivity of elemental iron materials. *J. Hazard. Mater.* **2009**, *172*, 943–951. [CrossRef]
176. Phukan, M.; Noubactep, C.; Licha, T. Characterizing the ion-selective nature of Fe0-based filters using azo dyes. *Chem. Eng. J.* **2015**, *259*, 481–491. [CrossRef]
177. Tepong-Tsindé, R.; Phukan, M.; Nassi, A.; Noubactep, C.; Ruppert, H. Validating the efficiency of the MB discoloration method for the characterization of Fe0/H2O systems using accelerated corrosion by chloride ions. *Chem. Eng. J.* **2015**, *279*, 353–362. [CrossRef]
178. Mitchell, G.; Poole, P.; Segrove, H.D. Adsorption of methylene blue by high-silica sands. *Nature* **1955**, *176*, 1025–1026. [CrossRef]
179. Morrison, S.J.; Mushovic, P.S.; Niesen, P.L. Early Breakthrough of Molybdenum and Uranium in a Permeable Reactive Barrier. *Environ. Sci. Technol.* **2006**, *40*, 2018–2024. [CrossRef] [PubMed]
180. Li, L.; Benson, C.H.; Lawson, E.M. Impact of mineral fouling on hydraulic behavior of permeable reactive barriers. *Groundwater* **2005**, *43*, 582–596. [CrossRef] [PubMed]
181. Lai, K.C.K.; Lo, I.M.C.; Birkelund, V.; Kjeldsen, P. Field monitoring of a permeable reactive barrier for removal of chlorinated organics. *J. Environ. Eng.* **2006**, *132*, 199–210. [CrossRef]
182. McGeough, K.L.; Kalin, R.M.; Myles, P. Carbon disulfide removal by zero valent iron. *Environ. Sci. Technol.* **2007**, *41*, 4607–4612. [CrossRef]
183. Bi, E.; Devlin, J.F.; Huang, B. Effects of mixing granular iron with sand on the kinetics of trichloroethylene reduction. *Groundw. Monit. Remediat.* **2009**, *29*, 56–62. [CrossRef]
184. Tang, H.; Zhu, D.; Li, T.; Kong, H.; Chen, W. Reductive dechlorination of activated carbon-adsorbed trichloroethylene by zero-valent iron: Carbon as electron shuttle. *Environ. Qual.* **2011**, *40*, 1878–1885. [CrossRef]
185. Huang, Y.H.; Tang, C.L.; Zeng, H. Removing molybdate from water using a hybridized zero-valent iron/magnetite/Fe(II) treatment system. *Chem. Eng. J.* **2012**, *200*, 205–263. [CrossRef]
186. Btatkeu-K, B.D.; Olvera-Vargas, H.; Tchatchueng, J.B.; Noubactep, C.; Caré, S. Characterizing the impact of MnO2 on the efficiency of Fe0-based filtration systems. *Chem. Eng. J.* **2014**, *250*, 416–422. [CrossRef]
187. Westerhoff, P.; James, J. Nitrate removal in zero-valent iron packed columns. *Water Res.* **2003**, *37*, 1818–1830. [CrossRef]
188. Song, D.-I.; Kim, Y.H.; Shin, W.S. A simple mathematical analysis on the effect of sand in Cr(VI) reduction using zero valent iron. *Korean J. Chem. Eng.* **2005**, *22*, 67–69. [CrossRef]
189. Notter, J.L. The purification of water by filtration. *Br. Med. J.* **1878**, *12*, 556–557. [CrossRef]
190. Chen, Z.; Wang, X.; Ge, Q.; Guo, G. Iron oxide red wastewater treatment and recycling of iron-containing sludge. *J. Clean. Prod.* **2015**, *87*, 558–566. [CrossRef]
191. Chaudhari, S.; Banerji, T.; Kumar, P.R. Domestic- and community-scale arsenic removal technologies suitable for developing countries. *Water Reclam. Sustain.* **2014**, *2014*, 155–182.
192. Marwa, J.; Lufingo, M.; Noubactep, C.; Machunda, R. Defeating fluorosis in the East African Rift Valley: Transforming the Kilimanjaro into a rainwater harvesting park. *Sustainability* **2018**, *10*, 4194. [CrossRef]
193. Ndé-Tchoupé, A.I.; Tepong-Tsindé, R.; Lufingo, M.; Pembe-Ali, Z.; Lugodisha, I.; Mureth, R.I.; Nkinda, M.; Marwa, J.; Gwenzi, W.; Mwamila, T.B.; et al. White teeth and healthy skeletons for all: The path to universal fluoride-free drinking water in Tanzania. *Water* **2019**, *11*, 131. [CrossRef]

194. Qi, Q.; Marwa, J.; Mwamila, T.B.; Gwenzi, W.; Noubactep, C. Making Rainwater Harvesting a Key Solution for Water Supply: The Universality of the Kilimanjaro Concept. *Sustainability* **2019**, *11*, 5606. [CrossRef]
195. Latrach, L.; Ouazzani, N.; Hejjaj, A.; Mahi, M.; Masunaga, T.; Mandi, L. Two-stage vertical flow multi-soil-layering (MSL) technology for efficient removal of coliforms and human pathogens from domestic wastewater in rural areas under arid climate. *Int. J. Hyg. Environ. Health* **2018**, *22*, 64–80. [CrossRef]
196. Allred, B.J.; Racharaks, R. Laboratory comparison of four iron-based filter materials for drainage water phosphate treatment. *Water Environ. Res.* **2014**, *86*, 852–862. [CrossRef]
197. Murphy, A.P. Removal of selenate from water by chemical reduction. *Ind. Eng. Chem. Res.* **1988**, *27*, 181–191. [CrossRef]
198. Oldright, G.L.; Keyes, H.E.; Miller, V.; Sloan, W.A. *Precipitation of Lead and Copper from Solution on Sponge Iron*; BuMines B 281; Washington, DC, USA, 1928; 131 p, Available online: https://searchworks.stanford.edu/view/7623313 (accessed on 27 February 2020).
199. Wu, S.; Vymazal, J.; Brix, H. Critical Review: Biogeochemical Networking of Iron in Constructed Wetlands for Wastewater Treatment. *Environ. Sci. Technol.* **2019**, *53*, 7930–7944. [CrossRef]
200. Rahman, M.A.; Wiegand, B.A.; Badruzzaman, A.B.M.; Ptak, T. Hydrogeological analysis of the upper Dupi Tila Aquifer, towards the implementation of a managed aquifer-recharge project bei Dhaka City, Bangladesh. *Hydrogeology* **2013**, *21*, 1071–1089. [CrossRef]
201. Fronczyk, J.; Markowska-Lech, K.; Bilgin, A. Treatment assessment of road runoff water in zones filled with ZVI, activated carbon and mineral materials. *Sustainability* **2020**, *12*, 873. [CrossRef]
202. Davis, F. *An Elementary Handbook on Potable Water*; Silver, Burdett & Co: New York, NY, USA; Boston, MA, USA; Chicago, IL, USA, 1891; 118 p.
203. Lee, E.W. Drainage Water Treatment and Disposal. In *Management of Water Use in Agriculture. Advanced Series in Agricultural Sciences*; Tanji, K.K., Yaron, B., Eds.; Springer: Berlin/Heidelberg, Germany, 1994; Volume 22.
204. Hussam, A. Contending with a development disaster: Sono filters remove arsenic from well water in Bangladesh. *Innovations* **2009**, *4*, 89–102. [CrossRef]
205. Bartzas, G.; Komnitsas, K.; Paspaliaris, I. Laboratory evaluation of Fe0 barriers to treat acidic leachates. *Miner. Eng.* **2006**, *19*, 505–514. [CrossRef]
206. Tansel, B. New technologies for water and wastewater treatment: A survey of recent patents. *Recent Pat. Chem. Eng.* **2008**, *1*, 17–26. [CrossRef]
207. Obiri-Nyarko, F.; Grajales-Mesa, S.J.; Malina, G. An overview of permeable reactive barriers for in situ sustainable groundwater remediation. *Chemosphere* **2014**, *111*, 243–259. [CrossRef]

Review

Application of Ionizing Radiation in Wastewater Treatment: An Overview

Rehab O. Abdel Rahman [1,*] and Yung-Tse Hung [2]

1 Hot Lab. Center, Atomic Energy Authority of Egypt, Cairo P.O. No. 13759, Egypt
2 Department of Civil and Environmental Engineering, Cleveland State University, Cleveland, OH 44115, USA; yungtsehung@gmail.com
* Correspondence: alaarehab@yahoo.com; Tel.: +2-010-6140-4462

Received: 28 October 2019; Accepted: 18 December 2019; Published: 19 December 2019

Abstract: Technological applications of nuclear science and technology in different sectors have proved their reliabilities and sustainability over decades. These applications have supported various human civilization needs, ranging from power generation to industrial, medical, and environmental applications. Environmental applications of radiation sources are used to support decision making processes in many fields; including the detection and analysis of pollutant transport, water resources management, and treatment of municipal and industrial wastewaters. This work reviewed recent advances in the research and applications of ionizing radiation in treating different wastewater effluents. The main objective of the work is to highlight the role of ionizing radiation technology in the treatment of complex wastewater effluents generated from various human activities and to address its sustainability. Results of both laboratory and industrial scale applications of this treatment technology have been reviewed, and information on operational safety of industrial irradiators, which affect the sustainability of this technology, has been summarized.

Keywords: ionizing radiation; agricultural effluents; dye treatment; pharmaceutical effluents; disinfection

---

1. Introduction

Providing clean water and sanitation is one of the sustainable development goals that were proposed by the United Nation (UN). One of the problems that affect this goal is the reduction of freshwater quality. This reduction is attributed to the continuous increase in untreated wastewaters volumes and poor management practices, which led to the introduction of hazardous materials into freshwater sources [1]. Wastewater is defined, as indicated in UNEP/UN-Habitat, as a combination of one or more of the following effluents: domestic, commercial, industrial, horticultural, aquaculture, and storm water [2]. Recent advances in wastewater management helped in addressing some of the problems in water supply, pollution control, water recycling, and environmental protection. Now wastewaters are proposed as a resource, where many wastewater treatment plants are operated by the biogas generated from the anaerobic digestion of sludge, and the effluents from these plants could be used after appropriate treatment to meet the industrial, agricultural, and potable water requirements. Currently, wastewater management in developing countries is characterized by the discharge of large quantity of wastewater into surface water bodies without proper treatment. It is also challenged by the difficulties to sustain financing, operating and maintaining infrastructure for wastewater treatment.

Conventional wastewater treatment plants (WWTP) aims at reducing the contamination levels to acceptable limits required by the national regulatory agencies and at complying with international guidelines, which allow its safe discharge or reuse. Multi-stages of treatment processes are used, where pre-treatment stage is applied to remove coarse and large solids from the waste stream, using physical treatment technologies such as screens and grit chambers [3]. Primary treatment methods are then applied to remove suspended solids that could be settled using gravity sedimentation with or without coagulation and flocculation [3,4]. The effluent from the primary treatment is directed to the secondary treatment stage to remove residual suspended solids and organic materials by using biological treatment processes [3]. The effluents from secondary stage contain some heavy metals, synthetic bio-refractory organic pollutants, and soluble microbial products derived during biological treatment [5,6]. The synthetic bio-refractory organic pollutants may include emerging micro-pollutants and disinfection by-products. In the tertiary stage of treatment, the effluents from the secondary stage are polished by removing persistence organic containments and heavy metals using advanced wastewater treatment technologies [3,4,7]. These technologies include filtration, sorption, gas stripping, ion-exchange, advanced oxidation processes (AOP), and distillation [7]. Finally, disinfection could be applied depending on the potential use of the treated effluents and the effluents characteristics [3].

Regulations on discharge/reuse indicators varied from country to another. In Denmark, Belgium, Spain, Germany, France, and Netherlands, the regulation covers heavy metals concentrations, total suspended solids (TSS), chemical oxygen demand (COD), 5 days biochemical oxygen demand (BOD_5), total nitrogen (TN), total phosphorus (TP), and quantity of effluent discharge [8]. On the other hand, toxicity is only considered as reuse/discharge indicator in Germany, France, and United Kingdom [8]. Table 1 lists some international guidelines on the maximum limits for the reuse of effluents containing heavy metals, and U.S. regulation for their reuse and discharge [9–12]. Table 2 presents effluent discharge and reuse quality requirements in USA, Canada, and EU [11,13–15]. It should be noted that three classes of effluents are listed in the Canadian regulation based on the degree of treatment processes used in the treatment plants, namely, A, B and C, that refers to effluent from tertiary and disinfection treatment, tertiary treatment, and secondary treatment, respectively. For reuse regulations in EU, the limit of each class is determined based on the processes used in the treatment plants, irrigation methods, and the type of corps. Class A represents effluents from tertiary and disinfection treatment used any irrigation method and to produce all food corps. Classes B and C are effluent from secondary treatment and disinfection, which are used to produce food and processed food and non-food corps using all irrigation methods, and drip irrigation, respectively. Finally class D in EU regulation represents effluents from secondary treatment and disinfection, where all irrigation methods are allowed to produce industrial, energy, and seed corps [13]. Discharges from un-complied conventional WWTP can lead to the introduction of persistent chemicals and eco-toxic micro-pollutants into the aquatic systems [16]. The incomplete removal of these containments (even in 10^{-9}–10^{-6} g/L concentration range) was reported to induce potential long-term detrimental impact on the environment and the human health [17]. Recent research studies reported in the literature supports the application of advanced oxidation processes (AOP) for wastewater treatment to remove these contaminants [16,17].

Table 1. Guidelines of maximum limits for discharge and reuse of some inorganic pollutants in treated wastewater.

	WHO, Reuse, ppb [11]	FAO, Reuse, ppb [12]	US, ppb [9]			WHO, Reuse, ppb [11]	FAO, Reuse, ppb [12]	US, ppb [9]	
			Discharge	Reuse				Discharge	Reuse
As	50	100	3000	50	Hg	1	NA	2000	2
Cd	5	100	15,000	5	Ni	0	200	12,000	100
Cr	NA	100	10,000	50	Ag	NA	NA	5000	50
Cu	1	200	15,000	NA	Zn	5000	200	25,000	5000
CN	NA	NA	10,000	5.2	pH	6.5–8.5	6.5–8.4	6–9	
Pb	50	5000	40,000	50					

NA: Not Available.

Table 2. Effluent reuse and discharge limits in different regulations [11,13–15].

Indicator	Reuse				Discharge				
	EU [14,15]				EU [14,15]	US [11]	Canada [13]		
	A	B	C	D			A	B	C
BOD_5, ppm	10	25				45	10	10	45
TSS, ppm	10	35				45	10	10	45
COD, ppm	NA	NA	NA	NA	125	NA	NA	NA	NA
E. coli, (CFU/100 mL)	10	10^2	10^3	10^4	NA	2.2	2.2	400	NA
Turbidity (NTU)	5	NA	NA	NA	NA	2	2	NA	NA
TN, ppm	NA	NA	NA	NA	15	10	20	NA	NA

NA: Not Available.

AOP use in-situ generated hydroxyl or sulfate radicals for organic pollutants degradation and heavy metals toxicity reduction [5,18,19]. AOP convert synthetic organic pollutants and soluble microbial products into simple biodegradable and harmless products, which lead to the reduction of COD and BOD in the treated effluents. AOP technologies employ various activation methods, where induced oxidation is achieved via exposure to photochemical or ionizing radiations, or chemicals [18,20,21]. Several studies were conducted to assess the feasibility of using ionizing radiation, in the form of gamma (γ) rays or electrons (e^-), to remove persistence contaminants and to disinfect treated water and sludge. The results of these studies indicated that the ionizing radiation treatment is technically and economically promising. The efficiency of this technology was also proved in the reduction of persistence heavy metals under different treatment conditions. Fifteen pilot plants and several full scale irradiation treatment facilities that employ electron accelerator or γ irradiators were established with varying capacities. International Atomic Energy Agency (IAEA) reported that there is a need to continue research in this area to increase the general awareness of these processes in the environmental engineering community [22]. This awareness could be achieved by presenting integrated reviews on the advances in ionizing radiation applications in industrial effluents treatment from technical, operational safety, and economical aspects. Some review papers were published that reviewed the role of ionizing radiation in the degradation of azo dyes [23], summarized the results of the IAEA coordination research project [24], and have reported the optimized doses and procedures [25]. There is a lack in review papers that summarize recent advances in this field and that provide integrated insights into the role of this technology in eliminating hazardous biodegradation and disinfection products. In this work, the effort is directed to summarize the current understanding of the decomposition and removal mechanisms for organic and inorganic pollutants, respectively. Principles and advances in investigating the scientific basis of the applicability of this technique in the treatment and disinfection of agriculture, dyes, pharmaceutical, and petrochemical effluents will be presented. Operational safety and economical factors that affect the sustainability of this technology will be summarized. Finally, knowledge gaps will be identified and research areas that need to be addressed will be highlighted.

2. Decomposition and Removal Mechanisms

Generally any emitted radiation is characterized by its ability to deposit some of its energy in the surrounding media. This energy excites the media atoms by striping their electrons or break the chemical bonds between its molecules [26]. This particular characteristic leads to the application of ionizing radiation in treating wastewater, where the radiation imparts some of its energy in the radiolysis of water molecules. The effectiveness of the irradiation process is evaluated by calculating the radiation chemical yield (G-value, μmol·J⁻) that quantifies the number of formed species due to the absorption of 100 eV and is given by:

$$G = \frac{6.023 \times 10^{23} C}{D \times 6.24 \times 10^{16}} \quad (1)$$

where C is the formed species concentration (mol/L) and D is the absorbed dose (Gy). Figure 1 illustrates the two stages radiolysis process for water molecules and its corresponding time scale, radiolysis products (primary intermediates), and their G-values. Primary intermediate (PI) then reacts with the pollutants (A) leading to the formation of secondary intermediates (SI) that are less persistence in a bi-molecular reaction to produce the degradation products (C + D), as follows [25,27,28]:

$$A + PI \underset{=}{\rightarrow} SI \rightarrow C + D \quad (2)$$

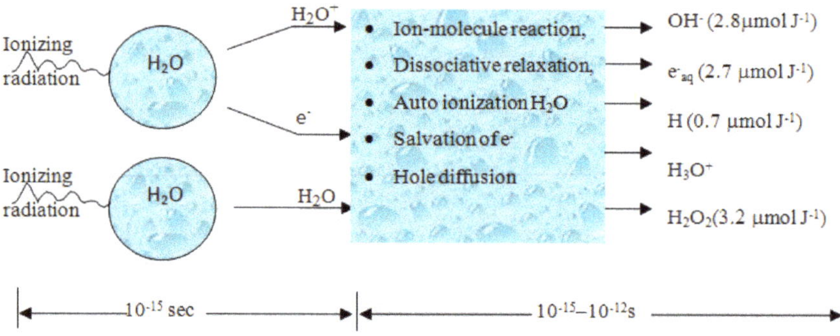

Figure 1. Two stages radiolysis process.

In actual wastewater, there is a competition between the pollutants and the anions in the solution, i.e., Cl^-, CO_3^{2-}, HCO_3^-, SO_4^{2-}, for the reaction with the primary intermediates. This competition can affect the efficiency of the overall treatment process [27]. Gehringer studied the competition kinetics of two pollutants (A, B) in wastewater. The reaction rate was attributed to the bi-molecular rate constant (k_A) between the solute concentration ([A] or [B]) and the primary intermediates concentration [OH] as follow:

$$-\frac{dA}{dt} = k_A[A][OH] \quad (3)$$

Table 3 lists the bi-molecular reaction rate constants for some pollutants [22,25,27,29]. The amount of the specific radicals available for interaction with certain solute could be calculated using the reaction probability (P_A) according to the following equation:

$$P_A = \frac{k_A[A]}{k_A[A] + k_B[B]} \quad (4)$$

Figure 2 illustrates the reactions between the primary intermediates and the available anions in the wastewater effluents. Wojnarovits and Takacs [30] indicated that intermediate reaction with chloride ions is dominant at pH < 5, whereas the reaction with carbonate and bicarbonate ions are dominant at neutral or slightly alkaline pH. The produced free radicals will subsequently react with dissolved organic matters in the wastewater via direct electron transfer (outer sphere electron), addition to double bonds (inner sphere electron) or abstraction of H-atoms from C-H bonds. The last reaction takes place between oxidizing radicals and saturated molecules. It should be noted that in some cases, it is very hard to determine if the reaction is electron transfer or radical addition/elimination.

Table 3. Bimolecular reaction rate constants for some pollutants [22,25,27,29].

Compound	Rate Constant, M^{-1} s^{-1}, $\times 10^9$			Compound	Rate Constant, M^{-1} s^{-1}, $\times 10^9$		
	OH	e^-	H		OH	e^-	H
Perchloroethylene	1.7	13		Bicarbonate	0.0085		4×10^{-3}
Trichloroethylene	2.6	19		Nitrogen		9.7	
Dichloroethylene	7	7.5		Oxygen		19	
Vinylchloride	12	0.25		Methyl t-butyl ether	2	0.018	
1,1,1-Trichloroethane	0.04			Ethyl t-butyl Ether	1.8	0.01	7×10^{-3}
Chloroform	0.005	30		Diisopropyl Ether	2.5	7×10^{-3}	0.067
T-Amyl Methyl Ether	2.4	3×10^{-3}	3×10^{-3}	Acetylenes	0.1–1		
Alcohols	0.1–1			Aldehydes	1		
Alkanes	0.001–1			Aromatics	0.1–100		
Carboxyl Acids	0.01–1			Ketones	1		
Organo-Nitrogen	0.1–100			Olefins	1–10		
Phenols	1–10			Organo-Sulfure	1–10		
Trichloronitromethane (TCNM), Chloropicrin	0.0497 ± 0.28	21.3 ± 0.03		Carbofuran	6.6		
Benzene	7.8			Carbendazim	2.2		

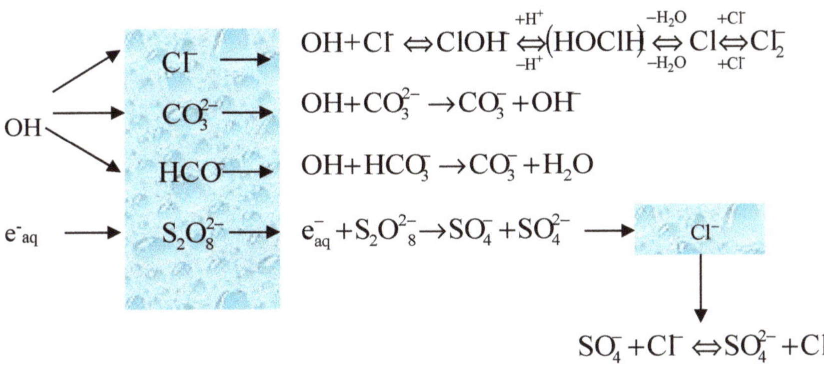

Figure 2. Schematic diagram of intermediate reaction with free radicals.

Ionizing radiation can reduce various forms of mercury via reaction with e^-_{aq} (Equation (5)) and H (Equation (6)) to form unstable compound that dimerized (Equation (7)) to produce insoluble form [31,32].

$$HgCl_2 + e^-_{aq} \rightarrow HgCl + Cl^- \quad (5)$$

$$HgCl_2 + H \rightarrow HgCl + Cl^- + H^+ \quad (6)$$

$$2HgCl_2 \rightarrow Hg_2Cl_2 \quad (7)$$

The presence of hydroxyl radicals inhibited the production of insoluble mercury (Hg_2Cl_2) through re-oxidation of HgCl [31]. To enhance the removal performance of this technology the use of organic radical to act as hydroxyl scavenger was proposed. For example, ethanol may be used (Equations (8) and (9)) with the following reactions [9]:

$$CH_3CH_2OH + OH \rightarrow CH_3CHOH + H_2O \quad (8)$$

$$HgCl_2 + CH_3CHOH \rightarrow HgCl + Cl^- + CH_3CHO + H^+ \quad (9)$$

Successive reduction of Cr species from (VI) state to (III) state is achieved via reaction with H [32]. On the other hand, cadmium (II) and lead (II) are reduced via reaction with H or/and e^-_{aq} (Equation (10)) to produce Cd(I) that undergoes disprotonation (Equation (11)) or oxidization through a reaction with OH^- or H_2O_2 (Equation (12)), or react with hydrogen to produce unstable MH^+ species that decay (Equation (13)) [18].

$$M^{2+} + e^-_{aq} \rightarrow M^+ \quad (10)$$

$$2M^+ \rightarrow M + M^{2+} \quad (11)$$

$$M^+ + OH \rightarrow M^{2+} + OH^- \quad (12)$$

$$2H + 2M^+ \rightarrow 2MH^+ \rightarrow H_2 + M^{2+} + M \quad (13)$$

To reduce the oxidation effect and subsequently to enhance cadmium and lead precipitation, an organic OH^- scavenger could be used or the process could be operated in the absence of oxygen [31–33].

3. Advances in Treating Agricultural Wastewaters

Despite chemical pesticides, herbicides, and fungicides are applied according to national agricultural guidelines to enhance the agricultural production efficiency. Residues of these persistence pollutants and their toxic byproducts exist in agricultural wastewater, and could migrate to surface-and ground-waters. Ionizing radiation treatment technology proved its effectiveness in decomposing a varied number of these pollutants using tertiary treatment method. Most of the researches conducted in this area focused on optimizing the irradiation conditions, investigating the effect of other waste components on the degradation process, and the possibility of using combined treatment technologies to enhance the overall efficiency of the treatment process. Table 4 summarizes the performance of ionizing radiation in removing some pollutants in agricultural wastewaters.

Table 4. Performance of agricultural wastewater pollutants degradation using irradiation.

Pollutant	Origin	Degradation Performance	Initial Contaminant Concentration, ppm	Dose, KGy	References
Polychlorinated Biphenyls (PCB)	Pesticide	96%	NA	<0.1	[34]
Alkali Halides	Herbicides, fungicides, insecticide	98%	100	1	[25]
Trihalomethanes	Disinfection by products	87.4%	NA	2	[25]
		95%	145–780	6	[35]
		87.8%	264	8	[35]
Nitrophenols	Degradation products		0.139	5	[25]
Carbofuran, Dimethoate, Imidacloprid, Procloraz	Pesticide	99%	50, (pH 5.5)	5	[36]
Methiocarb	Pesticide	67%			[36]
2,4-Dichlorophenol	Pesticide	Complete degradation	50	10	[22]

NA: Not Available.

The optimal irradiation conditions for six commercial pesticides, i.e., diazinon, dimethoate, procloraz, metiocarb, imidacloprid, and carbofuran, using electron-beam facility, were determined for different wastewater compositions and at different operational conditions [36]. The initial pesticide concentrations (C_i) varied in the range $40 < C_i < 400$ ppm, the initial pH in the range ($4.5 < pH < 8$) and the irradiation doses were varied from 2–10 KGy. The application of 5 KGy was found to be the most efficient for the treatment of procloraz, dimethoate, imidacloprid, and carbofuran (99% degradation), but not efficient for the treatment of metiocarb. Low irradiation doses (<1 KGy) led to the formation of hydroxylated intermediates that build-up in the solution with the progress of the irradiation process. By increasing the irradiation period, a complete removal of the pollutants is achieved as a result of hydroxylation of the accumulated intermediates. The effect of the combined process of aeration and irradiation treatment was evaluated. It was found that the use of aeration can improve the metiocarb removal by 18%. In general, the hydroxylation of the aromatic ring to produce hydroxyl-cyclohexadienyl radicals is reversible [37]. Using aeration to inhibit the formation of these reversible radicals by peroxidation will lead to more efficient degradation [37]. The optimum irradiation conditions for the degradation of three herbicide and one fungicide, i.e., 2,4-dichlorophenoxyacetic acid (2,4-D), 3,6-dichloro-2-methoxy-benzoic acid (dicamba), 4-chloro-2-methylphenoxyacetic acid (MCPA), and carbendazim were determined. It was found that complete radiolytic degradation of dicamba ($C_i = 110$ ppm) is achieved using an irradiation dose of 5 KGy and the process is insensitive to the presence of other waste constituents, i.e., NO_3^-. 2,4-D-degradation was found to be affected by the presence of other waste constituents [35]. Another study addressed the use of ionizing radiation in the treatment of 2,4-D, and MCPA at lower concentration levels (<50 ppm), in the presence Cl^-, Br^- and NO_3^-. The results confirmed the role of oxidative radicals (OH^-) in the degradation of these pollutants and the sensitivity of the irradiation process to the presence of the anions, where the secondary intermediate formed at low doses (<1 KGy) are more toxic than the original pollutants [38,39].

The feasibility of using ionizing radiation in the treatment of agricultural effluents containing chlorinated organic pesticide, i.e., (4-chloro phenoxyacetic acid (4-CPA), 2.4-dichlorophenoxyacetic acid (2,4-D), 2.4-dichlorophenoxyacetic propionic acid (2,4-DP), and 2.4-dichlorophenoxyacetic butanoic acid (2,4-DB) was studied. The results indicated good efficiency of the irradiation process at 1 KGy in decomposing these pollutants. It was observed that chlorine was released as a result of their degradation. To reduce the disinfection byproducts in the treated wastewater prior to its use in chicken and fish livestock, an irradiation of the wastewater stream at 16.2 KGy dose was proposed, and 27 KGy was proposed for sludge treatment [40].

Swine wastewaters, which are alkaline agricultural wastewaters, that contain different pollutants, i.e., carbohydrates, proteins, lipids nitrates, nitrite, phosphate, and ammonia, were treated using a combined process of irradiation and ion-exchange biological treatment method. The use of electron-beam at 75 KGy achieved 85.1% removal efficiency of chemical oxygen demand (COD) at organic loading rate of 1.41 kg/m^3.day and achieved 75% removal efficiency of total nitrogen. The nitrogen removal was found to be sensitive to variation in current density [41].

4. Advance in Dyes Treatments

The presence of organic dyes in industrial effluents can lead to serious health and environmental problems. Ionizing radiation treatment was proposed to remove these contaminants. Research efforts were directed to study model compounds, single polluted solution, simulated wastewater, and real effluents. Table 5 summarizes the kinetic reactions of model compounds with their primary intermediates, their corresponding rate coefficients and their secondary intermediates [23]. Most of these compounds, except Azobenzene, decompose as a result of hydroxylation reaction, where the rate coefficients are slightly varied depending on the molecular structure of the contaminant. On the other hand, reactions with e^- and H have small contribution to the decomposition process [23]. G value for phenol degradation was found to be inversely proportional to the irradiation dose. Degradation reaction is first order and favors neutral pH. Phenol degradation is enhanced with the addition of oxidants, i.e., O_3, or $S_2O_8{}^{2-}$. $S_2O_8{}^{2-}$ was more efficient due to the selectivity of $SO_4{}^{2-}$ radical to the formed by-products (carboxylic acids) [42].

Table 5. Degradationkinetics of model compounds [23].

Model Compound	Primary Intermediate	Rate Coefficient, $\times 10^{10}$ $mol^{-1}{\cdot}L{\cdot}s^{-1}$	Secondary Intermediate
Aniline	OH^-	0.86–1	-
	H^+	0.2	Anilino radical (directly and after water elimination.)
Phenol	OH^-	0.66–1.4	Cyclohexadienyl-type radical
	H^+	0.17	
Atrazine	OH^-	0.24	
Azobenzene	e^-	1–3.3	Hydrazyl radicals
	OH^-	2	
4-amino-5-hydroxynaphthalene-2,7-disulphonic acid	OH^-	-	Anilino-type

65% de-coloration of aqueous solution containing Alizarin Yellow GG (AY-GG, C_i = 100 ppm) was achieved using 9 KGy dose. Irradiation pre-biodegradation was proposed to improve the efficiency of treatment by 30% due to the formation of heterocyclic aromatic amines and cyanides. Increasing the irradiation dose can enhance the biodegradation due to the elimination of toxic secondary intermediates [43]. The aqueous solutions of Reactive Blue 15 (RB15) and Reactive Black 5 (RB5) dyes were irradiated with doses 0.1–15 KGy and at 2.87 and 0.14 KGy/h dose rates. Complete de-coloration was observed at 1 and 15 KGy doses for RB5 and RB15, respectively [25]. The de-coloration mechanisms for Apollofix Red (AR) and RB5 in aqueous solutions were investigated using γ-rays irradiation [44]. It was found that the de-coloration that was due to reactions with e^- and H, and increased linearly with increasing the irradiation dose, whereas de-coloration that was due to reaction with OH^- radical increase logarithmically. De-coloration was attributed to the destruction of the color bearing part of the dye as a consequence of OH^- addition to the aromatic ring [25]. Aqueous solutions containing AR and Apollofix Yellow (AY) were irradiated with doses of 1.0–8.0 KGy and at 0.14 KGy/h dose rate. The complete de-coloration was observed using 3.0 and 1.0 KGy doses for AR and AY, respectively [25].

The use of ionizing radiation to decompose wastewater effluents of varying pH ($1.6 < pH < 11.5$) and COD concentration ($650 < C_i < 2210$ ppm) from dye manufacturing factory was evaluated [45]. The use of 15 KGy decreased the optical density by 95% and removed COD by 72%. Use of aeration and H_2O_2 enhanced the degradation. The treatment of simulated wastewater contains four dyes, i.e., direct and reactive azo and anthraquinone dyes, using electron beam and γ radiation at different dyes concentration and irradiation doses in the presence and absence of H_2O_2 was studied [46]. The highest de-coloration performance ($\approx$100%) was obtained via electron beam irradiation using 7 KGy for direct and reactive azo dye (C_i = 1000 ppm). A combined process of electron-beam and biological treatment was used in a pilot plant study in treating two waste streams. The first stream was generated during the operation of dying process and the second stream was from polyester fiber production enriched with ethylene glycol and terephthalic acid in 1998. An industrial facility was later commissioned and operated based on the same technology in 2005 [24,25]. The use of electron-beam technology to de-color and detoxify three effluents that represent chemical, final, and standard textile effluents was conducted. The effluents were de-colored (96, 55, and 90%) using 40, 2.5, and 2.5 KGy, respectively. The use of irradiation pre-biological treatment was found to reduce the toxic effect to the subsequent biological treatment process [47]. The potential use of combined treatment for alkaline and nearly neutral textile effluent of COD (632 < COD < 127 ppm), BOD (311 < BOD < 490 ppm), and turbidity (75–77%) at irradiation doses (<3 KGy) was investigated. It was found that the application of coagulation prior to irradiation enhanced the de-coloration performance [48]. Electron beam irradiation of simulated effluents that contains reactive dye (reactive yellow 15), size (starch), synthetic size (PVA), alkali, color, and pigment (pigment red 139) was studied [49]. Application of 1 KGy irradiation for post-biodegradation treatment enhanced the quality of the treated effluents.

5. Advances in Wastewater and Sludge Disinfection

Disinfection of treated wastewaters using ionizing radiation has been studied extensively. The application of 4 KGy reduced the BOD_5, COD, and total organic carbon (TOC) to acceptable limits [50]. In another study, anaerobic digested sludge irradiation at 1 KGy eliminated 98% of the total and fecal coliform, whereas BOD_5 was reduced to an acceptable limit at 4 KGy and COD was not affected by irradiation up to 20 KGy [44]. The disinfection of acidic industrial effluents and sludge (2 < pH < 3) of high concentration of BOD_5 and COD (7093, and 32,664 ppm, respectively) using γ irradiation was investigated [17]. Sludge and treated water disinfection could be achieved using 7 and 4 KGy, respectively, whereas the reduction of the BOD_5 and COD concentration to acceptable limits was obtained at 18 KGy. A comparative study was conducted to evaluate the use of different advanced oxidation methods in the disinfection of the municipal wastewater, re-growth control, and the associated operating costs [51]. The results revealed that UV efficiency was affected by the seasonal variation in the wastewater composition, whereas ionizing radiation efficiency was respectively unaffected by this factor. Ionizing radiation provided high stable disinfection efficiency (95%) for the total colony count and total coliform at radiation doses >0.25 KGy and inhibited the re-growth. From economical point of view, the electric power consumption for UV and ozone is three orders of magnitude higher than that required for ionizing radiation. Electron beam irradiation used to disinfect a municipal wastewater at irradiation doses (<3 KGy) and removed 90% of coliforms [51].

6. Advances in Pharmaceutical and Petrochemical Wastewater Treatments

The possibility of treating pharmaceutical effluents was addressed by studying the effect of ionizing radiation on the biodegradability and toxicity of individual drugs. Changes in biodegradability and toxicity induced in aqueous solutions containing sulfamethoxazole (SMX, C_i = 0.1 mmol/L) using ionizing radiation treatment revealed that SMX biodegradability was improved by applying 0.4 KGy dose. At 2.5 KGy dose, SMX conversion to biologically treatable substances was noted [52]. The degradation of carbamazepine (CBZ) by ionizing radiation was enhanced by the application of oxidant, 10 mM H_2O_2 [20]. The decomposition of mutagenic and carcinogenic secondary intermediates, i.e., acridine (ACIN), was enhanced in the presence of H_2O_2. The ionization of aqueous solutions containing ciprofloxacin (CIR) and norfloxacin (NOR) ($C_i = 10^{-4}$ M) was investigated. The degradation reaction proceeds via OH^- and e^- reactions with comparable rate constant ($\approx 10^9$ $mol^{-1} \cdot L \cdot s^{-1}$). At low irradiation doses, the antibacterial activities of the secondary intermediates vanished. Pollutants hydroxylation during γ irradiations proceeds on the hydroxylated molecules and desethylene derivatives and during pulse radiolysis is attributed to absorbance of hydroxyl-cyclohexadienyl radicals. In hydrated electron reactions, electron adduct is formed then it underwent protonation yielding cyclohexadienyl type radical [53]. The treatment of real pharmaceutical effluent using combined process of coagulation, biological treatment, and γ irradiation was investigated [54]. Two tested neutral effluents, namely, low organic strength (LSW; BOD < 6730 ppm, COD < 12,715 ppm) and high organic strength (HSW; BOD < 27,242, COD < 51,223 ppm). The use of irradiation led to maximum reduction in COD of 45% in acidic media at 50 KGy (LSW) and 30% in acidic media at 100 KGy (HSW). The application of coagulation pre-treatment was found to affect the efficiency, where 55% and 50% could be achieved using 100 KGy, for LSW and HSW, respectively. The use of H_2O_2 led to enhanced COD and TOC removal efficiency, when compared to $S_2O_8{}^{2-}$. The combined treatment led to overall 92.7% ± 2.3% and 90.2% ± 2.9% removal of COD from LSW and HSW, respectively. In a separate study, combined process of coagulation, electron beam treatment, and biological treatment was performed. An overall reduction in COD of 94% and 89% was achieved LSW and HSW, respectively [55]. The radiation doses were varied from 25–100 KGy at different pH that represent acidic, neutral and alkaline media. The slightly improved performance of the electron irradiation process was related to the reaction of $e^{-}{}_{aq}$ with H through parallel reaction with the organic contaminants to generate H^{+}, which subsequently inhabits the recombination of $e^{-}{}_{aq}$ and OH^{-}. The consideration of the total cost of electron beam irradiation facility (20 MeV, 100 kW), its shielding and maintenance, and capital costs was estimated and the cost of the treatment was estimated to be 0.6 USD/m^{-3}.

The decomposition of naphthalene (C_i = 5–32 ppm) in aqueous solution was studied using γ irradiation combined with both H_2O_2 and TiO_2nano-particles. The application of 3 KGy dose led to high naphthalene removal performance (>98%) and TOC reduction (28–31%) due to hydroxylation reaction. This performance is enhanced by 35% due to the presence of 40 ppm of H_2O_2 and 48% due to the presence of 0.8 g/L TiO_2 [56].

7. Sustainability of the Technology

In general, the sustainability of the nuclear industry is governed by its technical competitive performance, economical feasibility, and safe operational practice. As presented in the previous sections, the technical performance of the application of ionizing radiation technology in wastewater treatment is effective in disinfection and reduces the bio-refractory nature of several persistence organic pollutants. Several studies proved the economical feasibility of the e-beam and γ irradiation technologies [21,32,41,48,51,54–63]. For e-beam technology, the capital costs include the accelerator price, building shield, conveyer, cooling and ventilation systems, and monitoring system [57]. The capital cost of the e-beam accelerator is dependent on the power (P, kW) consumed to produce optimum dose (D, KGy) for specified plant flow rate (Q, m^3/h) taking into account the utilization factor (φ) [48,60]:

$$p = \frac{DQ}{\varphi} \quad (14)$$

The cost of the accelerator installation (K_i, k$) is determined using the applied electron energy (E, MeV), power, accelerator type and manufacturer (b, d), and installation coefficient (a) as follows [48,57]:

$$K_i = a \cdot b(1 \pm d)E\sqrt{p} \quad (15)$$

The values of the coefficients b and d varied with time due to the evolution and advances in the manufacturing process [48]. On the other hand, the costs of γ irradiator are not accompanied by power consumption [54], where γ source irradiate spontaneously. The average cost (ATC, k$/$m^3$) of this irradiation technique is determined based on the price of the used radioactive source (R, k$/$C_i$), required activity (I, C_i), irradiation time (t, h) and half life ($t_{0.5}$, h), and the irradiation chamber volume (v, m^3)

$$A\,(C_i) = R \times I \times t/(t_{0.5} \times v) \quad (16)$$

To facilitate the cost comparison between the use of e-beam technology and other treatment technologies the relative treatment costs estimated at different time is presented in Figure 3 [48,64]. In terms of cost, only disinfection process, which uses chlorine, is better than the use of e-beam technology. Similar data are not available for γ irradiation. To compare the cost of this technique alone and cost of this technology combined with coagulation, the data reported by Changotra et al. [54] were plotted for two types of industrial effluents (Figure 4).

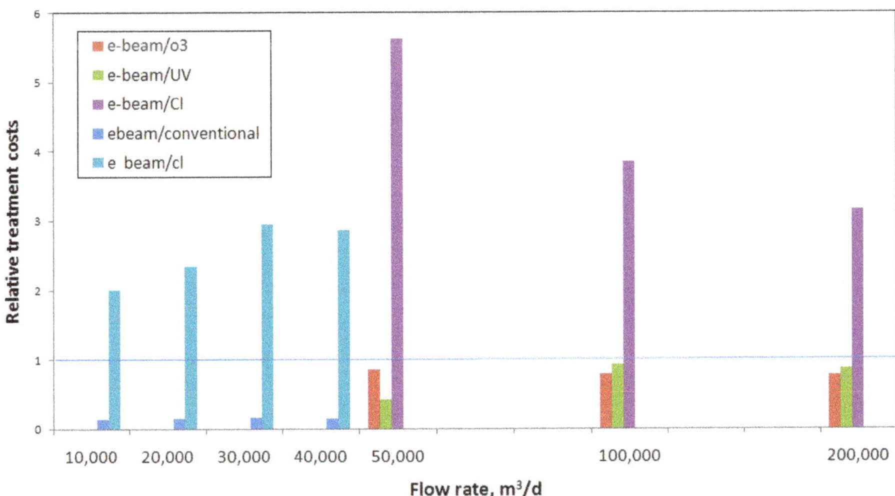

Figure 3. Relative treatment costs for wastewater using e-beam and other disinfection and conventional treatment methods [48,64].

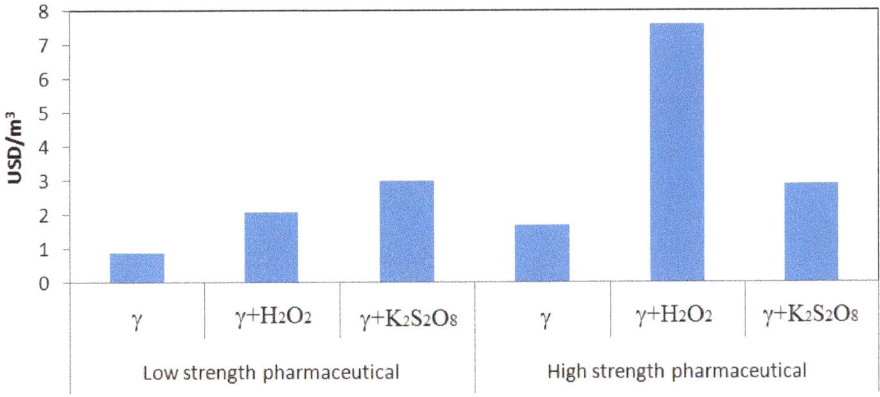

Figure 4. Treatment costs for γ irradiation and combine coagulation-γ irradiation [54].

There are some technical and non-technical issues that need to be resolved to ensure the sustainability of using this technology. These issues are listed below [22]

1. The usable output power of the electron accelerator limits its wide scale use as a safe substitute for chlorination for medium scale wastewater treatment plants.
2. Capital costs of electron accelerators needs to be reduced to ensure economical feasibility, where capital cost of installing an electron accelerator of 1MeV/400KW, could be attributed to half the accelerator price, whereas the design, construction, transportation, and installation of the facility contribute to 37.5% of capital cost.

From safety point of view, radiological accidents associated with different application of industrial irradiator, either γ irradiators or accelerators are limited, and could be classified as level 4 on the international nuclear and radiological event scale (INES). This level is used to describe accident with local consequence; Figure 5 summarizes the radiological exposures in these historical accidents [65]. Except the 1967 accelerator accident, one worker was exposed to radiation in each accident with a probability 0.092 and 0.12 a^{-1} for accelerators and γ irradiators, respectively. It should be noted that due to the more stringent regulatory safety requirements that were issued later, similar accidents were not reported over three decades ago. These regulatory requirements are not only related to safety aspects but also to the security aspects to ensure the application of 3S concept (Safety, Security, and Safeguards) [66,67]. To reduce these accidents, and based on the national regulation, the following measures should be considered [68–70]:

1. Appropriate safety measures, i.e., shielding requirements, operational procedures, provision of safety assessment documents, should be applied to ensure radiological containments.
2. Appropriate security measures should be applied to ensure safety of workers and public.

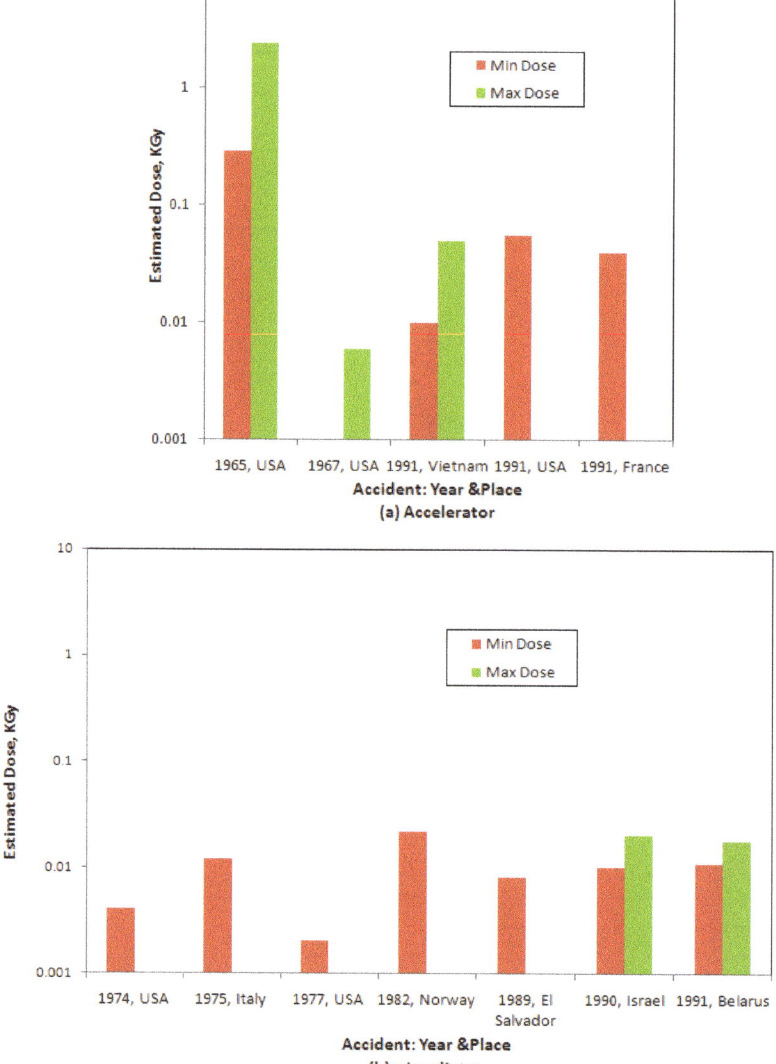

Figure 5. Summary of the historical industrial irradiator accidents; (**a**) accelerators, (**b**) γ irradiator.

Currently, the control strategy in these facilities is based on the combination of physical protection means, i.e., shields, barriers, and interlocks, and operational procedures [65,71]. This strategy ensured an accident probability for industrial irradiator in the order of 2×10^{-4} a^{-1}.

8. Conclusions

The review of ionizing radiation technology in decomposing bio-refractory organic contaminants and disinfecting different wastewater effluents were presented in this study. Factors that affect the sustainability of this technology were summarized. From this review the following conclusions could be drawn.

1. Most of the published work focused on the quantification of ionizing radiation effects on primary pollutants and the secondary intermediates toxicity, and determination of optimum irradiation conditions in different effluents. Research efforts are needed to study the feasibility of ionizing radiation treatment for petrochemical wastewater effluents,
2. Despite real industrial wastewater effluents were studied, there is a need to study the kinetics reactions in these complex systems to enable a better understanding and a better design for combined treatment schemes.
3. Compared to other disinfection technologies, the ionizing radiation technology provides economical, reliable, and safer operations that are not affected by the seasonal variation in the effluent composition, and reduces the generation of secondary toxic intermediates.
4. Operational safety of industrial irradiators has been improved due to the increasing stringent regulatory requirements and the updating the operational procedures, which lead to the reduction of accidents probability from 10^{-2} to 10^{-4} a^{-1}.

Author Contributions: Conceptualization, R.O.A.R. and Y.-T.H.; methodology, R.O.A.R.; resources, R.O.A.R. and Y.-T.H.; writing—original draft preparation, R.O.A.R.; writing—review and editing, R.O.A.R.; supervision, Y.-T.H.; funding acquisition, Y.-T.H. All authors have read and agreed to the published version of the manuscript.

Funding: This research received no external funding.

Conflicts of Interest: The authors declare no conflict of interest.

References

1. Allaoui, M.; Schmitz, T.; Campbell, D.; Porte, C.A.D.L. *Good Practices, For Regulating Wastewater Treatment, Legislations, Policies and Standards*; United Nations Environment Programme: Nairobi, Kenya, 2015.
2. UNESCAP. *Policy Guidance Manual on Wastewater Management, with a Special Emphasis on Decentralized Wastewater Treatment Systems*; UNESCAP: Bangkok, Thailand, 2015.
3. FAO. *Wastewater Treatment and Use in Agricultural-FAO Irrigation and Drainage Paper 47*; FAO: Rome, Italy, 1992; Available online: http://www.fao.org/3/t0551e/t0551e00.htm#Contents (accessed on 20 November 2019).
4. Quach-Cu, J.; Herrera-Lynch, B.; Marciniak, C.; Adams, S.; Simmerman, A.; Reinke, R.A. The Effect of Primary, Secondary, and Tertiary Wastewater Treatment Processes on Antibiotic Resistance Gene (ARG) Concentrations in Solid and DissolvedWastewater Fractions. *Water* **2018**, *10*, 37. [CrossRef]
5. Deng, Y.; Zhao, R. Advanced Oxidation Processes (AOPs) in Wastewater Treatment. *Curr. Pollut. Rep.* **2015**, *1*, 167–176. [CrossRef]
6. Shon, H.; Vigneswaran, S.; Snyder, S. Effluent organic matter (EfOM) in wastewater: Constituents, effects, and treatment. *Crit. Rev. Environ. Sci. Technol.* **2006**, *36*, 327–374. [CrossRef]
7. Tchobanoglous, G.; Burton, F.L.; Stensel, H.D. *Waste Water Engineering: Treatment and Reuse*; Metcalf and Eddy: Wakefield, UK, 2003.
8. EU. *Effluent Charging Systems in the EU Member States*; European Parliament: Luxembourg, 2001.
9. Wastewater Treatment Requirement. Available online: https://www.lacsd.org/civicax/filebank/blobdload.aspx?blobid=3325 (accessed on 18 December 2019).

10. Munter, R. Industrial wastewater characteristics. In *Sustainable Water Management in the Baltic Sea Basin: Water Use and Management*; Lundin, L.C., Ed.; The Baltic University Programme, Uppsala University: Uppsala, Sweden, 2000; pp. 185–194.
11. Kumar, M.; Puri, A. A review of permissible limits of drinking water. *Indian J. Occup. Environ. Med.* **2012**, *16*, 40–44. [PubMed]
12. WHO. *A Compendium of Standards for Wastewater Reuse in the Eastern Mediterranean Region*; WHO-EM/CEH/142/E; WHO: Cairo, Egypt, 2006.
13. Environmental Management Act. Municipal Wastewater Regulation B.C. Reg. 87/2012, Consolidation Current to 20 June2018. Available online: http://archives.leg.bc.ca/civix/document/id/crbc/crbc/87_2012 (accessed on 20 August 2019).
14. Councel Directive of 21 May 1991 Concerning Urban Wastewater Treatment (91/271/EEC). Available online: https://eur-lex.europa.eu/LexUriServ/LexUriServ.do?uri=OJ:L:1991:135:0040:0052:EN:PDF%20 (accessed on 18 December 2019).
15. EU. *Minimum Quality Requirements for Water Reuse in Agricultural Irrigation and Aquifer Recharge*; Publications Office of the European Union: Luxembourg, 2017.
16. Luo, Y.; Guo, W.; Ngo, H.H.; Nghiem, L.D.; Hai, F.I.; Zhang, J.; Liang, S.; Wang, X.C. A review on the occurrence of micropollutants in the aquatic environment and their fate and removal during wastewater treatment. *Sci. Total Environ.* **2014**, *473*, 619–641. [CrossRef]
17. Naing, T.T.; Lay, K.K. Utilization of gamma radiation in industrial wastewater treatment. *Int. J. Mech. Prod. Eng.* **2015**, *3*, 1–5.
18. Miklos, D.B.; Remy, C.; Jekel, M.; Linden, K.G.; Drewes, J.E.; Hübner, U. Evaluation of advanced oxidation processes for water and wastewater treatment—A critical review. *Water Res.* **2018**, *139*, 118–131. [CrossRef]
19. Babu, D.S.; Srivastava, V.; Nidheesh, P.V.; Kumar, M.S. Detoxification of water and wastewater by advanced oxidation processes. *Sci. Total Environ.* **2019**, *696*, 133961. [CrossRef]
20. Liu, N.; Lei, Z.D.; Wanga, T.; Wang, J.J.; Zhang, X.D.; Xu, G.; Tang, L. Radiolysis of carbamazepine aqueous solution using electron beam irradiation combining with hydrogen peroxide: Efficiency and mechanism. *Chem. Eng. J.* **2016**, *295*, 484–493. [CrossRef]
21. Meng, M.; Pellizzari, F.; Boukari, S.O.B.; Leitner, N.K.V.; Teychene, B. Impact of e-beam irradiation of municipal secondary effluent on MF and RO membranes performances. *J. Membr. Sci.* **2014**, *471*, 1–8. [CrossRef]
22. IAEA. *Radiation Treatment of Polluted Water and Wastewater*; IAEA: Vienna, Austria, 2008.
23. Wojnarovits, L.; Takacs, E. Irradiation treatment of azo dye containing wastewater: An overview. *Radiat. Phys. Chem.* **2008**, *77*, 225–244. [CrossRef]
24. Sampa, M.H.O.; Takacs, E.; Gehringer, P.; Rela, P.R.; Ramirez, T.; Amro, H.; Trojanowicz, M.; Botelho, M.L.; Han, B.; Solpan, D.; et al. Remediation of polluted waters and wastewater by radiation processing. *Nukleonika* **2007**, *52*, 137–144.
25. Jan, S.; Kamili, A.N.; Parween, T.; Hami, R.; Parray, J.A.; Siddiqi, T.O.; Ahmad, M.P. Feasibility of radiation technology for wastewater Treatment. *Desalin. Water Treat.* **2015**, *55*, 2053–2068. [CrossRef]
26. Abdel Rahman, R.O.; Kozak, M.W.; Hung, Y.T. Radioactive pollution and control. In *Handbook of Environment and Waste Management*; Hung, Y.T., Wang, L.K., Shammas, N.K., Eds.; World Scientific Publishing Co: Singapore, 2014; pp. 949–1027. [CrossRef]
27. Gehringer, P. Advances in radiation processing of wastewater—Basics of the process. In *Status of Industrial Scale Radiation Treatment of Wastewater and Its Future*; IAEA-TECDOC-1407; IAEA: Vienna, Austria, 2003; pp. 7–17.
28. Shojaie, F. Comparison of the rate constants of a bimolecular reaction using two methods. *Arab. J. Chem.* **2017**, *10*, S148–S156. [CrossRef]
29. IAEA. *Status of Industrial Scale Radiation Treatment of Wastewater and Its Future*; IAEA-TECDOC-1407; IAEA: Vienna, Austria, 2003.
30. Wojnárovits, L.; Takacs, E. Wastewater treatment with ionizing radiation. *J. Radioanal. Nucl. Chem.* **2017**, *311*, 973–981. [CrossRef]
31. Chaychian, M.; Al-Sheikhly, M.; Silverman, J.; McLaughlin, W.L. The mechanisms of removal of heavy metals from water by ionizing radiation. *Radiat. Phys. Chem.* **1998**, *53*, 145–150. [CrossRef]

32. Chmielewski, A.G. Application of ionizing radiation to environment protection. *Nukleonika* **2005**, *50*, s17–s24. [CrossRef]
33. Dalalia, N.; Kazeraninejada, M.; Akhavan, A. Removal of heavy metal ions from wastewater samples using electron beam radiation in the presence of TiO_2. *Desalin. Water Treat.* **2017**, *71*, 136–144. [CrossRef]
34. El-Motaium, R.A. Application of nuclear techniques in Environmental studies and pollution Control. In Proceedings of the 2nd Environmental Physics Conference, Alexandria, Egypt, 18–22 February 2006; pp. 169–182.
35. Rela, P.R.; Sampa, M.H.O.; Duarte, C.L.; Costa, F.E.D.; Rela, C.S.; Borrely, S.I.; Mori, M.N.; Somessari, E.S.R.; Silveira, C.G. The status of radiation process to treat industrial effluents in Brazil. In *Radiation Treatment of Polluted Water and Waste Water*; IAEA TECDOC-1598; IAEA: Vienna, Austria, 2008; pp. 27–41.
36. Ramirez, T.; Armas, M.; Uzcategui, M. Effect of accelerated electron beam on pesticides removal of effluents from flower plantations. In *Radiation Treatment of Polluted Water and Waste Water*; IAEA TECDOC-1598; IAEA: Vienna, Austria, 2008; pp. 43–58.
37. Emmi, S.S.; De Paoli, G.; Takács, E.; Caminati, S.; Pálfi, T. The e-induced decomposition of pesticides in water: A gamma and pulse radiolysis investigation on carbofuran. In *Radiation Treatment of Polluted Water and Waste Water*; IAEA TECDOC-1598; IAEA: Vienna, Austria, 2008; pp. 77–88.
38. Şolpan, D. The degradation of some pesticides in aqueous solutions by gamma radiation. In *Radiation Treatment of Polluted Water and Waste Water*; IAEA TECDOC-1598; IAEA: Vienna, Austria, 2008; pp. 177–199.
39. Utrill, J.R.; Abdel Daiem, M.M.; Polo, M.S.; Pérez, R.O.; Peñalver, J.J.L.; Gala, I.V.; Mota, A.J. Removal of compounds used as plasticizers and herbicides from water by means of gamma irradiation. *Sci. Total Environ.* **2016**, *569*, 518–526. [CrossRef]
40. Sabbagh, S.; El Mahmoudi, A.S.; Al-Dakheel, Y.Y. Waste Water Sterilization by Cobalt Co-60 for the Agricultural Irrigation: A Case Study. *Int. J. Water Resour. Arid Environ.* **2014**, *3*, 11–18.
41. Lim, S.J.; Kim, T.H. Combined treatment of swine wastewater by electron beam irradiation and ion-exchange biological reactor system. *Sep. Purif. Technol.* **2015**, *146*, 42–49. [CrossRef]
42. Alkhuraiji, T.S.; Boukari, S.O.B.; Alfadhl, F.S. Gamma irradiation-induced complete degradation and mineralization of phenol in aqueous solution: Effects of reagent. *J. Hazard. Mater.* **2017**, *328*, 29–36. [CrossRef] [PubMed]
43. Sun, W.; Chen, L.; Tian, J.; Wang, J.; He, S. Degradation of a monoazo dye Alizarin Yellow GG in aqueous solutions by gamma irradiation: Decolorization and biodegradability enhancement. *Radiat. Phys. Chem.* **2013**, *83*, 86–89. [CrossRef]
44. Takacs, E.; Wojnarovits, L.; Palfi, T. Azo dye degradation by high-energy irradiation: Kinetics and mechanism of destruction. *Nukleonika* **2007**, *52*, 69–75.
45. Muneer, M.; Bhatti, I.A.; Bhatti, H.N.; Rehman, K.U. Treatment of Dyes Industrial Effluents by Ionizing Radiation. *Asian J. Chem.* **2011**, *23*, 2392–2394.
46. Abdou, L.A.W.; Hakeim, O.A.; Mahmoud, M.S.; El-Naggar, A.M. Comparative study between the efficiency of electron beam and gamma irradiation for treatment of dye solutions. *Chem. Eng. J.* **2011**, *168*, 752–758. [CrossRef]
47. Borrely, S.I.; Morais, A.V.; Rosa, J.M.; Pedroso, C.B.; Pereira, M.D.C.; Hig, M.C. Decoloration and detoxification of effluents by ionizing radiation. *Radiat. Phys. Chem.* **2016**, *124*, 98–202. [CrossRef]
48. Meibodi, M.E.; Parsaeian, M.R.; Amraei, R.; Banaei, M.; Anvari, F.; Tahami, S.M.R.; Vakhshoor, B.; Mehdizadeh, A.; FallahNejad, N.; Shirmardi, S.P.; et al. An experimental investigation of wastewater treatment using electron beam irradiation. *Radiat. Phys. Chem.* **2016**, *125*, 82–87. [CrossRef]
49. Deogaonkar, S.C.; Wakode, P.; Rawat, K.P. Electron beam irradiation post treatment for degradation of non biodegradable contaminants in textile wastewater. *Radiat. Phys. Chem.* **2019**, *165*, 108377. [CrossRef]
50. Basfar, A.A.; Abdel Rehim, F. Disinfection of wastewater from a Riyadh Wastewater Treatment Plant with ionizing radiation. *Radiat. Phys. Chem.* **2002**, *65*, 527–532. [CrossRef]
51. Lee, O.M.; Kim, H.Y.; Park, W.; Kim, T.H.; Yu, S. A comparative study of disinfection efficiency and regrowth control of microorganism in secondary wastewater effluent using UV, ozone, and ionizing irradiation process. *J. Hazard. Mater.* **2015**, *295*, 201–208. [CrossRef] [PubMed]
52. Sági, G.; Kovács, K.; Bezsenyi, A.; Csay, T.; Takács, E.; Wojnárovits, L. Enhancing the biological degradability of sulfamethoxazole by ionizing radiation treatment in aqueous solution. *Radiat. Phys. Chem.* **2016**, *124*, 179–183. [CrossRef]

53. Tegze, A.; Sági, G.; Kovács, K.; Tótha, T.; Takács, E.; Wojnárovits, L. Radiation induced degradation of ciprofloxacin and norfloxacin: Kinetics and product analysis. *Radiat. Phys. Chem.* **2019**, *158*, 68–75. [CrossRef]
54. Changotra, R.; Rajput, H.; Guin, J.P.; Varshney, L.; Dhir, A. Hybrid coagulation, gamma irradiation and biological treatment of real pharmaceutical wastewater. *Chem. Eng. J.* **2019**, *370*, 595–605. [CrossRef]
55. Changotra, R.; Rajput, H.; Guin, J.P.; Khader, S.A.; Dhir, A. Techno-Economical Evaluation of Coupling Ionizing Radiation and Biological Treatment Process for the Remediation of Real Pharmaceutical Wastewater. *J. Clean. Prod.* **2020**, *242*, 118544. [CrossRef]
56. Chu, L.; Yu, S.; Wang, J. Gamma radiolytic degradation of naphthalene in aqueous solution. *Radiat. Phys. Chem.* **2016**, *123*, 97–102. [CrossRef]
57. Economical Aspects of Radiation Sterilization with Electron Beam. Available online: https://www.osti.gov/etdeweb/servlets/purl/644045 (accessed on 18 December 2019).
58. Maruthi1, Y.A.; Das, N.L.; Hossain, K.; Rawat, K.P.; Sarma, K.S.S.; Sabharwal, S. Application of electron beam technology in improving sewage water quality: An advance technique. *Eur. J. Sustain. Dev.* **2013**, *2*, 1–18.
59. Hossain, K.Y.; Maruthi, A.; Das, N.L.; Rawat, K.P.; Sarma1, K.S.S. Irradiation of wastewater with electron beam is a key to sustainable smart/green cities: A review. *Appl. Water Sci.* **2018**, *8*, 6. [CrossRef]
60. Gala, I.V.; Peñalver, J.J.L.; Polo, M.S.; Utrilla, J.R. Degradation of X-ray contrast media diatrizoate in different water matrices by gamma irradiation. *Chem. Technol. Biotechnol.* **2013**, *88*, 1336–1343. [CrossRef]
61. Maruthi, A.Y.; Das, N.L.; Hossain, K.; Rawat, K.P.; Sarma, K.S.S.; Sabharwal, S. Disinfection and reduction of organic load of sewage water by electron beam radiation. *Appl. Water Sci.* **2011**, *1*, 49–56. [CrossRef]
62. Ndong, J.E.; Uribe, R.M.; Gregory, R.; Gangoda, M.; Nickelsen, M.G.; Loar, P. Effect of electron beam irradiation on bacterial and Ascaris ova loads and volatile organiccompounds in municipal sewage sludge. *Radiat. Phys. Chem.* **2015**, *112*, 6–12. [CrossRef]
63. IAEA. *Radiation Technology for Cleaner Products and Processes*; IAEA-TECDOC-1786; IAEA: Vienna, Austria, 2016.
64. Pereira, L.S.; Cordery, I.; Iacovides, I. *Coping with Water Scarcity*; IHP VI, Technical Documents in Hydrology 58; UNSCO: Paris, France, 2002.
65. Ortiz, P.; Oresegun, M.; Weatley, J. Lessons from Major Radiation Accidents. In Proceedings of the 10th International Congress of the International Radiation Protection Association—IRPA 10, Hiroshima, Japan, 14–19 May 2000.
66. IAEA. *Management for the Prevention of Accidents from Disused Sealed Radioactive Sources*; IAEA, TECDOC-1205; IAEA: Vienna, Austria, 2001.
67. Abdel Rahman, R.O. Introductory Chapter: Introduction to Current Trends in Nuclear Material Research and Technology. In *Nuclear Material Performance*; Abdel Rahman, R.O., Saleh, H.E.M., Eds.; Intech: Rijeka, Croatia, 2016; pp. 3–14. Available online: http://www.intechopen.com/books/nuclear-material-performance/introductory-chapter-introduction-to-current-trends-in-nuclear-material-research-and-technology (accessed on 18 December 2019).
68. USDOE. *Safety of Accelerator Facilities*; DOE O440.C; USDOE: Washington, DC, USA, 2011.
69. IAEA. *Directory of Gamma Processing Facilities in Member States*; IAEA-DGPF/CD; IAEA: Vienna, Austria, 2004.
70. IAEA. *Radiation Safety of Gamma, Electron and X Ray Irradiation Facilities*; SSG No. 8; IAEA: Vienna, Austria, 1996.
71. Janžekovič, H. Overview of Radiation Accidents at Industrial Accelerator Facilities. In Proceedings of the 26th International Conference on Nuclear Energy for New Europe, NENE, Bled, Slovenia, 11–14 September 2017; p. 7.

MDPI
St. Alban-Anlage 66
4052 Basel
Switzerland
Tel. +41 61 683 77 34
Fax +41 61 302 89 18
www.mdpi.com

Water Editorial Office
E-mail: water@mdpi.com
www.mdpi.com/journal/water

www.ingramcontent.com/pod-product-compliance
Lightning Source LLC
LaVergne TN
LVHW071533170726
843515LV00008B/2132

9783036510224